Advances in the Biology of Phototrophic Bacteria

Advances in the Biology of Phototrophic Bacteria

Editor

Johannes F. Imhoff

MDPI • Basel • Beijing • Wuhan • Barcelona • Belgrade • Manchester • Tokyo • Cluj • Tianjin

Editor
Johannes F. Imhoff
Helmholtz Centre for Ocean
Research GEOMAR
Germany

Editorial Office
MDPI
St. Alban-Anlage 66
4052 Basel, Switzerland

This is a reprint of articles from the Special Issue published online in the open access journal *Microorganisms* (ISSN 2076-2607) (available at: https://www.mdpi.com/journal/microorganisms/special_issues/phototrophic_bacteria).

For citation purposes, cite each article independently as indicated on the article page online and as indicated below:

LastName, A.A.; LastName, B.B.; LastName, C.C. Article Title. *Journal Name* **Year**, *Volume Number*, Page Range.

ISBN 978-3-0365-2269-2 (Hbk)
ISBN 978-3-0365-2270-8 (PDF)

Cover image courtesy of Johannes F. Imhoff.

© 2021 by the authors. Articles in this book are Open Access and distributed under the Creative Commons Attribution (CC BY) license, which allows users to download, copy and build upon published articles, as long as the author and publisher are properly credited, which ensures maximum dissemination and a wider impact of our publications.

The book as a whole is distributed by MDPI under the terms and conditions of the Creative Commons license CC BY-NC-ND.

Contents

About the Editor vii

Johannes F. Imhoff
Editorial for the Special Issue: Advances in the Biology of Phototrophic Bacteria
Reprinted from: *Microorganisms* **2021**, *9*, 2119, doi:10.3390/microorganisms9102119 1

Johannes F. Imhoff, Tanja Rahn, Sven Künzel and Sven C. Neulinger
Phylogeny of Anoxygenic Photosynthesis Based on Sequences of Photosynthetic Reaction Center Proteins and a Key Enzyme in Bacteriochlorophyll Biosynthesis, the Chlorophyllide Reductase
Reprinted from: *Microorganisms* **2019**, *7*, 576, doi:10.3390/microorganisms7110576 7

John A. Kyndt, Jozef J. Van Beeumen and Terry E. Meyer
Simultaneous Genome Sequencing of *Prosthecochloris ethylica* and *Desulfuromonas acetoxidans* within a Syntrophic Mixture Reveals Unique Pili and Protein Interactions
Reprinted from: *Microorganisms* **2020**, *8*, 1939, doi:10.3390/microorganisms8121939 25

Ana Giraldo-Silva, Vanessa M. C. Fernandes, Julie Bethany and Ferran Garcia-Pichel
Niche Partitioning with Temperature among Heterocystous Cyanobacteria (*Scytonema* spp., *Nostoc* spp., and *Tolypothrix* spp.) from Biological Soil Crusts
Reprinted from: *Microorganisms* **2020**, *8*, 396, doi:10.3390/microorganisms8030396 43

Emma D. Dewey, Lynn M. Stokes, Brad M. Burchell, Kathryn N. Shaffer, Austin M. Huntington, Jennifer M. Baker, Suvarna Nadendla, Michelle G. Giglio, Kelly S. Bender, Jeffrey W. Touchman, Robert E. Blankenship, Michael T. Madigan and W. Matthew Sattley
Analysis of the Complete Genome of the Alkaliphilic and Phototrophic Firmicute *Heliorestis convoluta* Strain HH^T
Reprinted from: *Microorganisms* **2020**, *8*, 313, doi:10.3390/microorganisms8030313 59

Johannes F. Imhoff, Tanja Rahn, Sven Künzel, Alexander Keller and Sven C. Neulinger
Osmotic Adaptation and Compatible Solute Biosynthesis of Phototrophic Bacteria as Revealed from Genome Analyses
Reprinted from: *Microorganisms* **2021**, *9*, 46, doi:10.3390/ microorganisms9010046 83

Steven B. Kuzyk, Elizabeth Hughes and Vladimir Yurkov
Discovery of Siderophore and Metallophore Production in the Aerobic Anoxygenic Phototrophs
Reprinted from: *Microorganisms* **2021**, *9*, 959, doi:10.3390/microorganisms9050959 115

Daniel Roush and Ferran Garcia-Pichel
Succession and Colonization Dynamics of Endolithic Phototrophs within Intertidal Carbonates
Reprinted from: *Microorganisms* **2020**, *8*, 214, doi:10.3390/microorganisms8020214 135

Akira Hiraishi, Nobuyoshi Nagao, Chinatsu Yonekawa, So Umekage, Yo Kikuchi, Toshihiko Eki and Yuu Hirose
Distribution of Phototrophic Purple Nonsulfur Bacteria in Massive Blooms in Coastal and Wastewater Ditch Environments
Reprinted from: *Microorganisms* **2020**, *8*, 150, doi:10.3390/microorganisms8020150 151

Mathieu K. Licht, Aaron M. Nuss, Marcel Volk, Anne Konzer, Michael Beckstette, Bork A. Berghoff and Gabriele Klug
Adaptation to Photooxidative Stress: Common and Special Strategies of the Alphaproteobacteria *Rhodobacter sphaeroides* and *Rhodobacter capsulatus*
Reprinted from: *Microorganisms* **2020**, *8*, 283, doi:10.3390/microorganisms8020283 . . . . **171**

Xin Nie, Bernhard Remes and Gabriele Klug
Multiple Sense and Antisense Promoters Contribute to the Regulated Expression of the *isc-suf* Operon for Iron-Sulfur Cluster Assembly in *Rhodobacter*
Reprinted from: *Microorganisms* **2019**, *7*, 671, doi:10.3390/microorganisms7120671 . . . . **193**

Sonja Koppenhöfer and Andrew S. Lang
Interactions among Redox Regulators and the CtrA Phosphorelay in *Dinoroseobacter shibae* and *Rhodobacter capsulatus*
Reprinted from: *Microorganisms* **2020**, *8*, 562, doi:10.3390/microorganisms8040562 . . . . **213**

Karel Kopejtka, Yonghui Zeng, David Kaftan, Vadim Selyanin, Zdenko Gardian, Jürgen Tomasch, Ruben Sommaruga and Michal Koblížek
Characterization of the Aerobic Anoxygenic Phototrophic Bacterium *Sphingomonas* sp. AAP5
Reprinted from: *Microorganisms* **2021**, *9*, 768, doi:10.3390/microorganisms9040768 . . . . **229**

Shigeru Kawai, Joval N. Martinez, Mads Lichtenberg, Erik Trampe, Michael Kühl, Marcus Tank, Shin Haruta, Arisa Nishihara, Satoshi Hanada and Vera Thiel
In-Situ Metatranscriptomic Analyses Reveal the Metabolic Flexibility of the Thermophilic Anoxygenic Photosynthetic Bacterium *Chloroflexus aggregans* in a Hot Spring Cyanobacteria-Dominated Microbial Mat
Reprinted from: *Microorganisms* **2021**, *9*, 652, doi:10.3390/microorganisms9030652 . . . . **243**

About the Editor

Johannes F. Imhoff is a retired professor of Marine Microbiology at the GEOMAR Helmholtz Centre for Ocean Research in Kiel, Germany. He gained his Ph.D. in microbiology in 1980 at the Institute for Microbiology of the University of Bonn under the supervision of Prof. Dr. H.G. Trüper. In 1993, he was appointed as the head of the Marine Microbiology Department at the Institut für Meereskunde in Kiel (now the GEOMAR Helmholtz Centre for Ocean Research Kiel), where his main research themes dealt with marine microbial diversity with an emphasis on communities in deep-sea hot vents and in marine sponges. The introduction of appropriate molecular tools enabled studies of functional groups in microbial communities by sequence analysis of genes characteristic of photosynthesis and processes of sulfur oxidation, ammonia oxidation, and nitrate reduction. One of his major activities was establishing a research center for marine natural products, the "Kieler Wirkstoffzentrum" in 2005, of which he was founding director, and which was later transformed into the Marine Natural Product Chemistry research unit of GEOMAR. During these years, he laid the basis of large collections of marine bacteria and fungi, published the chemical structures and biological activities of numerous new compounds, and filed several patents on antitumoral active substances. He retired in 2017.

He published more than 350 papers, including chapters in several editions of "The Prokaryotes" and "Bergey´s Manual of Systematic Bacteriology". Since his first contact with microbiology in 1972, he has been fascinated by the culture of phototrophic bacteria, their ecology, diversity, and phylogeny. His studies focused on phototrophic bacteria from marine and hypersaline habitats, particularly salt and soda lakes, and their osmotic adaptation and compatible solute biosynthesis. He described more than 50 new taxa on the level of species, genus, family, and order of phototrophic and marine bacteria. A large culture collection of anoxygenic phototrophic bacteria, part of which was obtained from N. Pfennig in 1991, was maintained over the years and recently served to compile a large number of genomes of phototrophic bacteria. These data serve as a basis for studies on phylogenetic aspects, including photosynthesis.

Editorial

Editorial for the Special Issue: Advances in the Biology of Phototrophic Bacteria

Johannes F. Imhoff

GEOMAR Helmholtz Centre for Ocean Research, D-24105 Kiel, Germany; jimhoff@geomar.de

Citation: Imhoff, J.F. Editorial for the Special Issue: Advances in the Biology of Phototrophic Bacteria. *Microorg* 2021, *9*, 2119. https://doi.org/10.3390/microorganisms9102119

Received: 30 September 2021
Accepted: 2 October 2021
Published: 8 October 2021

Publisher's Note: MDPI stays neutral with regard to jurisdictional claims in published maps and institutional affiliations.

Copyright: © 2021 by the author. Licensee MDPI, Basel, Switzerland. This article is an open access article distributed under the terms and conditions of the Creative Commons Attribution (CC BY) license (https://creativecommons.org/licenses/by/4.0/).

Phototrophic bacteria represent a very ancient phylogenetic and highly diverse metabolic type of bacteria that diverged early into several major phylogenetic lineages with quite different properties. First of all, they differ in the structure of the photosynthetic apparatus, the light reactions, bacteriochlorophyll structure and biosynthesis. They are different in the relation to oxygen, i.e., are oxygen producers, are strict anaerobes or have a wide range of tolerance of oxygen. Some species can make use of oxygen for energy generation. They have adapted to all kinds of ecological niches and representatives are found in a wide range of environments from cold waters to hot springs, from freshwater to saturated salt brines, from acidic to alkaline habitats, in microbial mats and as inhabitants inside rocks. Among a number of books on phototrophic bacteria, two compendia that highlight major topics of early studies shall be mentioned [1,2].

Over the past decades, genomic and transcriptomic studies have pushed our knowledge on phototrophic bacteria in various aspects, especially in regard to phylogenetic relationships of species and evolution of physiological pathways, but also in the analysis of environmental communities and the metabolic flexibility and adaptation of communities and individual strains to specific ecological niches and to changing environmental conditions.

This Special Issue highlights recent advances in these aspects. It includes results specifically on green sulfur bacteria (*Chlorobi* [3]), heliobacteria (*Firmicutes* [4]), *Chloroflexi* [5], *Cyanobacteria* [6,7] and phototrophic purple bacteria (*Proteobacteria* [8–13]). In addition, phylogenetic studies regarding photosynthesis and osmotic adaptation consider representatives of all of these phyla [14,15]. This Special Issue includes results on identification and genomic characterization of new isolates [9], the adaptation to specific ecological niches in regard to temperature [6], salinity [14] and pH [4]. It also considers aerobic phototrophic purple bacteria and the relations to oxygen and oxidative stress [8–11,13] as well as environmental interactions [5,7,12] and syntrophic relations [3].

Photosynthesis is a key process in the development of life on earth, and over roughly 3.5 billion years of phototrophic life on earth, not only a number of significantly different phylogenetic lineages diverged but also different ecological niches were conquered. The analysis of gene sequences of a key enzyme in bacteriochlorophyll biosynthesis, the light-independent chlorophyllide reductase BchXYZ which is common to all anoxygenic phototrophic bacteria, including those with a type-I and those with a type-II photosynthetic reaction center, highlights their phylogenetic relationship [15]. The phylogenetic relations of more than 150 species demonstrate that bacteriochlorophyll biosynthesis had evolved in ancestors of phototrophic green bacteria (*Chlorobi*, *Heliobacteriaceae*, *Chloracidobacterium*) much earlier as compared to phototrophic purple bacteria and also that multiple events independently formed different lineages of aerobic phototrophic purple bacteria, some of which have ancient roots [15].

Apparently, phylogenetically distinct strains of the *Chlorobi* have specifically adapted to form syntrophic associations [16,17]. It has been the careful analysis of a mixed culture considered to be a green sulfur bacterium and named "Chloropseudomonas ethylica" that led to the discovery of a syntrophic relationship of a green sulfur bacterium and a chemotrophic sulfur-reducer realizing a short type of sulfur cycle [17]. Now, the simultaneous determination of the genome sequences of the green and colorless components of three mixtures named "Chloropseudomonas ethylica" N3 and N2 originating from E.N. Kondrat'eva many years ago and of the 2-K mixture (DSM 1685) revealed the identity of the green component as a distinct *Prosthecochloris* species and of the colorless component as a distinct cluster within *Desulfuromonas* strains [3]. *Prosthecochloris ethylica* is proposed as a new species name for the green component. Tight adhesion (Tad) types of pili are suggested to play a role in the syntrophic relationship of these bacteria by forming cell–cell interactions and a gene cluster encoding these pili is characteristic for those green sulfur bacteria (N2, N3, and DSM1685) that are involved in the interactions [3]. The formation of pili in both partners of the syntrophic association suggests an evolutionarily gained specific property of the syntrophic partners [3].

Over evolutionary times, phototrophic bacteria that have conquered different ecological niches have given rise to the evolution of phylogenetic groups of species that are living today in these niches. This is depicted in specific ecological niches of phylogenetic distinct groups of phototrophic bacteria with examples of some heliobacteria in alkaline waters [4], of different groups of phototrophic purple bacteria adapted to marine and hypersaline habitats [14] and of certain genera of cyanobacteria with different temperature responses [6]. The last aspect was studied by comparison of temperature responses of 30 isolates of these genera and by a meta-analysis of 84 locations around the world [6] and revealed advantages of *Tolypothrix* strains at lower temperatures, *Scytonema* strains at higher temperatures and *Nostoc* strains at moderate temperatures [6]. The complex situation in the habitat was demonstrated by an expanded upper temperature range for growth if fixed nitrogen sources are available.

Alkaliphilic heliobacteria of the genus *Heliorestis* live in soda lakes, grow optimally between pH 8.0 and 9.5 and form a phylogenetic distinct group among heliobacteria [4]. One of these, *Hrs. convoluta* is the first heliobacterium isolated from a soda lake in the Wadi-el-Natrun (Egypt), which is known as a habitat of alkaliphilic and extremely halophilic eubacteria and archaea, e.g., *Halorhodospira* and *Natronomonas* species [18–20]. The analysis of the complete genome sequence of *Hrs. convoluta* provided insight into the molecular adaptation to alkaline conditions, the photoheterotrophic metabolism, nitrogen utilization, sulfur assimilation, and pigment biosynthesis pathways of heliobacteria [4]. Recent genome analyses of a larger number of heliobacteria have led to reconsidering their phylogeny and systematic treatment and confirmed the distinct phylogenetic position of the alkaliphilic *Heliorestis* species [21].

Comparative genomic analyses have consolidated the phylogenetic relationships of phototrophic bacteria living in marine and hypersaline environments. Halophilic phototrophic bacteria have glycine betaine and ectoine as major compatible solutes [22,23] and their ability to transport and synthesize these compatible solutes has been found to correlate well with the occurrence of these bacteria in saline and hypersaline habitats [14]. Furthermore, phylogenetic relations of key genes of these pathways define different phylogenetic groups of halophilic phototrophic bacteria [14].

Another property that correlates with the occurrence in marine or freshwater habitats is the formation of siderophores by aerobic phototrophic bacteria [8]. As important iron chelators, siderophores participate in chelating and uptake of iron from the environment. Interestingly, a high proportion of siderophore producers among aerobic phototrophic purple bacteria was found in isolates originating from freshwater sources (hot springs, freshwater lakes, and biological soil crusts) in comparison to those from marine, meromictic lake, and saline spring habitats [8]. Halotolerant or halophilic aerobic phototrophic purple bacteria do not produce siderophores of equal activity or quantity as compared to bacteria that do not depend on NaCl for growth [8].

A special ecological niche of cyanobacteria as the primary settlers is the endolithic habitat. In this study, the primary colonization and following succession dynamics in intertidal carbonate rocks were studied over a period of nine months [7]. Based on 16S rRNA gene libraries, an "unknown boring cluster" of so far uncultured cyanobacteria was identified as the dominant primary settler [7]. With time, these primary settlers were replaced by other endolithic cyanobacteria and significant populations of anoxygenic *Chloroflexi* occurred in the mature endolithic intertidal ecosystem [7].

Prior to the establishment of oxygenic photosynthesis, phototrophic life had depended on anoxic and strongly reducing conditions. The appearance of oxygen primarily caused severe stress on all strictly anaerobes and triggered adaptation processes to tolerate and eventually use oxygen for energy generation. Today, for many phototrophic purple bacteria, the chemocline and boundary between oxic and anoxic/sulfidic parts of the environment offer conditions for significant developments if light is available. Different strategies have been realized to adapt to these dynamic ecological niches.

The relations to sulfide and oxygen remain important environmental factors determining the distribution and competition of anaerobic phototrophic purple bacteria in the environment. While purple sulfur bacteria have a preference for anoxic and sulfidic parts of the gradients, purple nonsulfur bacteria, to a different degree, have adapted to less sulfidic/sulfide-free, anoxic, microoxic or even oxic parts. In colored blooms and microbial mats, phototrophic purple nonsulfur bacteria and purple sulfur bacteria regularly occur together with a clear preference for the purple sulfur bacteria in sulfidic niches. The different niches of phototrophic purple bacteria were highlighted in a comparison of colored blooms in a coastal environment and in wastewater ditches [12]. The sulfidic coastal marine habitat contained purple sulfur bacteria as the major populations, and smaller but significant densities of purple nonsulfur bacteria, with members of *Rhodovulum* predominating. The freshwater/wastewater habitat exclusively yielded purple nonsulfur bacteria, with species of *Rhodobacter*, *Rhodopseudomonas*, and/or *Pararhodospirillum* as the major constituents. As important environmental factors affecting purple nonsulfur bacteria populations, organic matter, sulfide concentrations and the oxidation-reduction potential were identified [12]. Light-exposed, sulfide-deficient water bodies with high content of organic matter and within a limited range of oxidation-reduction potentials provide favorable conditions for the significant growth of purple nonsulfur bacteria [12].

Many phototrophic purple bacteria that perform photosynthesis under anoxic conditions have alternative ways of energy generation in the dark. The relation to oxygen is a crucial point in their life and oxygen is expected to determine the metabolic activities of these bacteria. They can switch between aerobic and anaerobic lifestyles and responses to these changes are regulated by a complex network of regulators.

Rba. sphaeroides and *Rba. capsulatus* are two of these facultative phototrophic bacteria which are able to adjust their lifestyle. They can perform photosynthesis under anoxic conditions but can also perform aerobic or anaerobic respiration or fermentation. Oxygen is a major regulatory factor for the formation of photosynthetic complexes, and several proteins involved in oxygen-mediated gene regulation have been identified [11]. In the presence of light and oxygen, they may be exposed to photooxidative stress by the formation of reactive singlet oxygen. In this study, the responses to photooxidative stress of *Rba. sphaeroides* and *Rba. capsulatus* are compared by transcriptomic and proteomic analyses. Although both species have quite a similar lifestyle, they show different responses to photooxidative stress [11].

Functional iron–sulfur clusters have diverse and important functions in phototrophic bacteria. They are essential for bacteriochlorophyll synthesis and photosynthetic electron transport and are involved in the defense of oxidative stress. A complex regulatory network of several promoters and regulatory proteins is supposed to adjust iron–sulfur cluster assembly to changing conditions in *Rba. sphaeroides* to avoid destruction by oxidative stress [13]. A model is proposed for the regulation of iron–sulfur cluster biosynthesis in which IscR is the main regulator and multiple promotors and regulators are involved in adjusting *ics-suf* expression to environmental conditions [13].

Another study with two facultative phototrophic bacteria, *Dinoroseobacter shibae* and *Rba. capsulatus*, demonstrates that adaptation processes to changes between oxic and anoxic conditions are controlled by a complex regulatory network with different transcriptional regulators in the two species [10]. It is shown that regulation of the CtrA regulon in

Dinoroseobacter shibae is controlled by the aerobic-anaerobic regulators Crp/Fnr and by FnrL/RegA in *Rba. capsulatus* [10].

An increasing number of phototrophic purple bacteria have been recently identified to be well adapted to the oxic environment. A new isolate, closely related to *Sphingomonas glacialis* and *Sphingomonas paucimobilis*, is characterized on the basis of its complete genome with detailed phenotypic, phylogenetic, genomic, physiological, and biochemical characterization [9].

Counteracting activities of oxygenic photosynthesis and respiration cause changes in stratified environments with moving oxygen horizons during the daily cycles. These are most pronounced in microbial mats. Bacteria living in these gradients have to deal with these regular changes. Prominent examples of bacteria adapted to microbial mats in hot springs are the *Chloroflexus* species [24]. The changes between oxic and anoxic conditions as well as between light and dark conditions force these gliding bacteria to move up and down in the mat to meet optimal conditions and to change key metabolic reactions to optimize energy generation under the variable conditions. A particularly detailed study on the adaptation to these dynamic environmental conditions, especially to daily moving gradients in microbial mats, has been performed with *Chloroflexus aggregatus* by metatranscriptomic and proteomic studies [5], in which the in situ metabolic activity of *Chloroflexus aggregans* in microbial mats of Nakabusa Hot Springs was analyzed [5]. This study reveals a well-coordinated regulation of key metabolic processes to assure that *Cfl. aggregans* uses its metabolic flexibility and capability for both phototrophic and chemotrophic growth to optimize its performance under the varying environmental conditions in its natural habitat [5]. During daytime, light is the main energy source supporting phototrophic growth of *Cfl. aggregans*. During the afternoon, under microoxic low-light conditions, chemoheterotrophic growth is based on aerobic respiration, while fermentation takes place under anoxic conditions at night. During early morning hours before sunrise, chemoautotrophic growth with oxygen as the terminal electron acceptor takes place [5]. During the daily cycle, *Cfl. aggregans* obviously also makes use of both forms of the Mg-protoporphyrin monomethylester cyclase to synthesize bacteriochlorophyll via the aerobic AcsF-dependent enzyme and via the anaerobic BchlE-dependent enzyme [5].

Altogether, the papers of this Special Issue demonstrate the exiting advances that have been made in our knowledge on various aspects of the biology of phototrophic bacteria by the application of genomic, transcriptomic and proteomic analyses. They bring new dimensions to our understanding of metabolic processes and their regulation and to environmental adaptation and place these into a phylogenetic context.

Funding: This work received no funding.

Acknowledgments: The author is grateful to all authors for their exciting contributions to this Special Issue. He also thanks the reviewers for their helpful recommendations and the staff of the editorial office of *Microorganisms* for support to prepare this Special Issue.

Conflicts of Interest: The author declares no conflict of interest.

References

1. Clayton, R.K.; Sistrom, W.R. (Eds.) *The Photosynthetic Bacteria*; Plenum Press: New York, NY, USA, 1978.
2. Blankenship, R.E.; Olson, J.M.; Miller, M. Antenna complexes from green photosynthetic bacteria. In *Anoxygenic Photosynthetic Bacteria*; Blankenship, R.E., Madigan, M.T., Bauer, C.E., Eds.; Kluwer Academic Publ.: Dordrecht, The Netherlands, 1995; pp. 399–435.
3. Kyndt, J.A.; Van Beeumen, J.J.; Meyer, T.E. Simultaneous Genome Sequencing of *Prosthecochloris ethylica* and *Desulfuromonas acetoxidans* within a Syntrophic Mixture Reveals Unique Pili and Protein Interactions. *Microorganisms* **2020**, *8*, 1939. [CrossRef] [PubMed]
4. Dewey, E.D.; Stokes, L.M.; Burchell, B.M.; Shaffer, K.N.; Huntington, A.M.; Baker, J.M.; Nadendla, S.; Giglio, M.G.; Bender, K.S.; Touchman, J.W.; et al. Analysis of the Complete Genome of the Alkaliphilic and Phototrophic Firmicute *Heliorestis convoluta* Strain HH^T. *Microorganisms* **2020**, *8*, 313. [CrossRef] [PubMed]

5. Kawai, S.; Martinez, J.N.; Lichtenberg, M.; Trampe, E.; Kühl, M.; Tank, M.; Haruta, S.; Nishihara, A.; Hanada, S.; Thiel, V. In-Situ Metatranscriptomic Analyses Reveal the Metabolic Flexibility of the Thermophilic Anoxygenic Photosynthetic Bacterium *Chloroflexus aggregans* in a Hot Spring Cyanobacteria-Dominated Microbial Mat. *Microorganisms* 2021, 9, 652. [CrossRef] [PubMed]
6. Giraldo-Silva, A.; Fernandes, V.M.C.; Bethany, J.; Garcia-Pichel, F. Niche Partitioning with Temperature among Heterocystous Cyanobacteria (*Scytonema* spp., *Nostoc* spp., and *Tolypothrix* spp.) from Biological Soil Crusts. *Microorganisms* 2020, 8, 396. [CrossRef] [PubMed]
7. Roush, D.; Garcia-Pichel, F. Succession and Colonization Dynamics of Endolithic Phototrophs within Intertidal Carbonates. *Microorganisms* 2020, 8, 214. [CrossRef] [PubMed]
8. Kuzyk, S.B.; Hughes, E.; Yurkov, V. Discovery of Siderophore and Metallophore Production in the Aerobic Anoxygenic Phototrophs. *Microorganisms* 2021, 9, 959. [CrossRef] [PubMed]
9. Kopejtka, K.; Zeng, Y.; Kaftan, D.; Selyanin, V.; Gardian, Z.; Tomasch, J.; Sommaruga, R.; Koblížek, M. Characterization of the Aerobic Anoxygenic Phototrophic Bacterium *Sphingomonas* sp. AAP5. *Microorganisms* 2021, 9, 768. [CrossRef] [PubMed]
10. Koppenhöfer, S.; Lang, A.S. Interactions among Redox Regulators and the CtrA Phosphorelay in *Dinoroseobacter shibae* and *Rhodobacter capsulatus*. *Microorganisms* 2020, 8, 562. [CrossRef] [PubMed]
11. Licht, M.K.; Nuss, A.M.; Volk, M.; Konzer, A.; Beckstette, M.; Berghoff, B.A.; Klug, G. Adaptation to Photooxidative Stress: Common and Special Strategies of the Alphaproteobacteria *Rhodobacter sphaeroides* and *Rhodobacter capsulatus*. *Microorganisms* 2020, 8, 283. [CrossRef] [PubMed]
12. Hiraishi, A.; Nagao, N.; Yonekawa, C.; Umekage, S.; Kikuchi, Y.; Eki, T.; Hirose, Y. Distribution of Phototrophic Purple Nonsulfur Bacteria in Massive Blooms in Coastal and Wastewater Ditch Environments. *Microorganisms* 2020, 8, 150. [CrossRef] [PubMed]
13. Nie, X.; Remes, B.; Klug, G. Multiple Sense and Antisense Promoters Contribute to the Regulated Expression of the isc-suf Operon for Iron-Sulfur Cluster Assembly in *Rhodobacter*. *Microorganisms* 2019, 7, 671. [CrossRef] [PubMed]
14. Imhoff, J.F.; Rahn, T.; Künzel, S.; Keller, A.; Neulinger, S.C. Osmotic Adaptation and Compatible Solute Biosynthesis of Phototrophic Bacteria as Revealed from Genome Analyses. *Microorganisms* 2021, 9, 46. [CrossRef] [PubMed]
15. Imhoff, J.F.; Rahn, T.; Künzel, S.; Neulinger, S.C. Phylogeny of Anoxygenic Photosynthesis Based on Sequences of Photosynthetic Reaction Center Proteins and a Key Enzyme in Bacteriochlorophyll Biosynthesis, the Chlorophyllide Reductase. *Microorganisms* 2019, 7, 576. [CrossRef] [PubMed]
16. Overmann, J. Phototrophic consortia. a tight cooperation between non-related eubacteria. In *Symbiosis, Cellular Origin, Life in Extreme Habitats and Astrobiology*; Seckbach, J., Ed.; Springer: Dordrecht, The Netherlands, 2001; Volume 4, pp. 239–255. [CrossRef]
17. Pfennig, N.; Biebl, H. *Desulfuromonas acetoxidans* gen. nov. and sp. nov., a new anaerobic, sulfur-reducing, acetate-oxidizing bacterium. *Arch. Microbiol.* 1976, 110, 3–12. [CrossRef] [PubMed]
18. Imhoff, J.F.; Hashwa, F.; Trüper, H.G. Isolation of extremely halophilic phototrophic bacteria from the alkaline Wadi Natrun, Egypt. *Arch. Hydrobiol.* 1978, 84, 381–388.
19. Imhoff, J.F.; Sahl, H.G.; Soliman, G.S.H.; Trüper, H.G. The Wadi Natrun: Chemical composition and microbial mass developments in alkaline brines of eutrophic desert lakes. *Geomicrobiol. J.* 1979, 1, 219–234. [CrossRef]
20. Soliman, G.S.H.; Trüper, H.G. *Halobacterium pharaonis* sp. nov., a new, extremely haloalkaliphilic archaebacterium with low magnesium requirement. *Zent. Bakteriol. Hyg. Abt. I Orig.* 1982, C3, 318–329. [CrossRef]
21. Kyndt, J.A.; Salama, D.M.; Meyer, T.E.; Imhoff, J.F. Phylogenetic relationship of phototrophic heliobacteria and systematic reconsideration of species and genus assignments based on genome sequences of eight species. *Int. J. Syst. Evol. Microbiol.* 2021, 71, 4729. [CrossRef]
22. Imhoff, J.F. True marine and halophilic anoxygenic phototrophic bacteria. *Arch. Microbiol.* 2001, 176, 243–254. [CrossRef]
23. Imhoff, J.F. Anoxygenic phototrophic bacteria from extreme environments. In *Modern Topics in the Phototrophic Prokaryotes: Environmental and Applied Aspects*; Hallenbeck, P.C., Ed.; Springer: Cham, Switzerland, 2017; pp. 427–480.
24. Pierson, B.K.; Castenholz, R.W. A phototrophic gliding filamentous bacterium of hot springs, *Chloroflexus aurantiacus*, gen. and sp. nov. *Arch. Microbiol.* 1974, 100, 5–24. [CrossRef]

Article

Phylogeny of Anoxygenic Photosynthesis Based on Sequences of Photosynthetic Reaction Center Proteins and a Key Enzyme in Bacteriochlorophyll Biosynthesis, the Chlorophyllide Reductase

Johannes F. Imhoff [1,*], Tanja Rahn [1], Sven Künzel [2] and Sven C. Neulinger [3]

1 GEOMAR Helmholtz Centre for Ocean Research, 24105 Kiel, Germany; trahn@geomar.de
2 Max Planck Institute for Evolutionary Biologie, 24306 Plön, Germany; kuenzel@evolbio.mpg.de
3 omics2view.consulting GbR, 24118 Kiel, Germany; s.neulinger@omics2view.consulting
* Correspondence: jimhoff@geomar.de

Received: 29 October 2019; Accepted: 15 November 2019; Published: 19 November 2019

Abstract: Photosynthesis is a key process for the establishment and maintenance of life on earth, and it is manifested in several major lineages of the prokaryote tree of life. The evolution of photosynthesis in anoxygenic photosynthetic bacteria is of major interest as these have the most ancient roots of photosynthetic systems. The phylogenetic relations between anoxygenic phototrophic bacteria were compared on the basis of sequences of key proteins of the type-II photosynthetic reaction center, including PufLM and PufH (PuhA), and a key enzyme of bacteriochlorophyll biosynthesis, the light-independent chlorophyllide reductase BchXYZ. The latter was common to all anoxygenic phototrophic bacteria, including those with a type-I and those with a type-II photosynthetic reaction center. The phylogenetic considerations included cultured phototrophic bacteria from several phyla, including *Proteobacteria* (138 species), *Chloroflexi* (five species), *Chlorobi* (six species), as well as *Heliobacterium modesticaldum* (Firmicutes), *Chloracidobacterium acidophilum* (Acidobacteria), and *Gemmatimonas phototrophica* (Gemmatimonadetes). Whenever available, type strains were studied. Phylogenetic relationships based on a photosynthesis tree (PS tree, including sequences of PufHLM-BchXYZ) were compared with those of 16S rRNA gene sequences (RNS tree). Despite some significant differences, large parts were congruent between the 16S rRNA phylogeny and photosynthesis proteins. The phylogenetic relations demonstrated that bacteriochlorophyll biosynthesis had evolved in ancestors of phototrophic green bacteria much earlier as compared to phototrophic purple bacteria and that multiple events independently formed different lineages of aerobic phototrophic purple bacteria, many of which have very ancient roots. The *Rhodobacterales* clearly represented the youngest group, which was separated from other *Proteobacteria* by a large evolutionary gap.

Keywords: phylogeny; photosynthetic reaction center proteins; bacteriochlorophyll biosynthesis; phototrophic purple bacteria; evolution of anoxygenic photosynthesis

1. Introduction

Anoxygenic photosynthesis is widely distributed among eubacteria and involves a number of genes for the photosynthetic reaction center and for the biosynthesis of photosynthetic pigments, bacteriochlorophylls, and carotenoids, which are essential elements to enable photosynthesis. While the biosynthesis of bacteriochlorophylls is common to all of them, the different structure of the photosynthetic reaction center clearly separates two groups of anoxygenic phototrophic bacteria, those having a type-I and those having a type-II photosystem [1–3].

Those bacteria employing a photosystem type-II photosynthetic apparatus include the phototrophic purple bacteria (*Proteobacteria*), as well as *Gemmatimonas* and *Chloroflexus*, with their photosynthetic relatives [1,3–5]. Essential components of the type-II photosynthetic apparatus are represented by two membrane-spanning photosynthetic reaction center proteins that are common to all of these bacteria. These PufLM proteins are binding bacteriochlorophyll molecules and are crucial components of the type-II photosynthetic apparatus. Together with an additional protein (PufH = PuhA), they form the core structure of the type-II photosynthetic reaction center in all phototrophic purple bacteria (*Proteobacteria* and *Gemmatimonas*). The PufH protein is absent from the *Chloroflexi* that have chlorosomes attached to the reaction center. In addition, a cytochrome c (PufC) is associated with the reaction center proteins in the majority of phototrophic purple bacteria but is lacking in a number of species [6]. While *pufLMC* genes form a stable genomic cluster (sometimes lacking the *pufC* gene), *pufH* (*puhA*) is located at a different place within the genome, associated with genes of bacteriochlorophyll biosynthesis [6]. It has been demonstrated that sequences of PufLM are excellent tools to study the phylogeny of anoxygenic phototrophic purple bacteria, as well as their diversity and environmental distribution [2,7–9]. In a comprehensive study based on the phylogeny of PufLM, it was shown that distinct lineages of *Proteobacteria* contained phototrophic representatives in 10 orders, including anaerobic as well as aerobic anoxygenic phototrophic purple bacteria [3].

Bacteriochlorophyll biosynthesis is common to all phototrophic bacteria, including those with a type-I and those with a type-II photosynthetic reaction center. A key enzyme in this pathway is the light-independent chlorophyllide reductase BchXYZ. Consequently, this protein enables a broad view on the phylogeny of anoxygenic photosynthetic bacteria with a capacity to synthesize bacteriochlorophyll [10].

In the present work, the phylogeny of anoxygenic phototrophic bacteria was analyzed on the basis of sequences of key proteins of the type-II photosynthetic reaction center PufHLM and of chlorophyllide reductase BchXYZ and was compared with the phylogeny of the 16S rRNA gene (Figure 1). The phylogenetic tree of BchXYZ (Figure 2) gave an overview of all considered strains, while that of combined sequences of PufHLM-BchXYZ (Figure 3) covered all considered phototrophic purple bacteria. In addition, phylogenies of combined PufHLM-BchXYZ sequences and 16S rRNA gene sequences were compared (Figure 4).

2. Material and Methods

2.1. Cultivation, Sequencing, and Assembly of DNA Sequences

Cells were grown in the appropriate media, as described for the purple sulfur bacteria [11,12] and several groups of phototrophic purple bacteria [13]. Extraction and sequencing of DNA and the assembly of sequences were described earlier [14].

2.2. Sequences

Sequences of PufL, PufM, PufH, PufC, BchXYZ were retrieved from the annotated genomes. Genome sequences were annotated using the rapid annotations using subsystems technology (RAST) [15,16]. All sequences were deposited in the EMBL database. Accession numbers, together with species and strain designations, as well as the corresponding higher taxonomic ranks, are included in Supplementary Table S1.

2.3. Phylogenetic Analyses

Multiple sequence alignments (MSAs) were produced with MAFFT v7.313 [17,18] from all sequences and were visually inspected for consistency. MAFFT was run with parameters '- globalpair - maxiterate 1000'. Alignment positions with >25% gaps were trimmed from MSAs. Maximum likelihood (ML) phylogenetic trees were calculated from MSAs with IQ-TREE v1.6.1 [19] using the best substitution models inferred from MSAs. For trees calculated from combined alignments ('bchXYZ' and 'bchXYZpufHLM'), substitution models were used as so-called partition models [20]. Ultrafast bootstrap approximation (UFBoot) [21] was used to provide branch support values with 1000 replicates based on the same substitution models as the original ML tree. Branch support values were assigned onto the original ML tree as the number of times each branch in the original tree occurred in the set of bootstrap replicates (IQ-TREE option '-sup').

Phylogenetic trees were midpoint-rooted and formatted using functionality from R packages ape v5.0.1 [22], phangorn v2.3.2 [23], and phytools v0.6.45 [24]. Bootstrap values within a range of 80–100% were visualized as filled circles. The circle area is a linear function of the respective bootstrap value. The scale bar beneath a tree indicates the number of substitutions per alignment site. A co-phylogenetic plot was produced to facilitate the comparison of selected phylogenies. Nodes of compared trees were rotated to optimize tip matching.

3. Results and Discussion

3.1. Strain and Sequence Selection

Representatives of phototrophic *Proteobacteria* (10 orders, 21 families, 86 genera, 138 species, 159 strains + five unclassified strains) together with five representatives of *Chloroflexi* (one order, three families, three genera, five species) and six selected *Chlorobi* (one order, one family, four genera, six species), as well as *Gemmatimonas phototrophica*, *Chloracidobacterium acidophilum*, and *Heliobacterium modesticaldum* were included in the phylogenetic analyses of this study.

Depending on the availability of gene and genomic information, primarily sequence information from the type and reference strains was considered. In order to avoid any incongruity due to strain-dependent sequence variation, sequences from identical strains were used for all phylogenetic trees. All species and strain numbers are presented in Supplementary Table S1.

3.2. Phylogeny According to 16S rRNA Gene Sequences

As the 16S rRNA gene is established as a phylogenetic reference since the pioneering work of Carl Woese [25], we included the phylogenetic tree of this gene showing the relationship of all strains selected for the present study (RNA tree, Figure 1) and later compared this phylogenetic relationship with that of key proteins of photosynthesis (Figure 4). Clearly separated and distinct major groups with the deepest branching points in the tree were represented by *Chlorobi*, *Chloroflexi*, as well as *Heliobacterium modesticaldum* (representative of *Firmicutes* phylum), *Chloracidobacterium thermophilum* (representative of *Acidobacteria* phylum), and *Gemmatimonas phototrophica* (representative of *Gemmatimonadetes* phylum) (Figure 1). Quite remarkable was the isolated position of *Gemmatimonas*, which encodes a typical proteobacterial photosynthetic apparatus [26,27].

The *Proteobacteria* formed two distinct major branches with all *Alphaproteobacteria* in one branch and the *Gammaproteobacteria* and *Betaproteobacteria* in another branch. In the *Gammaproteobacteria* branch, distinct lineages were represented by *Chromatiaceae*, the *Ectothiorhodospira* group, the *Halorhodospira* group, the *Cellvibrionales* (aerobic anoxygenic phototrophic *Gammaproteobacteria*), and the *Betaproteobacteria*.

A much more complex situation existed within the *Alphaproteobacteria*, with a number of small groups with larger phylogenetic distance. The *Rhodobacterales* and also core groups of *Rhodospirillales*, *Rhizobiales*, and *Sphingomonadales* formed well-supported branches, which were, however, poorly resolved in their relationship to each other. Supported branches were formed by the members of the following genera:

(i) *Rhodospirillum, Roseospirillum/Caenispirillum, Rhodospira, Pararhodospirillum,*
(ii) *Phaeospirillum, Oceanibaculum, Rhodocista, Skermanella,*
(iii) *Rhodopila, Rubritepida, Paracraurococcus, Acidiphilum, Acidisphaera,*
(iv) *Erythrobacter, Porphyrobacter, Novosphingobium, Sphingomonas,*
(v) *Rhodopseudomonas/Bradyrhizobium, Methylobacterium.*

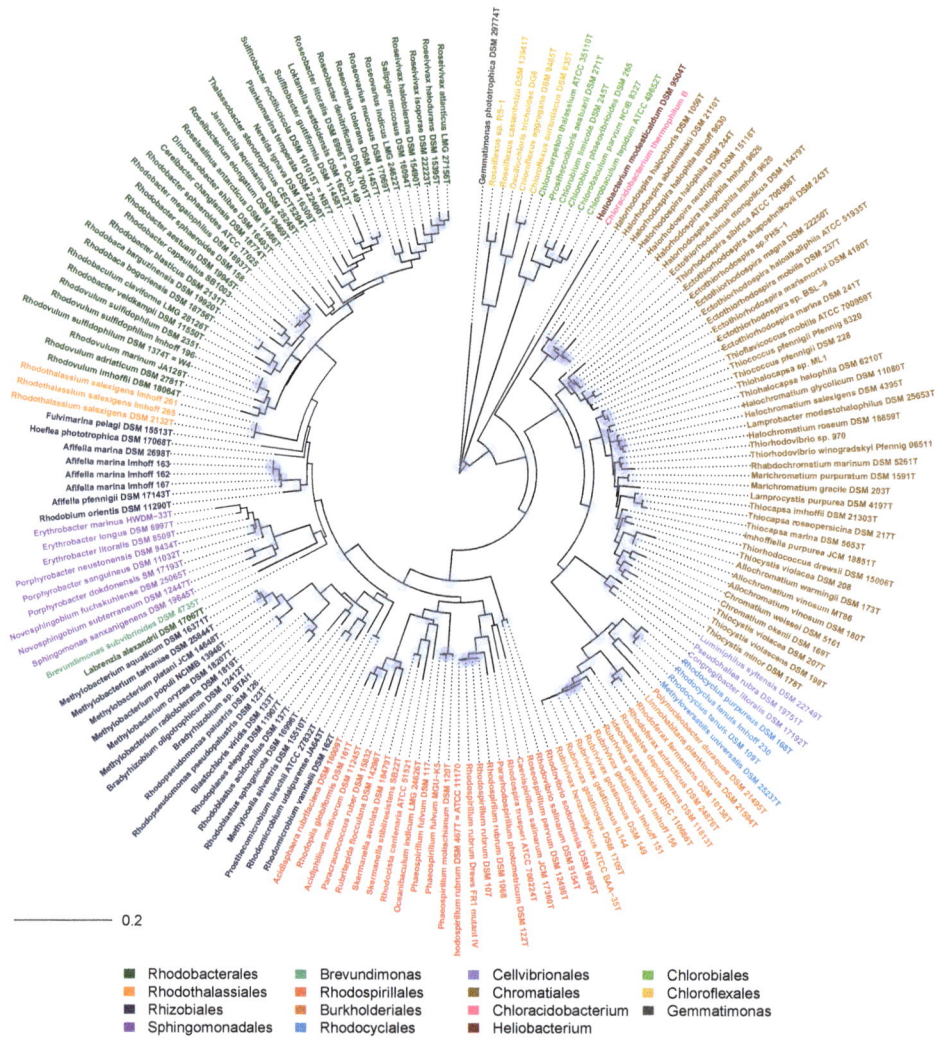

Figure 1. Phylogenetic tree (RNA tree) of phototrophic bacteria according to 16S rRNA gene sequences.

Most remarkable were the isolated positions of representatives of *Fulvimarina*, *Hoeflea*, *Labrenzia*, *Rhodothalassium*, and *Afifella-Rhodobium*. Though distant to other phototrophic bacteria, *Brevundimonas* clearly was linked to the *Sphingomonadales* branch. In addition, *Rhodovibrio* species appeared as clear outsiders and formed the most deeply branching lineage within the *Alphaproteobacteria*. In addition, several small groups were formed by single species or a few species only. These included species of *Blastochloris*, *Rhodoplanes*, *Rhodoblastus*, *Methylocella*, *Prosthecomicrobium*, and *Rhodomicrobium* (Figure 1).

It should be emphasized that *Roseospirillum parvum* was associated with the *Rhodospirillaceae* and, in particular, with the *Rhodospirillum/Pararhodospirillum* group as also *Caenispirillum* and *Rhodospira trueperi* do (Figure 1), supporting the current taxonomic classification [28].

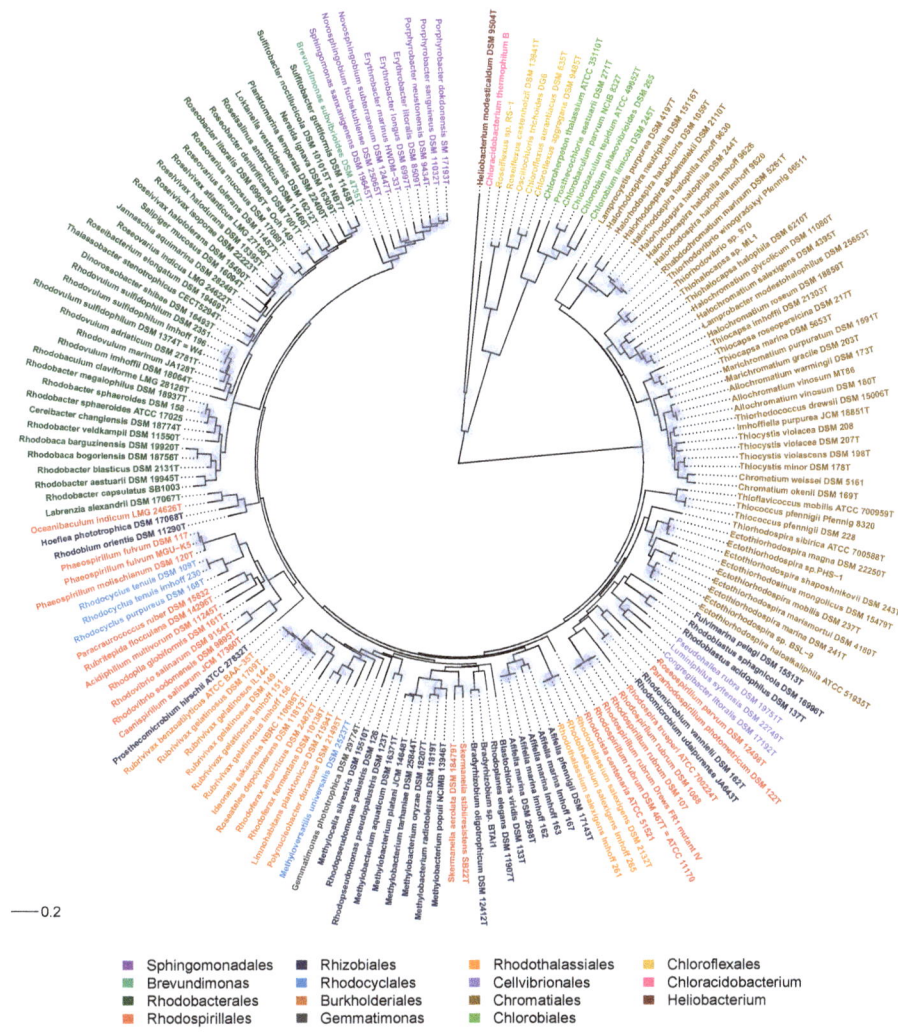

Figure 2. Phylogenetic tree of phototrophic bacteria, according to BchXYZ sequences.

3.3. Phylogeny of Photosynthesis

In order to evaluate the phylogeny of the photosynthetic apparatus, sequences of essential proteins for photosynthesis were analyzed. These included the bacteriochlorophyllide reductase BchXYZ and the photosynthetic reaction center proteins PufHLM and PufC. While the phylogenetic tree of BchXYZ (Figure 2) gave an overview of all considered strains and included all of the phototrophic green bacteria, the tree with combined sequences of PufHLM-BchXYZ (Figure 3) covered all phototrophic purple bacteria. PufC sequences were not considered in these trees because this component was absent from a number of representative species. All sequences and their accession numbers are presented in Supplementary Table S1.

3.3.1. Phylogeny according to BchXYZ Sequences

The phylogeny of BchXYZ allows the widest view on the phylogeny of photosynthesis in phototrophic bacteria, including PS-I and PS-II bacteria. The chlorophyllide reductase BchXYZ catalyzes the first step in bacteriochlorophyll biosynthesis that differentiates this pathway from the biosynthesis of chlorophyll. It is present in all phototrophic bacteria producing different forms of bacteriochlorophyll [10].

The deepest and likewise most ancient roots according to BchXYZ sequences (Figure 2) were found in the phototrophic green bacteria that employ a type-I photosystem, the *Chlorobi*, *Heliobacterium modesticaldum* and relatives, and *Chloracidobacterium thermophilum*, as well as in *Chloroflexi* that employ a type-II photosystem (like all *Proteobacteria*). The large sequence differences to the phototrophic purple bacteria pointed out that bacteriochlorophyll biosynthesis had evolved in ancestors of green bacteria much earlier as compared to phototrophic purple bacteria. This relationship quite well correlated to the phylogeny of the 16S rRNA gene (RNA tree) (Figure 1), with the exception of *Gemmatimonas phototrophica*, which, according to BchXYZ, was distantly associated with the *Betaproteobacteria*, specifically the *Burkholderiales* with *Rubrivivax* and *Rhodoferax* as representative genera. The phylogeny of photosynthesis in *Proteobacteria* was discussed on the basis of more comprehensive information of the BchXYZ-PufHLM sequences below (Figure 3).

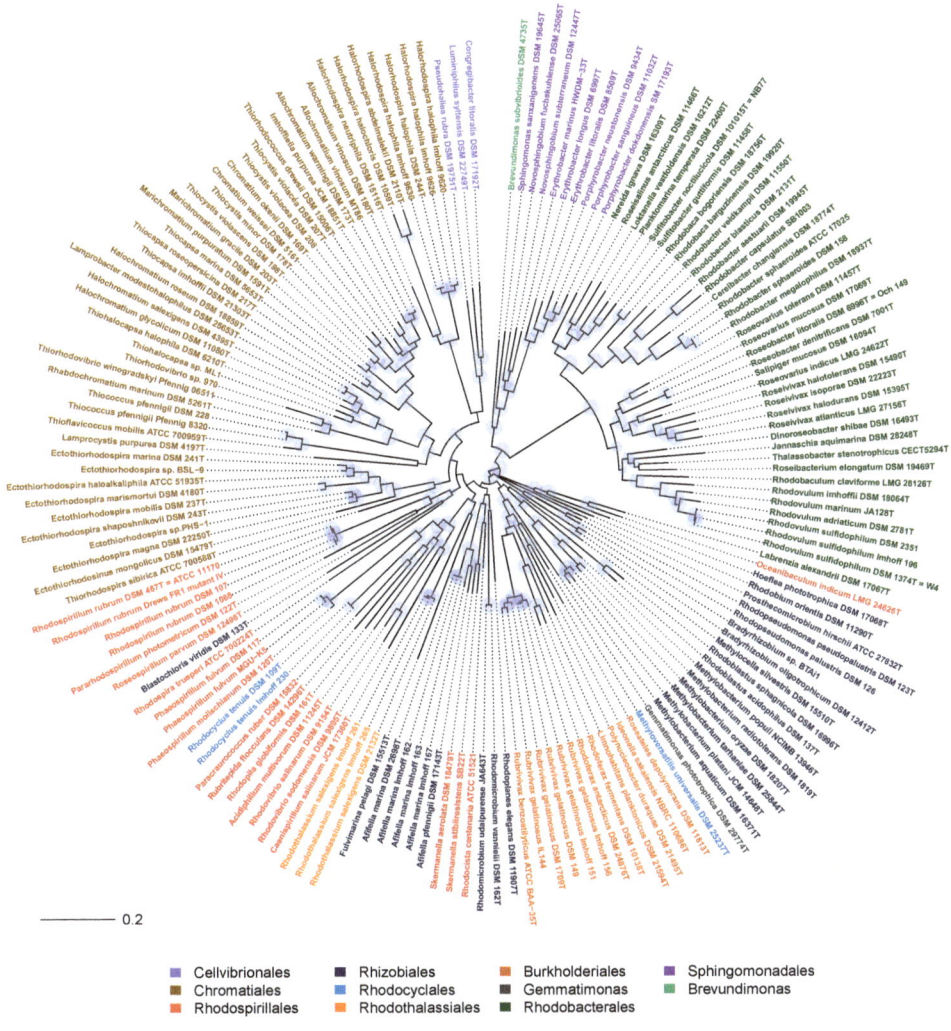

Figure 3. Phylogenetic tree (PS tree) of phototrophic bacteria, according to BchXYZ-PufHLM sequences.

3.3.2. Phylogeny of BchXYZ-PufHLM and Comparison with 16S rRNA Phylogeny

The combined sequence information of the key proteins of the photosynthetic reaction center in photosystem-II bacteria (PufHLM) and of the bacteriochlorophyll biosynthesis with the subunits of the chlorophyllide reductase (BchXYZ) gave a solid basis (alignment length, 2458 aa) to trace back the phylogeny of photosynthesis within the phototrophic purple bacteria (Figure 3). The consideration of PufHLM excluded the *Chloroflexi* (they lack PufH) in this consideration and restricted the view to *Proteobacteria* and *Gemmatimonas*. A direct comparison of the comprehensive phylogeny of anoxygenic photosynthesis, including sequences of BchXYZ-PufHLM (PS tree), with the phylogenetic relations according to 16S rRNA gene sequences (RNA tree) enlightened the evolution of photosynthesis as compared to that of the protein-producing machinery (Figure 4).

Gammaproteobacteria (Chromatiales and Cellvibrionales)

The phototrophic *Gammaproteobacteria* represented a well-established major phylogenetic branch with four major sub-branches, which were well supported within both PS tree and RNA tree. The sub-branches included

(i) the Halorhodospira species,
(ii) the *Ectothiorhodospiraceae*, including *Ectothiorhodospira* and *Ectothiorhodosinus* species, with *Thiorhodospira* being associated more distantly, but excluding the *Halorhodospira* species,
(iii) the *Chromatiaceae* with subgroups of a) the *Thiococcus* group of bacteriochlorophyll-b containing *Chromatiaceae*, including species of *Thiococcus* and *Thioflavicoccus*; b) the *Halochromatium* group with halophilic species of the genera *Halochromatium*, *Lamprobacter*, *Rhabdochromatium*, *Thiorhodovibrio*, and *Thiohalocapsa*; c) the *Chromatium* group with species of *Chromatium*, *Thiocapsa*, *Marichromatium*, *Allochromatium*, *Thiorhodococcus*, *Imhoffiella*, and *Thiocystis*; d) *Lamprocystis purpurea* as an outsider among the *Chromatiaceae* with distant relationship to others and no statistical support of its position. Most significantly, *Lamprocystis purpurea* formed a deeply branching line in the *Chromatium*-group according to both 16S rRNA phylogeny and PS phylogeny. Therefore, it is likely to be an ancient representative of the *Chromatiaceae*,
(iv) the Cellvibrionales (Haliaceae) with Congregibacter litoralis, Luminiphilus syltensis, and Pseudohaliea rubra (most likely including Chromatocurvus halotolerans [3]), which were linked with low confidence to the Halorhodospira group. The Cellvibrionales formed a group distant to other Gammaproteobacteria according to both trees. In the RNA tree, they were linked to the Betaproteobacteria (in this tree within the frame of the Gammaproteobacteria), and in the PS tree, associated with the Halorhodospira group. Apparently, they represent an ancient phylogenetic lineage of the Gammaproteobacteria without clearly resolved roots.

It was remarkable that the species with bacteriochlorophyll-b, according to the PS tree, formed different deeply rooted lineages associated with the corresponding bacteriochlorophyll-a containing relatives, *Hlr. abdelmalekii* and *Hlr. halochloris* were associated with the *Halorhodospira* branch, *Thiorhodococcus* and *Thioflavicoccus* species with the *Chromatiaceae*, and *Rhodospira trueperi* and *Blastochloris viridis* with the *Rhodospirillaceae*, specifically with the *Rhodospirillum* group though with a low significance (Figure 3). The incorporation of all bacteriochlorophyll-b-containing bacteria within one common cluster is restricted to the phylogeny of the reaction center proteins PufLM [3]. This has been previously explained by the congruent evolution of the reaction center proteins with respect to the specific binding requirements of the bacteriochlorophyll-b molecule [3] and implicates the independent evolution of the photosystems with bacteriochlorophyll-b in the different phylogenetic lineages.

Betaproteobacteria (Burkholderiales and Rhodocyclales)

One of the most obvious differences between PS and RNA trees was in the position of the *Betaproteobacteria*. In the RNA tree, *Rhodocyclales* and the *Burkholderiales* formed two related lineages of a major branch within the frame of the *Gammaproteobacteria* (Figure 1). In the PS tree, both groups formed clearly separated clusters, which were associated with different branches of the *Alphaproteobacteria* (Figure 3). The *Burkholderiales* formed a deep and not safely rooted branch, including separate lineages of *Rubrivivax*, *Ideonella/Roseateles*, *Rhodoferax/Limnohabitans*, *Polynucleobacter*, and *Methyloversatilis*. The deep roots identify the photosynthesis of these bacteria as very ancient and, despite the poorly supported branches, could indicate a possible acquisition of photosynthesis by gene transfer from an early phototrophic alphaproteobacterium, as supposed earlier [6,29] (Igarashi et al., 2001; Nagashima and Nagashima, 2013). The whole group was also visible in the RNA tree but associated with the *Gammaproteobacteria*. In the PS tree, *Rhodocyclus* was linked to *Phaeospirillum* and the *Rhodospirillales* (Figures 3 and 4), contrasting its link to the *Burkholderiales* in the RNA tree (Figure 1). This change might be indicative of a single event of a transfer of the photosynthesis genes from an ancient alphaproteobacterium within the *Rhodospirillales* frame to a *Rhodocyclus* ancestor.

Gemmatimonas (Gemmatimonadales)

Most significantly, *Gemmatimonas phototrophica* was found at the deepest branching point in the RNA tree, which placed this bacterium apart from all other phototrophic purple bacteria and also the phototrophic green bacteria. However, with the *Proteobacteria*, it shared the type-2 photosystem. In the PS tree, it formed a distinct line that split off at the deepest branching point from the *Burkholderiales* and was distantly linked to *Rubrivivax* (*Betaproteobacteria*). This was an indication that the photosynthetic roots of *Gemmatimonas* were associated with the ancient roots of the phototrophic *Burkholderiales*. If we exclude the acquisition of a foreign 16S rRNA, the most likely explanation for this discrepancy would be the acquirement of the photosynthesis genes by an early ancestor of *Gemmatimonas*, as suggested by Zeng et al. [26]. This event should have preceded the branching divergence of the *Burkholderiales*.

Alphaproteobacteria

The phototrophic *Alphaproteobacteria* formed the most fragmented and diverse array of groups in the PS tree with representatives of the six orders *Rhodospirillales*, *Rhizobiales*, *Sphingomonadales*, *Rhodobacterales*, *Caulobacterales*, and *Rhodothalassiales*. Most significantly, the *Rhodobacterales*, together with *Sphingomonadales* and *Brevundimonas* (*Caulobacterales*), formed a major branch, according to BchXYZ-PufHLM, which was clearly distinct from all other phototrophic *Alphaproteobacteria* (Figures 1, 3 and 4). A deep branching point separated the *Sphingomonadales* and *Brevundimonas* from the *Rhodobacterales*. The relations of other *Alphaproteobacteria*, however, were more problematic because most of the species had long-distance lines with deep branching points and only a few species arranged in stable groups that could be recognized in both PS tree and RNA tree.

Rhodobacterales. The most recent and shallow divergences were seen in the phylogeny of the *Rhodobacterales*, which, in contrast to most other phototrophic *Alphaproteobacteria*, appeared as a young group that had differentiated later than others and was well established as a group in PS tree and RNA tree. It diversified quite fast in evolutionary terms and now represents one of the largest orders of phototrophic bacteria known. The following groups of *Rhodobacterales* were formed in the PS tree. With the exception of the *Rhodobacter* group and the *Rhodovulum* group they represent aerobic phototrophic bacteria.

- *Rhodovulum* group: According to BchXYZ and BchXYZ-PufHLM, the *Rhodovulum* group was well recognized. *Rhodobaculum claviforme* appeared distantly associated with this group and, like the *Rhodovulum* species, had PufC (Supplementary Table S1). However, in the RNA tree, *Rhodobaculum claviforme* clustered with *Rhodobacter* species.
- *Rhodobacter/Rhodobaca* group: This group contained *Rhodobacter* and *Rhodobaca* species together with *Cereibacter changlensis* and was supported by all considered trees (BchXYZ-PufHLM, BchXYZ, RNA tree). The reaction center cytochrome PufC was absent (Supplementary Table S1). Quite remarkable *Rhodobaculum claviforme* was included in this group according to the RNA tree only.
- *Loktanella/Sulfitobacter* group: This group included species of *Loktanella*, *Sulfitobacter*, *Planktomarina*, and *Roseisalinus* and distantly linked also *Nereida ignava*. It was supported by BchXYZ-PufHLM and lacked PufC (Supplementary Table S1). According to the RNA tree, this group was not well supported, and *Roseobacter* but not *Roseisalinus* was included.
- *Roseobacter/Roseivivax* group: This group contained species of *Roseobacter*, *Roseivivax*, *Salipiger*, and *Roseovarius*. In line with the PS tree, PufC was present in all species, including *Roseobacter*. The RNA tree excluded *Roseobacter* from this group.
- *Dinoroseobacter/Jannaschia* group: *Dinoroseobacter shibae*, *Jannaschia aquamarina*, *Thalassobacter stenotrophicus*, and *Roseibacterium elongatum* formed a group of poorly linked bacteria, which did not fit into any of the aforementioned groups. All four species had PufC. Within the RNA tree, there was only weak support for this group (Figure 1).

Sphingomonadales. The *Sphingomonadales* formed a consistent lineage of aerobic phototrophic bacteria within all considered phylogenetic trees. *Sphingomonadaceae* with *Sphingomonas* and *Novosphingobium* species (likely also *Blastomonas*, see [3]) were forming one sub-branch and the *Erythrobacteraceae* with *Erythrobacter* and *Porphyrobacter* species (likely also *Erythromicrobium*, see [3]) a second one. There was support for the inclusion of *Erythrobacter marinus* into the *Sphingomonas* group from BchXYZ and BchXYZ-PufHLM phylogeny. In addition, *Erythrobacter marinus* contained PufC like *Sphingomonas* and *Novosphingobium* species but unlike other *Erythrobacteraceae*. According to the RNA tree, *Erythrobacter marinus* clustered with other *Erythrobacter* species, however, with low confidence in its position.

Brevundimonas (*Caulobacterales*). *Brevundimonas subvibrioides* represented an aerobic phototrophic bacterium, which clearly but distantly was linked to the *Sphingomonadales* branch according to the PS tree and RNA tree. *Brevundimonas* lacked PufC as the *Erythrobacteraceae* did. The deep branching point of *Brevundimonas* in the PS tree indicated that it was closest to the common ancestor of this branch.

The *Rhodobium/Hoeflea* group. A most deeply branching stable lineage in the PS tree was found within the radiation of the *Alphaproteobacteria* and was represented by the *Rhodobium/Hoeflea* group with *Rhodobium orientis, Hoeflea phototrophica, Labrenzia alexandrii,* and *Oceanibaculum indicum* (Figures 2 and 3). Despite the formation of a coherent group according to the PS tree, the species had different, though unsupported positions in the RNA tree (Figures 1 and 4). According to 16S rRNA, *Hoeflea phototrophica* (*Rhizobiales, Phylobacteriaceae*) had a deeply branching unsupported position; *Labrenzia alexandrii* (*Rhodobacterales, Rhodobacteraceae*) also had an unsupported position that was linked at the basis to *Brevundimonas* and the *Spingomonadales*; *Rhodobium orientis* (*Rhizobiales, Rhodobiaceae*) was found together with *Afifella* in a poorly rooted distinct branch; *Oceanibaculum indicum* (*Rhodospirillales, Rhodospirillaceae*) appeared distantly associated with *Rhodocista, Skermanella,* and the *Acetobacteraceae* (Figure 1). The photosynthesis of the *Rhodobium/Hoeflea* group represented one of the most ancient lines among the purple bacteria, and the most recent divergence (between *Labrenzia* and *Oceanibaculum*) was rooted much deeper as the basic divergence of the *Rhodobacterales* branch (Figure 3). In addition, there is no close relative to the photosynthesis system among other known phototrophic bacteria, which is a clear indication of the very ancient origin of photosynthesis in this lineage of phototrophic bacteria. If we do trust the phylogenetic reliability of the 16S rRNA system, we should assume quite early genetic transfers of major parts or the complete photosystem from an ancient ancestor within the *Rhodobium* lineage to the other bacteria. Alternatively, as the species of this branch formed poorly rooted lines in the RNA tree, the differences between PS and RNA tree might be explained by unresolved relationships and not correctly rooted positions of these bacteria in the RNA tree.

The *Rhodopseudomonas/Bradyrhizobium* group. In the PS tree, the *Rhodopseudomonas/Bradyrhizobium* group formed one of the most deeply branching lines distinct from other *Rhizobiales*. Both *Rhodopseudomonas* and *Bradyrhizobium* lacked PufC. According to 16S rRNA phylogeny, *Rhodopseudomonas* and *Bradyrhizobium* formed a sister branch to the photosynthetic *Methylobacterium* species, distant to other *Rhizobiales* (*Blastochloris* and *Rhodoplanes, Methylocella* and *Rhodoblastus, Prosthecomicrobium* and *Rhodomicrobium*).

The *Rhodopila* group. Another distinct group was represented by the *Acetobacteraceae* and supported by both RNA tree and PS tree with species of *Rhodopila, Acidiphilum, Paracraurococcus,* and *Rubritepida*.

The *Rhodospirillum* group. According to the PS tree, species of *Rhodospirillum, Pararhodospirillum,* and *Roseospirillum parvum* formed a group to which *Rhodospira trueperi* appeared distantly linked. In the RNA tree, *Caenispirillum* was included in this group, while in the PS tree, it had a separate position and formed a branch together with *Rhodovibrio* species, which, in turn, appeared as an isolated line at the basis of the *Alphaproteobacteria* within the RNA tree.

In addition to these groups, several separate lineages were represented by single genera of *Fulvimarina, Rhodothalassium, Prosthecomicrobium,* and *Afifella* in both PS and RNA trees (Figures 1 and 3). Thus, their phylogenetic positions remained unclear. While *Methylocella* specifically associated with *Rhodoblastus* in both RNA tree and PS trees, the following groupings were not well supported or had different positions in PS tree and RNA tree:

- *Rhodocista centenaria* and *Skermanella* species showed up jointly in the RNA tree with the Acetobacteraceae as a sister branch, while both formed a deeply rooted unsupported branch in the PS tree.
- *Rhodomicrobium* formed a distinct lineage within the *Rhizobiales* in the RNA tree but, according to the PS tree, separated from other *Rhizobiales* together with *Rhodoplanes* in a distinct deeply branching but unsupported line.
- *Blastochloris* separated from other *Rhizobiales* in the PS tree and formed an unsupported isolated line together with the bacteriochlorophyll-b containing *Rhodospira trueperi*, while it was included in a major branch of *Rhizobiales* in the RNA tree.

3.4. Distribution of PufC

The cytochrome associated with the photosynthetic reaction center is an important component in many of the PS-II type photosynthetic bacteria. As a more peripheral part of the photosynthetic reaction center, the cytochrome may be more easily replaced by alternative electron transport systems, and this obviously happened in a number of phylogenetic lineages of the *Alphaproteobacteria* and *Betaproteobacteria* (Supplementary Table S1). The general presence of the reaction center cytochrome PufC in phototrophic purple bacteria and the absence in quite a few distinct groups of the *Alphaproteobacteria* and *Betaproteobacteria* strongly suggested that PufC independently has been lost several times. PufC was absent in *Rhodoferax fermentans* (but present in the related *Rhodoferax antarcticus*), in *Rhodospirillum rubrum* (but present in the related *Pararhodospirillum photometricum*), in *Bradyrhizobium* and *Rhodopseudomonas* species (but present in other *Rhizobiales*), in *Brevundimonas*, *Porphyrobacter*, and *Erythrobacter* species (but present in *Erythrobacter marinus* and *Sphingomonadaceae*). It was also absent in one of the major *Rhodobacterales* branches of the PS tree, including the *Rhodobacter/Rhodobaca* group and the *Loktanella/Sulfitobacter* group. The presence/absence of PufC in species of the *Alphaproteobacteria* was congruent with the photosynthesis phylogeny. The presence of PufC supported the inclusion of *Rhodobaculum* into the *Rhodovulum* group, and its absence in *Roseisalinus* was in accord with its inclusion into the *Loktanella/Sulfitobacter* group according to the PS tree. Following the PS tree and the presence of PufC, *Erythrobacter marinus* fitted into the *Sphingomonadaceae* (rather than into the *Erythrobacteraceae*).

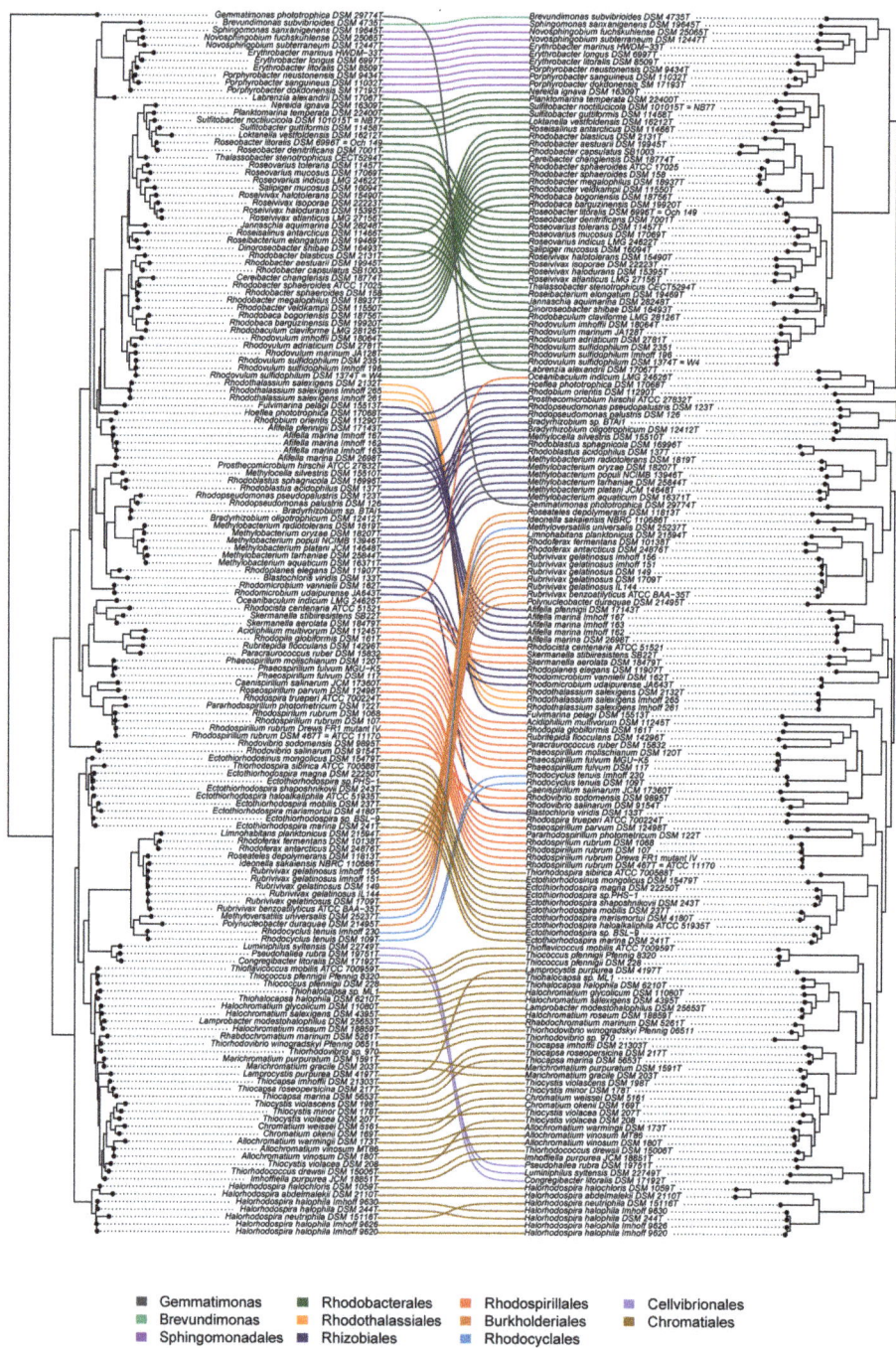

Figure 4. Linear comparison tree of 16S rRNA gene sequences (right) and BchXYZ-PufHLM sequences (left) of phototrophic bacteria.

3.5. Phylogenetic Aspects of Aerobic Anoxygenic Photosynthesis

As oxygen was absent from the earth during the first billion years of life, in which the basic concepts of photosynthesis are expected to have evolved and as oxygenic photosynthesis using two different consecutive photoreactions is considered a late event in the evolution of photosynthesis, the roots of anoxygenic, as well as oxygenic photosynthesis, are to be found in the ancient anoxic environments [30,31]. We assume that with the onset of oxygenic photosynthesis, the basic concepts of photosynthesis, as known today, have already been established. In fact, oxygen evolution by photosynthesis using water as electron source is dependent on the use of two different consecutive photosynthetic reactions that have evolutionary ancestors among anoxygenic phototrophic bacteria: a type-I photosynthesis in ancestors of *Heliobacteria*, *Chlorobi*, and *Chloracidobacterium* and a type-II photosynthesis in ancestors of *Chloroflexi* and all phototrophic purple bacteria [31].

The appearance of oxygenic photosynthesis approx. 3 billion years ago was a revolution in ecology. It drastically changed the environmental conditions on earth, and over approx. 2 billion of years caused the gradual increase of the atmospheric oxygen content to the actual level [32]. Quite likely, during this transition period, the radiation of the purple bacteria diverged to its full extension.

During adaptation to oxic conditions, quite a number of anoxygenic phototrophic purple bacteria may have gained the ability to perform under both anoxic and oxic conditions by maintaining the strict regulation of biosynthesis of the photosynthetic apparatus and its repression by oxygen. An example of such bacteria is found in *Rhodobacter* species performing anoxygenic photosynthesis under anaerobic conditions in the light and aerobic respiration under oxic conditions in the dark [33–36]. During further evolution, some of these phototrophic bacteria may have lost the ability to build up the photosynthetic apparatus in the absence of oxygen and, in contrast, required oxygen for the formation of the photosynthetic apparatus [37]. In bacteria, such as the anaerobic phototrophic *Rhodobacterales*, that have already been adapted to arrange themselves with certain levels of oxygen, this could have been a small step in modifying the oxygen-response in the formation of the photosynthetic apparatus. As a result, in various phylogenetic branches, aerobic anoxygenic phototrophic bacteria have evolved, of which *Erythrobacter* and *Roseobacter* species at present are the most well-known examples [37–39].

Aerobic representatives of phototrophic purple bacteria that require oxygen for the synthesis of bacteriochlorophyll and the photosynthetic apparatus were found in a number of well-defined groups. The *Haliaceae* of the *Gammaproteobacteria* and the *Erythrobacteraceae* and *Sphingomonadaceae* of the *Sphingomonadales* at present exclusively contain aerobic representatives. In addition, isolated lines of single representatives of aerobic phototrophic bacteria were found with species of *Fulvimarina*, *Gemmatimonas*, *Polynucleobacter*, and *Brevundimonas*. In the *Rhodobium/Hoeflea* group, several aerobic phototrophic bacteria joined the anaerobic phototrophic *Rhodobium*. In the *Acetobacteraceae*, several aerobic representatives were found together with the anaerobic phototrophic *Rhodopila*. A larger group of the aerobic phototrophic *Rhizobiales* with *Methylobacterium*, *Methylocella*, *Prosthecomicrobium*, and *Bradyrhizobium* was related to the anaerobic phototrophic *Rhodopseudomonas* and *Rhodoblastus* species. Among the *Rhodobacterales*, the groups around *Loktanella/Sulfitobacter*, *Dinoroseobacter/Jannaschia*, and *Roseivivax/Roseovarius* represented branches with aerobic species.

With the exception of the aerobic phototrophic *Rhodobacterales*, most of the aerobic phototrophic bacteria represented ancient phylogenetic lineages. This was especially the case for the *Sphingomonadales*, *Brevundimonas* (*Caulobacterales*), the *Haliaceae*, and those within the *Rhodobium/Hoeflea* group. As traces or small levels of oxygen already were present at the time when anaerobic phototrophic purple bacteria presumably diversified approx. 2.5 billion years ago, the aerobic phototrophic purple bacteria could have developed in parallel, which would explain the deep divergence of some of the lineages of aerobic phototrophic *Proteobacteria*.

The pattern of distribution of aerobic phototrophic bacteria among the phototrophic purple bacteria strongly suggested that aerobic phototrophic purple bacteria evolved from anaerobic ancestors in independent and multiple events. The deep branching points of some lineages indicated their early divergence from the anaerobic phototrophic ancestors. The phylogeny suggested that in the

Rhodobium/Hoeflea group, photosynthesis of aerobic representatives evolved from an anaerobic ancestor with a common root with *Rhodobium*. Known representatives of aerobic phototrophic *Sphingomonadales*, *Cellvibrionales (Haliaceae)*, *Caulobacterales (Brevundimonas)*, and *Gemmatimonas* could be present-day survivors of ancient anaerobic phototrophic relatives not known so far or extinct.

The development of aerobic phototrophic *Rhodobacterales* was considered to be a more recent event, as these bacteria and their photosynthesis were much younger compared to most other phototrophic bacteria (Figure 3). It has been calculated by molecular clock calculations using sequences of representative genes that the divergence of the last common ancestor of *Roseobacter* and *Rhodobacter* dates to approx. 900 Myr ago (+/- 200 Myr) [40]. At that time, the oxygen content of the earth's atmosphere almost had reached present-day levels [32], and it is tempting to assume that aerobic phototrophic lineages branched off from anaerobic phototrophic *Rhodobacterales* under these conditions. Though this is considered the more likely scenario, alternatively, the common ancestor of the *Rhodobacterales* could have been an aerobic phototrophic bacterium. This scenario could find support in the common roots of *Rhodobacterales* with the aerobic phototrophic *Sphingomonadales* and *Brevundimonas* but implies that the ancestors of *Sphingomonadales* and *Brevundimonas* also were aerobic phototrophic bacteria, which is not necessarily the case. It would also imply that aerobic phototrophic *Rhodobacterales* transformed back to perform anaerobic photosynthesis, which from an evolutionary and ecological perspective appears quite unlikely. Therefore, we assume that aerobic phototrophic representatives of *Rhodobacterales*, *Sphingomonadsales*, and *Brevundimonas* evolved independently from anaerobic phototrophic ancestors.

3.6. General Aspects

Traditionally and especially due to the pioneering work of Carl Woese [25], the 16S rRNA gene sequence has been established as the basic tool for the analysis of bacterial phylogenies. Though the sequence information contained in this molecule is of limited size (approx. 1400 nt), it is considered as particularly conservative in evolutionary terms. In consequence, 16S rRNA gene sequences still are used as a backbone for phylogenetic considerations, although limitations are to be expected due to the comparable small sequence information and restricted resolution. Further limitations may be due to multiple changes in individual sequence positions and insertions/deletions over time, which could blur the phylogenetic roots in particular of the deep branching lineages.

In this context, it was quite remarkable that some of those species/groups that revealed the most obvious differences between the PS tree and RNA tree also showed the deepest branching points within the PS tree. In fact, a number of those species and branches that were not congruent with respect to RNA and PS phylogeny had statistically poorly supported positions in either one or both of the trees. This was especially relevant for most of the *Alphaproteobacteria*, which appeared—with the remarkable exception of the *Rhodobacterales*—to be the most ancient group of phototrophic *Proteobacteria*.

Despite the uncertainty in the resolution of the very deeply branching lineages, the transfer of photosynthesis genes could explain several of the discrepancies between the PS and RNA trees. Such mechanisms have been postulated earlier [26,29], and genetic exchange could have occurred repeatedly in the early ages of photosynthesis in *Proteobacteria*. Examples of such possible exchange events during early diversification of the phototrophic purple bacteria may be found in *Gemmatimonas phototrophica*, the *Rhodobium/Hoeflea* group, and the *Betaproteobacteria* with different events of *Rhodocyclus* and *Rubrivivax* and their relatives.

The situation was different within the *Rhodobacterales*, which is a well-resolved group with a clear distinction from other *Proteobacteria*. It is the youngest diversification within the phototrophic *Proteobacteria*. For this group, lateral gene transfer has been demonstrated [41]. It could, in fact, be shown that the photosynthetic gene cluster in several genomes, including *Sulfitobacter* and *Roseobacter* species, is located on a plasmid, which enforces the genetic exchange of the whole cluster [41]. Several of such exchanges could explain the divergences between photosynthesis phylogeny and RNA phylogeny among the *Rhodobacterales* [41]. Despite the established gene transfer inside the *Rhodobacterales*, it appears highly unlikely that the whole group received the photosynthetic gene cluster by lateral transfer from an external donor. The long phylogenetic distance to most other photosynthesis systems and the basically good correlation of RNA and PS phylogeny in regard to relations of *Sphingomonadales*, *Brevundimonas*, and *Rhodobacterales* precludes the transfer from any other known phototrophic lineage. As we have no knowledge of the existence of similar gene transfer agents in other phototrophic bacteria, this kind of genetic exchange of the complete photosynthetic gene cluster could be a late acquisition and unique to the *Rhodobacterales*. However, it would be interesting to study the situation in *Erythrobacter marinus*, which could have similarly received its photosynthesis genes from a relative of the *Sphingomonadaceae* branch.

4. Conclusions

The immense phylogenetic diversity of photosynthetic prokaryotes was demonstrated by the wide systematic range of these bacteria. Bacteria considered in this communication were cultured representatives from six phyla (the cyanobacteria were not considered here) with 15 orders, 27 families, and 90 genera. The most ancient representatives of the phototrophic bacteria, the first that made bacteriochlorophyll (chlorophyllide reductase, BchXYZ) and performed photosynthesis were the phototrophic green bacteria, in particular, those with a type-I photosystem (*Chlorobi*, *Heliobacterium*, *Chloracidobacterium*) (Figures 1 and 2). Among those with a type-II photosystem, the *Chloroflexi* have by far the most ancient roots [3], and *Proteobacteria*, together with their photosystem, diversified to the present-day forms much later (Figures 1 and 3). There was an apparent large gap in the evolution of photosynthesis in the phototrophic green bacteria and in the *Proteobacteria*. This is an indication for the loss of early stages of the photosystem present in the Proteobacteria (Figures 1 and 2).

Today phototrophic *Proteobacteria* are by far the most diverse and the most abundant in the environment and have to be considered the most successful to adapt to the largely oxic environment. If we consider that the basic divergences within the *Rhodobacterales* (e.g., the separation of *Rhodobacter* and *Roseobacter*) have occurred approx. 1 billion years ago [40] and that the first photosynthetic prokaryotes have evolved approx. 3.2–3.5 billion years ago, it is reasonable to conclude from the phylogeny of photosynthesis that phototrophic *Proteobacteria* appeared around 2–2.5 billion years ago. If we use these rough estimates as a guide for the interpretation of the phylogenetic relations of photosynthesis, we can conclude that the ancestors of the green bacteria dominated the field over approx. a billion years and quite likely ancestors of the strictly anaerobic *Chlorobi* played a prominent role in the sulfur oxidation during this time. The *Chlorobi* maintained their strict phototrophic and also an anaerobic way of life up to today and consequently are pushed back to the few anoxic/sulfidic ecological niches that receive light. In the early ages also, the photosystem type-II originated and presumably soon separated into a system represented by our present-day *Chloroflexi* and a system that developed later within the phototrophic *Proteobacteria*. If we assume a common origin of the photosystem type-II in *Chloroflexi* and *Proteobacteria*, the system, as we know it from the *Proteobacteria*, is an advanced stage of a parallel development that diversified together with these bacteria much later. Ancient forms that could represent a link between the two type-II photosystems, apparently, were extinct or survivors have not yet been detected.

The most ancient roots of photosynthesis among *Proteobacteria* are found in the *Alphaproteobacteria* (excluding *Rhodobacterales*) and *Betaproteobacteria* with often unsupported deep divergences and long lines to the present-day representatives, the species/strains studied. Photosynthesis in *Gammaproteobacteria* diversified significantly later with the origin of the *Ectothiorhodospira* group, predating that of the others (*Halorhodospira*, *Chromatiaceae*, *Cellvibrionales*). As the photosynthesis phylogeny in general terms was congruent with the RNA phylogeny (Figures 1–4), it was concluded that this type-II photosystem diversified together with the *Proteobacteria*.

Compared to the phylogenetic depth and the systematic width found in the radiation of the *Alphaproteobacteria*, relatively few genera are known of these bacteria, with the exception of *Rhodobacterales*. The *Rhodobacterales*, on the other hand, represent the most recently diverged group.

These bacteria apparently are the most successful to live in our mostly oxic world today, are most versatile in their metabolism, are well adapted to live in the oxic environment, and represent one of the largest orders of phototrophic bacteria living today. A second large gap in the evolution of photosynthesis is, in fact, seen between the *Rhodobacterales* and all other *Proteobacteria* (Figure 3). Another large group, which is clearly separated from the others, but diversified earlier than the *Rhodobacterales* is represented by the *Chromatiales*. These bacteria characteristically are adapted to the borderline between the anoxic/sulfidic and the oxic environment, have found ecological niches over billions of years and survived successfully until today.

The phototrophic bacteria included in this investigation were representatives of most of those that are known and in laboratory culture today. Therefore, the presented data gave a comprehensive basis of the phylogeny of anoxygenic photosynthesis, although the view was limited due to the fact that all those that have escaped cultivation attempts or for other reasons have not been cultured could not be considered. As genetic studies with communities of phototrophic purple bacteria from marine coastal sediments based on the PufLM metagenomic diversity demonstrated that many of those present (or close relatives thereof) already had been cultivated [7], it was concluded that a great part of those out in nature have already been identified, at least from coastal marine sediments. Nevertheless, we will certainly continue finding new species and phylogenetic lines of phototrophic bacteria, in particular when unstudied or poorly studied locations and environments are investigated. Comprehensive metagenome studies on a great number of environmental communities might even detect missing links of photosynthesis evolution.

Supplementary Materials: The following are available online at http://www.mdpi.com/2076-2607/7/11/576/s1.

Author Contributions: Cultivation of bacterial cultures, DNA extraction, and purification was performed by T.R.; genomic sequencing and quality assurance by S.K.; sequence assembly, retrieval from databases, phylogenetic calculations, tree construction, and visualization by S.C.N. Sequence annotation, retrieval of sequences from annotated genomes and databases, design of the study, as well as the writing of the manuscript was made by J.F.I. All authors contributed to and approved the work for publication.

Conflicts of Interest: The authors declare no conflict of interest.

References

1. Madigan, M.T.; Schaaf, N.A.V.; Sattley, W.M. The Chlorobiaceae, Chloroflexaceae, and Heliobacteriaceae. In *Modern Topics in the Phototrophic Prokaryotes: Environmental and Applied Aspects*; Hallenbeck, P.C., Ed.; Springer: Cham, Switzerland, 2017; pp. 139–161.
2. Imhoff, J.F. Diversity of anaerobic anoxygenic phototrophic purple bacteria. In *Modern Topics in the Phototrophic Prokaryotes: Environmental and Applied Aspects*; Hallenbeck, P.C., Ed.; Springer: Cham, Switzerland, 2017; pp. 47–85.
3. Imhoff, J.F.; Rahn, T.; Künzel, S.; Neulinger, S.C. Photosynthesis is widely distributed among proteobacteria as revealed by the phylogeny of PufLM reaction center proteins. *Front. Microbiol.* **2018**, *8*, 2679. [CrossRef] [PubMed]
4. Zhang, H.; Sekiguchi, Y.; Hanada, S.; Hugenholtz, P.; Kim, H.; Kamagata, Y.; Nakamura, K. *Gemmatimonas aurantiaca* gen. nov., sp. nov., a Gram-negative, aerobic, polyphosphate-accumulating microorganism, the first cultured representative of the new bacterial phylum Gemmatimonadetes phyl. nov. *Int. J. Syst. Evol. Microbiol.* **2003**, *53*, 1155–1163. [CrossRef] [PubMed]
5. Zeng, A.; Koblížek, M. Phototrophic Gemmatimonadetes: A new "purple" branch of the bacterial tree of life. In *Modern Topics in the Phototrophic Prokaryotes: Environmental and Applied Aspects*; Hallenbeck, P.C., Ed.; Springer: Cham, Switzerland, 2017; pp. 163–192.
6. Nagashima, S.; Nagashima, K.V.P. Comparison of photosynthrtic gene clusters retrieved from total genome sequences of purple bacteria. In *Genome Evolution of Photosynthetic Bacteria*; Beatty, T.J., Ed.; Elsevier: Amsterdam, The Netherlands, 2013; pp. 151–178.
7. Tank, M.; Blümel, M.; Imhoff, J.F. Communities of purple sulfur bacteria in a Baltic Sea coastal lagoon analyzed by *pufLM* gene libraries and the impact of temperature and NaCl concentration in experimental enrichment cultures. *FEMS Microbiol. Ecol.* **2011**, *78*, 428–438. [CrossRef] [PubMed]

8. Tank, M.; Thiel, V.; Imhoff, J.F. Phylogenetic relationship of phototrophic purple sulfur bacteria according to *pufL* and *pufM* genes. *Intern. Microbiol.* **2009**, *12*, 175–185.
9. Thiel, V.; Tank, M.; Neulinger, S.C.; Gehrmann, L.; Dorador, C.; Imhoff, J.F. Unique communities of anoxygenic phototrophic bacteria in saline lakes of Salar de Atacama (Chile). Evidence for a new phylogenetic lineage of phototrophic Gammaproteobacteria from *pufLM* gene analyses. *FEMS Microbiol. Ecol.* **2010**, *74*, 510–522. [CrossRef]
10. Willows, R.D.; Kriegel, A.M. Biosynthesis of bacteriochlorophylls in purple bacteria. In *Advances in Photosynthesis and Respiration. The Purple Phototrophic Bacteria*; Hunter, C.N., Daldal, F., Thurnauer, M.C., Beatty, J.T., Eds.; Springer: Dordrecht, The Netherlands, 2009; pp. 57–79.
11. Pfennig, N.; Trüper, H.G. Isolation of members of the families Chromatiaceae and Chlorobiaceae. In *The Prokaryotes, A Handbook on Habitats, Isolation and Identification of Bacteria*; Starr, M.P., Stolp, H., Trüper, H.G., Balows, A., Schlegel, H.G., Eds.; Springer: New York, NY, USA, 1981; pp. 279–289.
12. Imhoff, J.F. The Chromatiaceae. In *The Prokaryotes. A Handbook on the Biology of Bacteria*, 3rd ed.; Dworkin, M., Falkow, S., Rosenberg, E., Schleifer, K.-H., Stackebrandt, E., Eds.; Springer: New York, NY, USA, 2006; Volume 6, pp. 846–873.
13. Imhoff, J.F. Anoxygenic phototrophic bacteria. In *Methods in Aquatic Bacteriology*; Austin, B., Ed.; John Wiley & Sons: Chichester, UK, 1988; pp. 207–240.
14. Imhoff, J.F.; Rahn, T.; Künzel, S.; Neulinger, S.C. New insights into the metabolic potential of the phototrophic purple bacterium *Rhodopila globiformis* DSM 161^T from its draft genome sequence and evidence for a vanadium-dependent nitrogenase. *Arch. Microbiol.* **2018**, *200*, 847–857. [CrossRef]
15. Aziz, R.K.; Bartels, D.; Best, A.A.; DeJongh, M.; Disz, T.; Edwards, R.A.; Formsma, K.; Gerdes, S.; Glass, E.M.; Kubal, M.; et al. The RAST Server: Rapid Annotations using Subsystems Technology. *BMC Genom.* **2008**, *9*, 75. [CrossRef]
16. Overbeek, R.; Olson, R.; Pusch, G.D.; Olsen, G.J.; Davis, J.J.; Disz, T.; Edwards, R.A.; Gerdes, S.; Parrello, B.; Shukla, M.; et al. The SEED and the Rapid Annotation of microbial genomes using Subsystems Technology (RAST). *Nucleic Acids Res.* **2014**, *42*, D206–D214. [CrossRef]
17. Katoh, K.; Misawa, K.; Kuma, K.-I.; Miyata, T. MAFFT: A novel method for rapid multiple sequence alignment based on fast Fourier transform. *Nucleic Acids Res.* **2002**, *30*, 3059–3066. [CrossRef]
18. Katoh, K.; Standley, D.M. MAFFT multiple sequence alignment software version 7: Improvements in performance and usability. *Mol. Biol. Evol.* **2013**, *30*, 772–780. [CrossRef]
19. Nguyen, L.-T.; Schmidt, H.A.; von Haeseler, A.; Minh, B.Q. IQ-TREE: A fast and effective stochastic algorithm for estimating maximum-likelihood phylogenies. *Mol. Biol. Evol.* **2015**, *32*, 268–274. [CrossRef] [PubMed]
20. Chernomor, O.; von Haeseler, A.; Minh, B.Q. Terrace aware data structure for phylogenomic inference from supermatrices. *Syst. Biol.* **2016**, *65*, 997–1008. [CrossRef] [PubMed]
21. Minh, B.Q.; Nguyen, M.A.T.; von Haeseler, A. Ultrafast approximation for phylogenetic bootstrap. *Mol. Biol. Evol.* **2013**, *30*, 1188–1195. [CrossRef] [PubMed]
22. Paradis, E.; Claude, J.; Strimmer, K. APE: Analyses of phylogenetics and evolution in R language. *Bioinformatics* **2004**, *20*, 289–290. [CrossRef] [PubMed]
23. Schliep, K.P. Phangorn: Phylogenetic analysis in R. *Bioinformatics* **2011**, *27*, 592–593. [CrossRef]
24. Revell, L.J. Phytools: An R package for phylogenetic comparative biology (and other things). *Methods Ecol. Evol.* **2012**, *3*, 217–223. [CrossRef]
25. Woese, C.R. Bacterial evolution. *Microbiol. Rev.* **1987**, *51*, 4673–4680.
26. Zeng, A.; Feng, F.; Medova, H.; Dean, J.; Koblizek, M. Functional type 2 photosynthetic reaction centers found in the rare bacterial phylum Gemmatimonadetes. *Proc. Natl. Acad. Sci. USA* **2014**, *111*, 7795–7800. [CrossRef]
27. Zeng, A.; Selyanin, V.; Lukes, M.; Dean, J.; Kaftan, D.; Feng, F.; Koblizek, M. Characterization of the microaerophilic, bacteriochlorophyll a-containing bacterium *Gemmatimonas phototrophica* sp. nov., and emended description of the genus *Gemmatimonas* and *Gemmatimonas auranttiaca*. *Int. J. Syst. Evol. Microbiol.* **2015**, *65*, 2410–2419. [CrossRef]
28. Imhoff, J.F. Genus *Roseospirillum*. In *Bergey's Manual of Systematic Bacteriology*, 2nd ed.; Brenner, D.J., Krieg, N.R., Staley, J.T., Eds.; Springer: New York, NY, USA, 2005; Volume 2, p. 39.

29. Igarashi, N.; Harada, J.; Nagashima, S.; Matsuura, K.; Shimada, K.; Nagashima, K.V.P. Horizontal transfer of photosynthesis gene cluster and operon rearrangement in purple bacteria. *J. Mol. Evol.* **2001**, *52*, 333–341. [CrossRef]
30. Holland, H.D. The oxygenation of the atmosphere and oceans. *Phil. Trans. R. Soc. B* **2006**, *361*, 903–915. [CrossRef] [PubMed]
31. Blankenship, R.E. Early evolution of photosynthesis. *Plant Physiol.* **2010**, *154*, 434–438. [CrossRef] [PubMed]
32. Canfield, D.E. The early history of atmospheric oxygen: Homage to Robert, M. Garrels. *Annu. Rev. Earth Planet. Sci.* **2005**, *33*, 1–36. [CrossRef]
33. Drews, G.; Imhoff, J.F. Phototrophic purple bacteria. In *Variations in Autotrophic Life*; Shively, J.M., Barton, L.L., Eds.; Academic Press: London, UK, 1991; pp. 51–97.
34. Imhoff, J.F. Genus *Rhodobacter*. In *Bergey's Manual of Systematic Bacteriology*, 2nd ed.; Brenner, D.J., Krieg, N.R., Staley, J.T., Eds.; Springer: New York, NY, USA, 2005; Volume 2, pp. 161–167.
35. Imhoff, J.F. Genus *Rhodovulum*. In *Bergey's Manual of Systematic Bacteriology*, 2nd ed.; Brenner, D.J., Krieg, N.R., Staley, J.T., Eds.; Springer: New York, NY, USA, 2005; Volume 2, pp. 205–209.
36. Imhoff, J.F. The phototrophic Alphaproteobacteria. In *The Prokaryotes. A Handbook on the Biology of Bacteria.*, 3rd ed.; Dworkin, M., Falkow, S., Rosenberg, E., Schleifer, K.-H., Stackebrandt, E., Eds.; Springer: New York, NY, USA, 2006; Volume 5, pp. 41–64.
37. Shimada, K. Aerobic anoxygenic phototrophs. In *Anoxygenic Photosynthetic Bacteria*; Blankenship, R.E., Madigan, M.T., Bauer, C.E., Eds.; Kluwer: Dordrecht, The Netherlands, 1995; pp. 105–122.
38. Shiba, T. *Roseobacter litoralis* gen. nov., sp. nov., and *Roseobacter denitrificans* sp. nov., aerobic pink-pigmented bacteria which contain bacteriochlorophyll a. *Syst. Appl. Microbiol.* **1991**, *14*, 140–145. [CrossRef]
39. Yurkov, V. Aerobic phototrophic Proteobacteria. In *The Prokaryotes. A Handbook on the Biology of Bacteria*, 3rd ed.; Dworkin, M., Falkow, S., Rosenberg, E., Schleifer, K.-H., Stackebrandt, E., Eds.; Springer: New York, NY, USA, 2006; Volume 5, pp. 562–584.
40. Koblížek, M.; Zeng, Y.; Horák, A.; Oborník, M. Regressive evolution of photosynthesis in the *Roseobacter* clade. In *Genome Evolution of Photosynthetic Bacteria*; Beatty, T.J., Ed.; Elsevier: Amsterdam, The Netherlands, 2013; pp. 1385–1405.
41. Brinkmann, H.; Göker, M.; Koblížek, M.; Wagner-Döbler, I.; Petersen, J. Horizontal operon transfer, plasmids, and the evolution of photosynthesis in Rhodobacteraceae. *ISME J.* **2018**, *12*, 1994–2010. [CrossRef]

© 2019 by the authors. Licensee MDPI, Basel, Switzerland. This article is an open access article distributed under the terms and conditions of the Creative Commons Attribution (CC BY) license (http://creativecommons.org/licenses/by/4.0/).

Article

Simultaneous Genome Sequencing of *Prosthecochloris ethylica* and *Desulfuromonas acetoxidans* within a Syntrophic Mixture Reveals Unique Pili and Protein Interactions

John A. Kyndt [1,*], Jozef J. Van Beeumen [2] and Terry E. Meyer [3,†]

1. College of Science and Technology, Bellevue University, Bellevue, NE 68005, USA
2. Department of Biochemistry and Microbiology, Ghent University, 9000 Gent, Belgium; jozef.vanbeeumen@ugent.be
3. Department of Chemistry and Biochemistry, University of Arizona, Tucson, AZ 85721, USA; temeyer@email.arizona.edu
* Correspondence: jkyndt@bellevue.edu; Tel.: +1-402-557-7551
† Passed away while this manuscript was in the final stages of preparation.

Received: 1 November 2020; Accepted: 4 December 2020; Published: 7 December 2020

Abstract: Strains of *Chloropseudomonas ethylica*, 2-K, N2, and N3 are known to be composed of a syntrophic mixture of a green sulfur bacterium and a sulfur-reducing colorless component. Upon sequence analysis, the green sulfur photosynthetic bacterial component of strain N3 was dominant and was readily sequenced, but the less abundant sulfur-reducing bacterial component was apparent only when analyzed by metagenomic binning. Whole-genome comparison showed that the green bacterium belonged to the genus *Prosthecochloris* and apparently was a species for which there was no genome sequence on file. For comparison, we also sequenced the genome of *Prosthecochloris* sp. DSM 1685, which had previously been isolated from the 2-K mixture in pure culture and have shown that all three *Prosthecochloris* genomes belong to a new species, which we propose to be named *Prosthecochloris ethylica* comb. nov. Whole genomes were also sequenced for the isolated *Desulfuromonas* strains DSM 1675 (from strain 2-K) and DSM 1676 (from strain N2) and shown to be nearly identical to the genome found in the N3 mixture. The genome of the green sulfur bacterium contains large genes for agglutination proteins, similar to the ones proposed to be involved in larger photosynthetic consortia of *Chlorochromatium aggregatum*. In addition, we also identified several unique "tight adhesion (tad)" pili genes that are presumably involved in the formation of cell–cell interactions. The colorless component, on the other hand, contained a unique large multiheme cytochrome C and unique genes for e-pili (geopilin) formation, genetically clustered with a conserved ferredoxin gene, which are all expected to play an electron transfer role in the closed sulfur cycle in the syntrophic mixture. The findings from the simultaneous genome sequencing of the components of *Cp. ethylica* have implications for the phenomenon of direct interspecies interactions and coupled electron transfer in photosynthetic symbionts. The mechanisms for such interactions appear to be more common in the environment than originally anticipated.

Keywords: green sulfur bacteria; syntrophy; e-pili; adhesion protein; photosynthetic symbionts; large multiheme cytochrome; metagenomic binning

1. Introduction

Chloropseudomonas ethylica was originally described as a motile green sulfur bacterium capable of utilizing ethanol [1]. However, no pure cultures of green sulfur bacteria were previously known that would utilize ethanol or that were motile. Strains of *Cp. ethylica* were isolated from the Krujalnik

Estuary, near Odesa, Russia (strain 2-K) and from Lake Saksky in Crimea, Russia [2], which in hindsight is a remarkable coincidence considering the approximately 200-mile distance between these habitats. However, doubts were raised about the purity of the strains when two different cell morphologies were observed in growing cultures [2,3]. Subsequently, the colorless, motile, ethanol-oxidizing, and sulfur-reducing bacterium *Desulfuromonas acetoxidans* strains DSM 1675 and DSM 1676 were isolated from either the 2-K or the N2 syntrophic mixtures [4]. A pure culture of *Ds. acetoxidans* DSM 684^T was also isolated from South Orkney Island, Antarctica [4]. The green component was either described as a species of *Chlorobium* [2] or *Prosthecochloris* [3]. The amino acid sequences of cytochrome C-555, alternatively known as c5, from the *Cp. ethylica* 2-K mixture was clearly related to those of *Prosthecochloris* species and not to *Chlorobium* [5], whereas the cytochrome C-551.5, alternatively known as c_7 [6] came from the *Desulfuromonas* component [7]. It is unclear whether there is a specificity to the connection between the two species in the *Cp. ethylica* mixture. Agar shake cultures with ethanol, sulfide, and bicarbonate show large green colonies (faster-growing when involved in sulfur cycling) and small green colonies. The large ones were verified to be in mixed culture and the smaller ones as a pure culture [8] and our unpublished observations.

Mutualistic relationships that involve close cell–cell interactions are most studied between bacterial and eukaryotic interactions, e.g., between nitrogen-fixing *Rhizobium* species and legumes, or bacterial pathogens and eukaryotic hosts. Symbiotic interactions amongst archaea and bacteria can be found in microbial mats where nutrient exchange and waste removal roles are crucial, in anaerobic methane-oxidizing communities of marine environments or in human digestive systems [9–13]. These interactions have only recently been studied in more detail and appear to be more common than historically expected. Larger bacterial and archaeal consortia that are formed through cell–cell interactions of two or more microorganisms have been observed to form a high degree of interdependence between taxonomically unrelated species [14–16]. Phototrophic consortia of this sort were first reported over a century ago [17]. The nature of the *Cp. ethylica* syntrophy appears to be centered around a closed sulfur cycle [18], similar to what has been shown in syntrophic cocultures of *Chlorobium phaeovibrioides* and *Desulfuromonas acetoxidans*, where acetate is oxidized by *Ds. acetoxidans*, with sulfur as an electron acceptor [19]. The process leads to the recycling of the sulfide that can then be used for anoxygenic photosynthesis by *Cb. phaeovibrioides*. Although these syntrophic cocultures appear to be more widespread than commonly expected and the nutrient cycle and mutual benefits are clear in some cases, very little is known about the physical interaction and formation of these cell–cell interactions and the specific components of the electron chemistry involved to establish such a mutually beneficial nutrient cycle. Studies with larger phototrophic consortia, such as *Chlorochromatium aggregatum*, have shown that the green sulfur bacteria involved in this complex are likely preadapted to a symbiotic lifestyle, and specific ultrastructures (periplasmic extruding tubules) can be formed between the central bacterium and the epibiont [14,20]. *Cp. ethylica* was not described to form larger aggregates and appears to form a simpler model of syntrophy. However, the formation of such larger consortia, even in *Chlorochromatium aggregatum*, is dependent on the cultivation strategy. To gain further insight into the possible physical interactions and the electron transfer mechanism involved in *Cp. ethylica*, we set out to determine the genomes of the syntrophic mixtures.

We now report the simultaneous determination of the genome sequences of the green and colorless components of the *Cp. ethylica* N3 mixture. Although the genome sequence of *Desulfuromonas acetoxidans* DSM 684^T was previously determined [21], we have now also established the genome sequences of the *Desulfuromonas* strains 2-K (DSM 1675) and N2 (DSM 1676), previously isolated from two of the *Cp. ethylica* mixtures. We attempted to simultaneously determine the sequences of both organisms from strain N2, but obtained only the genome belonging to the green sulfur bacterial genome, likely due to slightly different culture conditions that may have led to a lower abundance of the colorless component in the mixture.

2. Materials and Methods

2.1. Cultures and DNA Preparation

Cultures of *Chloropseudomonas ethylica* N3 and N2 were originally obtained by one of us (Terry E. Meyer) directly from the laboratory of E.N. Kondrat'eva, and had been grown in our laboratory over the years and kept in a lyophilized state. Cultures were grown according to [8] and harvested by centrifugation. DNA extracted from decades-old frozen cultures of the N3 and N2 syntrophic mixtures were the source for the genomic analysis presented here. Genomic DNA was extracted using the GeneJET DNA purification kit (Thermo Scientific, Waltham, MA, USA). The quantity and purity of DNA, determined using Qubit and Nanodrop instruments, showed an A260/280 ratio of 1.75.

Genomic DNA for *Prosthecochloris* sp. DSM 1685, *Desulfuromonas acetoxidans* DSM 1675 and DSM 1676 was obtained from DSMZ (Deutsche Sammlung von Mikroorganismen und Zellkulturen, GmbH). The quantity and purity of DNA, determined using Qubit and Nanodrop instruments, showed A260/280 ratios between 1.8 and 1.9.

2.2. DNA Sequencing, Assembly, and Annotation

The DNA libraries were prepared with the Nextera DNA Flex Library Prep Kit (Illumina, Inc., San Diego, CA, USA). All five genomes were sequenced, using 500 µL of a 1.8 pM library, with an Illumina MiniSeq instrument, using paired-end sequencing (2 × 150 bp). Quality control of the reads was performed using FASTQC within BaseSpace (Illumina, version 1.0.0), using a k-mer size of 5 and contamination filtering. The data for each of the genomes of *Chloropseudomonas ethylica* N3 and N2, *Prosthecochloris* sp. DSM 1685, and *Desulfuromonas acetoxidans* DSM 1675 and DSM 1676 was assembled de novo using Unicycler within PATRIC (Pathosystems Resource Integration Center) [22]. The genome sequences were annotated using RAST (Rapid Annotations using Subsystem Technology; version 2.0) [23].

Average percentage nucleotide identity (ANIb) between the whole genomes was calculated using JSpecies [24]. A whole-genome-based phylogenetic tree was generated applying the CodonTree method within PATRIC [22], which used PGFams as homology groups. Moreover, 519 PGFams were found among these selected genomes using the CodonTree analysis to construct the *Prosthecochloris* tree, while 598 unique PGFams were identified for the *Desulfuromonas* tree. The aligned proteins and coding DNA from single-copy genes were used for RAxML analysis [25,26] for the trees. The support values for the phylogenetic tree were generated using 100 rounds of the "Rapid bootstrapping" option of RAxML. Tree visualization was performed with iTOL [27].

This Whole-Genome Shotgun project has been deposited at DDBJ/ENA/GenBank, and the accession numbers for all of the sequenced genomes are listed in Table 1.

2.3. Metagenomic Binning

The sequencing reads of *Cp. ethylica* N3 were used to perform a metagenomic binning using the Metagenomic Binning service within PATRIC [22]. Paired-end reads were used as input, and default parameters were used. Sets of contig bins were constructed, with hits against contigs that have less than fourfold coverage or are less than 400 bp in length being removed. The contig pool was split into bins using reference genomes. Quality control of each bin was performed using checkM [28]. Each bin was automatically annotated using RAST within PATRIC [22], and consistency checks of the annotation were performed, producing a coarse score (percentage of roles that are correctly present or absent) and a fine score (percentage of roles that are correctly absent or present in the correct number). Identified genomes were ranked based on their coarse score, fine score, and completeness.

2.4. Synteny Analysis

For synteny analysis, comparative genome regions were generated using global PATRIC PGFam families to determine a set of genes that match a focus gene. All genomes were used in the search and compared to the reference genome. The gene set was compared to the focus gene using BLAST and sorted by BLAST scores within PATRIC [22]. The *Prosthecochloris ethylica* N3 TadZ/CpaE (PGFam_00109911) and agglutination protein (PGFam_02064367) were used as focus genes to analyze synteny of the Tad pili and adhesion protein gene clusters, respectively. For the *Desulfuromonas acetoxidans* synteny comparisons, the Type IV major assembly protein PilA (PGFam_00056426) and cytochrome C-551.5 (PGFam_10701576) were used as focus genes.

3. Results

The results of the genome sequences that were determined in this work, i.e., N2 and N3 syntrophic mixtures, those of the pure cultures DSM 1675 and DSM 1676 (isolated from the N2 and 2-K mixtures), and the pure culture DSM 1685 (from the 2-K mixture) are shown in Table 1. Knowing that *Cp. ethylica* N2 and N3 consists of a mixture of two species, we attempted to separate the raw genome data by metagenomic binning. In the case of *Cp. ethylica* N3, we obtained two valuable bins, one genome related to *Prosthecochloris* sp. (97.0% coarse consistency and 96.7% fine consistency), the other to *Desulfuromonas acetoxidans* (98.4% coarse consistency and 97.3% fine consistency), both with 100 % completeness (included in Table 1). From the *Cp. ethylica* N2 data, we only obtained the genome of the *Prosthecochloris* sp.

Table 1. Genome characteristics of the *Prosthecochloris* and *Desulfuromonas* genomes used in this study.

Species	Genome Size	GC Content	Contigs	Coverage	CDS	tRNAs	Reference	Genbank Accession #
Prosthecochloris (N3 mix)	2.45 Mb	55.1	72	223x	2480	45	this study	JADGII000000000
Prosthecochloris (N2 mix)	2.47 Mb	55.1	60	187x	2340	46	this study	JADGIH000000000
Prosthecochloris (2K, DSM 1685)	2.44 Mb	55.1	66	123x	2348	45	this study	JABVZQ010000000
Desulfuromonas (N3 mix)	3.78 Mb	51.9	121	84x	3557	51	this study	JADGIJ000000000
Desulfuromonas (N2, DSM 1676)	3.81 Mb	51.9	52	38x	3567	55	this study	JABWTG010000000
Desulfuromonas (2K, DSM 1675)	3.80 Mb	51.9	59	39x	3550	55	this study	JABWTF010000000
Desulfuromonas DSM 684	3.83 Mb	51.8	51	N/A	3573	49	unpublished	NZ_AAEW00000000

3.1. Prosthecochloris Genome from the Green Sulfur Component

Average nucleotide identity (ANIb) comparison showed that the genome sequences of the green component of the N2 and N3 mixtures were virtually the same as the one isolated from the 2-K mixture (DSM 1685) and both were similar to those of *Prosthecochloris* species, with an average nucleotide identity (ANI) of 75% to the nearest, previously determined, strain HL130 (Table 2). The ANI values of all the other *Prosthecochloris* species are well below 95%, which is the arbitrary cutoff value for species differentiation [24]. It is, therefore, likely that the genome sequence of the green component of the N2, N3, and 2-K mixture is from a new species, which we propose to be called *Prosthecochloris ethylica* comb. nov.

Table 2. Percentage Average Nucleotide Identity (ANIb) of *Prosthecochloris* green bacteria.

Pr. ethylica DSM 1685, isolated in pure culture from the *Cp. ethylica* 2K mixture									
99.91	*Pr. ethylica* N2 in the *Cp. ethylica* N2 mixture								
99.95	99.93	*Pr. ethylica* N3 in the *Cp. ethylica* N3 mixture							
74.5	74.7	74.6	*Prosthecochloris* sp HL130						
71	70.9	70.9	70.3	*Prosthecochloris* sp Ty Vent					
72.8	73	72.9	72.2	70.8	*Pr. aestuarii* DSM 271				
69.5	69.5	69.5	69.5	70.5	70.2	*Pr. marina* VI			
69.8	69.9	69.8	69.6	71.1	70.5	**77.2**	*Prosthecochloris* sp A305		
70.3	70.3	70.3	70.5	71.5	71.4	**74.6**	**76.3**	*Pr. phaeobacteroides* BS1	
72.2	72.3	72.3	72	70	71.2	68.9	69	70.4	*Pr. vibrioforme* DSM 260
72	72.2	72.1	71.8	69.9	71	68.9	69	70.3	**98.3** *Pr. phaeum* 2401

Bold means: ANI values above 95% which is the arbitrary cutoff value for species differentiation.

As expected, these genomes contained a gene for cytochrome C-555, for which the translated protein sequences of strains N2 and N3 were identical (100% identity) to that previously reported for the green component of the *Cp. ethylica* strain 2-K mixture [5], as shown in Figure 1.

A whole-genome phylogenetic tree for *Prosthecochloris* placed the green component of the N2 and N3 mixtures as nearly similar to that of strain DSM 1685, isolated from the 2-K mixture (red clade in Figure 2). This is consistent with the ANI comparisons mentioned above and clearly distinguishes this clade as separate from the other sequenced *Prosthecochloris* strains, with strain HL130 as the closest relative. This further supports the proposal of a new *Prosthecochloris* species.

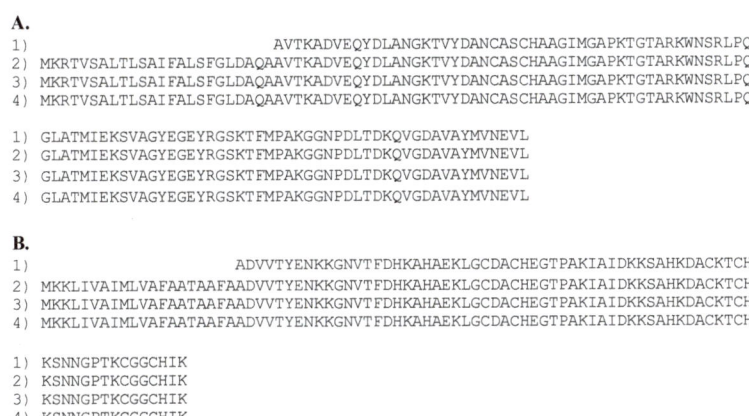

Figure 1. (**A**) Sequence alignment of cytochromes c5 from *Prosthecochloris*. (1) Van Beeumen et al. [5] soluble protein from 2-K mix, (2) translated genome from pure 2-K DSM 1685, (3) translated genome from N2 mix, and (4) translated genome from N3 mix. (**B**) Sequence alignment of cytochromes c-551.5 from *Desulfuromonas*. (1) Ambler [6] soluble protein from 2-K mix, (2) translated genome from pure 2-K DSM1675, (3) translated genome from pure N2 DSM1676, and (4) translated genome from N3 mix.

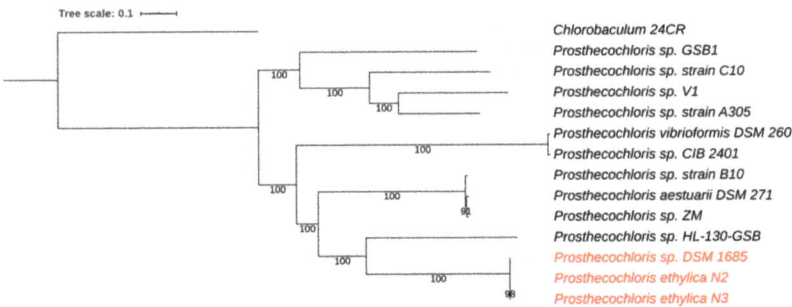

Figure 2. Whole-genome-based phylogenetic tree of all sequenced *Prosthecochloris* species. The phylogenetic tree was generated using the CodonTree method within PATRIC [22], which used PGFams as homology groups. The support values for the phylogenetic tree are generated using 100 rounds of the "Rapid bootstrapping" option of RaxML [25]. *Chlorobaculum* sp. 24CR was used as an outgroup [29]. The branch length tree scale is defined as the mean number of substitutions per site, which is an average across both nucleotide and amino acid changes. Species marked in red belong to the newly proposed *Prosthecochloris ethylica* comb. nov.

A search for protein families that are unique amongst the *Prosthecochloris* strains using PATRIC revealed a Tad (Tight Adhesion) pili gene cluster that is found exclusively in strains N2, N3, and DSM1685 and with lower homology in strains GSB1 and CIB2401, but appears to be absent from all other *Prosthecochloris* strains. This Tad pili gene cluster consists of at least 10 genes related to Type II/IV Flp pili assembly and secretion. The synteny of the gene cluster is preserved in all these species (Figure 3), indicating an evolutionary conservation of this cluster. The closest relatives with a similar gene cluster were found to be *Pelodictyon phaeoclathratiforme* BU-1 and *Chlorobium luteolum* DSM 273. Interestingly, *P. phaeoclathratiforme* is a brown-colored *Chlorobiaceae* that was described to form net-like colonies [30].

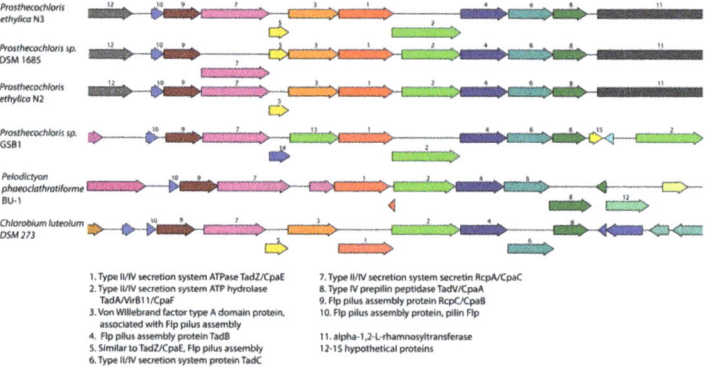

Figure 3. Gene cluster organization and synteny comparison of the Tad (Tight Adhesion) genes identified in the Prosthecochloris genomes. Genes are colored according to protein family (PGFam).

Tad pili gene clusters encode a macromolecular transport system (type II secretion system). They are present in the genomes of a wide variety of both Gram-negative and Gram-positive bacteria and are involved in close adhesion of cells within biofilm formation, colonization, and, sometimes, pathogenesis [31–34]. The long filamentous fibrils are formed by bundles of individual pilus strands, consisting of the fimbrial low-molecular-weight protein Flp [35–37]. Their attachment to surfaces and other cells are expected to create an optimized environment for nutrient, metal, and electron exchange between cells [31,33]. Given the presence of these conserved genes and the synteny in the *Prosthecochloris* strains isolated from the three *Cp. ethylica* mixtures, these Tad-encoded pili likely play a similar role in the syntrophic relationship of these strains by forming cell–cell interactions.

It has been described earlier that in larger multicellular phototrophic consortia of *Chlorochromatium aggregatum*, a few large genes encode proteins that are anticipated to play an important role in the formation of close interspecies interactions [14]. This was suggested based on in silico analysis, but the exact physical role of these proteins in forming these interactions have not been described. The four involved putative ORFs showed similarities to hemagglutinins (2 ORFs), an RTX toxin and hemolysin, and were found to be some of the largest genes detected in prokaryotes and to contain multiple characteristic repeats [13,14]. When searching for large ORFs in the *Prosthecochloris* genomes that we sequenced, we found the largest ORF (coding for 2417 aa.) to be annotated as "hypothetical protein." When performing an NCBI BLASTP search using this sequence, we found it to be homologous to an "outer membrane adhesin-like protein" from *Pelodictyon phaeoclathratiforme* and a "tandem-95 repeat" protein from *Prosthecochloris aestuarii*, albeit with low protein identity (<39%). Further comparison showed that all three of these proteins are homologous to hemagglutinin/adhesin-like proteins, similar to outer membrane adhesin proteins of the RTX toxin family, which contain numerous, internally repeated, calcium-binding domains [38].

Although highly diverse in terms of structure and/or adhesives properties, outer membrane adhesins in Gram-negative bacteria are usually grouped into two main categories: the adhesins secreted through a type 1 secretion system (T1SS) and the adhesins secreted through one of the type 5 secretion systems (T5SS) [39]. The most studied of these secreted adhesins are the biofilm-associated family of proteins (Bap), which are high-molecular-weight multidomain proteins containing an N-terminal secretion signal, a core domain of highly repeated motifs, and a glycine-rich C-terminal domain. Bap family members have been shown to be involved in cell adhesion to abiotic surfaces and biofilm formation in both Gram-positive and Gram-negative bacteria (for reviews, see [40,41]). However, only a few of these proteins have been characterized experimentally. A closer look at the gene region surrounding the large *Prosthecochloris* putative adhesion protein revealed, first, that the gene is located at the end of a contig in the genomes of strains N3, N2, and DSM1685, indicating that the full size of the encoded protein might be larger than 2417 amino acids and, second, that the gene is followed by a gene encoding an agglutination protein (TolC family type I secretion outer membrane protein), an ABC transmembrane transporter (type I secretion system ATPase), and a HlyD homologous protein (type I secretion membrane fusion protein) (Figure 4). Comparison of this gene region to closely related genomes showed that *Pr. aestuarii* DSM 271 contains a complete sequence for the adhesin-like gene (encoding 4748 amino acids) and is preceded by a large protein (4983 amino acids) that is homologous to the structural toxin protein RtxA (which is a T1SS-143 repeat domain-containing protein) in a similar gene cluster. The presence of the large RtxA homologue and the adhesin-like protein in a gene cluster with other T1SS-related proteins indicates that these genes indeed encode adhesins, secreted through a type 1 secretion system.

The occurrence of the adhesin-like gene at the end of a contig in our *Prosthecochloris* genomes is likely due to the fact that these genes contain multiple sequence repeats, which are known to be a potential challenge for next-generation sequencing-based genome assembly programs [42]. When searching the N3 genome with the adhesin and RtxA toxin-like proteins from *Pr. aestuarii* DSM 271 (using BLASTP in PATRIC), we did find partial genes (annotated as hypothetical proteins) located at the ends of two other contigs (contigs 008 and 029). These partial gene-containing contigs from strain N3 align very well with the gene cluster identified in strain DSM 271 (see Figure 4), which supports the hypothesis that the missing partial adhesin gene is due to assembly software limitations. The same was true for the 2-K (DSM1685) and N2 genomes where the RtxA homologue was also found on separate contigs. We can conclude that the *Prosthecochloris* genomes from the 2-K, N2, and N3 mixtures all contain large genes similar to the putative ORFs (hemagglutinin and RTX toxin) that were described in *C. aggregatum* to be important for the formation of close interspecies interactions [13,14].

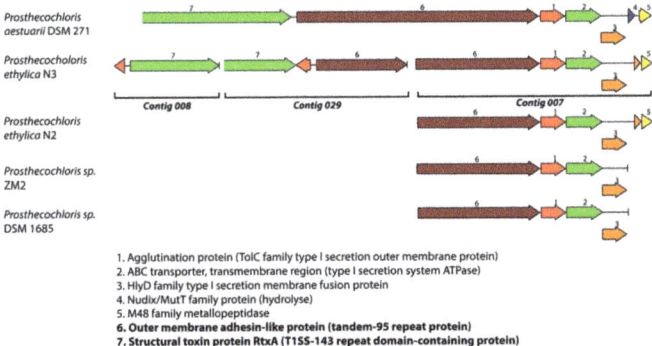

Figure 4. Gene cluster organization and synteny comparison of the large adhesin-like protein-encoded genes, identified in the *Prosthecochloris* N3 genome and homologues. The genes for large extracellular outer membrane adhesion protein and RtxA discussed in the text are marked in bold.

3.2. Desulfuromonas Genome from the Colorless Sulfur-Reducing Component

Based on ANIb comparisons (Table 3) and whole-genome phylogenetic analysis (Figure 5) of the colorless component in the N3 mixture, we can conclude that it is very similar to *Ds. acetoxidans* DSM

1675 and DSM 1676, previously isolated from the mixtures 2-K and N2, and that they are nearly the same as the type strain DSM 684.

The *Cp. ethylica* 2-K cytochrome C-551.5 protein [6] was identical to that of the translated gene from the *Desulfuromonas* component of the N3 mixture and from the DSM 1675 and DSM 1676 pure culture, as shown in Figure 1B, but was apparently absent from the type strain DSM 684. The C-551.5 gene synteny is conserved in the N3 mixture, DSM 1675, and DSM 1676 genomes (Figure 6), and the gene is surrounded in all of the strains by a gene for cytochrome C (PGFam_04122568; located downstream) and a Mg/Co/Ni transporter MgtE (PGFam_04560429; upstream and antisense). These surrounding genes are both present in strain DSM 684; however, they are each located near the end of separate contigs (Figure 6). It is, therefore, possible that the C-551.5 gene was lost during assembly of the DSM 684 genome. Further studies will be needed to conclude the presence of C-551.5 in the type strain.

Table 3. Percentage average nucleotide identity of some *Desulfuromonas* sulfur-reducing bacteria.

								Ds. acetoxidans DSM 684^T pure culture
99.92								*Ds. acetoxidans* 2K DSM 1675 pure culture
99.91	99.99							*Ds. acetoxidans* N2 DSM 1676 pure culture
99.86	99.92	99.93						*Ds. acetoxidans* in the *Cp. ethylica* N3 mixture
69.9	69.9	70	70					*Ds. thiophila* DSM 8987
67.2	67.4	67.4	67.4	68.9				*Ds. acetexigens* null
67.4	67.5	67.5	67.4	68.7	72.1			*Ds. sp.* WTL
67.2	67.3	67.3	67.2	69.5	71.5	72.3		*Ds. sp.* DDH 964

Bold: ANI values above 95% which is the arbitrary cutoff value for species differentiation.

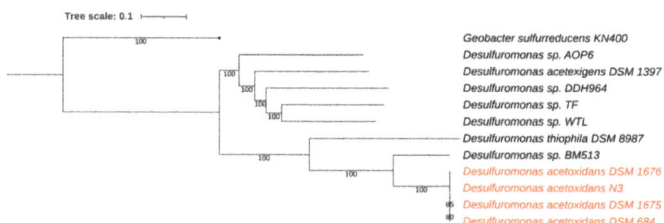

Figure 5. Whole-genome-based phylogenetic tree of all sequenced *Desulfuromonas* species. The phylogenetic tree was generated using the CodonTree method within PATRIC [22], which used PGFams as homology groups. The support values for the phylogenetic tree are generated using 100 rounds of the "Rapid bootstrapping" option of RaxML [25]. *Geobacter sulfurreducens* KN400 was used as an outgroup [43]. The branch length tree scale is defined as the mean number of substitutions per site, which is an average across both nucleotide and amino acid changes.

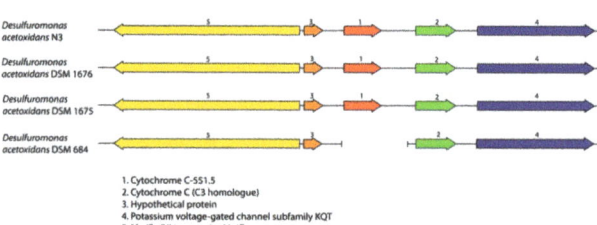

Figure 6. Gene organization comparison around the cytochrome *c*-551.5 gene in *Desulfuromonas* genomes. Genes are colored according to protein family (PGFam).

The genomes of *Desulfuromonas* strains DSM 1675, DSM 1676, DSM 684^T, and the genome isolated from the N3 mixture contain many genes from unique protein families (PGFams identified in PATRIC) that are apparently not found in the other sequenced *Desulfuromonas* strains. At least 18 unique PGFams were related to the synthesis of Type IV pili of two different classes. Eight of them encoded mannose-sensitive hemagglutinin (MSHA) type pili, and the other 10 encode the elements of a different Type IV pili group (Table 4). Type IV pili are found on the surface of a variety of Gram-negative bacteria [44] and have been demonstrated to be important as host colonization factors, bacteriophage receptors, mediators of DNA transfer and, more recently, also as electron transfer factors over longer distances (so called e-pili) [45–47].

Table 4. Overview of some of the PGFams uniquely identified in *Ds.* strains DSM 1675 (2-K), DSM 1676 (N2), N3, and DSM 684 (red clade in Figure 5).

Gene	PGFam	Protein Description	Size (aa.)
PilA	PGF_00056426	Tfp pilus major pilin assembly protein	314
PilZ	PGF_00414431	Type IV pilus assembly protein	121
PilX	PGF_03889320	Type IV fimbrial biogenesis protein	174
	PGF_04686021		204
PilV	PGF_04940117	Type IV fimbrial biogenesis protein	141
	PGF_05800309		126
PilP	PGF_06326322	Type IV fimbrial biogenesis protein	171
PilW	PGF_10370792	Type IV fimbrial biogenesis protein	330
FimT	PGF_10544072	Type IV fimbrial biogenesis protein	143
PglB	PGF_05008343	Pilin glycosylation protein	206
Ferro	PGF_00004340	Ferrodoxin domain protein	260
MshA	PGF_06702067	MSHA pilin protein	169
	PGF_12682451		178
MshD	PGF_01602851	MSHA pilin protein	152
MshI	PGF_05907321	MSHA biogenesis protein	198
MshJ	PGF_02203893	MSHA biogenesis protein	230
MshK	PGF_00018410	MSHA biogenesis protein	119
MshN	PGF_00938251	MSHA biogenesis protein	430
MshP	PGF_01166820	MSHA biogenesis protein	130
CytC	PGF_12883110	Multiheme Cytochrome C family protein	624

The type IV major pilin assembly protein PilA found exclusively in the four *Desulfuromonas* strains is significantly larger (314 residues) than the common PilA homologue found in many other bacteria (only ~60–70 aa.). The typical shorter PilA homologue was also found in all of the sequenced *Desulfuromonas* strains. This larger PilA protein (PGFam_00056426) contains a transmembrane region (from ~residues 70–160), and a BLASTP search revealed a geopilin domain membrane protein (247 aa.; PGFam_10038571) from *Pelobacter carbinolicus* DSM 2380 as the closest relative (49% protein identity and 67% similarity). Geopilin proteins have been implicated in direct interspecies electron transfer (e-pili) within syntrophic aggregates [48–50] and have also been shown to enhance current production in fuel cells [51].

The synteny of the geopilin-PilA protein is conserved in all four of our sequenced *Desulfuromonas* strains (Figure 7). The gene is preceded by a ferredoxin domain protein (PGFam_00004340) and followed by DUF419 (a protein of unknown function; PGFam_00038332), which are both also unique to these four *Desulfuromonas* strains. Ferredoxins are small proteins containing iron-sulfur clusters and function as biological "capacitors" that can accept or discharge electrons and are involved in electron transfer reactions in many organisms (for review see [52]). The presence of the ferredoxin protein directly upstream of the geopilin, which is proposed to be an electron-transfer pili (e-pili), is certainly intriguing and points towards a functional coupling of these proteins. In addition, unique PGFams for PilZ, PilX, PilV, PilP, and PilW, as well as FimT biogenesis proteins and the Pilin glycosylation protein PglB were found at other locations in our selected genomes. Using BLASTP (within PATRIC), we were able to identify one other *Desulfuromonas* strain, BM513, which contains a distant homologue of the larger PilA protein (66% sequence identity); however, the gene synteny is less conserved

(Figure 7). This latter genome was assembled from a metagenomic sample of an environmental isolate, and nothing is currently known about potential symbiotic relationships of this strain. The presence of this larger geopilin-PilA homologue and several of the type IV pili biogenesis proteins indicates that the *Desulfuromonas* strains N3, N2, 2-K, and possibly the environmental strain BM513, are able to produce a unique set of type IV pili (e-pili) that could play a role in syntrophy and electron transfer.

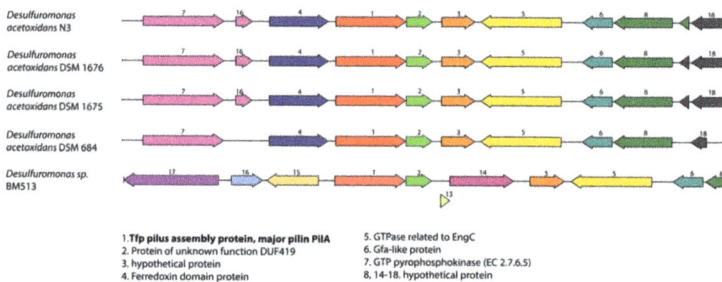

Figure 7. Gene organization comparison around the geopilin-PilA pilus assembly protein in *Desulfuromonas* genomes. Genes are colored according to protein family (PGFam). The gene for the large major pilin protein PilA, discussed in the text, is marked in bold.

The mannose-sensitive hemagglutinin (MSHA; Table 4) is likewise a member of the family of type IV pili. While the exact function of MSHA is unknown, studies have shown that *msh*A mutants of *Vibrio cholerae* are unable to produce biofilms on abiotic surfaces, and these pili might have an environmental role of survival outside the host [53]. Two homologues of MshA were found to be present in DSM 1675, DSM 1676, DSM 684, and the N3 strain, in addition to homologues for MshD, I, J, K, N, and P, which are essential for MSHA pili biosynthesis (Table 4). Although these Msh PGFams were not found by PATRIC in the other *Desulfuromonas* strains, we also identified, by performing a BLASTP search within all *Desulfuromonas* strains in PATRIC, more distant homologues (<40% aa. identity) of MshA, and found the same conserved gene cluster in the genomes of *Desulfuromonas* sp. AOP6, BM513, and *Ds. thiophila*. This indicates that the *msh* pili gene cluster might be more widespread amongst the *Desulfuromonas* species.

Besides these pili genes, the *Desulfuromonas* genomes also contain genes for several flagellar proteins, including the flagellar assembly protein FliH, flagellar basal body P-ring formation protein FlgA, the basal body rod protein FlgF, flagellar protein FlgJ (2 copies), and a flagellar regulatory protein FleQ. These were initially found as unique families by PATRIC in strains N3, DSM 1675, DSM 1676, and DSM 684. However, when performing a BLAST search, we found homologues of them (<45% protein identity) in several of the other strains, e.g., DSM1397, AOP6, and *Ds. thiophila*. This is consistent with the earlier observations that the *Desulfuromonas acetoxidans* species in the syntrophic mixtures are motile and able to produce functional flagella [8].

3.3. Comparison of the Geobacter sulfurreducens Genomes to the Desulfuromonas Genomes

It has been shown that *Prosthecochloris* could grow by direct interspecies electron transfer from *Geobacter sulfurreducens*, a close relative of *Desulfuromonas* [54]. To elucidate potential unique features in the species that have been shown to form the *Prosthecochloris* syntrophy, we compared the *Geobacter sulfurreducens* genomes of strains PCA and KN400 to the *Desulfuromonas* genomes. This revealed a unique large cytochrome C family protein (624 aa; PGFam_12883110) that is only present in the genomes of *Ds.* DSM1675 (2-K), DSM1676 (N2), N3, 684, AOP6, and the two *Gb. sulfurreducens* strains. An EXPASY-BLASTP search revealed that this is a multiheme cytochrome, with the closest relative being an uncharacterized cytochrome C from *Geothermobacter* sp. HR-1 (67% identity and 80% similarity), and several homologues in other *Geobacter* and *Thermodesulfovibrio* strains, but missing from all the other *Desulfuromonas* strains. The protein contains an N-terminal signal peptide and, as shown in the alignment in Figure 8, at least 9 possible heme-binding sites (identified as CXXCH) were found to be present. The function of this large multiheme cytochrome is currently unknown and the surrounding genes consist of mainly hypothetical genes in all of the *Desulfuromonas* strains where it was found, so it does not appear to be genetically associated with any known specific pathway. The recently published genome of *Ds.* strain AOP6 showed that this strain has genes for a large c-type cytochrome and unique Type IV pili [55], although no further analysis was performed. We now found this to be the case for our four genomes of *Desulfuromonas* as well. Since the *Geobacter* species can produce type IV pili and cytochromes that directly transport electrons through the pili to crystalline Fe(III) and Mn(IV) oxides [47,56], it is possible that this related multiheme cytochrome *c* and the Type IV pili play similar electron transport roles in the *Desulfuromonas* strains that contain these specific proteins.

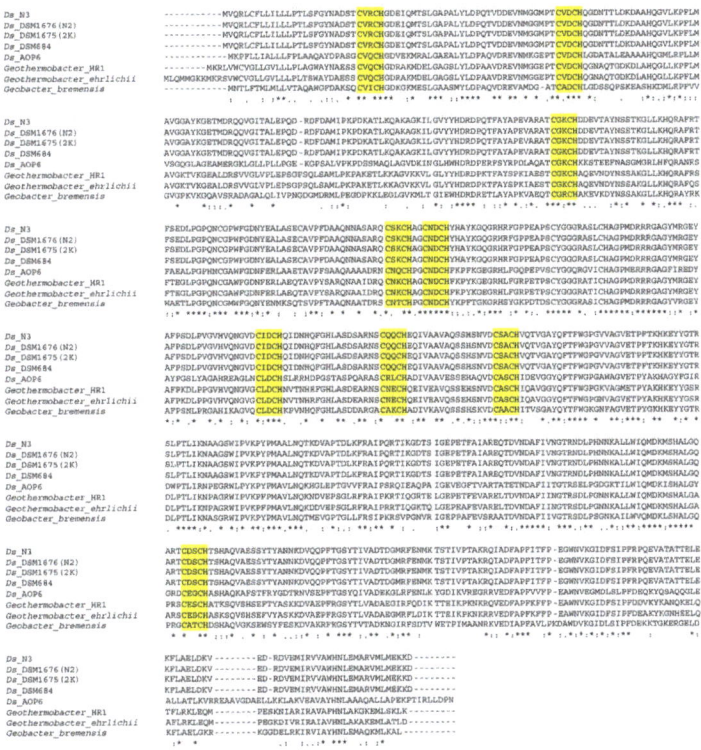

Figure 8. Alignment of the large unique multiheme cytochrome C sequences from *Desulfuromonas* genomes and their closest relatives from *Geothermobacter* and *Geobacter* species. Potential heme-binding sites are marked in yellow. Ds = *Desulfuromonas acetoxidans*.

4. Discussion

It is apparent that all three syntrophic mixtures analyzed, *Cp. ethylica* N2, N3, and 2-K, are virtually the same in both the green and colorless components. That is, the interaction between the two components of the mixtures appears to be highly specific, although no physical connection has been described and the mixtures were isolated from different habitats hundreds of miles apart. However, the finding that both the green component and the colorless component of the three syntrophic mixtures are virtually the same asks for some explanation. It would require to be quite a coincidence considering that the mixtures were isolated from habitats geographically separated by hundreds of miles in Russia and that the type strain of *Desulfuromonas* was isolated from the Antarctic on the far side of the Earth. In support of coincidence, Biebl and Pfennig [18] reported that there is no specificity in the syntrophic growth of *Desulfuromonas* DSM 1675 with green bacteria. As far as growth yield is concerned, *Prosthecochloris* DSM 1685 could be replaced by *Chlorobium* strain DSM 258, and *Desulfuromonas* DSM 1675 could be replaced by *Desulfuromonas* DSM 684^T in the syntrophic mixture. However, the doubling times were not measured, which might have shown a difference. A far better explanation of the identity of the three isolates, therefore, is that there is some undetected physical interaction between the two species in the mixtures.

A cell–cell complex, however fleeting, may require surface recognition proteins as previously postulated for the green bacterial consortia of *C. aggregatum* [13]. In that larger consortium, a single central, nonpigmented, heterotrophic bacterium forms close cell–cell interactions with multiple green-sulfur bacteria (15–40 cells) [13,14]. Even though the metabolic interactions in the *Cp. ethylica* symbiotic relationship are possibly different from the *C. aggregatum* and other large consortia, and are likely driven by sulfur cycling in the case of *Cp. ethylica*, the fact that similar large adhesin/agglutination proteins are detected in *Prosthecochloris ethylica* strains indicates that similar interactions may be formed in this more simple symbiont as those in the larger green sulfur consortia. The fact that this gene cluster appears to be present in other *Prosthecochloris* genomes as well (Figure 4) indicates that the capability to produce large outer membrane adhesin structures might be more widespread among green sulfur bacteria than anticipated.

The presence of unique Tad (tight adhesion) pili genes, only found in the *Prosthecochloris* strains that are known to form a symbiotic relationship, suggests that the structural formation of the cell–cell interactions occurs through specific pili and that large agglutination proteins are expected to help maintain this interaction. Both in the larger *C. aggregatum* and in a similar archaeal consortium with *Nanoarchaeum equitans*, fibers are observed that form periplasmic tubules surrounding the entire cell surface [14,20]. The exact nature of these fibers is still unresolved. However, tight adhesion pili, like the ones we found in the *Prosthecochloris ethylica* genomes, may be involved in the initial formation of the connecting tubules. Possibly, the pili are forming the initial connection before wider periplasmic tubules are established. This process would be similar to the better-known process of how conjugation pili (F pili) help to establish conjugation bridges during the process of DNA conjugation in many bacteria. It is interesting that both the Tad pili and the adhesion protein products described here are best known from studies of bacterial virulence factors. However, based on their presence in other nonpathogenic species, they seem to be more widespread amongst bacteria and are likely involved in many environmentally important syntrophic interactions between bacteria.

In the *Chloropseudomonas ethylica* syntrophy, the colorless *Ds. acetoxidans* bacterium likely reduces sulfur compounds and returns the resulting H_2S to the *Prosthecochloris* green sulfur component, as proposed by Biebl and Pfennig [18]. This would allow the green sulfur bacteria to use sulfide for their anoxygenic photosynthesis. Studies have shown that various cytochrome c_7s can directly reduce metal ions or sulfur, and the *Desulfuromonas* C-551.5 is no exception in that respect [57,58]. Thus, this cytochrome may be responsible for sulfur cycling in the mixtures. In fact, it has previously been shown that *Ds. acetoxidans* cytochrome C-551 is capable of reducing polysulfides and is suggested to be the terminal reductase [59]. It was also shown that another strain of *Prosthecochloris* can grow by direct interspecies electron transfer with *Geobacter sulfurreducens* (a close relative of *Desulfuromonas*) as the electron donor and that they could form a cell–cell complex [54]. These results suggest that the *Desulfuromonas* c_7 (C-551.5) in the *Chloropseudomonas* mixtures may transfer electrons directly to *Prosthecochloris* without reducing sulfur to sulfide, although that too is a possibility, as shown above. Based on our findings of geopilin-type pili genes and of a larger multiheme cytochrome C gene found exclusively in the genomes of the strains that form these syntrophic mixtures, it is likely that a more complex system of electron transfer through e-pili and multiheme cytochromes is involved, with cytochrome C-551.5 functioning as the terminal sulfur reducing agent in this complex.

Large multiheme cytochromes with monomeric molecular masses of 50 and 65 kDa (containing 6 and 8 hemes, respectively) were previously observed in Ds. *acetoxidans*, but no sequence data to compare with our current multiheme cytochrome C are available [59]. These multiheme cytochromes were found to cover an extremely wide range of reduction potential, but did not show any hydroxylamine oxidoreductase nor polysulfide reductase activity. Multiheme cytochromes of *Geobacter* and *Desulfuromonas* have been shown to play critical roles in the processing of many metals [57,60], although the monomers were much smaller in size (70–80 residues). A recent review on the role of multiheme cytochromes in anaerobic bacterial respiration [61] reports that multiheme cytochromes with a large molecular ratio of weight/heme ratio (7kDa/heme, or higher) appear to be most common in so-called electroactive organisms that are involved in the reduction of extracellular substrates, such as *Geobacteraceae* and *Shewanellaceae*. The mechanism used by *Geobacter* to transfer electrons onto solid extracellular substrates is still poorly understood, but involves both a pool of periplasmatic cytochromes and several outer membrane multiheme cytochromes. At least some of these form larger polymers [61,62]. The exact electron transfer process is still undetermined, but it is likely a complex multicomponent system where at least some of the multiheme cytochromes have overlapping functions covering the wide range of reduction potentials. By sequence homology alone we cannot state the function of the unique larger multiheme cytochrome C from the *Desulfuromonas acetoxidans* strains, but its high molecular weight/heme ratio (estimated to be 7.5 kDa/heme), N-terminal signal sequence, and unique homology to *Geobacter* large multiheme cytochromes point to a role in extracellular electron transfer.

Our findings from the simultaneously sequencing and comparison of the genomes in the *Cp. ethylica* syntrophic mixtures, in combination with observations described in related organisms, allow us to propose a basic model by which Tad pili and large agglutination proteins from the green sulfur *Prosthecochloris* are key elements in the formation of the syntrophic complex (Figure 9). This may be in conjunction with the formation of periplasmic tubules that were observed in larger photosynthetic consortia. The formation of larger consortia is dependent on cultivation conditions and is not easily reproduced under laboratory conditions. Although this has not been observed yet in *Cp. ethylica*, given the detection of similar adhesion proteins and pili, it is possible that it could produce similar larger consortia under the right cultivation conditions. Once close cell–cell interactions are formed, the closed sulfur cycle can be established by electron transfer through specialized e-pili and several cytochromes produced by the *Desulfuromonas* component (Figure 9). Further biochemical studies will be needed to determine the exact electron transfer chain and the function of the cytochromes involved, but it is likely that cytochrome C-551.5 plays an essential role in the sulfur reduction. The sulfur metabolism of green sulfur bacteria involves the oxidation of sulfide and the deposition of elemental sulfur globules outside the cells [63]. The e-pili produced by *Desulfuromonas* are therefore expected to be involved in extracellular electron transfer to the sulfur acceptor deposited extracellularly by the green sulfur bacterium (Figure 9). The expression of both specific geopilin and MSHA pili (particularly the major pilin *msh*A gene) in the related organism *Pelobacter carbinolicus* was found to be substantially upregulated during ethanol oxidation, presumably for improved attachment or electron transfer to the extracellular electron acceptor S^0 [50]. It requires further functional analysis, but the conserved ferredoxin protein directly upstream of the geopilin gene in Ds. *acetoxidans* may potentially be involved in the electron transfer through the e-pili.

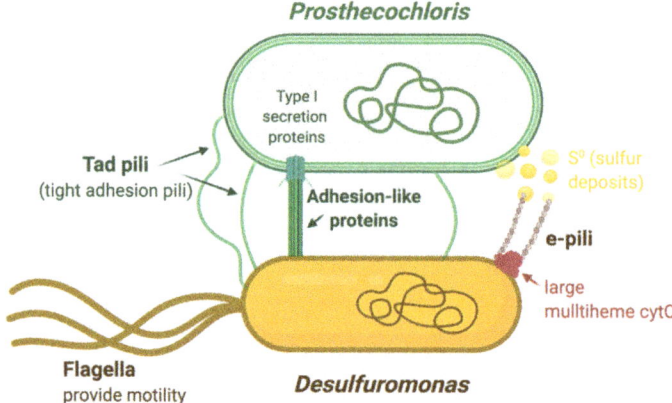

Figure 9. Schematic overview of proposed interactions in the *Chloropseudomonas ethylica* syntrophic mixture. Unique protein structures identified in the genomes are indicated in green for *Prosthecochloris ethylica* and in brown/red for *Desulfuromonas acetoxidans*. Image created in BioRender.com.

Irrespective of the electron transfer proteins involved, the mutualistic metabolic benefits to each of the components are clear from the sulfur and electron cycling. In addition, the colorless *Desulfuromonas* also contains several genes for flagella production, and, therefore, provides the additional benefit of motility. The potential advantage of motility was previously proposed for the larger *C. aggregatum* consortium [14] and appears to be a part of the *Cp. ethylica* syntrophy as well. None of the green sulfur bacteria observed in these consortia produce flagella and obtaining motility provided by the nonpigmented component is likely to result in a competitive advantage to orient themselves much faster towards light and sulfide gradients in stratified or meromictic lakes.

Author Contributions: Conceptualization, J.A.K. and T.E.M.; methodology, J.A.K. and T.E.M.; software, J.A.K.; validation, J.A.K., J.J.V.B. and T.E.M.; formal analysis, J.A.K., J.J.V.B. and T.E.M.; investigation, J.A.K., J.J.V.B. and T.E.M.; resources, J.A.K. and T.E.M.; data curation, J.A.K., J.J.V.B. and T.E.M.; writing—original draft preparation, J.A.K. and T.E.M.; writing—review and editing, J.A.K., J.J.V.B. and T.E.M.; visualization, J.A.K.; project administration, J.A.K., J.J.V.B. and T.E.M.; funding acquisition, J.A.K. All authors have read and agreed to the published version of the manuscript.

Funding: This research received no external funding.

Conflicts of Interest: The authors declare no conflict of interest.

References

1. Shaposhnikov, V.V.; Kondratieva, E.N.; Federov, V.D. A new species of green sulfur bacteria. *Nature* **1960**, *187*, 167–168. [CrossRef]
2. Olson, J.M. Historical note on *Chloropseudomonas ethylica* strain 2-K. *Int. J. Syst. Bacteriol.* **1973**, *23*, 265–266. [CrossRef]
3. Olson, J.M. Confused history of *Chloropseudomonas ethylica* 2-K. *Int. J. Syst. Bacteriol.* **1978**, *28*, 128–129. [CrossRef]
4. Pfennig, N.; Biebl, H. *Desulfuromonas acetoxidans* gen. nov. and sp. nov., a new anaerobic, sulfur-reducing, acetate-oxidizing bacterium. *Arch. Microbiol.* **1976**, *110*, 3–12. [CrossRef] [PubMed]
5. Van Beeumen, J.; Ambler, R.P.; Meyer, T.E.; Kamen, M.O.; Olson, J.M.; Shaw, E.K. The amino acid sequences of the cytochromes c-555 from two green sulphur bacteria of the genus *Chlorobium*. *Biochem. J.* **1976**, *159*, 757–769. [CrossRef]
6. Ambler, R.P. The amino acid sequence of cytochrome c-551.5 (Cytochrome c(7)) from the green photosynthetic bacterium *Chloropseudomonas ethylica*. *FEBS Lett.* **1971**, *18*, 351–353. [CrossRef]

7. Gray, B.H.; Fowler, C.F.; Nugent, N.A.; Fuller, R.C. A reevaluation of the presence of low midpoint potential cytochrome C-551.5 in the green photosynthetic bacterium *Chloropseudomonas ethylica*. *Biochem. Biophys. Res. Commun.* **1972**, *47*, 322–327. [CrossRef]
8. Gray, B.H.; Fowler, C.F.; Nugent, N.A.; Rigopoulos, N.; Fuller, R.C. Reevaluation of *Chloropseudomonas ethylica* strain 2-K. *Int. J. Syst. Bacteriol.* **1973**, *23*, 256–264. [CrossRef]
9. Overmann, J. Phototrophic Consortia: A Tight Cooperation Between Non-Related Eubacteria. In *Symbiosis. Cellular Origin, Life in Extreme Habitats and Astrobiology*; Seckbach, J., Ed.; Springer: Dordrecht, The Netherlands, 2001; Volume 4.
10. Schink, B.; Stams, A.J.M. Syntrophism among Prokaryotes. In *The Prokaryotes*; Dworkin, M., Falkow, S., Rosenberg, E., Schleifer, K.H., Stackebrandt, E., Eds.; Springer: New York, NY, USA, 2006.
11. McInerney, M.J.; Sieber, J.R.; Gunsalus, R.P. Syntrophy in anaerobic global carbon cycles. *Curr. Opin. Biotechnol.* **2009**, *20*, 623–632. [CrossRef]
12. Klatt, C.; Wood, J.; Rusch, D.; Bateson, M.M.; Hamamura, N.; Heidelberg, J.F.; Grossman, A.R.; Bhaya, D.; Cohan, F.M.; Kühl, M.; et al. Community ecology of hot spring cyanobacterial mats: Predominant populations and their functional potential. *ISME J.* **2011**, *5*, 1262–1278. [CrossRef]
13. Liu, Z.; Müller, J.; Li, T.; Alvey, R.M.; Vogl, K.; Frigaard, N.U.; Rockwell, N.C.; Boyd, E.S.; Tomsho, L.P.; Schuster, S.C.; et al. Genomic analysis reveals key aspects of prokaryotic symbiosis in the phototrophic consortium "*Chlorochromatium aggregatum*". *Genome Biol.* **2013**, *14*, R127. [CrossRef] [PubMed]
14. Müller, J.; Overmann, J. Close interspecies interactions between prokaryotes from sulfureous environments. *Front. Microbiol.* **2011**, *2*, 146. [CrossRef] [PubMed]
15. Ishii, S.; Kosaka, T.; Hori, K.; Hotta, Y.; Watanabe, K. Coaggregation facilitates interspecies hydrogen transfer between *Pelotomaculum thermopropionicum* and *Methanothermobacter thermautotrophicus*. *Appl. Environ. Microbiol.* **2005**, *71*, 7838–7845. [CrossRef] [PubMed]
16. Huber, H.; Hohn, M.J.; Stetter, K.O.; Rachel, R. The phylum Nanoarchaeota: Present knowledge and future perspectives of a unique form of life. *Res. Microbiol.* **2003**, *154*, 165–171. [CrossRef]
17. Lauterborn, R. Zur Kenntnis der sapropelischen Flora. *Allg. Bot. Z.* **1906**, *12*, 196–197.
18. Biebl, H.; Pfennig, N. Growth yields of green sulfur bacteria in mixed cultures with sulfur and sulfate reducing bacteria. *Arch. Microbiol.* **1978**, *117*, 9–16. [CrossRef]
19. Warthmann, R.; Cypionka, H.; Pfennig, N. Photoproduction of H_2 from acetate by syntrophic cocultures of green sulfur bacteria and sulfur-reducing bacteria. *Arch. Microbiol.* **1992**, *157*, 343–348. [CrossRef]
20. Wanner, G.; Vogl, K.; Overmann, J. Ultrastructural characterization of the prokaryotic symbiosis in "*Chlorochromatium aggregatum*". *J. Bacteriol.* **2008**, *190*, 3721–3730. [CrossRef]
21. Copeland, A.; Lucas, S.; Lapidus, A.; Barry, K.; Detter, J.C.; Glavina del Rio, T.; Hammon, N.; Israni, S.; Dalin, E.; Tice, H.; et al. Sequencing of the draft genome and assembly of *Desulfuromonas acetoxidans* DSM 684. *Jt. Genome Inst.* **2005**, unpublished.
22. Wattam, A.R.; Davis, J.J.; Assaf, R.; Boisvert, S.; Brettin, T.; Bun, C.; Conrad, N.; Dietrich, E.M.; Disz, T.; Gabbard, J.L.; et al. Improvements to PATRIC, the all-bacterial Bioinformatics Database and Analysis Resource Center. *Nucleic Acids Res.* **2017**, *45*, D535–D542. [CrossRef]
23. Aziz, R.K.; Bartels, D.; Best, A.A.; DeJongh, M.; Disz, T.; Edwards, R.A.; Fromsma, K.; Gerdes, S.; Glass, E.M.; Kabul, M.; et al. The RAST server: Rapid annotations using subsystems technology. *BMC Genom.* **2008**, *9*, 75. [CrossRef]
24. Richter, M.; Rosselló-Móra, R.; Glöckner, F.O.; Peplies, J. JSpeciesWS: A web server for prokaryotic species circumscription based on pairwise genome comparison. *Bioinformatics* **2016**, *32*, 929–931. [CrossRef]
25. Stamatakis, A.; Hoover, P.; Rougemont, J.J.S.B. A rapid bootstrap algorithm for the RAxML web servers. *Syst. Biol.* **2008**, *57*, 758–771. [CrossRef]
26. Stamatakis, A.J.B. RAxML version 8: A tool for phylogenetic analysis and post-analysis of large phylogenies. *Bioinformatics* **2014**, *30*, 1312–1313. [CrossRef]
27. Letunic, I.; Bork, P. Interactive Tree Of Life (iTOL) v4: Recent updates and new developments. *Nucleic Acids Res.* **2019**, *47*, 256–259. [CrossRef] [PubMed]
28. Parks, D.H.; Imelfort, M.; Skennerton, C.T.; Hugenholtz, P.; Tyson, G.W. Assessing the quality of microbial genomes recovered from isolates, single cells, and metagenomes. *Genome Res.* **2014**, *25*, 1043–1055. [CrossRef] [PubMed]

29. Freed, S.; Robertson, S.; Meyer, T.; Kyndt, J. Draft Whole-Genome Sequence of the Green Sulfur Photosynthetic Bacterium *Chlorobaculum* sp. Strain 24CR, Isolated from the Carmel River. *Microbiol. Resour. Announc.* **2019**, *8*, e00116-19. [CrossRef] [PubMed]
30. Overmann, J.; Pfennig, N. *Pelodictyon phaeoclathiforme* sp. nov., a new brown-colored member of the Chlorobiaceae forming net-like colonies. *Arch. Microbiol.* **1989**, *152*, 401–406. [CrossRef]
31. Tomich, M.; Planet, P.J.; Figurski, D.H. The tad locus: Postcards from the widespread colonization island. *Nat. Rev. Microbiol.* **2007**, *5*, 363–375. [CrossRef]
32. Bodenmiller, D.; Toh, E.; Brun, Y.V. Development of surface adhesion in *Caulobacter crescentus*. *J. Bacteriol.* **2004**, *186*, 1438–1447. [CrossRef]
33. Entcheva-Dimitrov, P.; Spormann, A.M. Dynamics and control of biofilms of the oligotrophic bacterium *Caulobacter crescentus*. *J. Bacteriol.* **2004**, *186*, 8254–8266. [CrossRef] [PubMed]
34. Sangermani, M.; Hug, I.; Sauter, N.; Pfohl, T.; Jenal, U. Tad Pili play a dynamic role in *Caulobacter crescentus* surface colonization. *mBio* **2019**, *10*, e01237-19. [CrossRef] [PubMed]
35. Wu, H.; Fives-Taylor, P.M. Molecular strategies for fimbrial expression and assembly. *Crit. Rev. Oral Biol. Med.* **2001**, *12*, 101–115. [CrossRef] [PubMed]
36. Fernandez, L.A.; Berenguer, J. Secretion and assembly of regular surface structures in Gram-negative bacteria. *FEMS Microbiol. Rev.* **2000**, *24*, 21–44. [CrossRef]
37. Mattick, J.S. Type IV pili and twitching motility. *Annu. Rev. Microbiol.* **2002**, *56*, 289–314. [CrossRef]
38. Satchell, K.J. Structure and function of MARTX toxins and other large repetitive RTX proteins. *Annu. Rev. Microbiol.* **2011**, *65*, 71–90. [CrossRef]
39. Berne, C.; Ducret, A.; Hardy, G.G.; Brun, Y.V. Adhesins Involved in Attachment to Abiotic Surfaces by Gram-Negative Bacteria. *Microbiol. Spectr.* **2015**, *3*. [CrossRef]
40. Lasa, I.; Penades, J.R. Bap: A family of surface proteins involved in biofilm formation. *Res. Microbiol.* **2006**, *157*, 99–107. [CrossRef]
41. Yousef, F.; Espinosa-Urgel, M. In silico analysis of large microbial surface proteins. *Res. Microbiol.* **2007**, *158*, 545–550. [CrossRef]
42. Treangen, T.J.; Salzberg, S.L. Repetitive DNA and next-generation sequencing: Computational challenges and solutions. *Nat. Rev. Genet.* **2011**, *13*, 36–46. [CrossRef]
43. Nagarajan, H.; Butler, J.E.; Klimes, A.; Qiu, Y.; Zengler, K.; Ward, J.; Young, N.D.; Methé, B.A.; Palsson, B.Ø.; Lovley, D.R.; et al. De novo assembly of the complete genome of an enhanced electricity-producing variant of *Geobacter sulfurreducens* using only short reads. *PLoS ONE* **2010**, *5*, e10922. [CrossRef] [PubMed]
44. Parge, H.E.; Forest, K.T.; Hickey, M.J.; Christensen, D.A.; Getzoff, E.D.; Tainer, J.A. Structure of the fibre-forming protein pilin at 2.6 Å resolution. *Nature* **1995**, *378*, 32–38. [CrossRef] [PubMed]
45. Reguera, G.; McCarthy, K.D.; Mehta, T.; Nicoll, J.S.; Tuominen, M.T.; Lovley, D.R. Extracellular electron transfer via microbial nanowires. *Nature* **2005**, *435*, 1098–1101. [CrossRef] [PubMed]
46. Liu, X.; Wang, S.; Xu, A.; Zhang, L.; Liu, H.; Ma, L.Z. Biological synthesis of high-conductive pili in aerobic bacterium *Pseudomonas aeruginosa*. *Appl. Microbiol. Biotechnol.* **2019**, *103*, 1535–1544. [CrossRef] [PubMed]
47. Lovley, D.R.; Walker, D.J.F. Geobacter Protein Nanowires. *Front. Microbiol.* **2019**, *10*, 2078. [CrossRef] [PubMed]
48. Summers, Z.M.; Fogarty, H.E.; Leang, C.; Franks, A.E.; Malvankar, N.S.; Lovley, D.R. Direct exchange of electrons within aggregates of an evolved syntrophic coculture of anaerobic bacteria. *Science* **2010**, *330*, 1413–1415. [CrossRef]
49. Malvankar, N.S.; Lovley, D.R. Microbial nanowires: A new paradigm for biological electron transfer and bioelectronics. *ChemSusChem* **2012**, *5*, 1039–1046. [CrossRef]
50. Aklujkar, M.; Haveman, S.A.; DiDonato, R.; Chertov, O.; Han, C.S.; Land, M.L.; Brown, P.; Lovley, D.R. The genome of *Pelobacter carbinolicus* reveals surprising metabolic capabilities and physiological features. *BMC Genom.* **2012**, *13*, 690. [CrossRef]
51. Reguera, G.; Nevin, K.P.; Nicoll, J.S.; Covalla, S.F.; Woodard, T.L.; Lovley, D.R. Biofilm and nanowire production leads to increased current in *Geobacter sulfurreducens* fuel cells. *Appl. Environ. Microbiol.* **2006**, *72*, 7345–7348. [CrossRef]
52. Zanetti, G.; Pandini, V. Ferredoxin. In *Encyclopedia of Biological Chemistry*, 2nd ed.; William, J., Lennarz, M., Lane, D., Eds.; Academic Press: Cambridge, MA, USA, 2013; pp. 296–298. ISBN 9780123786319.

53. Marsh, J.W.; Taylor, R.K. Genetic and Transcriptional Analyses of the *Vibrio cholerae* Mannose-Sensitive Hemagglutinin Type 4 Pilus Gene Locus. *J. Bacteriol.* **1999**, *181*, 1110–1117. [CrossRef]
54. Ha, P.T.; Lindemann, S.R.; Shi, L.; Dohnalkova, A.C.; Frederickson, J.K.; Madigan, M.T.; Beyenal, H. Syntrophic Anaerobic Photosynthesis via Direct Interspecies Electron Transfer. *Nat. Commun.* **2017**, *8*, 13924. [CrossRef] [PubMed]
55. Guo, Y.; Aoyagi, T.; Inaba, T.; Sato, Y.; Habe, H.; Hori, T. Complete Genome Sequence of *Desulfuromonas* sp. Strain AOP6, an Iron(III) Reducer Isolated from Subseafloor Sediment. *Microbiol. Res. Announc.* **2020**, *9*, e01325-19. [CrossRef] [PubMed]
56. Childers, S.E.; Ciufo, S.; Lovley, D.R. Geobacter metallireducens accesses insoluble Fe(III)oxide by chemotaxis. *Nature* **2002**, *416*, 767–769. [CrossRef] [PubMed]
57. Pokkuluri, P.R.; Londer, Y.Y.; Duke, N.E.; Long, W.C.; Schiffer, M. Family of cytochrome c7-type proteins from *Geobacter sulfurreducens*: Structure of one cytochrome c7 at 1.45 Å resolution. *Biochemistry* **2004**, *43*, 849–859. [CrossRef] [PubMed]
58. Probst, I.; Bruschi, M.; Pfennig, N.; Le Gall, J. Cytochrome c-551.5 (c7) from *Desulfuromonas acetoxidans*. *Biochim. Biophys. Acta* **1977**, *460*, 58–64. [CrossRef]
59. Pereira, I.A.; Pacheco, I.; Liu, M.Y.; Legall, J.; Xavier, A.V.; Teixeira, M. Multiheme cytochromes from the sulfur-reducing bacterium *Desulfuromonas acetoxidans*. *Eur. J. Biochem.* **1997**, *248*, 323–328. [CrossRef]
60. Mathews, F.S. The structure, function and evolution of cytochromes. *Prog. Biophys. Mol. Biol.* **1985**, *45*, 1–56. [CrossRef]
61. Edwards, M.J.; Richardson, D.J.; Paquete, C.M.; Clarke, T.A. Role of multiheme cytochromes involved in extracellular anaerobic respiration in bacteria. *Protein Sci.* **2020**, *29*, 830–842. [CrossRef]
62. Leang, C.; Qian, X.; Mester, T.; Lovley, D.R. Alignment of the c-type cytochrome OmcS along pili of *Geobacter sulfurreducens*. *Appl. Environ. Microbiol.* **2010**, *76*, 4080–4084. [CrossRef]
63. Imhoff, J.F. Biology of Green Sulfur Bacteria. In *eLS*; John Wiley & Sons, Ltd.: Hoboken, NJ, USA, 2014. [CrossRef]

Publisher's Note: MDPI stays neutral with regard to jurisdictional claims in published maps and institutional affiliations.

© 2020 by the authors. Licensee MDPI, Basel, Switzerland. This article is an open access article distributed under the terms and conditions of the Creative Commons Attribution (CC BY) license (http://creativecommons.org/licenses/by/4.0/).

Article

Niche Partitioning with Temperature among Heterocystous Cyanobacteria (*Scytonema* spp., *Nostoc* spp., and *Tolypothrix* spp.) from Biological Soil Crusts

Ana Giraldo-Silva [1,2], Vanessa M. C. Fernandes [1,2], Julie Bethany [1,2] and Ferran Garcia-Pichel [1,2,*]

- [1] School of Life Sciences, Arizona State University, Tempe, AZ 85287, USA; amgirald@asu.edu (A.G.-S.); vanessa.fernandes@asu.edu (V.M.C.F.); jabethan@asu.edu (J.B.)
- [2] Center for Fundamental and Applied Microbiomics (CFAM), Biodesing Institute, Arizona State University, Tempe, AZ 85287, USA
- [*] Correspondence: ferran@asu.edu, Tel.: +1-4807270498

Received: 14 February 2020; Accepted: 10 March 2020; Published: 12 March 2020

Abstract: Heterocystous cyanobacteria of biocrusts are key players for biological fixation in drylands, where nitrogen is only second to water as a limiting resource. We studied the niche partitioning among the three most common biocrust heterocystous cyanobacteria sts using enrichment cultivation and the determination of growth responses to temperature in 30 representative isolates. Isolates of *Scytonema* spp. were most thermotolerant, typically growing up to 40 °C, whereas only those of *Tolypothrix* spp. grew at 4 °C. *Nostoc* spp. strains responded well at intermediate temperatures. We could trace the heat sensitivity in *Nostoc* spp. and *Tolypothrix* spp. to N_2-fixation itself, because the upper temperature for growth increased under nitrogen replete conditions. This may involve an inability to develop heterocysts (specialized N_2-fixing cells) at high temperatures. We then used a meta-analysis of biocrust molecular surveys spanning four continents to test the relevance of this apparent niche partitioning in nature. Indeed, the geographic distribution of the three types was clearly constrained by the mean local temperature, particularly during the growth season. This allows us to predict a potential shift in dominance in many locales as a result of global warming, to the benefit of *Scytonema* spp. populations.

Keywords: biological soil crust; drylands; niche partitioning; nitrogen fixing cyanobacteria

1. Introduction

In drylands, where plant growth is limited by water and nutrients, the soil surface can be occupied by communities of microorganisms known as biological soil crusts (biocrusts; see [1] for a primer, and [2] for a monograph), which play crucial roles for the fertility and stability of drylands. Their presence enhances resistance to erosion caused by water [3] or wind [4,5], modifies soil surface temperature [6], and influences water retention and runoff [7–9]. Colonization of bare soils, typically pioneered by highly motile filamentous cyanobacteria like *Microcoleus vaginatus* and *Microcoleus steenstrupii* [10] results in the formation of incipient communities. Once the surface is stabilized, sessile, heterocystous cyanobacteria colonize secondarily. The community also hosts a variety of populations of heterotrophic bacteria [11,12], archaea [13], and fungi [14] as well as lichens and mosses [15], which are typical of the most developed crusts. Once established, these heterocystous cyanobacteria are significant contributors to dinitrogen inputs in soils crusts [16], taking over this role from the heterotrophic diazotrophic bacteria [17] that enter in C for N symbioses with *M. vaginatus* in early succession stages [18]. Three phylogenetically well-defined clades, *Scytonema* spp., *Nostoc* spp. and *Tolypothrix/Spirirestis* spp., have been identified as the most abundant diazotrophic cyanobacteria in biocrusts communities of the Southwestern US [19]. Soil crusts are typically in a perennial state of N deficiency because the internal

N cycle is broken (denitrification is apparently absent from most biocrusts; [20,21]). Biological fixation thus remains a necessity for continued growth. Fixed atmospheric C and N [20,22,23], along with other elements [24] can then be exported to underlying soils, improving landscape soil fertility. Because drylands cover nearly 45% of the total Earth continental area [25], and aridity is predicted to increase due to global warming [26–28], this N export activity of biocrusts matters not only locally, but also globally. In fact, the global N_2-fixation of cryptogamic covers, much of which are biocrusts, has been estimated at 49 Tg/yr, accounting for nearly 50% of the biological N_2-fixation on land [29].

N_2-fixation activity has been determined experimentally to be optimal in the range of 15–30 °C regardless of the biocrusts origin or successional stage assayed in the US Southwest [30,31], with rates decreasing significantly between 30 and 35 °C [31]. This sensitivity has been ascribed to possible deleterious effects of temperature on N_2-fixing cyanobacteria [31]. Thermophysiological studies using laboratory isolates [31–33] or geographical distribution in molecular tallies [34] have shown that the three main clades of biocrust heterocystous cyanobacteria are characterized by different temperature ranges for growth: the *Scytonema* spp. clade tends to be more thermotolerant, whereas the *Tolypothrix* spp. clade shows psychrophilic preferences, and strains in the *Nostoc* spp. clade shows a preference for mild temperatures (15 to 30 °C). However, these results come from the evaluation of a restricted number of sites or strains, and the patterns were not always robust. Clearly, however, the results point to a potential for differential sensitivity of these cyanobacteria to environmental warming, a future scenario with which biocrust will have to contend. Drylands at large will likely become warmer and drier in response to global warming. In particular, the US Southwest is predicted to experience an increase in temperature of about 1 °C per decade [26], accompanied by alterations in precipitation frequency [35–37].

In this contribution we wanted to evaluate in detail the thermophysiology of biocrust heterocystous cyanobacteria using cultivated isolates, and to test if their niche differentiation is regulated by N_2-fixation. Finally, we wanted to test if the physiological data obtained from cultures, can explain the current biogeographic distribution of each clade, and hence potentially help us predict their fate in the face of global warming. Our results show that these cyanobacteria show markedly different thermophysiological patterns in culture and consistent world-wide distributions in nature. This points to a potential for differential sensitivity among them to global warming, allowing us to predict a microbial replacement that biocrusts will have to contend with in future climate change scenarios.

2. Materials and Methods

2.1. Enrichment Cultures

Field biocrusts were collected from the (cold) Great Basin Desert (Utah, USA), and from the (warm) Chihuahuan desert (New Mexico, USA), and from two soil textural types in each, Great Basin: sandy clay loam and clay loam, and Chihuahuan: clay loam and loamy sand. Locations and soil types details are given in [38]. Three enrichment cultures were prepared from each site and incubation temperature by randomly placing small biocrust crumbles and spreading it over 1.5 % (*w/v*) agar-solidified minimal medium without combined nitrogen (BG11$_0$; [39,40]) in Petri dishes. They were incubated at 4, 25 and 30 °C, for 20 days, under 20 to 27 μmoL m^{-2} s^{-1} light from fluorescent bulbs under a 14 h photoperiod. After incubation, colonies were counted, sampled and observed under the compound microscope (labophot-2, Nikon, Tokyo, Japan) to be assigned to one of the three morphotypes. Differences in the relative proportions were assessed via permutational multivariate analysis of variance (PERMANOVA). PERMANOVAS were performed on the Bray-Curtis distance matrices of relative proportions derived from colonies counts and used 999 permutations. PERMANOVAS were run on PRIMER 6 software with PERMANOVA+ add on [41,42].

2.2. Experimental Organisms and Growth Conditions

Thirty cyanobacterial strains: 12 *Scytonema* spp., 10 *Nostoc* spp., and eight *Tolypothrix* spp. previously isolated as a part of our "microbial biocrust nurseries" protocols (see [38] as well as a description of the cyanobacterial community structure of the biocrust communities of origin), were used in our experiments. Briefly, strains were isolated from enrichment cultures in agar-solidified BG11$_0$ Petri plates followed by multiple streaking of colonies on fresh agar plates. Strain identity was first assessed by microscopy, and then confirmed by PCR amplification of the V4 region of the 16S rRNA gene using cyanobacteria specific primers CYA359F/CYA781R [43] (PCR protocol therein), blast comparisons, and by placing the sequences on the cyanobacterial tree Cydrasil (https://itol.embl.de/tree/1491698589270801574806192). PCR products were sequenced using Sanger sequencing. All strains were unicyanobacterial, are kept in our local culture collection, and are available upon request. Strain accession numbers along with their denomination coding for site of origin can be found in Table S3. Stock cultures were grown in 175 mL cell culture flasks containing 100 mL of medium free of combined nitrogen (BG11$_0$). Cultures were maintained at 25 ± 2 °C, under a 14 h photoperiod, illuminated at 20–27 µmoL (photon) m^{-2} s^{-1} provided by white fluorescent tubes.

2.3. Delineation of Temperature Range for Growth and Survival of Isolates

Prior to inoculation, stock liquid cultures of each strain were homogenized by repeatedly forcing biomass through a 60 mL sterile syringe, and immediately washed with fresh BG11$_0$ medium by five consecutive centrifugations (8 min, 8437 g, 25 °C). Aliquots of this homogenized cultures served as inoculum (5% *v/v*) for experimental cultures, which were run on 50 mL cell culture flasks filled to the 20 mL mark. Each strain was incubated at 4, 15, 25, 30, 35, 40 and 45 °C in triplicate, exposed to a light intensity of 20–27 µmoL (photon) m^{-2} s^{-1} provided by white fluorescent tubes, in a 12 h photoperiod regime. Growth was estimated visually after 30 days as either positive for growth (there was an obvious increase in biomass at the end of the incubation period compared to initial inoculum) or negative for growth (either no-growth (stasis) or patent death). Assays assigned to "no-growth" looked healthy, with brightly pigmented cells, but did not show appreciable biomass increase during the incubation, whereas assays assigned to patent death exhibited a total loss of pigmentation. The whole experiment was replicated a second time in full, and growth in any of the trials was reported as positive.

2.4. Influence of Diazotrophy on the Upper Temperature Limit for Growth

A homogenized, cleaned culture mix was prepared for each of the strains as detailed above, and inoculated (5% *v/v*) in 50 mL cell culture flasks containing either medium without combined nitrogen (BG110) or nitrogen-containing medium (BG11). Triplicate cultures were incubated at 35 and 40 °C, illuminated with 20–27 µmol (photon) m^{-2} s^{-1} provided by white fluorescent tubes, in a 12 h photoperiod regimen, for 30 days.

2.5. Heterocyst and Vegetative Cell Counts

To determine the frequency of heterocysts we conducted microscopic cell counts on fresh wet mounts under bright field illumination in a Nikon labophot-2 compound microscope. At least 200 cells were counted in each determination. To determine the effect of nitrogen source and incubation temperature on heterocyst frequency we examined triplicate cultures of each strain at 25, 35 and 40 °C, all at day 7 after inoculation, time at which all tested strains appeared healthy. The full experiment was replicated for a total $n = 6$.

2.6. Chlorophyll a Determination

Chlorophyll *a* (Chl *a*) was measured as a proxy for phototrophic biomass. Chl *a* was extracted in triplicate, in 90% acetone, according to [44], vortexed for 30 s. and allowed to extract for 24 h at

4 °C in the dark. Extracts were clarified by centrifugation (5 m at 8437 g). Absorbance spectra of the clarified extracts was recorded on a UV-visible spectrophotometer (UV-1601, Shimadzu, Kyoto, Japan). Interference from scytonemin and carotenoids was corrected using the trichromatic equation of [45].

2.7. Meta-analysis of Temperature Niches

In an attempt to look for a temperature segregation pattern among the studied taxa in the natural biocrust environment, we performed a meta-analysis of all bacterial 16S rRNA tallies available publicly. We performed a literature search, and either downloaded from public databases or directly requested raw sequence data from authors from multiple environmental biocrust surveys conducted at different locations around the world. We collected data from different arid and semiarid regions in USA [6,34,46–48], Mexico [33] and Australia [49], from arid, semiarid and alpine regions in Europe [32,50], from the arid Gurbantunggut desert in China [51], and from the Brazilian savannah (Cerrado) [52]. A complete list of the biocrust surveys with locations, environmental variables, and other relevant information can be found in Table S4.

For all but the dataset from [34], forward reads obtained with pyrosequencing [51] and paired-end reads obtained with Illumina were demultiplexed, and quality controlled using the DADA2 plugin [53] available in Qiime 2018.6 [54], creating a feature table containing representative sequences (features) and their frequency of occurrence. Highly variable positions were removed using MAFFT [55], and phylogenetic trees were generated using FastTree [56]. Preliminary taxonomic assignment was done using the Naïve Bayes classifier [57] trained on the Greengenes 13.8 release database [58]. For the [34] dataset, because quality files (.fastq) were not available, and in an effort to control for sequence quality before preforming any downstream analysis, raw sequences were first filtered using USEARCH 7 [59] to remove all sequences with less than 210 bp. Overall this step filtered out up to 5% of the total sequences in some but not all samples. Additionally, the first and last 10 bp of each sequence were trimmed using Fastx (http://hannonlab.cshl.edu/fastx_toolkit/). Quality controlled sequences were assigned to individual samples and barcodes were removed using Qiime 1.8 [54] using the *multiple_split_librairies_fastq.py* script. Operational taxonomic units (OTUs) were defined with a threshold of 97% similarity and clustered using UCLUST [59] using the *pick_open_reference_otus.py* script in Qiime. Potential chimeras, and singleton OTUs were removed from further consideration. Preliminary taxonomic assignments were done with the RDP (Ribosomal Database Project) classifier [60], and representative sequences were then aligned against the Greengenes database core reference alignment [58].

Cyanobacterial sequences (features) and OTUs were filtered out from the master file, and a more refined taxonomic assignment at the genus and species level was further informed throughout phylogenetic placements. Query cyanobacterial sequences (and OTUs) were phylogenetically placed in our cyanobacteria reference tree CYDRASIL version-0.22a (https://github.com/FGPLab/cydrasil/tree/0.22a, accessed in July, 2019), by aligning sequences to the cyanobacterial tree alignment using PaPaRa [61], and then placing them into the reference tree using the RaxML8 Evolutionary Placement Algorithm [62]. The resulting trees were imported and visualized in the iTOL4 server [63]. Accession numbers of representative strains of the clades in which *Scytonema* spp., *Nostoc* spp. and *Tolypothrix* spp. were assigned according to CYDRASIL are included in Table S1.

The proportion of *Scytonema* spp., *Nostoc* spp. and *Tolypothrix* spp. within the heterocystous cyanobacterial community was calculated by dividing the number of reads of either *Scytonema* spp., *Nostoc* spp. or *Tolypothrix* spp., by the sum of the number of reads of all N_2-fixing cyanobacteria found at each location. Resulting proportions were plotted against the mean annual temperature (MAT) and the mean temperature of the wettest quarter of the year (growth season) in each location of origin. A total of 25 (out of 109) locations at which the total relative abundance of N_2-fixing cyanobacteria was lower than 0.5 % of all reads were excluded from final plots. Mean annual temperature and mean temperature of the wettest quarter of the year were calculated from environmental variables of monthly climate data for minimum, mean, and maximum temperature and for precipitation for 1970–2000. Data was downloaded from WorldClim -Global Climate Data -version 2 (http://www.worldclim.org; [64].

Linear regressions between the proportion of sequence reads (arcsine transformed) of each taxon among heterocystous cyanobacteria and climatic parameters (MAT and MTempWetQ) were used to test significance of environmental patterns.

3. Results

3.1. Encrichment Cultivation

Enrichment cultures for diazotrophic photoautotrophs carried out at different temperatures using inoculum from four different biocrusts were very revealing. Only heterocystous cyanobacteria were enriched for in our medium free of nitrate and ammonium, and all 994 colonies examined belonged to one of the three major clades known from biocrusts: *Nostoc* spp., *Tolypothrix* spp., and *Scytonema* spp. [19], as determined by microscopic inspection. The relative proportions obtained, however, were strongly dependent on the temperature of incubation (Figure 1). The composition of the enrichments at 4 °C was significantly different from those growing at 25 °C (PERMANOVA pseudo-F: 6.22 df: 22 $p \geq 0.001$) and 30 °C (PERMANOVA pseudo-F: 9.36 df: 22 $p \geq 0.001$); the same was true for the comparison of 25 and 30 °C (PERMANOVA pseudo-F: 6.43 df: 22 $p \geq 0.001$). *Scytonema* spp. made up the majority of the colonies at 30 °C, whereas *Tolypothrix* spp. was preferentially selected for at 4 °C. *Nostoc* spp. had a slight advantage at lower temperatures as well. This was so regardless of the origin of the crusts used for inoculation, in that there was no significant effect on outcomes by location (PERMANOVA, $p \leq 0.2$; full dataset presented in Table S2).

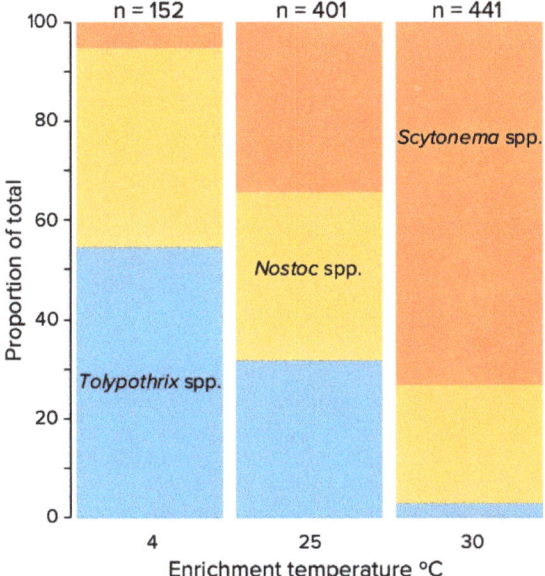

Figure 1. Relative proportion of colonies assignable to *Scytonema* spp., *Nostoc* spp., and *Tolypothrix* spp. in enriched cultures obtained on medium without combined nitrogen as a function of incubation temperature.

3.2. Temperature Range for Growth (or Survival) of Isolated Strains

All cyanobacterial strains (tested in medium without combined N) showed robust growth at 15 and 25 °C, while none grew at 45 °C (Figure 2), the lower limit of moderate thermophilly. Formally then, all these strains were mesophiles with respect to temperature. At 4 °C, all *Tolypothrix* spp. strains grew well, while only one *Scytonema* sp. strain did. At this temperature, three *Nostoc* spp. strains did not grow, while five strains were in apparent stasis (they neither grew nor show signs of cellular degradation). At 30 °C four out of eight *Tolypothrix* spp., nine out of ten *Nostoc* spp., and eleven out of twelve *Scytonema* spp. strains grew well. At 35 and 40 °C, no *Nostoc* spp. or *Tolypothrix* spp. strains grew, while eleven out of twelve *Scytonema* spp. did.

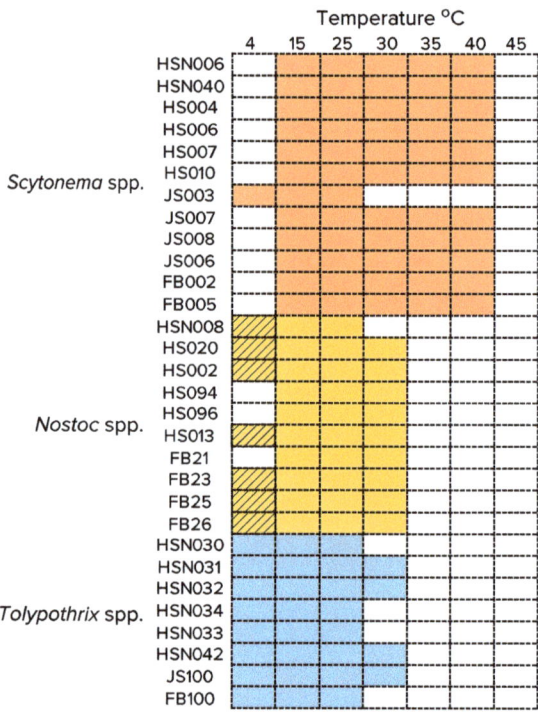

Figure 2. Temperature range at which the studied cyanobacterial strains can grow or survive under diazotrophic conditions. Colored rectangles indicate positive growth; hatched rectangles indicate stasis (no growth, but no obvious deterioration).

3.3. Upper Temperature Limit for Growth and N_2-Fixation

We looked at growth (and survival) responses more in detail as a function of nitrogen source (N_2-fixing vs. non N_2-fixing conditions) in the upper range of temperature (35 and 40 °C) in an effort to infer if N_2-fixation was the most sensitive cellular process determining the observed outcomes. Figure 3 shows the biomass yield of the 30 cyanobacterial strains after 30 days of growth cultivated in medium without combined nitrogen (nitrogen-free) and nitrogen replete media.

Our results show that providing a source of fixed nitrogen expanded the range for growth in many of them to 35 °C (*Scytonema* spp. JS003; *Nostoc* spp. HS002, HS094, HS013, *Tolypothrix* spp., HSN032, HSN033, HSN034) and in some cases, strains survived at 40 °C (*Nostoc* spp. HSN008, HS020, HS002, HS096, FB23, FB26; *Tolypothrix* sp. HSN042). The last column in Figure 3. shows the biomass yield in nitrogen replete minus that attained in medium without combined nitrogen at 35 °C, indicating a generalized positive effect on growth under nitrogen-replete conditions. For sixteen out of thirty strains this difference in growth was significant. This gives support to the contention that the upper temperature for growth may be determined by the sensitivity of N_2-fixation in *Nostoc* spp. and *Tolypothrix* spp.; whereas it is not nearly as determinant for *Scytonema* spp.

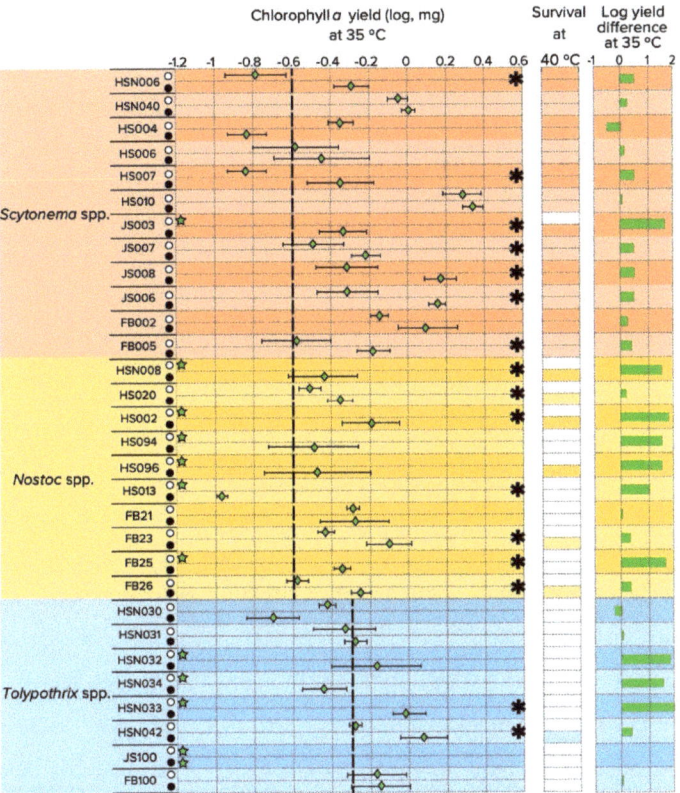

Figure 3. Growth yield of N_2-fixing cyanobacterial strains in the upper range of temperature for growth in nitrogen free (o) vs. nitrogen replete media (•). Rhombuses indicate the mean and error bars indicate ± 1 SE, with $n = 3$. Vertical dashed lines indicate the amount of inoculum provided. At 40 °C, only observational data were recorded: colored rectangles indicate survival and white rectangles indicate death. * Denotes statistically significant differences between growing conditions according to Wilcox's test.

3.4. Heterocyst Frequency

To determine if this effect on N_2-fixation was perhaps due to an inability to develop heterocysts (a developmentally specialized cell type dedicated to this process), we conducted microscopic counts of vegetative cells and heterocysts in strains incubated for seven days at different temperatures (Table 1). Counts were performed only on apparently healthy filaments, but at 35 and 40 °C, biomass from all replicates of *Nostoc* sp. HSN008 and *Tolypothrix* sp. HSN042 looked yellowish, and microscopy revealed high cell mortality as well. In fact, in one occasion, one set of replicates of *Nostoc* sp. HSN008 did not survive to day 7 (Table 1). All strains grown at 25 °C looked healthy when counts were performed. Those caveats aside, the frequency of heterocysts declined precipitously for *Nostoc* spp. strains above 35 degrees, and above 30 degrees for *Tolypothrix* spp. strains. In *Scytonema* spp., there were only slight decreases in this frequency in the temperature range tested. This is consistent with a cell developmental basis for the sensitivity of N_2-fixation to high temperatures in *Nostoc* spp. and *Tolypothrix* spp.

Table 1. Frequency of heterocysts (number of vegetative cells per heterocyst) in representative cyanobacterial strains after incubation at 30, 35 and 40 °C for 7 days. Averages of $n = 6$ determinations ± standard deviation are given. H: heterocystous, VG: Vegetative cells.

Strain	Incubation Temperature °C								
	25			35			40		
	H	VG	Ratio H:VG	H	VG	Ratio H:VG	H	VG	Ratio H:VG
Nostoc spp.	26	230	1:9	10	207	1:21	18	569	1:32
	23	205	1:9	12	258	1:22	28	773	1:28
	32	206	1:6	11	223	1:20	21	642	1:31
	24	208	1:9	13	246	1:10	23	730	1:32
	26	207	1:8	14	236	1:17	26	701	1:27
	22	212	1:10	-	-	-	27	689	1:26
Tolypothrix spp.	35	467	1:13	6	628	1:105	3	900	1:300
	38	507	1:13	3	444	1:148	4	900	1:225
	40	538	1:13	2	250	1:125	2	900	1:450
	37	557	1:15	5	576	1:115	6	900	1:150
	43	613	1:14	6	553	1:92	4	900	1:225
	36	582	1:15	3	324	1:108	4	900	1:224
Scytonema spp.	27	534	1:20	15	352	1:23	12	526	1:44
	29	553	1:19	5	150	1:20	12	600	1:50
	32	669	1:21	7	225	1:32	10	550	1:55
	37	844	1:23	9	276	1:32	14	708	1:44
	18	308	1:17	8	242	1:30	16	641	1:50
	25	412	1:16	7	233	1:33	41	1087	1:55

3.5. Thermal Niche of Biocrust Heterocystous Cyanobacteria through Meta-Analyses of Molecular Surveys

A total of 84 locations from eleven different biocrust surveys conducted in different arid and semiarid regions in North and South America, Europe, Australia, and China, and in the Brazil Savannah (see Table S4), were used in a meta-analysis to assess the relative contribution of the three main clades of heterocystous cyanobacteria along temperature related parameters. Figure 4 shows the relative proportion of *Scytonema* spp., *Nostoc* spp. and *Tolypothrix* spp., plotted against the mean annual temperature (MAT) of origin and the corresponding mean temperature during the wettest quarter of the year (MTempWetQ). MTempWetQ was used as a proxy for growth season since biocrust organisms are metabolically active only when water is available [65] and are relatively insensitive to heat stress when dry. The relative abundance of *Scytonema* spp. was positively correlated with MAT ($p = 4 \times 10^{-4}$) and with MTempWetQ ($p = 10^{-8}$); *Nostoc* spp. was negatively correlated with MAT ($p = 5 \times 10^{-3}$) and with MTempWetQ ($p = 10^{-7}$). The relative abundance of *Tolypothrix* spp. was also negatively correlated with MAT ($p = 0.035$) and with MTempWetQ ($p = 3 \times 10^{-3}$). Using linear regression on arcsine transformed data, MAT explained 15, 9 and 1 of the variability in *Scytonema* spp., *Nostoc* spp. and *Tolypothrix* spp., respectively. The explanatory power of MTempWetQ was much higher in all cases, rising to 32, 28 and 11%, respectively. Based on MTempWetQ, *Scytonema* spp. could attain dominance at warmer temperatures (Figure 4A), while at lower temperatures, *Tolypothrix* spp. (Figure 4C), followed by *Nostoc* spp. (Figure 4B) attain higher maximal relative abundances. Detailed statistics are in Supplementary Materials (Figure S1, Tables S5–S10).

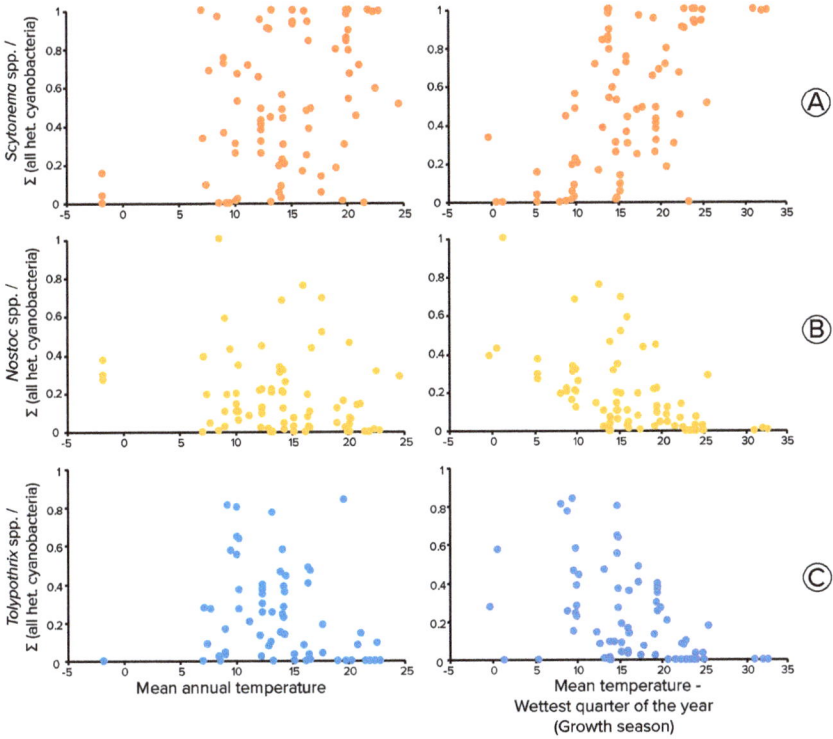

Figure 4. Proportion of sequence reads assignable to *Scytonema* spp. (**A**; orange), *Nostoc* spp. (**B**; yellow), and *Tolypothrix* spp. (**C**; blue) to those assignable to all heterocystous (Order Nostocales) cyanobacteria, in 16S rRNA molecular survey datasets, as a function of climate temperature indicators. Data are from biocrust communities surveyed at 84 locations around the world (see Table S4). Each dot represents a different location.

4. Discussion

The cyanobacteria *Scytonema* spp., *Nostoc* spp. and *Tolypothrix* spp. are secondary colonizers in the ecological succession of biocrust communities [6], where they are among the most common heterocystous organisms [6,19,32,38,50], and contribute much of the nitrogen inputs to the community at this stage of development [66]. Therefore, it is logical to assume that their presence and relative abundance have direct effects on the N_2-fixation capability of late successional biocrusts. Using quantitative enrichment cultures we could clearly demonstrate differential fitness in these cyanobacteria at different temperatures, in a pattern that confirms the preferences inferred in prior field [32,34] and cyanobacterial cultures thermophysiological assays [32–34], where the biocrust cyanobacteria *M. steenstrupii* complex and *Scytonema* spp. were found to be more thermotolerant than *M. vaginatus*, and *Tolypothrix* spp. and *Nostoc* spp., respectively.

Using a set of cultivated strains (12 *Scytonema* spp., 10 *Nostoc* spp. and eight *Tolypothrix* spp.) isolated from cold and hot desert locations of the Southwestern US, the temperature range for growth revealed a pattern of niche differentiation according to temperature: *Tolypothrix* spp. strains having an advantage at the lower temperatures, and *Scytonema* spp. strains at higher temperatures. *Nostoc* spp. strains occupied only the mesic part of the temperature range. This niche separation is similar to that found in non-heterocystous filamentous cyanobacteria of soil crusts [34], and parallels the much more conspicuous niche differentiation of cyanobacteria known from hot springs at temperatures between 45-73 °C [67]. Similar niche separation in cyanobacterial genera are found as a function of salinity [68,69] or desiccation frequency in marine intertidal systems [70]. We could also show that the upper temperature limit for growth (and survival) under N_2-fixing conditions is more constrained than that under non N_2-fixing conditions (Figure 3), implicating N_2-fixation as a possible driver of the effective upper limit of temperature range in nature. Although measurements of N_2-fixation rates (by acetylene reduction assay or ^{15}N isotopes) would give a much more direct result, the observed thermophysiological responses of the tested strains at 35 °C, coincide with more dramatic decreases in N_2-fixation rates (above 30 °C) in cold than in hot biocrusts locations [31], and are congruent with the fact that *Nostoc* spp. and particularly *Tolypothrix* spp. are more abundant in biocrusts from colder locations, while *Scytonema* spp. typically dominate in warmer ones [32–34].

In an effort to better understand the basis for this effect on N_2-fixation we determined the ratio of heterocyst frequency at different temperatures in a selected set of strains, which were responsive to our experimental conditions (*Scytonema* sp. JS006, *Nostoc* sp. HSN008 and *Tolypothrix* pp. HSN042, Figure 3). The results suggest that in *Nostoc* spp. and *Tolypothrix* spp., the impossibility of these strains to grow under N_2-fixation conditions at temperatures above 30 °C may be determined by an inability to carry out the sophisticated developmental cycle leading to the differentiation of heterocysts [71]. While *Scytonema* spp. may have overcome such developmental problems (Table 1), nitrogenase denaturation, which has been reported to happen at temperatures above 39 °C [72] could be the basis for the observed differences in *Scytonema* spp. strains' biomass yield at 35 °C (Figure 3). It is also possible that the observed inability of *Nostoc* spp. and *Tolypothrix* spp. strains to differentiate heterocysts at higher temperatures is the result of a resource allocation constraint to obtain the energy required to differentiate these specialized cells. However, N_2-fixation and heterocyst differentiation at temperatures above 40 °C is not a problem in principle, in that the freshwater thermophilic cyanobacterium *Mastigocladus laminosus* performs N_2-fixation at 45 °C [73], and is able to grow at temperatures as high as 57 °C [74]. Whether the observed heterocyst frequency decrease in *Nostoc* spp. and *Tolypothrix* spp. is a direct effect of temperature rather than a side effect due to stress on other physiological processes will need further investigation.

We tested the relevance of this temperature-based niche differentiation in nature by studying the distribution of the three cyanobacterial types as a function of climate parameters in a meta-analysis of a large dataset of biocrust surveys. Indeed, we found that the maximal relative proportion of *Scytonema* spp. among all heterocystous cyanobacteria increased along the temperature gradient with increasing temperatures (Figure 4A), when the average temperatures of the growth (wet) season was considered. Clearly, however, the results point to a potential for differential sensitivity of these cyanobacteria to environmental warming, a future scenario with which biocrust will have to contend. Drylands at large will likely become warmer and drier in response to global warming. In particular, the US Southwest is predicted to experience an increase in temperature of about 1 °C per decade [26], accompanied by alterations in precipitation frequency [35,36].

Given the observed differential response of biocrust N_2-fixing cyanobacteria to temperature, and in agreement with Muñoz-Martín et al., (2018), it is reasonable to forecast that a microbial replacement within biocrust heterocystous cyanobacteria may indeed be in store as a result of global warming. *Scytonema* spp. may replace more cold- and mesic-temperature adapted *Tolypothrix* spp. and *Nostoc* spp. In places such as the Colorado Plateau, the Mojave desert, the north part of the Chihuahuan Desert (Sevilleta LTER) in the USA, Alicante in Spain, Western Australia [49], temperate areas in Mexico [33], and the Brazilian savannah (Cerrado) [52], where the mean annual temperature during the growth season falls between the 17 and 23 °C range, this microbial replacement will likely happen faster than at those locations exhibiting mean average temperatures below 17 °C, that are not projected to reach sensitive temperature ranges for decades to centuries, or locations with average temperatures above 24 °C, which already exhibit a dominance of *Scytonema* spp. (Figure 4). This microbial replacement could have implications for drylands and biocrust nitrogen inputs beyond a mere compositional change. *Scytonema* spp. have been shown to be one of the most sensitive taxa in biocrust to changes in precipitation patterns [47]. In this scenario, the N_2-fixing cyanobacteria taxa that seem to be better adapted to withstand increases in temperature, ironically, seem to be among the least adapted to withstand drought. Although it makes sense that cyanobacterial distribution patterns with increasing temperature became more apparent when mean temperature during the wettest quarter of the year was used as an explanatory variable, we were surprised by the fact that plots using MAT did not show clearer patterns (Figure 4). This highlights the need to take into account the ecophysiology of microorganisms when seeking to find important climatic drivers.

These results can also serve to improve strategies to restore biological soil crust communities, of much recent interest in conservation ecology [46,75], by providing information to optimize inoculation season and microbial inoculum formulations.

Supplementary Materials: The following are available online at http://www.mdpi.com/2076-2607/8/3/396/s1, Table S1: Accession numbers for the main generic groups for *Scytonema* spp. *Nostoc* spp. and *Tolypothrix* spp. according to our taxonomic assignment using our own cyanobacterial reference tree CYDRASIL., Table S2: Outcome of enrichment cultures for nitrogen-fixing photoautotrophs (nitrogen and organic carbon free medium, in the light) using variously sourced biocrusts as inoculum as a function of the incubation temperature. Given are the number of colonies containing each cyanobacterial taxa of interest, as identified morphologically by microscopy inspection. "S" stands for *Scytonema* spp., "N" for *Nostoc* spp., and "T" for *Tolypothrix* spp., Table S3: Cyanobacterial strains and their accession number in NCBI of their partial 16S rRNA sequence. Strain denominations include coding for the site of origin (HSN: cold desert sandy clay loam soil; HS: cold desert clay loam soil; FB: warm desert loamy sandy soil; JS: warm desert clay loam soil), Table S4: Environmental biocrust surveys conducted at different locations around the world used in the meta-analysis, and the corresponding climate data. Raw sequences were downloaded from bacterial 16S rRNA tallies available publicly (see references). Environmental data was downloaded from WorldClim. "MAT" stands for mean annual temperature and "MTemWetQ" for mean temperature during the wettest quarter of the year (growth season)., Table S5: Full results for linear regression between relative proportions (arcsine transformed) of *Scytonema* spp. and mean annual temperature (MAT)., Table S6: Full results for linear regression between relative proportions (arcsine transformed) of *Scytonema* spp. and mean temperature during the wettest quarter of the year (MTemWetQ)., Table S7: Full results for linear regression between relative proportions (arcsine transformed) of *Nostoc* spp. and mean annual temperature (MAT)., Table S8: Full results for linear regression between relative proportions (arcsine transformed) of *Nostoc* spp. and mean temperature during the wettest quarter of the year (MTemWetQ)., Table S9: Full results for linear regression between relative proportions (arcsine transformed) of *Tolypothrix* spp. and mean annual temperature (MAT)., Table S10: Full results for linear regression between relative proportions (arcsine transformed) of *Tolypothrix* spp. and mean temperature during the wettest quarter of the year (MTemWetQ)., Figure S1. Linear regressions between the proportion of sequence reads (arcsine transformed) of each taxon among heterocystous cyanobacteria and climatic parameters (MAT and MTempWetQ).

Author Contributions: Conceptualization, A.G.-S. and F.G.-P.; methodology, A.G.-S., V.M.C.F. and J.B.; validation, A.G.-S and F.G.-P.; formal analysis, A.G.-S; investigation, A.G.-S.; data curation, A.G.-S. and F.G.-P.; writing—original draft preparation, A.G.-S.; writing—review and editing, A.G.-S. and F.G.-P.; visualization, A.G.-S.; supervision, F.G.-P.; project administration, A.G.-S.; funding acquisition, F.G.-P. All authors have read and agreed to the published version of the manuscript.

Funding: This research was partly funded by the Barret-Honors College at Arizona State University and by the Center for Bio-mediated & Bio-inspired Geotechnics (CBBG) at Arizona State University.

Acknowledgments: We would like to thank the late Richard W. Castenholz for his pioneering work on the role of temperature on the ecology of cyanobacteria, which showed us the path forward.

Conflicts of Interest: The authors declare no conflict of interest.

References

1. Garcia-Pichel, F. Desert Environments: Biological Soil Crusts. In *Encyclopedia of Environmental Microbiology 6 Volume*; Bitton, G., Ed.; Set. Wiley-Interscience: New York, NY, USA, 2003; pp. 1019–1023.
2. Weber, B.; Caldwell, M.M.; Jayne, B.; Bettina, W.; Büdel, B.; Belnap, J.; Weber, B.; Büdel, B. Biological Soil Crusts: An Organizing Principle in Drylands. In *Biological Soil Crusts: An Organizing Principle in Drylands*; Springer: Basel, Switzerland, 2016; pp. 3–14, ISBN 978-3-319-30212-6.
3. Gaskin, S.; Gardner, R. The role of cryptogams in runoff and erosion control on Bariland in the Nepal middle hills of the Southern Himalaya. *Earth Surf. Process. Landf.* **2001**, *26*, 1303–1315. [CrossRef]
4. Belnap, J.; Gillette, D.A. Disturbance of Biological Soil Crusts: Impacts on Potential Wind Erodibility of Sandy Desert Soils in Southeastern Utah. *Land Degrad. Dev.* **1997**, *8*, 355–362. [CrossRef]

5. Zhang, Y.M.; Wang, H.L.; Wang, X.Q.; Yang, W.K.; Zhang, D.Y. The microstructure of microbiotic crust and its influence on wind erosion for a sandy soil surface in the Gurbantunggut Desert of Northwestern China. *Geoderma* **2006**, *132*, 441–449. [CrossRef]
6. Couradeau, E.; Karaoz, U.; Lim, H.C.; Nunes da Rocha, U.; Northen, T.; Brodie, E.; Garcia-Pichel, F. Bacteria increase arid-land soil surface temperature through the production of sunscreens. *Nat. Commun.* **2016**, *7*, 1–7. [CrossRef] [PubMed]
7. Verrecchia, E.; Yair, A.; Kidron, G.J.; Verrecchia, K. Physical properties of the psammophile cryptogamic crust and their consequences to the water regime of sandy softs, north-western Negev Desert, Israel. *J. Arid Environ.* **1995**, 427–437. [CrossRef]
8. Rodríguez-Caballero, E.; Cantón, Y.; Chamizo, S.; Afana, A.; Solé-Benet, A. Effects of biological soil crusts on surface roughness and implications for runoff and erosion. *Geomorphology* **2012**, *145–146*, 81–89. [CrossRef]
9. Faist, A.M.; Herrick, J.E.; Belnap, J.; Zee, J.W.V.; Barger, N.N. Biological soil crust and disturbance controls on surface hydrology in a semi-Arid ecosystem. *Ecosphere* **2017**, *8*. [CrossRef]
10. Garcia-Pichel, F.; Wojciechowski, M.F. The evolution of a capacity to build supra-cellular ropes enabled filamentous cyanobacteria to colonize highly erodible substrates. *PLoS ONE* **2009**, *4*, e7801. [CrossRef]
11. Nunes da Rocha, U.; Cadillo-Quiroz, H.; Karaoz, U.; Rajeev, L.; Klitgord, N.; Dunn, S.; Truong, V.; Buenrostro, M.; Bowen, B.P.; Garcia-Pichel, F.; et al. Isolation of a significant fraction of non-phototroph diversity from a desert biological soil crust. *Front. Microbiol.* **2015**, *6*, 1–14. [CrossRef]
12. Nagy, M.L.; Pérez, A.; Garcia-Pichel, F. The prokaryotic diversity of biological soil crusts in the Sonoran Desert (Organ Pipe Cactus National Monument, AZ). *FEMS Microbiol. Ecol.* **2005**, *54*, 233–245. [CrossRef]
13. Soule, T.; Anderson, I.J.; Johnson, S.L.; Bates, S.T.; Garcia-Pichel, F. Archaeal populations in biological soil crusts from arid lands in North America. *Soil Biol. Biochem.* **2009**, *41*, 2069–2074. [CrossRef]
14. Bates, S.T.; Nash, T.H.; Garcia-Pichel, F. Patterns of diversity for fungal assemblages of biological soil crusts from the southwestern United States. *Mycologia* **2012**, *104*, 353–361. [CrossRef] [PubMed]
15. Ullmann, L.; Büdel, B. Ecological Determinants of Species Composition of Biological Soil Cruts on a Landscape Scale.pdf. In *Biological Soil Cruts: Structure, Function, and Managment*; Belnap, J., Lange, O.L., Eds.; Springer: Berlin/Heidelberg, Germany, 2001; pp. 203–213.
16. Johnson, S.L.; Budinoff, C.R.; Belnap, J.; Garcia-Pichel, F. Relevance of ammonium oxidation within biological soil crust communities. *Environ. Microbiol.* **2005**, *7*, 1–12. [CrossRef]
17. Pepe-Ranney, C.; Koechli, C.; Potrafka, R.; Andam, C.; Eggleston, E.; Garcia-Pichel, F.; Buckley, D.H. Non-cyanobacterial diazotrophs dominate dinitrogen fixation in biological soil crusts during early crust formation. *ISME J.* **2015**, *10*, 287–298. [CrossRef] [PubMed]
18. Couradeau, E.; Giraldo-Silva, A.; De Martini, F.; Garcia-Pichel, F. Spatial segregation of the biological soil crust microbiome around its foundational cyanobacterium, Microcoleus vaginatus, and the formation of a nitrogen-fixing cyanosphere. *Microbiome* **2019**, *7*, 1–12. [CrossRef] [PubMed]
19. Yeager, C.M.; Kornosky, J.L.; Morgan, R.E.; Cain, E.C.; Garcia-Pichel, F.; Housman, D.C.; Belnap, J.; Kuske, C.R. Three distinct clades of cultured heterocystous cyanobacteria constitute the dominant N2-fixing members of biological soil crusts of the Colorado Plateau, USA. *FEMS Microbiol. Ecol.* **2007**, *60*, 85–97. [CrossRef]
20. Johnson, S.L.; Neuer, S.; Garcia-Pichel, F. Export of nitrogenous compounds due to incomplete cycling within biological soil crusts of arid lands. *Environ. Microbiol.* **2007**, *9*, 680–689. [CrossRef]
21. Strauss, S.L.; Day, T.A.; Garcia-Pichel, F. Nitrogen cycling in desert biological soil crusts across biogeographic regions in the Southwestern United States. *Biogeochemistry* **2012**, *108*, 171–182. [CrossRef]
22. Thiet, R.K.; Boerner, R.E.J.; Nagy, M.; Jardine, R. The effect of biological soil crusts on throughput of rainwater and N into Lake Michigan sand dune soils. *Plant Soil* **2005**, *278*, 235–251. [CrossRef]
23. Thomazo, C.; Couradeau, E.; Garcia-Pichel, F. Possible nitrogen fertilization of the early Earth Ocean by microbial continental ecosystems. *Nat. Commun.* **2018**, *9*, 1–8. [CrossRef]
24. Beraldi-Campesi, H.; Hartnett, H.E.; Anbar, A.; Gordon, G.W.; Garcia-Pichel, F. Effect of biological soil crusts on soil elemental concentrations: Implications for biogeochemistry and as traceable biosignatures of ancient life on land. *Geobiology* **2009**, *7*, 348–359. [CrossRef] [PubMed]
25. Prăvălie, R. Drylands extent and environmental issues. A global approach. *Earth Sci. Rev.* **2016**, *161*, 259–278. [CrossRef]
26. Seager, R.; Vecchi, G.A. Greenhouse warming and the 21st century hydroclimate of southwestern North America. *Proc. Natl. Acad. Sci. USA* **2010**, *107*, 21277–21282. [CrossRef] [PubMed]

27. Petrie, M.D.; Collins, S.L.; Litvak, M.E. The ecological role of small rainfall events in a desert grassland. *Ecohydrology* **2015**, *8*, 1614–1622. [CrossRef]
28. Petrie, M.D.; Collins, S.L.; Gutzler, D.S.; Moore, D.M. Regional trends and local variability in monsoon precipitation in the northern Chihuahuan Desert, USA. *J. Arid Environ.* **2014**, *103*, 63–70. [CrossRef]
29. Elbert, W.; Weber, B.; Burrows, S.; Steinkamp, J.; Büdel, B.; Andreae, M.O.; Pöschl, U. Contribution of cryptogamic covers to the global cycles of carbon and nitrogen. *Nat. Geosci.* **2012**, *5*, 459–462. [CrossRef]
30. Barger, N.N.; Castle, S.C.; Dean, G.N. Denitrification from nitrogen-fixing biologically crusted soils in a cool desert environment, southeast Utah, USA. *Ecol. Process.* **2013**, *2*, 1–9. [CrossRef]
31. Zhou, X.; Smith, H.; Giraldo Silva, A.; Belnap, J.; Garcia-Pichel, F. Differential Responses of Dinitrogen Fixation, Diazotrophic Cyanobacteria and Ammonia Oxidation Reveal a Potential Warming-Induced Imbalance of the N-Cycle in Biological Soil Crusts. *PLoS ONE* **2016**, *11*, e0164932. [CrossRef]
32. Muñoz-Martín, M.Á.; Becerra-Absalón, I.; Perona, E.; Fernández-Valbuena, L.; Garcia-Pichel, F.; Mateo, P. Cyanobacterial biocrust diversity in Mediterranean ecosystems along a latitudinal and climatic gradient. *New Phytol.* **2018**, *221*, 123–141. [CrossRef]
33. Becerra-Absalón, I.; Muñoz-Martín, M.Á.; Montejano, G.; Mateo, P. Differences in the Cyanobacterial Community Composition of Biocrusts From the Drylands of Central Mexico. Are There Endemic Species? *Front. Microbiol.* **2019**, *10*, 1–21. [CrossRef]
34. Garcia-Pichel, F.; Loza, V.; Marusenko, Y.; Mateo, P.; Potrafka, R.M. Temperature Drives the Continental-Scale Distribution of Key Microbes in Topsoil Communities. *Science* **2013**, *340*, 1574–1577. [CrossRef] [PubMed]
35. Cable, J.M.; Huxman, T.E. Precipitation pulse size effects on Sonoran Desert soil microbial crusts. *Oecologia* **2004**, *141*, 317–324. [CrossRef] [PubMed]
36. Knapp, A.K.; Beier, C.; Briske, D.D.; Classen, A.T.; Luo, Y.; Reichstein, M.; Smith, M.D.; Smith, S.D.; Bell, J.E.; Fay, P.A.; et al. Consequences of More Extreme Precipitation Regimes for Terrestrial Ecosystems. *Bioscience* **2008**, *58*, 811–821. [CrossRef]
37. Sala, O.E.; Lauenroth, W.K. Small Rainfall Events: An Ecological Role in Semiarid Regions. *Oecologia* **1982**, *53*, 301–304. [CrossRef] [PubMed]
38. Giraldo-Silva, A.; Nelson, C.; Barger, N.; Garcia-Pichel, F. Nursing biocrusts: Isolation, cultivation and fitness test of indigenous cyanobacteria. *Restor. Ecol.* **2019**, *27*, 793–803. [CrossRef]
39. Allen, M.M.; Stanier, R. Growth and Division of Some Unicellular Blue-green Algae. *J. Gen. Microbiol.* **1968**, *51*, 199–202. [CrossRef]
40. Rippka, R.; Deruelles, J.; Waterbury, J.B. Generic assignments, strain histories and properties of pure cultures of cyanobacteria. *J. Gen. Microbiol.* **1979**, *111*, 1–61. [CrossRef]
41. Clarke, K.; Gorley, R. *Primer v6: User Manual/Tutorial*; PRIMER-E: Plymouth, UK, 2006.
42. Anderson, M.; Gorley, R.N.; Clarke, K.R. *PERMANOVA + for PRIMER: Guide to Software and Statistical Methods*; PRIMER-E: Plymouth, UK, 2008.
43. Nübel, U.; Muyzer, G.; Garcia-pichel, F.; Muyzer, G. PCR primers to amplify 16S rRNA genes from cyanobacteria PCR Primers To Amplify 16S rRNA Genes from Cyanobacteria. *Microbiology* **1997**, *63*, 3327–3332.
44. Castle, S.C.; Morrison, C.D.; Barger, N.N. Extraction of chlorophyll a from biological soil crusts: A comparison of solvents for spectrophotometric determination. *Soil Biol. Biochem.* **2011**, *43*, 853–856. [CrossRef]
45. Garcia-Pichel, F.; Castenholz, R.W. Characterization and biological implications of scytonemin, a cyanobacterial sheath pigment. *J. Phycol.* **1991**, *409*, 395–409. [CrossRef]
46. Velasco Ayuso, S.; Giraldo Silva, A.; Nelson, C.; Barger, N.N.; Garcia-pichel, F. Microbial Nursery Production of High- Quality Biological Soil Crust Biomass for Restoration of Degraded Dryland Soils. *Appl. Environ. Microbiol.* **2017**, *83*, 1–16. [CrossRef] [PubMed]
47. Fernandes, V.M.C.; Machado de Lima, N.M.; Roush, D.; Rudgers, J.; Collins, S.L.; Garcia-Pichel, F. Exposure to predicted precipitation patterns decreases population size and alters community structure of cyanobacteria in biological soil crusts from the Chihuahuan Desert. *Environ. Microbiol.* **2018**, *20*, 259–269. [CrossRef] [PubMed]
48. Bethany, J.; Giraldo-Silva, A.; Nelson, C.; Barger, N.N.; Garcia-Pichel, F. Optimizing the Production of Nursery-Based Biological Soil Crusts for Restoration of Arid Land Soils. *Appl. Environ. Microbiol.* **2019**, *85*, 1–13. [CrossRef] [PubMed]

49. Moreira-Grez, B.; Tam, K.; Cross, A.T.; Yong, J.W.H.; Kumaresan, D.; Nevill, P.; Farrell, M.; Whiteley, A.S. The Bacterial Microbiome Associated With Arid Biocrusts and the Biogeochemical Influence of Biocrusts Upon the Underlying Soil. *Front. Microbiol.* **2019**, *10*, 1–13. [CrossRef] [PubMed]
50. Williams, L.; Loewen-Schneider, K.; Maier, S.; Büdel, B. Cyanobacterial diversity of western European biological soil crusts along a latitudinal gradient. *FEMS Microbiol. Ecol.* **2016**, *92*, 1–9. [CrossRef] [PubMed]
51. Zhang, B.; Kong, W.; Nan, W.; Zhang, Y. Bacterial diversity and community along the succession of biological soil crusts in the Gurbantunggut Desert, Northern China. *J. Basic Microbiol.* **2016**, *56*, 670–679. [CrossRef]
52. Machado-de-Lima, N.M.; Fernandes, V.M.C.; Roush, D.; Velasco Ayuso, S.; Rigonato, J.; Garcia-Pichel, F.; Zanini Branco, L.H. The Compositionally Distinct Cyanobacterial Biocrusts From Brazilian Savanna and Their Environmental Drivers of Community Diversity. *Front. Microbiol.* **2019**, *10*, 1–10. [CrossRef]
53. Callahan, B.J.; Mcmurdie, P.J.; Rosen, M.J.; Han, A.W.; Johnson, A.J.; Holmes, S.P. DADA2: High resolution sample inference from amplicon data. *bioRxiv* **2015**, *13*, 581–583. [CrossRef]
54. Caporaso, J.G.; Kuczynski, J.; Stombaugh, J.; Bittinger, K.; Bushman, F.D.; Costello, E.K.; Fierer, N.; Peña, A.G.; Goodrich, J.K.; Gordon, J.I.; et al. QIIME allows analysis of high-thorughput community sequencing data. *Nat. Methods* **2010**, *7*, 335–336. [CrossRef]
55. Katoh, K.; Standley, D.M. MAFFT Multiple Sequence Alignment Software Version 7: Improvements in Performance and Usability Article Fast Track. *Mol. Biol. Evol.* **2013**, *30*, 772–780. [CrossRef]
56. Price, M.N.; Dehal, P.S.; Arkin, A.P. FastTree 2-Approximately maximum-likelihood trees for large alignments. *PLoS ONE* **2010**, *5*. [CrossRef] [PubMed]
57. Xu, S. Bayesian Naïve Bayes classifiers to text classification. *J. Inf. Sci.* **2016**, *44*, 48–59. [CrossRef]
58. McDonald, D.; Price, M.N.; Goodrich, J.; Nawrocki, E.P.; DeSantis, T.Z.; Probst, A.; Andersen, G.L.; Knight, R.; Hugenholtz, P. An improved Greengenes taxonomy with explicit ranks for ecological and evolutionary analyses of bacteria and archaea. *ISME J.* **2012**, *6*, 610–618. [CrossRef] [PubMed]
59. Edgar, R.C. Search and clustering orders of magnitude faster than BLAST. *Bioinformatics* **2010**, *26*, 2460–2461. [CrossRef] [PubMed]
60. Wang, Q.; Garrity, G.M.; Tiedje, J.M.; Cole, J.R. Naive Bayesian classifier for rapid assignment of rRNA sequences into the new bacterial taxonomy. *Appl. Environ. Microbiol.* **2007**, *73*, 5261–5267. [CrossRef] [PubMed]
61. Berger, S.A.; Stamatakis, A. Aligning short reads to reference alignments and trees. *Bioinformatics* **2011**, *27*, 2068–2075. [CrossRef]
62. Berger, S.A.; Krompass, D.; Stamatakis, A. Performance, accuracy, and web server for evolutionary placement of short sequence reads under maximum likelihood. *Syst. Biol.* **2011**, *60*, 291–302. [CrossRef]
63. Letunic, I.; Bork, P. Interactive tree of life (iTOL) v3: An online tool for the display and annotation of phylogenetic and other trees. *Nucleic Acids Res.* **2016**, *44*, 242–245. [CrossRef]
64. Fick, S.E.; Hijmans, R.J. WorldClim 2: New 1-km spatial resolution climate surfaces for global land areas. *Int. J. Climatol.* **2017**, *37*, 4302–4315. [CrossRef]
65. Rajeev, L.; Da Rocha, U.N.; Klitgord, N.; Luning, E.G.; Fortney, J.; Axen, S.D.; Shih, P.M.; Bouskill, N.J.; Bowen, B.P.; Kerfeld, C.A.; et al. Dynamic cyanobacterial response to hydration and dehydration in a desert biological soil crust. *ISME J.* **2013**, *7*, 2178–2191. [CrossRef]
66. Yeager, C.M.; Kornosky, J.L.; Housman, D.C.; Grote, E.E.; Belnap, J.; Kuske, C.R. Diazotrophic Community Structure and Function in Two Successional Stages of Biological Soil Crusts from the Colorado Plateau and Chihuahuan Desert. *Appl. Environ. Microbiol.* **2004**, *70*. [CrossRef] [PubMed]
67. Castenholz, R.W. Thermophilic blue-green algae and the thermal environment. *Bacteriol. Rev.* **1969**, *33*, 476–504. [CrossRef] [PubMed]
68. Nübel, U.; Garcia-Pichel, F.; Kuhl, M.; Muyzer, G. Spatial scale and diversity of benthic cyanobacteria and diatoms in a salina. In *Molecular Ecology of Aquatic Communities. Developments in Hydrology*; Zehr, J., Voytek, M.A., Eds.; Springer: Dordrecht, The Netherlands, 1999.
69. Garcia-pichel, F.; Ku, M.; Nübel, U.; Muyzer, G. Salinity-dependent limitation of photosynthesis and oxygen exchange in microbial mats. *J. Phycol.* **1999**, *35*, 227–238. [CrossRef]
70. Rothrock, M.J.; Garcia-Pichel, F. Microbial diversity of benthic mats along a tidal desiccation gradient. *Environ. Microbiol.* **2005**, *7*, 593–601. [CrossRef] [PubMed]
71. Adams, D.G.; Duggan, P.S. Heterocyst and akinete differentiation in cyanobacteria. *New Phytol.* **1999**, *144*, 3–33. [CrossRef]

72. Hennecke, H.; Shanmugam, K.T. Temperature control of nitrogen fixation in Klebsiella pneumoniae. *Arch. Microbiol.* **1979**, *123*, 259–265. [CrossRef]
73. Nierzwicki Bauer, S.A.; Balkwill, D.L.; Stevens, S.E. Heterocyst differentiation in the cyanobacterium Mastigocladus laminosus. *J. Bacteriol.* **1984**, *157*, 514–525. [CrossRef]
74. Miller, S.R.; Castenholz, R.W.; Pedersen, D. Phylogeography of the thermophilic cyanobacterium Mastigocladus laminosus. *Appl. Environ. Microbiol.* **2007**, *73*, 4751–4759. [CrossRef]
75. Bowker, M.A.; Antoninka, A.J. Rapid ex situ culture of N-fixing soil lichens and biocrusts is enhanced by complementarity. *Plant Soil* **2016**, *408*, 415–428. [CrossRef]

© 2020 by the authors. Licensee MDPI, Basel, Switzerland. This article is an open access article distributed under the terms and conditions of the Creative Commons Attribution (CC BY) license (http://creativecommons.org/licenses/by/4.0/).

Article

Analysis of the Complete Genome of the Alkaliphilic and Phototrophic Firmicute *Heliorestis convoluta* Strain HH[T]

Emma D. Dewey [1], Lynn M. Stokes [1], Brad M. Burchell [1], Kathryn N. Shaffer [1,†], Austin M. Huntington [1], Jennifer M. Baker [1,‡], Suvarna Nadendla [2], Michelle G. Giglio [2], Kelly S. Bender [3], Jeffrey W. Touchman [4,§], Robert E. Blankenship [5,||], Michael T. Madigan [3] and W. Matthew Sattley [1,*]

1. Division of Natural Sciences, Indiana Wesleyan University, Marion, IN 46953, USA; emma.dewey@myemail.indwes.edu (E.D.D.); lynn.stokes@myemail.indwes.edu (L.M.S.); brad.burchell@indwes.edu (B.M.B.); kshaffer@cedarville.edu (K.N.S.); austinm.huntington@gmail.com (A.M.H.); jennbak@umich.edu (J.M.B.)
2. Institute for Genome Sciences, University of Maryland School of Medicine, Baltimore, MD 21201, USA; snadendla@som.umaryland.edu (S.N.); mgiglio@som.umaryland.edu (M.G.G.)
3. Department of Microbiology, Southern Illinois University, Carbondale, IL 62901, USA; bender@micro.siu.edu (K.S.B.); madigan@siu.edu (M.T.M.)
4. School of Life Sciences, Arizona State University, Tempe, AZ 85287, USA; jefftouchman@icloud.com
5. Departments of Biology and Chemistry, Washington University in Saint Louis, St. Louis, MO 63130, USA; blankenship@wustl.edu
* Correspondence: matthew.sattley@indwes.edu; Tel.: +1-765-677-2128
† Present address: School of Pharmacy, Cedarville University, Cedarville, OH 45314, USA.
‡ Present address: Department of Microbiology and Immunology, University of Michigan, Ann Arbor, MI 48109, USA.
§ Present address: Intrexon Corporation, 1910 Fifth Street, Davis, CA 95616, USA.
|| Present address: 3536 South Kachina Drive, Tempe, AZ 85282, USA.

Received: 26 January 2020; Accepted: 22 February 2020; Published: 25 February 2020

Abstract: Despite significant interest and past work to elucidate the phylogeny and photochemistry of species of the *Heliobacteriaceae*, genomic analyses of heliobacteria to date have been limited to just one published genome, that of the thermophilic species *Heliobacterium* (*Hbt.*) *modesticaldum* str. Ice1[T]. Here we present an analysis of the complete genome of a second heliobacterium, *Heliorestis* (*Hrs.*) *convoluta* str. HH[T], an alkaliphilic, mesophilic, and morphologically distinct heliobacterium isolated from an Egyptian soda lake. The genome of *Hrs. convoluta* is a single circular chromosome of 3.22 Mb with a GC content of 43.1% and 3263 protein-encoding genes. In addition to culture-based observations and insights gleaned from the *Hbt. modesticaldum* genome, an analysis of enzyme-encoding genes from key metabolic pathways supports an obligately photoheterotrophic lifestyle for *Hrs. convoluta*. A complete set of genes encoding enzymes for propionate and butyrate catabolism and the absence of a gene encoding lactate dehydrogenase distinguishes the carbon metabolism of *Hrs. convoluta* from its close relatives. Comparative analyses of key proteins in *Hrs. convoluta*, including cytochrome c_{553} and the F_o alpha subunit of ATP synthase, with those of related species reveal variations in specific amino acid residues that likely contribute to the success of *Hrs. convoluta* in its highly alkaline environment.

Keywords: heliobacteria; *Heliorestis convoluta*; alkaliphilic bacteria; soda lake; anoxygenic phototroph; bacteriochlorophyll *g*

1. Introduction

Heliobacteria comprise a unique group of strictly anaerobic, anoxygenic phototrophs that have been isolated from a wide diversity of soil and aquatic habitats [1–4]. Unlike all other phototrophic bacteria, heliobacteria use bacteriochlorophyll (Bchl) *g* as the chief chlorophyll pigment for phototrophic growth [5], but despite their ability to use light as an energy source, heliobacteria are apparently incapable of autotrophic growth and, thus, are obligate heterotrophs [4,6]. Heliobacteria are the only phototrophs of the large bacterial phylum *Firmicutes* [4,7,8], and although they typically stain Gram-negatively, thin sections of cells of heliobacteria exhibit a Gram-positive cell wall morphology [9,10]. In addition to these distinctive properties, cells of heliobacteria are able to differentiate into heat-resistant endospores [4,11], and some heliobacteria have also demonstrated the ability to reduce toxic metals, such as Hg^{2+}, and therefore may be useful for applications in bioremediation [12,13].

Species of *Heliobacteriaceae* can be divided into two physiological groups—neutrophiles and alkaliphiles—that track closely with their phylogeny [7] (Figure 1). Included in the neutrophilic clade, the moderate thermophile *Hbt. modesticaldum* was the first heliobacterium to have its genome sequenced and, with its simple phototrophic machinery consisting of a type I reaction center (RC) and no peripheral antenna photocomplex, has been a model organism for studies of photosynthesis and related photochemistry [6,14]. Like other neutrophilic heliobacteria, *Hbt. modesticaldum* exhibits both phototrophic growth in the light and chemotrophic growth in the dark [3,15–17].

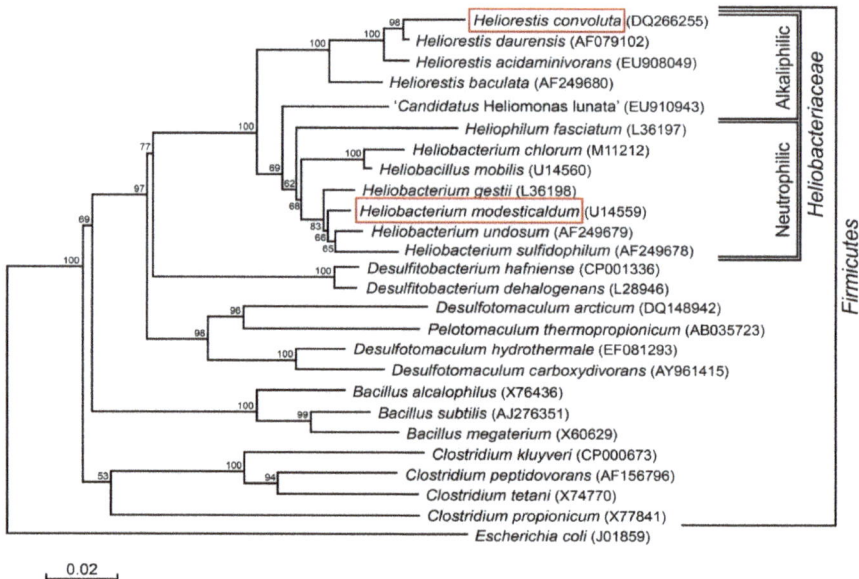

Figure 1. Phylogenetic (16S rRNA) tree of *Heliorestis convoluta* and related *Firmicutes*. Heliobacteria, the only phototrophic *Firmicutes*, are divided into alkaliphilic and neutrophilic species. *Heliobacterium modesticaldum* (boxed) is the model organism for physiological and biochemical studies of the heliobacteria; *Hrs. convoluta* (boxed) is the first alkaliphilic heliobacterium to have a described genome. Note that the branching pattern shown here suggests a possible alkaliphilic origin to the heliobacteria, as previously discussed by Sattley and Swingley [7]. The weighted neighbor-joining method [18] and Jukes-Cantor corrected distance model were used for tree construction. Nodes represent bootstrap values (≥50%) based on 100 replicates, and *Escherichia coli* was used to root the tree. GenBank accession numbers for each sequence used in the analysis are shown in parentheses, adapted from Sattley and Swingley [7], *Adv. Bot. Res.* **2013**, *66*, 67–97, Copyright 2013 Elsevier Ltd.

Species of alkaliphilic heliobacteria grow optimally between pH 8–9.5 and, unlike neutrophilic heliobacteria, are obligate photoheterotrophs, using light and organic compounds for growth but incapable of chemotrophic growth in darkness [19–22]. Consistent with other alkaliphilic heliobacteria originating from the soils and waters of soda lakes [19,20,22], *Hrs. convoluta* str. HH^T was isolated from the shore of the alkaline (pH 10) Lake El Hamra (Figure 2A), located in the Wadi El Natroun region of northern Egypt [21]. In the past, the saline lakes of the Wadi El Natroun have also been a fertile source of alkaliphilic purple bacteria, yielding many extremely alkaliphilic (and in some cases also extremely halophilic) species, including in particular, new species of the genus *Halorhodospira* [23–25]. However, *Hrs. convoluta* is the first heliobacterium to originate from these unusual lakes. Experimental work with *Hrs. convoluta* revealed motile cells having an unusual tightly coiled morphology (Figure 2B) and displaying a mesophilic (optimal growth at 33 °C) and alkaliphilic (optimal growth at pH 8.5–9) physiology [21].

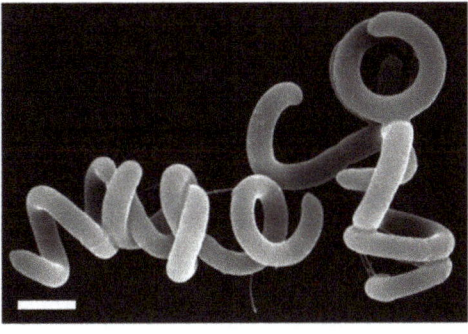

A B

Figure 2. Habitat and cells of *Heliorestis convoluta* strain HH^T. (**A**) Red bloom of alkaliphilic *Bacteria* and *Archaea* on the shore of Lake El Hamra, Wadi Natroun, Egypt. M.T.M. sampled this bloom in May 2001, and enrichments for heliobacteria yielded *Hrs. convoluta*. The bloom is about 2 m in diameter.; (**B**) Scanning electron micrograph of cells of *Hrs. convoluta* strain HH^T. A cell of *Hrs. convoluta* is about 0.5 μm in diameter and coils are of variable length. Scale bar = 1 μm.

To complement the analysis of the genome sequence of *Hbt. modesticaldum* [6,26], we present here a comparative analysis of the genome of *Hrs. convoluta*. Although a number of highly conserved genes encoding proteins that coordinate key processes in the cell (e.g., phototrophy and central carbon metabolism) are shared between these species, a close comparison of the two heliobacterial genomes revealed several genes encoding functions in carbon metabolism, biotin biosynthesis, nitrogen and sulfur assimilation, and carotenoid biosynthesis that are not held in common by these heliobacteria, which inhabit vastly different extreme environments. In addition, a comparative analysis of selected cytochrome and ATP synthase proteins in *Hrs. convoluta* revealed adaptations that likely facilitate its alkaliphilic lifestyle. The availability of a second heliobacterial genome, as well as the recent development of a genetic system in *Hbt. modesticaldum* [14], paves the way for increasing our understanding of the unique metabolism and physiology of heliobacteria.

2. Materials and Methods

Total genomic DNA from *Hrs. convoluta* str. HH^T (ATCC BAA-1281 and DSMZ 19787) [21] was isolated through proteinase K treatment and subsequent phenol extraction. Complete genome sequencing was performed using a random shotgun approach, and reads were assembled using Velvet v. 2010 [27]. Pyrosequencing on a Roche-454 GS20 sequencer (Hoffman-La Roche AG, Basel, Switzerland) provided 14-fold genome coverage, and an additional 35-fold coverage was generated by the Illumina GAIIx platform.

Annotation of the *Hrs. convoluta* genome was performed in accordance with the Prokaryotic Annotation Pipeline of the University of Maryland School of Medicine's Institute for Genome Sciences [28]. This pipeline employs Glimmer for gene identification and then searches the protein sequences with BLAST-extend-repraze (BER; a combination of BLAST and Smith–Waterman algorithms) to generate pairwise alignments, Hidden Markov Model (HMM), transmembrane (Tm) HMM, and SignalP predictions. An automated process employing the Pfunc evidence hierarchy is used to assign functional annotations. Manual verification of automated annotations was facilitated through the online tool Manatee [29] in conjunction with online databases including the Kyoto Encyclopedia of Genes and Genomes (KEGG), the Braunschweig Enzyme Database (BRENDA), MetaCyc, and Uniprot. The National Center for Biotechnology Information (NCBI) database was accessed to retrieve gene and protein sequences from related species for comparative analyses with corresponding genes in the genome of *Hrs. convoluta*.

The phylogenetic tree was generated as described in the legend to Figure 1. Genome statistics were compiled using the Pfam database v. 30.0 [30], the SignalP database v. 4.1 [31], the TMHMM database v. 2.0 [32], and CRISPRFinder v. 2.0 [33]. This complete genome sequence project has been deposited at DDBJ/EMBL/GenBank under accession number CP045875.

3. Results and Discussion

3.1. Genome Properties

The 3,218,981 base-pair (bp) genome of *Heliorestis convoluta* str. HH^T is organized into a single circular chromosome with no plasmids (Table 1). The 43.1% GC content of *Hrs. convoluta* is among the lowest of all heliobacteria (41%–57.7%) and is typical of alkaliphilic species of this group of phototrophs [4]. Nearly 87% of the *Hrs. convoluta* genome content is protein-encoding, with a total of 3263 protein coding genes at an average length of 855 nucleotides (Table 1). The genome contains nine ribosomal RNA (rRNA) genes, including multiple copies each of 5S, 16S (two full and one partial), and 23S rRNA, which are distributed randomly on the chromosome. Nearly 11% of the open reading frames (ORFs) were of unknown enzyme specificity or function, and 28% of genes were annotated as hypothetical. The role category breakdown of protein-encoding genes of *Hrs. convoluta* is shown in Table 2.

Table 1. Comparison of genome features of *Heliorestis convoluta* str. HH^T and *Heliobacterium modesticaldum* str. Ice1^T [6].

Characteristic	Hrs. convoluta	Hbt. modesticaldum
Chromosome size (bp)	3,218,981	3,075,407
G + C content (%)	43.1	56.0
Coding DNA (%)	86.9	87
Protein-encoding genes (no.)	3,263	3,138
Average gene length (bp)	855	882
ATG initiation codons (%)	63.5	62.1
GTG initiation codons (%)	15.7	19.1
TTG initiation codons (%)	20.8	18.8
rRNAs (no.)	9	24
tRNAs (no.)	105	104
Transposases (no.)	18	70
Putative pseudogenes (no.)	22	8
CRISPR repeats (no.)	1	Not determined

Genes encoding a total of 105 transfer RNAs (tRNAs) were identified in the *Hrs. convoluta* genome, as well as genes encoding all twenty common aminoacyl-tRNA synthetases except asparaginyl-tRNA synthetase, which could not be confirmed. However, genes encoding aspartyl/glutamyl-tRNA amidotransferase (*gatABC*) were identified in *Hrs. convoluta* and, as proposed for *Hbt. modesticaldum* [6], may encode a protein that compensates for the missing asparaginyl-tRNA synthetase by converting aspartyl-tRNA to asparaginyl-tRNA [34,35].

Table 2. Functional role categories of *Heliorestis convoluta* str. HH^T genes.

Characteristic	Genes	% of Genome Content *
Amino acid biosynthesis	119	3.64
Biosynthesis of cofactors, prosthetic groups, and carriers	142	4.35
Cell envelope and surface features	216	6.61
Cellular processes (cell division, motility, sporulation, etc.)	477	14.6
DNA metabolism	225	6.88
Energy and central intermediary metabolism	512	15.94
Fatty acid and phospholipid metabolism	66	2.02
Mobile and extrachromosomal element functions	76	2.33
Protein synthesis and fate	338	10.34
Purines, pyrimidines, nucleosides, and nucleotides	59	1.81
Regulatory functions	137	4.19
Signal transduction	80	2.45
Transcription	142	4.35
Transport and binding proteins	342	10.47
Hypothetical proteins	899	27.51

* Total exceeds 100%, as some genes are assigned to more than one role category.

3.2. Central Carbon Metabolism

Analysis of the *Hrs. convoluta* genome confirmed culture-based observations of the limited set of carbon sources able to support light-driven growth of this species [21]. As an obligate photoheterotroph, *Hrs. convoluta* grows only in anoxic, light conditions when supplied with mineral media containing CO_2 plus acetate, pyruvate, propionate, or butyrate as organic carbon sources [21]. Of the 12 described species of heliobacteria (Figure 1), only *Heliorestis acidaminivorans*, *Heliorestis daurensis*, and *Hrs. convoluta* are capable of propionate photoassimilation [4,19,21,22]. Genes encoding enzymes of the methylmalonyl pathway, which converts propionyl-coenzyme A (CoA) to succinyl-CoA for propionate assimilation, were identified in the *Hrs. convoluta* genome (Figure 3). Although a gene encoding propionyl-CoA carboxylase, which is thought to catalyze the first step in the proposed pathway [36,37], was not identified in the *Hrs. convoluta* genome, a gene predicted to encode methylmalonyl-CoA carboxyltransferase (FTV88_3237), which could circumvent this deficiency, was identified.

Unlike other *Heliorestis* species, *Hrs. convoluta* and a few other heliobacteria can use butyrate as a carbon source [19–21,38]. Analysis of the *Hrs. convoluta* genome revealed genes encoding enzymes that catabolize butyrate to acetyl-CoA for incorporation into the citric acid cycle (CAC) [39] (Figure 3). Genes encoding butyryl-CoA:acetate CoA transferase, which catalyzes the conversion of butyrate to butyryl-CoA in butyrate catabolism [39], and propionyl-CoA synthetase, which converts propionate to propionyl-CoA in propionate catabolism [37], were not identified in the genome of *Hrs. convoluta*. However, an experimentally characterized butyryl-CoA:acetate CoA transferase from *Desulfosarcina cetonica* [39] showed 47% amino acid sequence identity with 4-hydroxybutyrate CoA-transferase (FTV88_0224) from *Hrs. convoluta*. In addition, the product of a gene annotated as acetyl-coenzyme A synthetase (FTV88_0994) in *Hrs. convoluta* showed 37% sequence identity with propionyl-CoA synthetase from *Salmonella enterica* and contained the conserved lysine residue (Lys592) required in the initial reaction of propionate catabolism [40]. These findings suggest possible roles for 4-hydroxybutyrate CoA-transferase and acetyl-coenzyme A synthetase in butyrate and propionate catabolism, respectively, in *Hrs. convoluta*.

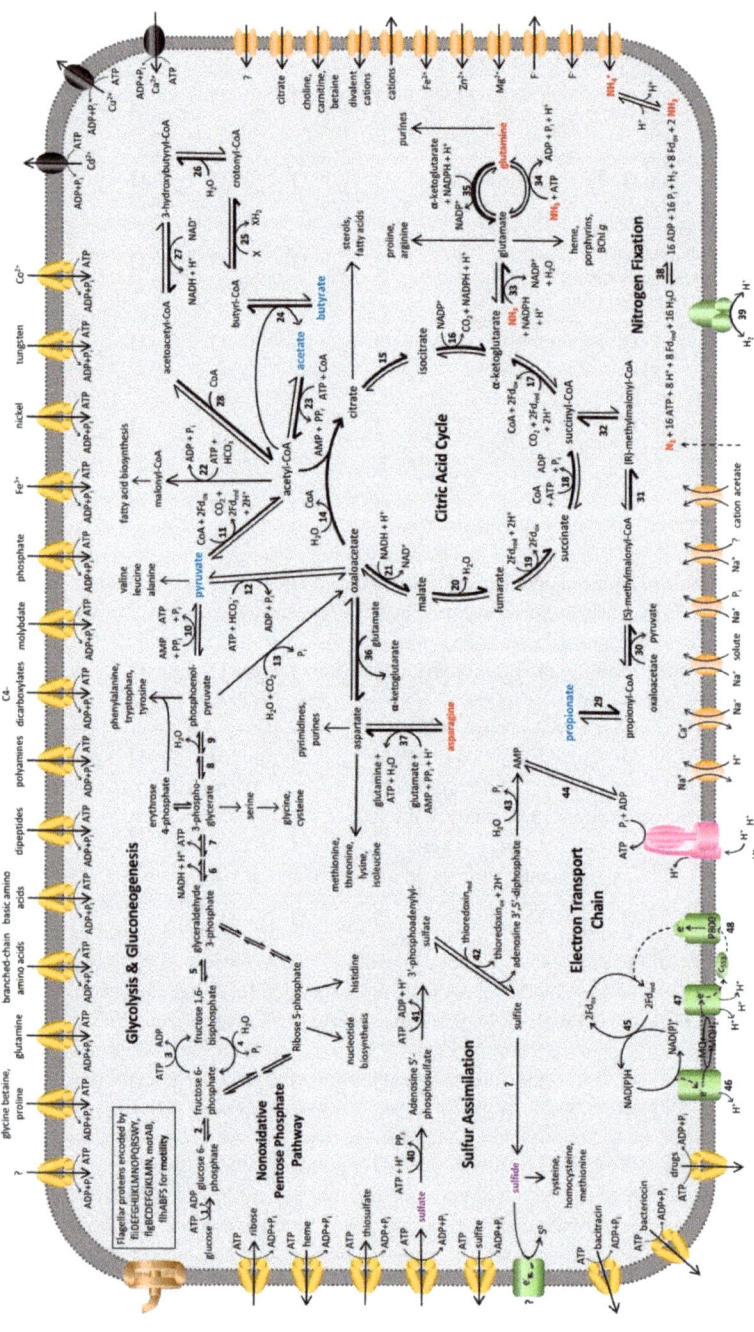

Figure 3. Overview of the proposed metabolic pathways and membrane transporters in *Helioresits convoluta*. Carbon (blue), nitrogen (red), and sulfur (purple) sources are catabolized or assimilated for phototrophic growth of *Hrs. convoluta*. The predicted dominant direction of metabolic flow is shown by bolded arrows. Numbers signify enzymes identified in the genome of *Hrs. convoluta*, whereas question marks indicate unidentified but anticipated enzymes catalyzing the respective reaction. Enzymes involved in glycolysis or gluconeogenesis include (1) glucokinase, (2) glucose-6-phosphate isomerase, (3) 6-phosphofructokinase, (4) fructose 1,6-bisphosphatase, (5) fructose-1,6-bisphosphate aldolase, (6) glyceraldehyde-3-phosphate dehydrogenase, (7) phosphoglycerate kinase, (8) phosphoglycerate mutase,

(9) enolase, and (10) pyruvate-phosphate dikinase. CAC enzymes are (11) pyruvate:ferredoxin oxidoreductase, (12) pyruvate carboxylase, (13) phosphoenolpyruvate carboxylase, (14) citrate (*re*)-synthase, (15) aconitate hydratase, (16) NADP$^+$-dependent isocitrate dehydrogenase, (17) 2-oxoglutarate synthase/2-oxoglutarate:ferredoxin oxidoreductase, (18) succinyl-CoA synthetase, (19) succinate dehydrogenase/fumarate reductase, (20) fumarate hydratase, and (21) NAD$^+$-dependent malate dehydrogenase. Acetyl-CoA metabolism is carried out by (22) acetyl-CoA carboxylase and (23) acetyl-CoA synthetase. Butyrate metabolism enzymes include (24) CoA transferase, (25) acyl-CoA dehydrogenase, (26) enoyl-CoA hydratase, (27) 3-hydroxybutyryl-CoA dehydrogenase, and (28) acetyl-CoA C-acetyltransferase. Propionate metabolism is catabolized by (29) CoA transferase, (30) methylmalonyl-CoA carboxytransferase, (31) methylmalonyl-CoA epimerase, and (32) methylmalonyl-CoA mutase. Amino acid metabolism enzymes are (33) NADP$^+$-specific glutamate dehydrogenase, (34) glutamine synthetase, (35) NADPH-dependent glutamate synthase, (36) pyridoxal phosphate-dependent aminotransferase, and (37) asparagine synthase. The enzymes (38) nitrogenase and (39) uptake [NiFe] hydrogenase catalyze nitrogen fixation and H$_2$ oxidation, respectively, and sulfur assimilation is performed by (40) sulfate adenyltransferase, (41) adenylyl-sulfate kinase, (42) phosophoadenylyl-sulfate reductase, (43) bifunctional oligoribonuclease and PAP phosphatase, and (44) adenylate kinase. Finally, the electron transport chain includes (45) ferredoxin:NADP$^+$ reductase, (46) NADH:quinone oxidoreductase, (47) cytochrome *bc* complex, and the (48) light-harvesting reaction center. Membrane proteins include ABC transporters (yellow), P-type ATPases (black), ATP synthase (pink), flagellar and motor proteins (brown), other transporters (orange), and other membrane proteins (green).

Although capable of growth on pyruvate, *Hrs. convoluta* str. HH^T is unable to grow photoheterotrophically on lactate [21], a phenotype distinct from that of most other heliobacteria and the result of an underlying genetic deficiency. In this connection, a gene encoding a putative L-lactate dehydrogenase in *Hbt. modesticaldum* [6] showed no meaningful similarity to any genes in *Hrs. convoluta*. In addition to lactate, no growth was detected when alcohols of any kind were used as sole carbon source in cultures of strain HH^T [21]. Despite this observation, genes encoding alcohol dehydrogenase and aldehyde dehydrogenase were annotated in the *Hrs. convoluta* genome and, thus, could potentially play a role in non-energetic processes, such as detoxification.

Although a full complement of genes encoding enzymes of the glycolytic and nonoxidative pentose phosphate pathways was present in the *Hrs. convoluta* genome (Figure 3), various common sugars did not support photoheterotrophic growth of strain HH^T [21]. An inability to use sugars was also originally reported for *Hbt. modesticaldum* [16], but later experimentation showed that *Hbt. modesticaldum* utilized the glycolytic pathway when D-ribose, D-glucose, or D-fructose were supplied with low levels of yeast extract [41]. Although no gene encoding a hexose transporter was annotated in the *Hrs. convoluta* genome, a putative ribose ABC transporter complex (FTV88_0053, FTV88_0054, FTV88_0055) was identified and may allow for carbohydrate transport [41]. As genes encoding glycolytic pathway enzymes are present in the *Hrs. convoluta* genome, it is tempting to speculate that the alkaliphile can utilize sugars in a manner similar to *Hbt. modesticaldum*. The absence of genes encoding glucose 6-phosphate dehydrogenase and 6-phosphogluconolactonase suggest incomplete Entner-Doudoroff and oxidative pentose phosphate pathways, which was also the case for *Hbt. modesticaldum* [6].

It is likely that *Hrs. convoluta* can catalyze many of the steps in the CAC based on biochemical studies of *Hbt. modesticaldum* [42] and high sequence similarity of key CAC enzymes between the two species (Figure 3). However, since both *Hrs. convoluta* and *Hbt. modesticaldum* lack a gene encoding pyruvate dehydrogenase for oxidizing pyruvate to acetyl-CoA, this reaction in heliobacteria is likely catalyzed by the enzyme pyruvate:ferredoxin oxidoreductase (PFOR); the gene encoding PFOR in *Hrs. convoluta* (FTV88_3370) shares 61% sequence identity to an orthologous gene in *Hbt. modesticaldum* [6]. Furthermore, an unusual citrate synthase, citrate (*re*)-synthase, which specifically catalyzes the addition of the acetyl moiety from acetyl-CoA to the *re* face of the ketone carbon of oxaloacetate [a stereospecificity opposite to that of citrate (*si*)-synthase], has been identified in several clostridia and other strictly anaerobic *Firmicutes*, including *Hbt. modesticaldum* [42]. In *Hrs. convoluta*, a gene (FTV88_1447) having high amino acid sequence identity (81%) to the gene encoding citrate (*re*)-synthase (HM1_2993) in *Hbt. modesticaldum* supports the presence of citrate (*re*)-synthase in *Hrs. convoluta* and suggests this unusual form of citrate synthase is common to all heliobacteria.

In regards to photoautotrophic capacity, no genes encoding enzymes of any form of the Calvin-Benson cycle, including ribulose 1,5-bisphosphate carboxylase and phosphoribulokinase, were identified in the *Hrs. convoluta* genome. In addition, the lack of genes encoding key enzymes of other autotrophic pathways, such as malyl-CoA lyase (3-hydroxypropionate/4-hydroxybutyrate pathway) and acetyl-CoA synthase (Wood-Ljungdahl pathway), also prevents *Hrs. convoluta* from assimilating CO_2 into organic carbon molecules for growth. The capacity for CO_2 fixation by the reverse CAC, as observed in green sulfur bacteria [43], is apparently disrupted by the absence of a gene encoding ATP-citrate lyase. Although an ORF identified as a citrate lyase family protein (FTV88_0308) was annotated in the *Hrs. convoluta* genome based on sequence identities of approximately 50% with corresponding genes from other *Firmicutes* (but having no similarity to genes in *Hbt. modesticaldum*), biochemical analysis of this gene product would be required to assess its activity and role, if any, in metabolic pathways of *Hrs. convoluta*. Although anapleurotic CO_2 assimilation has been shown in heliobacteria supplied with usable organic carbon sources [44], cultures of *Hrs. convoluta* strain HH^T, like all other cultured heliobacteria, were unable to grow using CO_2 as sole carbon source [21], thus supporting the premise that heliobacteria require an organic carbon source during phototrophic growth.

In addition to phototrophy, neutrophilic heliobacteria are able to grow chemotrophically in the dark by pyruvate fermentation [4]. Interestingly, however, the capacity for pyruvate fermentation has not been observed in any alkaliphilic heliobacterial isolate to date, including *Hrs. convoluta* [4,15,17,21]. Studies have suggested that the neutrophile *Hbt. modesticaldum* carries out substrate-level phosphorylation via acetyl-CoA conversion to acetate in dark, anoxic (fermentative) conditions through the activity of phosphotransacetylase (PTA) and acetate kinase (ACK) [15,17,41]. A gene encoding ACK (FTV88_2009) was annotated in the genome of *Hrs. convoluta* and has 67% sequence identity to a corresponding gene in *Hbt. modesticaldum*. However, a gene encoding PTA could not be identified in either *Hrs. convoluta* or *Hbt. modesticaldum*. Therefore, the genetic determinants that coordinate pyruvate fermentation in neutrophilic heliobacteria but are apparently absent from alkaliphilic heliobacteria remain unidentified.

Three *Hrs. convoluta* genes encoding acetyl-CoA synthetase (ACS) were identified in the genome, one of which showed 87% amino acid sequence identity with the corresponding gene in *Hbt. modesticaldum*. Activity of ACS in *Hbt. modesticaldum* cell extracts was detected only under phototrophic (light/anoxic) conditions, and expression levels of the ACS gene decreased when the bacterium was cultured in darkness [41], thus indicating that, although technically reversible, ACS activity is predominately skewed toward the production of acetyl-CoA from acetate (Figure 3). Activity of ACS therefore allows both *Hbt. modesticaldum* and *Hrs. convoluta* to grow photoheterotrophically using acetate as sole carbon source [16,21].

In contrast to all other heliobacteria, which require biotin for growth, *Hrs. convoluta* and close relative *Hrs. acidaminivorans* (Figure 1) have no growth factor requirements [21,22]. The presence of a full complement of genes (*bioABCDF*) encoding enzymes for biotin biosynthesis allows *Hrs. convoluta* to synthesize biotin, thereby supporting culture-based observations [21]. By contrast, analysis of the *Hbt. modesticaldum* genome revealed the absence of two key genes for biotin biosynthesis, *bioC* and *bioF*, thus explaining the absolute requirement for biotin in that species [16].

3.3. Nitrogen Metabolism

Hrs. convoluta is strongly diazotrophic [21], and as in *Hbt. modesticaldum*, genes for nitrogen fixation are grouped into a single *nif* gene cluster containing *nifI$_1$*, *nifI$_2$*, *nifH*, *nifD*, *nifK*, *nifE*, *nifN*, *nifX*, *fdxB*, *nifB*, and *nifV* [6]. Each of these genes shows between 63% and 93% sequence identity and analogous gene synteny to corresponding genes in *Hbt. modesticaldum*. A study with *Paenibacillus* sp. WLY78—also an endospore-former within the phylum *Firmicutes*—concluded that nine genes (*nifB*, *nifH*, *nifD*, *nifK*, *nifE*, *nifN*, *nifX*, *hesA*, *nifV*), which were grouped into a single gene cluster, are essential to synthesize a catalytically-active nitrogenase for dinitrogen assimilation [45]. All of these nitrogen fixation genes, except for *hesA*, were identified in the *Hrs. convoluta* and *Hbt. modesticaldum* genomes. Since HesA is proposed to play a role in metallocluster biosynthesis [45], it is possible that a gene (FTV88_2056) located outside of the *nif* gene cluster and encoding a putative dinitrogenase Fe/Mo cofactor biosynthesis protein fills this role in *Hrs. convoluta*. This encoded protein showed high (~64%) sequence identity to a corresponding protein in *Hrs. acidaminivorans* and over 50% sequence similarity to that from a variety of nonphototrophic *Firmicutes*, but it showed no significant similarity to proteins encoded by *Hbt. modesticaldum*.

Research on *Hbt. modesticaldum* revealed changes in expression levels of numerous genes essential for various metabolic, biosynthetic, and other cellular pathways when the organism was grown under N_2-fixing conditions [46]. This diazotrophic effect likely exists in other heliobacteria as well, including *Hrs. convoluta*. In terms of regulation of nitrogen fixation genes, however, it is interesting that neither *orf1* nor *nifA*, which encode regulatory proteins for the expression of *nif* structural genes [47,48], could be identified in the *Hrs. convoluta* genome. In *Hbt. modesticaldum*, the *orf1* gene product likely regulates the expression of *nif* genes when levels of fixed nitrogen are too low to support non-diazotrophic growth of the organism [16,48]. It is possible that *Hrs. convoluta* lacks the *orf1* and *nifA* regulatory genes and instead employs only *nifI$_1$* (FTV88_2453) and *nifI$_2$* (FTV88_2454) to coordinate post-translational regulation of nitrogenase [49].

Hrs. convoluta and *Hbt. modesticaldum* both contain gene clusters (*hypABCDEF* and *hupCDLS*) that encode an uptake [NiFe] hydrogenase that can putatively catalyze the oxidation of H_2 produced during nitrogen fixation [6] (Figure 3). The arrangement of these genes in *Hrs. convoluta* is identical to that reported for *Hbt. modesticaldum* [6], being organized into a single cluster instead of dispersed throughout different regions of the chromosome, as has been observed in the genomes of other *Firmicutes* (Figure 4).

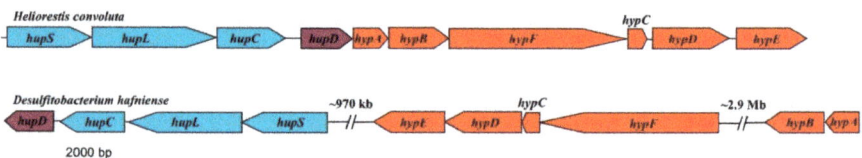

Figure 4. Comparison of uptake [NiFe]-hydrogenase genes in related *Firmicutes*. The genes are concatenated within a single region in *Heliorestis convoluta*, but they are dispersed in different regions of the *Desulfitobacterium hafniense* chromosome. Colors: blue, [NiFe]-hydrogenase structural genes; purple, hydrogenase expression/formation; red, hydrogenase assembly/maturation. Adapted from Sattley et al. [6]. *J. Bacteriol.* **2008**, *190*, 4687–4696. Copyright 2008 American Society for Microbiology.

In addition to performing N_2 fixation, cells of *Hrs. convoluta* strain HH^T could assimilate ammonia, glutamine, and asparagine as nitrogen sources [21]. Accordingly, genes encoding the ammonium transporter protein Amt (FTV88_2595) and enzymes of the glutamine synthetase-glutamate synthase pathway, which incorporates ammonia in the formation of glutamine from glutamate [50,51] (Figure 3), were identified in the *Hrs. convoluta* genome. Following transport, glutamine can then be used for purine biosynthesis or, through the activity of NADPH-dependent glutamate synthase, can be condensed with α-ketoglutarate to yield two molecules of glutamate for other biosynthetic pathways [50,51]. In addition, a gene encoding NADP-specific glutamate dehydrogenase (FTV88_2506) enables *Hrs. convoluta* to assimilate ammonia when synthesizing glutamate directly from α-ketoglutarate (Figure 3). Finally, genes encoding a glutamine-hydrolyzing asparagine synthetase (FTV88_1161 and FTV88_3319), which converts asparagine and glutamate into aspartate and glutamine, respectively (Figure 3), allow for the use of asparagine as a nitrogen source. In contrast, aspartate and glutamate cannot serve as nitrogen sources for strain HH^T [21]. Taken together, these findings suggest that, although the reactions are generally considered reversible, the enzymes catalyzing the conversion of asparagine to aspartate and glutamine to glutamate are physiologically unidirectional, strongly favoring the formation of aspartate and glutamate, respectively (shown as bolded arrows in Figure 3).

3.4. Assimilation of Sulfur

Growth studies indicate *Hrs. convoluta* is capable of assimilatory sulfate reduction [21]. Consistent with these observations, genomic analyses revealed that the pathway of assimilatory sulfate reduction in *Hrs. convoluta* begins with sulfate uptake using a sulfate/thiosulfate ABC transporter (*cysAWTP*). Typically, an enzyme encoded by *cysD* and *cysN*, sulfate adenyltransferase, catalyzes the assimilation of sulfate as adenosine phosphosulfate (APS) [52,53]. *Hrs. convoluta* lacks *cysD*, but genes encoding the bifunctional enzyme CysN/CysC (FTV88_1460 and FTV88_1458, respectively), which can also perform this function [54], are present. As shown in Figure 3, adenylyl-sulfate kinase (*cysC*) and phosphoadenylyl-sulfate reductase (*cysH*, FTV88_1461) catalyze the subsequent reaction to yield sulfite [52,53].

To produce sulfide for amino acid biosynthesis, sulfite must undergo further reduction through sulfite reductase [53]. However, a gene encoding sulfite reductase could not be identified, suggesting that *Hrs. convoluta* may employ an unusual reductase or an alternative mechanism to perform this reaction. Genes encoding all successive enzymes necessary to synthesize cysteine, homocysteine, and methionine from hydrogen sulfide were identified (data not shown). By comparison, the *Hbt.*

modesticaldum genome lacked *cysN*, *cysH*, and sulfite reductase, supporting physiological studies indicating that *Hbt. modesticaldum* requires a reduced sulfur source for biosynthetic purposes [16].

Interestingly, cultures of *Hrs. convoluta* strain HH^T were able to grow well in the presence of high levels of sulfide (10mM), with sulfide oxidation accompanied by the production of elemental sulfur globules during growth [21]. However, the pathway for this reaction remains unclear, as the *Hrs. convoluta* genome appears to lack genes encoding traditional sulfide oxidoreductases, such as the sulfide:quinone oxidoreductase (SQR) from the green sulfur bacterium *Chlorobaculum* (*Chlorobium*) *tepidum* that oxidizes H_2S to S^0 and reduces quinone [55], or sulfide:flavocytochrome *c* oxidoreductase from the purple sulfur bacterium *Allochromatium vinosum* that oxidizes sulfide to sulfur or polysulfides [56]. Thus, it is possible that *Hrs. convoluta* contains a novel sulfide oxidoreductase for this purpose.

3.5. Photosynthesis Genes and Pigment Biosynthesis

Heliobacteria synthesize bacteriochlorophyll (BChl) *g*, a pigment absorbing light maximally between 785 and 790 nm, for phototrophic growth [4]. Accordingly, genes encoding enzymes that catalyze the conversion of glutamic acid to divinyl protochlorophyllide (*gltX*, *hemALBCDEN*, and *bchIDHME*) for pigment biosynthesis (Figure 5) were annotated in *Hrs. convoluta*. However, as for *Hbt. modesticaldum*, neither of the genes encoding protoporphyrinogen oxidase (*hemY* or *hemG*), which catalyzes the oxidation of protoporphyrinogen to protoporphyrin, was identified in the *Hrs. convoluta* genome. Moreover, comparisons with *hemG* from *Escherichia coli* and *hemY* from *Bacillus subtilis* yielded no significant sequence identity to genes in the *Hrs. convoluta* genome. Due to the anaerobic nature of *Hrs. convoluta*, an alternative and unidentified enzyme likely acts as a dehydrogenase rather than an oxidase in this step of pigment biosynthesis. Studies with *Desulfovibrio gigas*, also a strict anaerobe, suggest that electron carriers, such as flavins and pyridine nucleotides, or electron-transport complexes, such as nitrite and fumarate reductases, do not use O_2 as the electron acceptor in the conversion of protoporphyrinogen to protoporphyrin [57,58]. More recently, however, an alternative pathway that does not use protoporphyrin to synthesize heme has been described in *Hbt. modesticaldum* [59], and a similar mechanism likely exists in *Hrs. convoluta*.

Following the synthesis of divinyl protochlorophyllide in *Hrs. convoluta*, genes encoding protochlorophyllide reductase (*bchLNB*), chlorophyllide reductase (*bchXYZ*), and bacteriochlorophyll synthase (*bchG*) are present to facilitate catalysis of subsequent reactions and produce BChl *g*. Previous work with *Hbt. modesticaldum* suggested the need for an isomerase in the interconversion between 8-vinyl bacteriochlorophyllide *a* and bacteriochlorophyllide *g* [6], but more recent experimental work with this species revealed the ability of chlorophyllide reductase to perform both reduction and isomerization of divinyl chlorophyllide *a* and circumvent the need for a separate isomerase in the biosynthesis of bacteriochlorophyllide *g* [60,61].

Heliobacteria also contain an alternative form of chlorophyll (Chl) *a*, 8^1-OH-Chl *a*, which was observed as a smaller absorption peak at 672 nm in spectrophotometric studies of *Hrs. convoluta* [21]. Whereas BChl *g*, a bacteriochlorin-type chlorophyll, is reduced at the C-7 and C-8 bond and has an ethylidene functional group at C-8 [5], 8^1-OH-Chl *a*, a chlorin, has a double bond connecting C-7 and C-8 with a hydroxyethyl group at C-8 [44]. BChl *g* and 8^1-OH-Chl *a* are putatively synthesized from a common precursor, divinyl chlorophyllide *a* [7,60].

Hydration of the C-8 vinyl group of divinyl chlorophyllide *a* is catalyzed by 8-vinyl chlorophyllide hydratase, and bacteriochlorophyll synthase catalyzes the addition of a farnesyl group to produce the mature 8^1-OH-Chl *a* [7,60]. However, a gene encoding chlorophyllide hydratase or an analogous enzyme was not identified in the genomes of *Hrs. convoluta* or *Hbt. modesticaldum* [6]. Hence, a possible alternative mechanism for 8^1-OH-Chl *a* synthesis includes steps of dehydrogenation and subsequent hydroxygenation of bacteriochlorophyllide *g* to produce 8^1-OH-chlorophyllide *a* [60], but genes encoding enzymes for this reaction were not identified in either *Hrs. convoluta* or *Hbt. modesticaldum*. Yet another possible mechanism for 8^1-OH-Chl *a* synthesis would require the irreversible conversion of

BChl g into 8^1-OH-Chl a upon exposure to O_2 and light [62]. However, as strict anaerobes, the viability of heliobacteria is compromised upon exposure to O_2, and therefore this mechanism is unlikely as the major pathway for 8^1-OH-Chl a production [3,62].

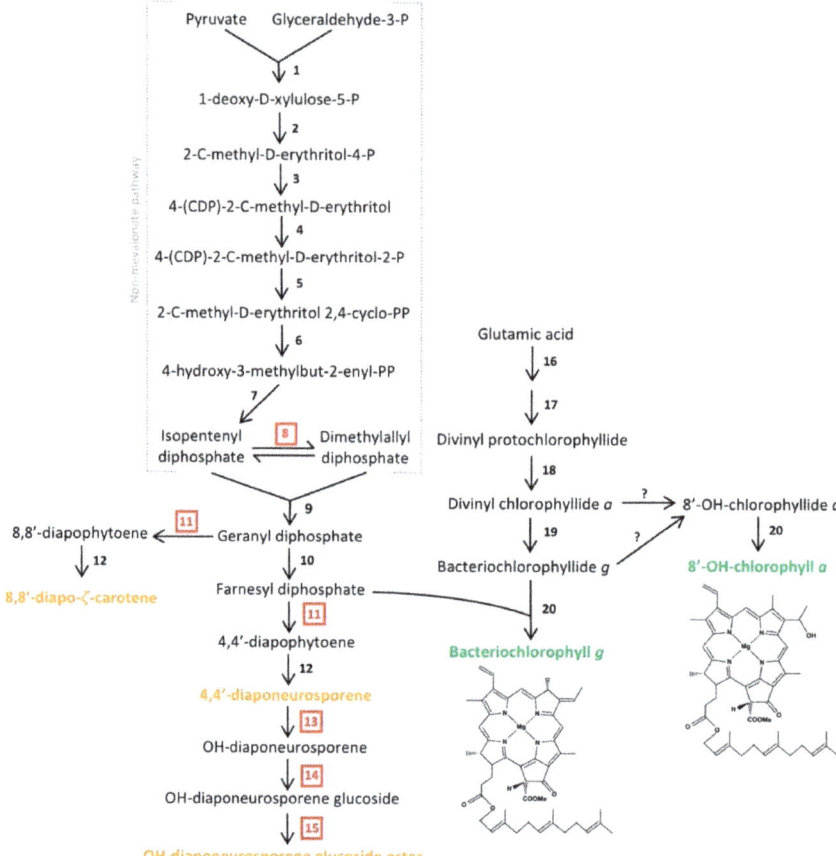

Figure 5. Predicted biosynthetic pathway of major pigments in *Heliorestis convoluta*. The non-mevalonate pathway shows the synthesis of farnesyl diphosphate for either conversion into carotenoids (orange) or incorporation into the final chlorophyll (green) structures. The enzymes that catalyze each individual numbered reaction are (1) 1-deoxy-D-xylulose-5-P synthase, (2) 1-deoxy-D-xylulose-5-P reductoisomerase, (3) 4-(CDP)-2-C-methyl-D-erythritol synthase, (4) 4-(CDP)-2-C-methyl-D-erythritol kinase, (5) 2-C-methyl-D-erythritol 2,4-cyclo-PP synthase, (6) 4-hydroxy-3-methylbut-2-enyl-PP synthase, (7) 4-hydroxy-3-methylbut-2-enyl-PP reductase, (8) isomerase, (9) geranyl diphosphate synthase, (10) farnesyl diphosphate synthase, (11) 4,4′-diapophytoene synthase, (12) diapophytoene dehydrogenase, (13) hydratase, (14) glucosyl transferase, (15) esterase, (16) enzymes encoded by *gltx* and *hemALBCDENYG* genes, (17) enzymes encoded by *bchIDHME* genes, (18) protochlorophyllide reductase (*bchLNB*), (19) chlorophyllide reductase (*bchXYZ*), and (20) bacteriochlorophyll synthase (*bchG*). Red, boxed numbers represent enzymes not yet identified in the *Hrs. convoluta* genome but are proposed based on the predicted pathway. Adapted from Takaichi et al. [63], *Arch. Microbiol.* **2003**, *179*, 95–100. Copyright 2002 Springer Nature; Dubey et al. [64] *J. Biosci.* **2003**, *28*, 637–646. Copyright 2003 Springer Nature; Sattley et al. [6] *J. Bacteriol.* **2008**, *190*, 4687–4696. Copyright 2008 American Society for Microbiology; Sattley and Swingley [7], *Adv. Bot. Res.* **2013**, *66*, 67–97, Copyright 2013 Elsevier Ltd.; and Tsukatani et al. [60], *Biochim. Biophys. Acta* **2013**, *1827*, 1200–1204. Copyright 2013 Elsevier Ltd.

Due to high sequence identity between genes allowing for phototrophic growth (data not shown), a mechanism similar to BChl g and 8^1-OH-Chl a biosynthesis in *Hbt. modesticaldum* [6,7,60] is predicted for *Hrs. convoluta* (Figure 5). Many of the genes encoding enzymes required for pigment biosynthesis are grouped into a single photosynthesis gene cluster (PGC) in heliobacteria. The PGCs of *Hrs. convoluta* and *Hbt. modesticaldum* were nearly identical and displayed a shared gene synteny in all key genes, including those associated with pigment and cofactor biosynthesis, electron transport, and light harvesting, suggesting that a common genetic architecture—one that differs substantially from the PGCs present in the genomes of purple bacteria—defines the heliobacterial PGC (Figure 6).

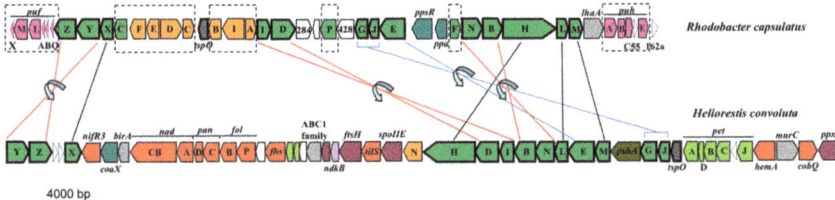

Figure 6. Photosynthesis gene clusters from *Heliorestis convoluta* and the purple bacterium *Rhodobacter capsulatus*. Shared genes are outlined in bold. Lines indicate gene synteny: black, single gene rearrangements; red, inverted genes; and blue, inverted genes with a gene insertion. Dashed boxes show *Rba. capsulatus* photosynthesis genes absent from *Hrs. convoluta*. Colors: green, bacteriochlorophyll biosynthesis (bch); orange, carotenoid biosynthesis (crt); pink, proteobacterial reaction centers (puf) and light harvesting complexes (puh); olive, heliobacterial reaction center (psh); teal, regulatory proteins; light green, electron transport (pet); red, cofactor biosynthesis; purple, cell division and sporulation; light blue, nitrogen fixation; grey, transcription; light grey, other nonphotosynthesis genes; and white, uncharacterized genes. Adapted from Sattley et al. [6]. *J. Bacteriol.* **2008**, *190*, 4687–4696. Copyright 2008 American Society for Microbiology.

Like BChls *c*, *d*, and *e* of green sulfur bacteria, both BChl g and 8^1-OH-Chl a of heliobacteria are esterified with farnesol [63]. A non-mevalonate pathway is employed to synthesize the esterifying alcohol, farnesyl diphosphate, of heliobacterial pigments [64]. As was noted for *Hbt. modesticaldum*, *Hrs. convoluta* contained the complete complement of genes for this pathway, beginning with pyruvate and glyceraldehyde-3-phosphate and proceeding to an unidentified but predicted isomerase that could catalyze the interconversion of isopentenyl diphosphate and dimethylallyl diphosphate [6,64] (Figure 5). Following this, farnesyl diphosphate can either be incorporated into the final structures of BChl g and 8^1-OH-Chl a or further transformed into the major carotenoids found in *Hrs. convoluta* [63] (Figure 5). The high specificity of BchG for incorporation of a farnesol moiety over longer alcohol groups, such as phytol, has been demonstrated in studies of pigment biosynthesis in *Hbt. modesticaldum* [61], and the high sequence identity (68%) of BchG from *Hrs. convoluta* to that of *Hbt. modesticaldum* suggests a similar activity in the alkaliphile.

Experimental work and pigment extraction from *Hrs. convoluta*, *Hrs. daurensis*, and *Hrs. baculata* revealed that the major carotenoid in alkaliphilic heliobacteria is OH-diaponeurosporene glucoside C16:0 ester, followed by 4,4′-diaponeurosporene, OH-diaponeurosporene glucoside C16:1 ester, and 8,8′-zeta-carotene [21,63]. These novel glucoside esters in alkaliphilic heliobacteria were not found in neutrophilic heliobacteria, in which 4,4′-diaponeurosporene was the major carotenoid [63,65]. The synthesis of these C_{30} carotenoids [66] is complicated by the apparent absence of a gene (*crtM*) encoding 4,4′-diapophytoene synthase in both *Hbt. modesticaldum* [6] and *Hrs. convoluta*. Presumably, the presence of an enzyme with 4,4′-diapophytoene synthase activity is essential in the proposed biosynthetic pathway for each of the carotenoids found in heliobacteria [63] (Figure 5). Although *crtM* was not identified, two nonidentical copies of *crtN* (FTV88_2648 and FTV88_3059) having a sequence identity of 71% were annotated in the *Hrs. convoluta* genome, and it is possible that one of their gene products exhibits CrtM-like activity.

In alkaliphilic heliobacteria, a proposed CrtC-like hydratase catalyzes the formation of OH-diaponeurosporene from 4,4′-diaponeurosporene, followed by synthesis of OH-diaponeurosporene glucoside by a CrtX-like glucosyl transferase, with a putative esterase making the final conversion to the mature glucoside ester [63]. Genes encoding the enzymes catalyzing the final three steps of OH-diaponeurosporene glucoside ester synthesis were not identified in *Hrs. convoluta* (Figure 5), but genes encoding two carotenoid biosynthesis proteins (FTV88_0301 and FTV88_0302) were annotated. These genes showed no significant sequence similarity to genes of the neutrophilic *Hbt. modesticaldum*, and they may be candidates for encoding proteins to perform the final steps of carotenoid biosynthesis in alkaliphilic heliobacteria.

3.6. Reaction Center and Electron Transport Chain

Heliobacteria possess a type I (Fe–S type) photosynthetic reaction center (RC) imbedded in the cytoplasmic membrane [4,67,68]. As the simplest known and perhaps most ancient extant (bacterio)chlorophyll-binding photochemical apparatus [69], the heliobacterial RC is a symmetrical homodimer consisting of the PshA polypeptide and the novel, single-transmembrane helix PshX polypeptide [70]. PshA of *Hrs. convoluta* (encoded by *pshA*, FTV88_2638) showed 71% sequence identity to PshA of *Hbt. modesticaldum* but nearly 96% identity to PshA of *Hrs. acidaminivorans* (GenBank accession WBXO01000000, unpublished). As is the case in *Hbt. modesticaldum*, *pshX* (FTV88_2551) is situated outside of the PGC in *Hrs. convoluta* and encodes a protein consisting of just 31 amino acids. The PshX RC subunit from *Hrs. convoluta* showed a 74% sequence identity to that of *Hbt. modesticaldum*.

The crystal structure of the *Hbt. modesticaldum* RC revealed the presence of 54 BChl *g* molecules, two 8^1-OH-Chl *a* molecules, two carotenoids (4,4′-diaponeurosporene), four BChl *g*′ molecules (a C-13 epimer of BChl *g* that functions as the primary electron donor, P_{800}) [68,71,72], two lipids, and one [4Fe–4S] cluster [70]. Experimental data on the structure of the *Hrs. convoluta* RC are not yet available. However, with their highly similar PshA and PshX proteins, the geometry and pigment composition of the *Hrs. convoluta* RC should closely resemble that of the *Hbt. modesticaldum* RC [69]. Nevertheless, some distinctions may materialize considering the alternative carotenoids produced by alkaliphilic heliobacteria and their inherently alkaline habitat [63].

Proteins of the electron transport chain (ETC) of *Hrs. convoluta* exhibited high sequence similarity to those from *Hbt. modesticaldum*, and thus the overarching mechanism of light-driven energy conservation is likely to be highly conserved across all heliobacterial taxa. Although not experimentally confirmed, it is likely that electrons first enter the chain by the activity of either NADH:quinone oxidoreductase (Figure 3), a 14-subunit protein complex embedded in the cytoplasmic membrane and encoded by *nuoABCDEFGHIJKLMN*, or perhaps a complex having ferredoxin:menaquinone oxidoreductase activity. As observed in *Hbt. modesticaldum*, the *nuoEFG* genes in *Hrs. convoluta* are not co-localized within the same operon as the other *nuo* genes. However, unlike in *Hbt. modesticaldum*, in which *nuoEF* are fused, *nuoE* and *nuoF* exist as individual genes (present in duplicate copies) in *Hrs. convoluta* (Figure 7). With the exception of this distinction, all *nuo* genes show high sequence identity (62–79%) between the two species. As for *Hbt. modesticaldum*, menaquinone is predicted to shuttle electrons from Complex I to the cytochrome *bc* complex (PetABCD), and electron transfer through these complexes drives translocation of H^+ to the periplasmic space (Figure 3), forming a proton motive force (PMF) [6,73].

Cytochrome bc_1 complexes, which are found in a variety of anoxygenic phototrophs and also in eukaryotic mitochondria, consist of a minimum of three protein subunits: cytochrome *b*, cytochrome c_1, and the Rieske iron-sulfur protein [74]. In contrast, the related cytochrome b_6f complex, which is present in cyanobacteria and chloroplasts, is comprised of cytochrome *f* (PetA), cytochrome b_6 (PetB), the Rieske iron-sulfur protein (PetC), and subunit IV (PetD) [74]. Having similar functions but distinct structural properties, cytochrome *b* contains eight transmembrane helices, whereas cytochrome b_6 and its associated subunit IV contain four and three transmembrane helices, respectively [74]. Cytochrome b_6 shows homology to the N-terminal half of cytochrome *b*, and subunit IV is homologous to the C-terminal half of cytochrome *b* [74].

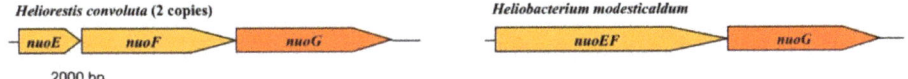

Figure 7. Comparison of *nuoEFG* genes in *Heliorestis convoluta* and *Heliobacterium modesticaldum*. Whereas in *Hbt. modesticaldum* (and *Heliobacillus mobilis*) *nuoE* and *nuoF* are fused, these genes are independent in *Hrs. convoluta*. In both species, however, *nuoEFG* are separated from other *nuo* genes on the chromosome. In addition, unlike in *Hbt. modesticaldum*, *Hrs. convoluta* contains two copies of *nuoEFG*, as well as the *hydEFG* maturase genes (FTV88_1003–1005) that may impart [FeFe] hydrogenase activity. Colors: gold, NADH dehydrogenase subunits; orange, structural genes.

An analysis of the cytochrome *bc* complex of *Hrs. convoluta* indicated that it resembles a hybrid of the cytochrome b_6f complex and the cytochrome bc_1 complex. A comparison of cytochrome b_6 and subunit IV proteins from *Hrs. convoluta* and the model cyanobacterium *Synechocystis* PCC 6803 showed 48% and 42% amino acid sequence identity, respectively. However, cytochrome b_6 and subunit IV from *Hrs. convoluta* also showed 36% and 30% amino acid identity, respectively, to the N-terminal and C-terminal halves of cytochrome *b* from the purple bacterium *Rhodobacter sphaeroides*. Furthermore, whereas subunit IV from *Hrs. convoluta* is predicted to contain the usual three transmembrane helices, cytochrome b_6 from *Hrs. convoluta* contained a predicted five transmembrane regions instead of the four typically observed in the b_6f complex. Notably, cytochrome b_6 from *Hbt. modesticaldum* is predicted to contain the conventional four transmembrane helices. Therefore, considering the above sequence analyses and their total of eight predicted transmembrane helices, the PetB and PetD proteins of *Hrs. convoluta* may represent a structural and evolutionary intermediate between cytochrome *b* and cytochrome b_6/subunit IV proteins, a distinction perhaps not shared with neutrophilic heliobacteria.

The PetA protein in heliobacteria is also of interest because it functions as a diheme cytochrome *c* (as opposed to the typical monoheme protein) and shows no sequence or structural similarity to cytochrome *f* [74]. Although unusual among the *Firmicutes*, the diheme cytochrome *c* has been identified in all heliobacteria studied thus far and is likely a universal feature of these phototrophs. PetA from *Hrs. convoluta* showed high sequence identity with PetA from *Hrs. acidaminivorans* (79%), and sequence identities to PetA from neutrophilic heliobacteria (e.g., *Hbt. modesticaldum*, *Hbt. gestii*, *Heliobacillus mobilis*, and *Heliophilum fasciatum*) were all near 50%. Based on similarities in its N- and C-terminal domains, the heliobacterial diheme cytochrome *c* may have been the result of a past gene duplication and subsequent fusion [75].

A single operon containing all eight genes encoding the subunits of ATP synthase (*atpABCDEFGH*) was identified in the genome of *Hbt. modesticaldum* [6,26], and the encoded ATP synthase itself has since been biochemically characterized [76]. The composition and arrangement of ATP synthase genes in *Hrs. convoluta* was identical to that in the *Hbt. modesticaldum* genome. Kinetic studies with *Hba. mobilis* and *Hbt. modesticaldum* and physiological similarity to photosystem I of cyanobacteria suggest that a PMF established by cyclic electron flow drives photophosphorylation in heliobacteria [73,77,78]. For overviews of electron transfer reactions in heliobacteria, see Sattley and Swingley [7], Kondo et al. [79], and, more recently, Kashey et al. [73].

3.7. Endosporulation

A likely universal trait of heliobacteria is the ability to form endospores [11], differentiated and largely dormant cells that are highly resistant to environmental extremes, such as heat and desiccation. Genomic comparisons of *Hrs. convoluta* and *Hbt. modesticaldum* revealed high similarity between endosporulation genes in each species. For example, genes encoding key sporulation sigma factors (σ^H, σ^E, σ^F, σ^G, σ^K) in *Hbt. modesticaldum* were also identified in the *Hrs. convoluta* genome. Like *Hbt. modesticaldum*, *Hrs. convoluta* lacked the *spo0M* gene functioning to regulate stage 0 development of endosporulation [80] and the *spoIIB* gene necessary for robust sporulation in *B. subtilis* [81]. This may help explain the sporadic (as opposed to consistent) production of endospores in serially subcultured cells of *Hrs. convoluta* strain HH^T [21], as the deletion of either *spo0M* or *spoIIB* in *B. subtilis* results in impairment of endosporulation [80,81]. Additionally, the 20 *cot* genes encoding proteins that comprise the protective spore coat for *B. subtilis*, including *cotH* required for spore coat assembly [82], did not show significant similarity to genes in *Hbt. modesticaldum* [6] or *Hrs. convoluta*. Likewise, key proteins that coordinate spore coat assembly and composition in *Clostridioides* (*Clostridium*) *difficile*, including CotA and CotB [83], showed no sequence similarity to genes in *Hrs. convoluta*. Despite these deficiencies, cells of *Hrs. convoluta* strain HH^T were still capable of forming heat-resistant endospores, even if sporadically [21]. These findings suggest shared biosynthetic and regulatory mechanisms governing endosporulation in *Hbt. modesticaldum* and *Hrs. convoluta* that differ in some respects from those that govern endosporulation in species of *Bacillus* and *Clostridium*.

3.8. Molecular Adaptations to Alkaliphily in Heliorestis convoluta

Alkaliphilic bacteria employ several mechanisms to maintain intracellular pH homeostasis in their highly alkaline environments. Experimental work conducted with alkaliphiles revealed that these organisms maintain a lower cytoplasmic pH than their external environment—up to a 2.3 pH unit difference—for optimal enzyme activity and cellular functioning [84,85]. Despite its optimal growth pH of 8.5–9 and ability to grow slowly at pH 10 [21], it is likely that *Hrs. convoluta* maintains a cytoplasmic pH at or below pH 8, as is true from studies of several alkaliphilic strains of *Bacillus* [84,86].

Cytoplasmic pH homeostasis in *Hrs. convoluta* is likely supported by the presence of a Na$^+$/H$^+$ antiporter encoded by *nhaA* (FTV88_0116). To maintain cytoplasmic pH at homeostatic levels, the Na$^+$/H$^+$ antiporter operates in an electrogenic manner, facilitating the import of twice as many H$^+$ as Na$^+$ exported [87,88]. The inward movement of protons through the antiporter acidifies the cytoplasm to maintain a pH closer to neutral [87–89]. The NhaA protein from *Hrs. convoluta* was found to be 87% identical in amino acid sequence to NhaA from *Heliorestis acidaminivorans*—also an alkaliphile—but only 50% identical to NhaA from the neutrophile *Hbt. modesticaldum*. The NhaA enzyme may, therefore, be a good candidate to study which amino acid residues facilitate antiporter activity in alkaline versus neutral environments.

In addition to its role in cytoplasmic pH maintenance, the Na$^+$/H$^+$ antiporter generates a sodium motive force (SMF) that has been shown to be important for secondary active transport of various substrates [89–91] (see Figure 3 for examples in *Hrs. convoluta*). The use of Na$^+$-coupling for transport is potentially more important in *Hrs. convoluta* than in its neutrophilic relative, *Hbt. modesticaldum*, as genes encoding multiple Na$^+$-dependent transporters (FTV88_2418 and FTV88_1400) and a Na$^+$/Ca$^+$ antiporter (FTV88_2739) in *Hrs. convoluta* showed little to no significant similarity with genes in *Hbt. modesticaldum*.

The more neutral cytoplasm compared to the alkaline extracellular milieu would seemingly create an outward-directed bulk PMF rather than the inward-directed PMF needed to drive ATP synthesis [89–91]. Despite this, most alkaliphiles, including *Hrs. convoluta*, still employ a PMF rather than a SMF to power ATP synthase [89,92]. Alkaliphilic bacteria must therefore have mechanisms in place to prevent H$^+$ equilibration with the external environment so that an effective local PMF can be established. To this end, carotenoids, which are produced in large quantities by alkaliphilic heliobacteria [93], have been proposed to play a role in organizing proton pumps close to ATP synthases in the membrane, thus facilitating more efficient ATP generation [91,94,95]. In addition, cardiolipin, a glycerophospholipid that assists in membrane domain organization, may also help prevent H$^+$ equilibration by functioning as a proton sink for the H$^+$-coupled ATP synthase [96–98]. By functioning in this capacity, cardiolipin allows for the retention of H$^+$ near the surface of the cell membrane so that they are unable to spontaneously diffuse into the alkaline environment. Notably, a gene encoding cardiolipin synthase was present in *Hrs. convoluta* (FTV88_2523), but no corresponding homolog was identified in *Hbt. modesticaldum*.

In addition to producing proteins and other molecules that counteract the pH difference between the cytoplasm and environment and the consequences thereof, homologous proteins also have amino acid substitutions that optimize the functioning of normal processes for the alkaline environment. In alkaliphiles, the portions of extracellular enzymes that are exposed to the external environment tend to have decreased numbers of basic residues (arginine, histidine, or lysine), with acidic amino acids (aspartate or glutamate) or neutral residues in their place [99–101]. In a noteworthy example, the amino acid sequence for cytochrome c_{553} (PetJ) of *Hrs. convoluta* contained 13 more acidic amino acid residues and 11 fewer basic residues than PetJ of *Hbt. modesticaldum* (Figure 8A). In line with previous discussion, the elevated number of acidic residues and corresponding decrease in basic residues in the externally-functioning *Hrs. convoluta* PetJ should contribute to OH$^-$ repulsion and H$^+$ attraction near the membrane surface and help maintain the PMF [89]. Although additional investigation of the cell surface of *Hrs. convoluta* is required to confirm its electrochemical nature, genomic data suggest that this phototroph can sequester H$^+$ near the cell surface to create an effective PMF for ATP synthesis and flagellar motility.

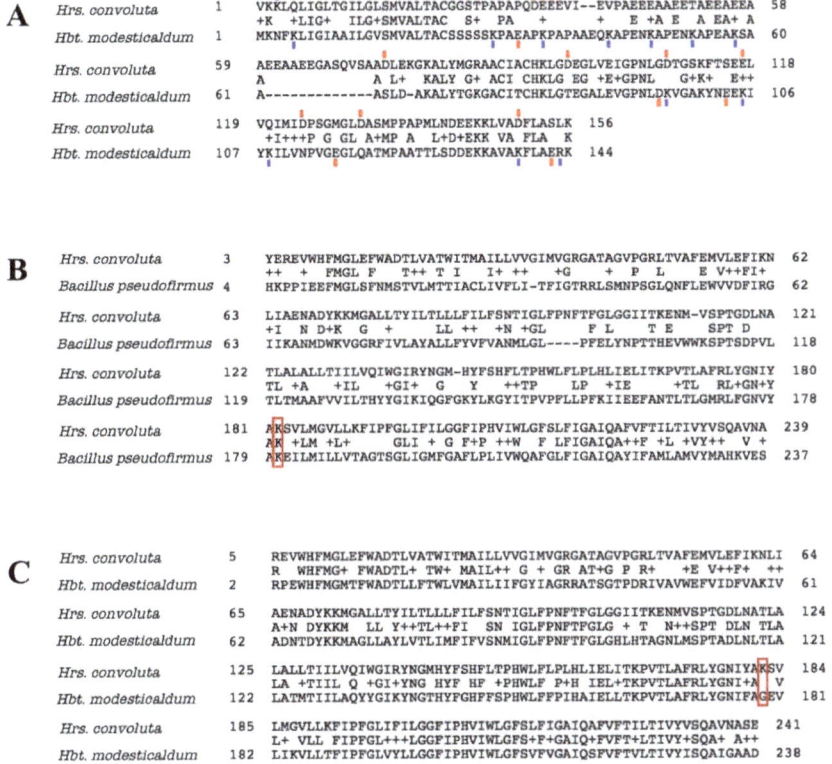

Figure 8. Amino acid sequence alignments for cytochrome c_{553} and ATP synthase F$_o$ alpha subunit of *Heliorestis convoluta* with related species. (**A**) The sequence alignment for cytochrome c_{553} of *Hrs. convoluta* and *Heliobacterium modesticaldum*. Acidic amino acid residues (red), aspartate (D) and glutamate (E), and basic amino acid residues (blue), arginine (R) and lysine (K), that differed between each species were indicated by a colored dash directly above or below the residue. Acidic or basic amino acids in the gap (–) regions were not marked. (**B**) Sequence alignment for ATP synthase F$_o$ alpha subunit of *Hrs. convoluta* and *Bacillus pseudofirmus* OF4. The lysine residue of interest at position 180 in *B. pseudofirmus* aligns with Lys^{182} in *Hrs. convoluta* (red box). (**C**) Sequence alignment for ATP synthase F$_o$ alpha subunit of *Hrs. convoluta* and *Hbt. modesticaldum*. The lysine residue of interest at position 182 in *Hrs. convoluta* aligns with Gly^{179} in *Hbt. modesticaldum* (red box). All sequence alignments were generated using the BLAST algorithm.

In a similar way, several key amino acid residues and motifs in ATP synthase have been found to contribute to optimal functioning of the enzyme at different pH levels [89,91,102,103]. For example, a lysine residue found at position 180 in the F_o alpha subunit of ATP synthase in *Bacillus pseudofirmus* OF4 was determined to favor H^+-powered ATP synthesis at an alkaliphilic pH due to its basic properties [102,104–106]. As expected, a corresponding Lys^{182} in the *Hrs. convoluta* F_o alpha subunit (Figure 8B) can presumably capture protons optimally from the alkaline environment and release them into the rotor subunit of ATP synthase at an external basic pH near the high pKa of the side chain [102]. The lysine residue would be detrimental to ATP synthesis in a neutral pH range, as H^+ would be retained on the residue side chain at ~pH 7 (below the side chain pKa). This highlights the significance of a glycine residue at the corresponding position in the F_o alpha subunit in neutrophilic bacteria, including *Hbt. modesticaldum* [102] (Figure 8C).

Several alkaliphilic bacteria use a SMF to power flagellar motor proteins, thus reserving the valuable and limited PMF for ATP production [107,108]. Research conducted with alkaliphilic *Bacillus* spp. concluded that a highly conserved valine residue is present in H^+-driven (MotB) flagellar motor protein sequences, whereas a leucine residue takes the place of this valine in Na^+-driven (MotS) motor protein sequences [107,108]. The alkaliphilic *Bacillus* spp. contained MotS with the conserved leucine amino acid, allowing these bacteria to use a SMF to power Na^+-coupled flagellar motility [107,108]. Interestingly, MotB—with its conserved valine—was identified in both *Hrs. convoluta* and *Hbt. modesticaldum*, suggesting that a PMF is used to power motility in both alkaliphilic and neutrophilic heliobacteria. Genomic analyses confirmed the presence of a core set of 24 genes (*fliCDEFGHIMNPQR, flgBCDEFGKL, motAB, flhAB*) in *Hrs. convoluta* that are essential for PMF-driven swimming motility in numerous flagellated bacteria [109].

4. Conclusions

The analysis of the complete genome sequence of *Hrs. convoluta* has provided further insight into the photoheterotrophic metabolism, nitrogen utilization, sulfur assimilation, and pigment biosynthesis pathways of heliobacteria, as well as molecular adaptations to an alkaliphilic existence. Further biochemical and genetic experimentation with alkaliphilic heliobacteria, including *Hrs. convoluta*, is necessary to confirm genomics-based predictions regarding the roles of specific genes and the apparent absence of specific enzyme activities.

Author Contributions: J.W.T., R.E.B., M.T.M., E.D.D., and W.M.S. conceived and designed the experiments. All authors performed the experiments and analyzed the data, with the bulk of the analysis done by E.D.D., L.M.S., B.M.B., K.N.S., J.M.B., and W.M.S. The culture of *Heliorestis convoluta* str. HH^T was provided by M.T.M. Submission and accessioning of the *Hrs. convoluta* genome with GenBank was initiated and overseen by S.N. and M.G.G. The manuscript was prepared by E.D.D., W.M.S., and M.T.M., with other authors providing reviews and edits for the final version. All authors have read and agreed to the published version of the manuscript.

Funding: This work was supported by the U.S. National Science Foundation Phototrophic Prokaryotes Sequencing Project as an award to J.W.T., R.E.B., and M.T.M. (Evolutionary Diversification of Photosynthesis and the Anoxygenic to Oxygenic Transition; NSF Grant #0950550).

Acknowledgments: Support for participation of E.D.D., L.M.S., B.M.B., K.N.S., and W.M.S. in this research was provided by Hodson Research Institute (Indiana Wesleyan University) grants to W.M.S. We thank Dr. Marie Asao for her detailed studies of the basic physiology of *Hrs. convoluta* str. HH^T.

Conflicts of Interest: The authors declare no conflict of interest.

References

1. Gest, H.; Favinger, J.L. *Heliobacterium chlorum*, an anoxygenic brownish-green photosynthetic bacterium containing a "new" form of bacteriochlorophyll. *Arch. Microbiol.* **1983**, *136*, 11–16. [CrossRef]
2. Stevenson, A.K.; Kimble, L.K.; Woese, C.R.; Madigan, M.T. Characterization of new phototrophic heliobacteria and their habitats. *Photo. Res.* **1997**, *53*, 1–12. [CrossRef]

3. Madigan, M.T.; Ormerod, J.G. Taxonomy, physiology, and ecology of heliobacteria. In *Anoxygenic Photosynthetic Bacteria*; Blankenship, R.E., Madigan, M.T., Bauer, C.E., Eds.; Kluwer Academic Publishers: Dordrecht, The Netherlands, 1995; Volume 2, pp. 17–30.
4. Sattley, W.M.; Madigan, M.T. The Family *Heliobacteriaceae*. In *The Prokaryotes—Firmicutes and Tenericutes*; Rosenberg, E., DeLong, E.F., Lory, S., Stackebrandt, E., Thompson, F., Eds.; Springer: Berlin/Heidelberg, Germany, 2014; pp. 185–196.
5. Brockmann, H.; Lipinski, A. Bacteriochlorophyll *g*. A new bacteriochlorophyll from *Heliobacterium chlorum*. *Arch. Microbiol.* **1983**, *136*, 17–19. [CrossRef]
6. Sattley, W.M.; Madigan, M.T.; Swingley, W.D.; Cheung, P.C.; Clocksin, K.M.; Conrad, A.L.; Dejesa, L.C.; Honchak, B.M.; Jung, D.O.; Karbach, L.E.; et al. The genome of *Heliobacterium modesticaldum*, a phototrophic representative of the *Firmicutes* containing the simplest photosynthetic apparatus. *J. Bacteriol.* **2008**, *190*, 4687–4696. [CrossRef] [PubMed]
7. Sattley, W.M.; Swingley, W.D. Properties and evolutionary implications of the heliobacterial genome. In *Genome Evolution of Photosynthetic Bacteria*; Beatty, T.J., Ed.; Academic Press, Elsevier Inc.: San Diego, CA, USA, 2013; Volume 66, pp. 67–98.
8. Woese, C.R.; Debrunner-Vossbrinck, B.A.; Oyaizu, H.; Stackebrandt, E.; Ludwig, W. Gram-positive bacteria: Possible photosynthetic ancestry. *Science* **1984**, *229*, 762–765. [CrossRef]
9. Miller, K.R.; Jacob, J.S.; Smith, U.; Kolaczkowski, S.; Bowman, M.K. *Heliobacterium chlorum*: Cell organization and structure. *Arch. Microbiol.* **1986**, *146*, 111–114. [CrossRef]
10. Pickett, M.W.; Weiss, N.; Kelly, D.J. Gram-positive cell wall structure of the A3γ type in heliobacteria. *FEMS Microbiol. Lett.* **1994**, *122*, 7–12. [CrossRef]
11. Kimble-Long, L.K.; Madigan, M.T. Molecular evidence that the capacity for endosporulation is universal among phototrophic heliobacteria. *FEMS Microbiol. Lett.* **2001**, *199*, 191–195. [CrossRef]
12. Grégoire, D.S.; Lavoie, N.C.; Poulain, A.J. Heliobacteria reveal fermentation as a key pathway for mercury reduction in anoxic environments. *Environ. Sci. Technol.* **2018**, *52*, 4145–4153. [CrossRef]
13. Lavoie, N.C.; Grégoire, D.S.; Stenzler, B.R.; Poulain, A.J. Reduced sulphur sources favour Hg^{II} reduction during anoxygenic photosynthesis by heliobacteria. *Geobiology.* **2020**, *18*, 70–79. [CrossRef]
14. Baker, P.L.; Orf, G.S.; Khan, Z.; Espinoza, L.; Leung, S.; Kevershan, K.; Redding, K.E. A molecular biology tool kit for the phototrophic Firmicute *Heliobacterium modesticaldum*. *Appl. Environ. Microbiol.* **2019**, *85*, e01287-19. [CrossRef] [PubMed]
15. Kimble, L.K.; Stevenson, A.K.; Madigan, M.T. Chemotrophic growth of heliobacteria in darkness. *FEMS Microbiol. Lett.* **1994**, *115*, 51–55. [CrossRef] [PubMed]
16. Kimble, L.K.; Mandelco, L.; Woese, C.R.; Madigan, M.T. *Heliobacterium modesticaldum*, sp. nov., a thermophilic heliobacterium of hot springs and volcanic soils. *Arch. Microbiol.* **1995**, *163*, 259–267. [CrossRef]
17. Pickett, M.W.; Williamson, M.P.; Kelly, D.J. An enzyme and ^{13}CNMR study of carbon metabolism in heliobacteria. *Photosynth. Res.* **1994**, *41*, 75–88. [CrossRef] [PubMed]
18. Bruno, W.J.; Nicholas, D.S.; Halpern, A.L. Weighted neighbor joining: A likelihood-based approach to distance-based phylogeny reconstruction. *Mol. Biol. and Evol.* **2000**, *17*, 189–197. [CrossRef] [PubMed]
19. Bryantseva, I.A.; Gorlenko, V.M.; Kompantseva, E.I.; Achenbach, L.A.; Madigan, M.T. *Heliorestis daurensis*, gen. nov. sp. nov., an alkaliphilic rod-to-coiled-shaped phototrophic heliobacterium from a Siberian soda lake. *Arch. Microbiol.* **1999**, *172*, 167–174. [CrossRef]
20. Bryantseva, I.A.; Gorlenko, V.M.; Kompantseva, E.I.; Tourova, T.P.; Kuznetsov, B.B.; Osipov, G.A. Alkaliphilic heliobacterium *Heliorestis baculata* sp. nov., and emended description of the genus *Heliorestis*. *Arch. Microbiol.* **2000**, *174*, 283–291. [CrossRef]
21. Asao, M.; Jung, D.O.; Achenbach, L.A.; Madigan, M.T. *Heliorestis convoluta* sp. nov., a coiled, alkaliphilic heliobacterium from the Wadi El Natroun, Egypt. *Extremophiles* **2006**, *10*, 403–410. [CrossRef]
22. Asao, M.; Takaichi, S.; Madigan, M.T. Amino acid-assimilating phototrophic heliobacteria from soda lake environments: *Heliorestis acidaminivorans* sp. nov. and 'Candidatus Heliomonas lunata'. *Extremophiles* **2012**, *16*, 585–595. [CrossRef]
23. Imhoff, J.F. Anoxygenic phototrophic bacteria from extreme environments. In *Modern Topics in the Phototrophic Prokaryotes: Environmental and Applied Aspects*; Hallenbeck, P.C., Ed.; Springer: Dordrecht, The Netherlands, 2017; Chapter 13; pp. 427–480.

24. Imhoff, J.F.; Hashwa, F.; Trüper, H.G. Isolation of extremely halophilic phototrophic bacteria from the alkaline Wadi Natrun, Egypt. *Arch. Hydrobiol.* **1978**, *84*, 381–388.
25. Imhoff, J.F.; Sahl, H.G.; Soliman, G.S.H.; Trüper, H.G. The Wadi Natrun: Chemical composition and microbial mass developments in alkaline brines of eutrophic desert lakes. *Geomicrobiol. J.* **1979**, *1*, 219–234. [CrossRef]
26. Sattley, W.M.; Blankenship, R.E. Insights into heliobacterial photosynthesis and physiology from the genome of *Heliobacterium modesticaldum*. *Photosynth. Res.* **2010**, *104*, 113–122. [CrossRef] [PubMed]
27. Zerbino, D.R. Using the Velvet *de novo* assembler for short-read sequencing technologies. *Curr. Protoc. Bioinform.* **2010**, *31*, 11.5.1–11.5.12. [CrossRef] [PubMed]
28. Galens, K.; Orvis, J.; Daugherty, S.; Creasy, H.H.; Angiuoli, S.; White, O.; Wortman, J.; Mahurkar, A.; Giglio, M.G. The IGS standard operating procedure for automated prokaryotic annotation. *Stand Genomic Sci.* **2011**, *4*, 244–251. [CrossRef] [PubMed]
29. Manatee. Available online: http://manatee.sourceforge.net (accessed on 24 February 2020).
30. Finn, R.D.; Coggill, P.; Eberhardt, R.Y.; Eddy, S.R.; Mistry, J.; Mitchell, A.L.; Potter, S.C.; Punta, M.; Qureshi, M.; Sangrador-Vegas, A.; et al. The Pfam protein families database: Towards a more sustainable future. *Nucleic Acids Res.* **2016**, *44*, D279–D285. [CrossRef] [PubMed]
31. Petersen, T.N.; Brunak, S.; von Heijne, G.; Nielsen, H. SignalP 4.0: Discriminating signal peptides from transmembrane regions. *Nat. Methods* **2011**, *8*, 785–786. [CrossRef]
32. Krogh, A.; Larrson, B.; von Heijne, G.; Sonnhammer, E.L. Predicting transmembrane protein topology with a hidden Markov model: Application to complete genomes. *J. Mol. Biol.* **2001**, *305*, 567–580. [CrossRef]
33. Grissa, I.; Vergnaud, G.; Pourcel, C. CRISPRFinder: A web tool to identify clustered regularly interspaced short palindromic repeats. *Nucleic Acids Res.* **2007**, *35*, W52–W57. [CrossRef]
34. Curnow, A.W.; Tumbula, D.L.; Pelaschier, J.T.; Min, B.; Söll, D. Glutamyl-tRNAGln amidotransferase in *Deinococcus radiodurans* may be confined to asparagine biosynthesis. *Proc. Natl. Acad. Sci. USA* **1998**, *95*, 12838–12843. [CrossRef]
35. Akochy, P.-M.; Bernard, D.; Roy, P.H.; Lapointe, J. Direct glutaminyl-tRNA biosynthesis and indirect asparaginyl-tRNA biosynthesis in *Pseudomonas aeruginosa* PAO1. *J. Bacteriol.* **2004**, *186*, 767–776. [CrossRef]
36. Savvi, S.; Warner, D.F.; Kana, B.D.; McKinney, J.D.; Mizrahi, V.; Dawes, S.S. Functional characterization of vitamin B_{12}-dependent methylmalonyl pathway in *Mycobacterium tuberculosis*: Implications for propionate metabolism during growth on fatty acids. *J. Bacteriol.* **2008**, *190*, 3884–3895. [CrossRef] [PubMed]
37. Suvorova, I.A.; Ravcheev, D.A.; Gelfand, M.S. Regulation and evolution of malonate and propionate catabolism in proteobacteria. *J. Bacteriol.* **2012**, *194*, 3234–3240. [CrossRef] [PubMed]
38. Ormerod, J.G.; Kimble, L.K.; Nesbakken, T.; Torgersen, T.A.; Woese, C.R.; Madigan, M.T. *Heliophilum fasciatum* gen. nov. et sp. nov., and *Heliobacterium gestii* sp. nov. endospore-forming heliobacteria from rice field soils. *Arch. Microbiol.* **1996**, *165*, 226–234.
39. Janssen, P.H.; Schink, B. Pathway of butyrate catabolism by *Desulfobacterium cetonicum*. *J. Bacteriol.* **1995**, *177*, 3870–3872. [CrossRef] [PubMed]
40. Horswill, A.R.; Escalante-Semerena, J.C. Characterization of the propionyl-CoA synthetase (PrpE) enzyme of *Salmonella enterica*: Residue Lys592 is required for propionyl-AMP synthesis. *Biochemistry* **2002**, *41*, 2379–2387. [CrossRef] [PubMed]
41. Tang, K.H.; Yue, H.; Blankenship, R.E. Energy metabolism of *Heliobacterium modesticaldum* during phototrophic and chemotrophic growth. *BMC Microbiol.* **2010**, *10*, 150. [CrossRef]
42. Tang, K.H.; Feng, X.; Zhuang, W.Q.; Alvarez-Cohen, L.; Blankenship, R.E.; Tang, Y.J. Carbon flow of heliobacteria is related more to clostridia than to the green sulfur bacteria. *J. Biol. Chem.* **2010**, *285*, 35104–35112. [CrossRef]
43. Evans, M.C.; Buchanan, B.B.; Arnon, D.I. New cyclic process for carbon assimilation by a photosynthetic bacterium. *Science* **1966**, *152*, 673. [CrossRef]
44. Sattley, W.M.; Asao, M.; Tang, J.K.H.; Collins, A.M. Energy conservation in heliobacteria: Photosynthesis and central carbon metabolism. In *The Structural Basis of Biological Energy Generation, Advances in Photosynthesis and Respiration*; Hohmann-Marriott, M.F., Ed.; Springer: Dordrecht, The Netherlands, 2014; Volume 39, pp. 231–247.
45. Wang, L.; Zhang, L.; Liu, Z.; Zhao, D.; Liu, X.; Zhang, B.; Xie, J.; Hong, Y.; Li, P.; Chen, S.; et al. A minimal nitrogen fixation gene cluster from *Paenibacillus* sp. WLY78 enables expression of active nitrogenase in *Escherichia coli*. *PLoS Genet.* **2013**, *9*, e1003865. [CrossRef]

46. Sheehy, D.; Lu, Y.-K.; Osman, F.; Alattar, Z.; Flores, C.; Sussman, H.; Zaare, S.; Dooling, M.; Meraban, A.; Baker, P.; et al. Genome-wide Transcriptional Response during the Shift to N_2-fixing Conditions in *Heliobacterium modesticaldum*. *J. Proteomics Bioinform.* **2018**, *11*, 143–160. [CrossRef]
47. Merrick, M.J. Regulation of nitrogen fixation in free-living diazotrophs. In *Genetics and regulation of nitrogen fixation in free-living bacteria*; Klipp, W.B., Masepohl, B., Gallon, J.R., Newton, W.E., Eds.; Springer: Dordrecht, The Netherlands, 2004; Volume 2, pp. 197–223.
48. Enkh-Amgalan, J.; Kawasaki, H.; Oh-oka, H.; Seki, T. Cloning and characterization of a novel gene involved in nitrogen fixation in *Heliobacterium chlorum*: A possible regulatory gene. *Arch. Microbiol.* **2006**, *186*, 327–337. [CrossRef] [PubMed]
49. Dodsworth, J.A.; Leigh, J.A. Regulation of nitrogenase by 2-oxoglutarate reversible, direct binding of a PII-like nitrogen sensor protein to dinitrogenase. *Proc. Natl. Acad. Sci. USA* **2006**, *103*, 9779–9784. [CrossRef] [PubMed]
50. Vanoni, M.A.; Curti, B. Glutamate synthase: A complex iron-sulfur flavoprotein. *Cell Mol. Life Sci.* **1999**, *55*, 617–638. [CrossRef] [PubMed]
51. Chen, J.S. Nitrogen fixation in the clostridia. In *Genetics and Regulation of Nitrogen Fixation in Free-living Bacteria*; Klipp, W., Masepohl, B., Gallon, J.R., Newton, W.E., Eds.; Springer: Dordrecht, The Netherlands, 2004; Volume 2, pp. 53–64.
52. Peck, H.D. Enzymatic basis for assimilatory and dissimilatory sulfate reduction. *J. Bacteriol.* **1961**, *82*, 933–939. [CrossRef] [PubMed]
53. Sekowska, A.; Kung, H.-F.; Danchin, A. Sulfur metabolism in *Escherichia coli* and related bacteria: Facts and fiction. *J. Mol. Microbiol. Biotechnol.* **2000**, *2*, 145–177. [PubMed]
54. Schwedock, J.S.; Liu, C.; Leyh, T.S.; Long, S.R. *Rhizobium meliloti* NodP and NodQ form a multifunctional sulfate-activating complex requiring GTP for activity. *J. Bacteriol.* **1994**, *176*, 7055–7064. [CrossRef]
55. Eisen, J.A.; Nelson, K.E.; Paulsen, I.T.; Heidelberg, J.F.; Wu, M.; Dodson, R.J.; Deboy, R.; Gwinn, M.L.; Nelson, W.C.; Haft, D.H.; et al. The complete genome sequence of *Chlorobium tepidum* TLS, a photosynthetic, anaerobic, green-sulfur bacterium. *Proc. Natl. Acad. Sci. USA* **2002**, *99*, 9509–9514. [CrossRef]
56. Weissgerber, T.; Zigann, R.; Bruce, D.; Chang, Y.-J.; Detter, J.C.; Han, C.; Hauser, L.; Jeffries, C.D.; Land, M.; Munk, A.C.; et al. Complete genome sequence of *Allochromatium vinosum* DSM 180^T. *Stand. Genomic Sci.* **2011**, *5*, 311–330. [CrossRef]
57. Klemm, D.J.; Barton, L.L. Oxidation of protoporphyrinogen in the obligate anaerobe *Desulfovibrio gigas*. *J. Bacteriol.* **1985**, *164*, 316–320. [CrossRef]
58. Klemm, D.J.; Barton, L.L. Purification and properties of protoporphyrinogen oxidase from an anaerobic bacterium, *Desulfovibrio gigas*. *J. Bacteriol.* **1987**, *169*, 5209–5215. [CrossRef]
59. Dailey, H.A.; Gerdes, S.; Dailey, T.A.; Burch, J.S.; Phillips, J.D. Noncanonical coproporphyrin-dependent bacterial heme biosynthesis pathway that does not use protoporphyrin. *Proc. Natl. Acad. Sci. USA* **2015**, *112*, 2210–2215. [CrossRef] [PubMed]
60. Tsukatani, Y.; Yamamoto, H.; Mizoguchi, T.; Fujita, Y.; Tamiaki, H. Completion of biosynthetic pathways for bacteriochlorophyll g in *Heliobacterium modesticaldum*: The C8-ethylidene group formation. *Biochim. Biophys. Acta* **2013**, *1827*, 1200–1204. [CrossRef] [PubMed]
61. Ortega-Ramos, M.; Canniffe, D.P.; Radle, M.I.; Hunter, C.N.; Bryant, D.A.; Golbeck, J.H. Engineered biosynthesis of bacteriochlorophyll g_F in *Rhodobacter sphaeroides*. *Biochim. Biophys. Acta Bioenerg.* **2018**, *1859*, 501–509. [CrossRef] [PubMed]
62. van de Meent, E.J.; Kobayashi, M.; Erkelens, C.; van Veelen, P.A.; Amesz, J.; Watanabe, T. Identification of 8^1-hydroxychlorophyll a as a functional reaction center pigment in heliobacteria. *Biochim. Biophys. Acta* **1991**, *1058*, 356–362. [CrossRef]
63. Takaichi, S.; Oh-oka, H.; Maoka, T.; Jung, D.O.; Madigan, M.T. Novel carotenoid glucoside esters from alkaliphilic heliobacteria. *Arch. Microbiol.* **2003**, *179*, 95–100. [CrossRef]
64. Dubey, V.S.; Bhalla, R.; Luthra, R. An overview of the non-mevalonate pathway for terpenoid biosynthesis in plants. *J. Biosci.* **2003**, *28*, 637–646. [CrossRef]
65. Takaichi, S.; Inoue, K.; Akaike, M.; Kobayashi, M.; Oh-oka, H.; Madigan, M.T. The major carotenoid in all known species of heliobacteria is the C30 carotenoid 4,4'-diaponeurosporene, not neurosporene. *Arch. Microbiol.* **1997**, *168*, 277–281. [CrossRef]

66. Takaichi, S. Carotenoids and carotenogenesis in anoxygenic photosynthetic bacteria. In *The Photochemistry of Carotenoids: Applications in Biology*; Frank, H.A., Cogdell, R.J., Young, A., Britton, G., Eds.; Springer: Dordrecht, The Netherlands, 1999; Volume 8, pp. 39–69.
67. Trost, J.T.; Brune, D.C.; Blankenship, R.E. Protein sequences and redox titrations indicate that the electron acceptors in reaction centers from heliobacteria are similar to Photosystem I. *Photosynth. Res.* **1992**, *32*, 11–22. [CrossRef]
68. Neerken, S.; Amesz, J. The antenna reaction center complex of heliobacteria: Composition, energy conversion and electron transfer. *Biochim. Biophys. Acta* **2001**, *1507*, 278–290. [CrossRef]
69. Orf, G.S.; Gisriel, C.; Redding, K.E. Evolution of photosynthetic reaction centers: Insights from the structure of the heliobacterial reaction center. *Photosynth. Res.* **2018**, *138*, 11–37. [CrossRef]
70. Gisriel, C.; Sarrou, I.; Ferlez, B.; Golbeck, J.H.; Redding, K.E.; Fromme, R. Structure of a symmetric photosynthetic reaction center–photosystem. *Science* **2017**, *357*, 1021–1025. [CrossRef] [PubMed]
71. Kobayashi, M.; van de Meent, E.J.; Erkelens, C.; Amesz, J.; Ikegami, I.; Watanabe, T. Bacteriochlorophyll g epimer as a possible reaction center component of heliobacteria. *Biochim. Biophys. Acta* **1991**, *1057*, 89–96. [CrossRef]
72. Kobayashi, M.; Watanabe, T.; Ikegami, I.; van de Meent, E.J.; Amesz, J. Enrichment of bacteriochlorophyll g' in membranes of *Heliobacterium chlorum* by ether extraction: Unequivocal evidence for its existence in vivo. *FEBS Lett.* **1991**, *284*, 129–131. [CrossRef]
73. Kashey, T.S.; Luu, D.D.; Cowgill, J.C.; Baker, P.L.; Redding, K.E. Light-driven quinone reduction in heliobacterial membranes. *Photosynth. Res.* **2018**, *138*, 1–9. [CrossRef] [PubMed]
74. Blankenship, R.E. *Molecular Mechanisms of Photosynthesis*, 2nd ed.; Wiley Blackwell: Oxford, UK, 2014.
75. Yue, H.; Kang, Y.; Zhang, H.; Gao, X.; Blankenship, R.E. Expression and characterization of the diheme cytochrome c subunit of the cytochrome bc complex in *Heliobacterium modesticaldum*. *Arch. Biochem. Biophys.* **2012**, *517*, 131–137. [CrossRef]
76. Yang, J.-H.; Sarrou, I.; Martin-Garcia, J.M.; Zhang, S.; Redding, K.E.; Fromme, P. Purification and biochemical characterization of the ATP synthase from *Heliobacterium modesticaldum*. *Protein. Expr. Purif.* **2015**, *114*, 1–8. [CrossRef]
77. Kramer, D.M.; Schoepp, B.; Liebl, U.; Nitschke, W. Cyclic electron transfer in *Heliobacillus mobilis* involving a menaquinol-oxidizing cytochrome bc complex and an RCI-type reaction center. *Biochemistry* **1997**, *36*, 4203–4211. [CrossRef]
78. Heinnickel, M.; Shen, G.; Golbeck, J.H. Identification and characterization of PshB, the dicluster ferredoxin that harbors the terminal electron acceptors F_A and F_B in *Heliobacterium modesticaldum*. *Biochemistry* **2007**, *46*, 2530–2536. [CrossRef]
79. Kondo, T.; Itoh, S.; Matsuoka, M.; Azai, C.; Oh-oka, H. Menaquinone as the secondary electron acceptor in the type I homodimeric photosynthetic reaction center of *Heliobacterium modesticaldum*. *J. Phys. Chem. B.* **2015**, *119*, 8480–8489. [CrossRef]
80. Han, W.-D.; Kawamoto, S.; Hosoya, Y.; Fujita, M.; Sadaie, Y.; Suzuki, K.; Ohashi, Y.; Kawamura, F.; Ochi, K. A novel sporulation-control gene (*spo0M*) of *Bacillus subtilis* with a σ^H-regulated promoter. *Gene* **1998**, *217*, 31–40. [CrossRef]
81. Margolis, P.S.; Driks, A.; Losick, R. Sporulation gene *spoIIB* from *Bacillus subtilis*. *J. Bacteriol.* **1993**, *175*, 528–540. [CrossRef] [PubMed]
82. Naclerio, G.; Baccigalupi, L.; Zilhao, R.; De Felice, M.; Ricca, E. *Bacillus subtilis* spore coat assembly requires *cotH* gene expression. *J. Bacteriol.* **1996**, *178*, 4375–4380. [CrossRef] [PubMed]
83. Permpoonpattana, P.; Phetcharaburanin, J.; Mikelsone, A.; Dembek, M.; Tan, S.; Brisson, M.C.; La Ragione, R.; Brisson, A.R.; Fairweather, N.; Hong, H.A.; et al. Functional characterization of *Clostridium difficile* spore coat proteins. *J. Bacteriol.* **2013**, *195*, 1492–1503. [CrossRef] [PubMed]
84. Horikoshi, K. Alkaliphiles. *Proc. Jpn. Acad. Ser. B* **2004**, *80*, 166–178. [CrossRef]
85. Hicks, D.B.; Liu, J.; Fujisawa, M.; Krulwich, T.A. F_1F_o-ATP synthases of alkaliphilic bacteria: Lessons from their adaptations. *Biochim. Biophys. Acta* **2010**, *1797*, 1362–1377. [CrossRef]
86. Sturr, M.G.; Guffanti, A.A.; Krulwich, T.A. Growth and bioenergetics of alkaliphilic *Bacillus firmus* OF4 in continuous culture at high pH. *J. Bacteriol.* **1994**, *176*, 3111–3116. [CrossRef]
87. Macnab, R.M.; Castle, A.M. A variable stoichiometry model for pH homeostasis in bacteria. *Biophys. J.* **1987**, *52*, 637–647. [CrossRef]

88. Padan, E.; Schuldiner, S. Bacterial Na$^+$/H$^+$ antiporters: Molecular biology, biochemistry and physiology. In *Handbook of Biological Physics*; Konings, W.N., Kaback, H.R., Lolkema, J.S., Eds.; Elsevier Science: Amsterdam, The Netherlands, 1996; Volume 2, pp. 501–531.
89. Krulwich, T.A.; Liu, J.; Morino, M.; Fujisawa, M.; Ito, M.; Hicks, D.B. Adaptive mechanisms of extreme alkaliphiles. In *Extremophiles Handbook*; Horikoshi, K., Ed.; Springer: Tokyo, Japan, 2011; pp. 119–139.
90. Peddie, C.J.; Cook, G.M.; Morgan, H.W. Sucrose transport by the alkaliphilic, thermophilic *Bacillus* sp. strain TA2.A1 is dependent on a sodium gradient. *Extremophiles* **2000**, *4*, 291–296. [CrossRef]
91. Preiss, L.; Hicks, D.B.; Suzuki, S.; Meier, T.; Krulwich, T.A. Alkaliphilic bacteria with impact on industrial applications, concepts of early life forms, and bioenergetics of ATP synthesis. *Front. Bioeng. Biotechnol.* **2015**, *3*, 1–16. [CrossRef]
92. von Ballmoos, C.; Cook, G.M.; Dimroth, P. Unique rotary ATP synthase and its biological diversity. *Annu. Rev. Biophys.* **2008**, *37*, 43–64. [CrossRef]
93. Asao, M.; Madigan, M.T. Family IV *Heliobacteriaceae* Madigan 2001, 625. In *Bergey's Manual of Systematic Bacteriology (The Firmicutes)*, 2nd ed.; De Vos, P., Garrity, G.M., Jones, D., Krieg, N.R., Ludwig, W., Rainey, F.A., Schleifer, K.-H., Whitman, W.B., Eds.; Springer: New York, NY, USA, 2009; Volume 3, pp. 923–931.
94. Aono, R.; Horikoshi, K. Carotenes produced by alkaliphilic yellow-pigmented strains of *Bacillus*. *Agric. Biol. Chem.* **1991**, *55*, 2643–2645. [CrossRef]
95. Steiger, S.; Perez-Fons, L.; Cutting, S.M.; Fraser, P.D.; Sandmann, G. Annotation and functional assignment of the genes for the C30 carotenoid pathways from the genomes of two bacteria: *Bacillus indicus* and *Bacillus firmus*. *Microbiology* **2015**, *161*, 194–202. [CrossRef] [PubMed]
96. Haines, T.H.; Dencher, N.A. Cardiolipin: A proton trap for oxidative phosphorylation. *FEBS Lett.* **2002**, *528*, 35–39. [CrossRef]
97. Dowhan, W.; Bogdanov, M.; Mileykovskaya, E. Functional roles of lipids in membranes. In *Biochemistry of Lipids. Lipoproteins and Membranes*, 5th ed.; Vance, D.E., Vance, J.E., Eds.; Elsevier Press: Amsterdam, The Netherlands, 2008; pp. 1–37.
98. Schlame, M. Cardiolipin synthesis for the assembly of bacterial and mitochondrial membranes. *J. Lipid Res.* **2008**, *49*, 1607–1620. [CrossRef]
99. Kang, S.-K.; Kudo, T.; Horikoshi, K. Molecular cloning and characterization of an alkalophilic *Bacillus* sp. C125 gene homologous to *Bacillus subtilis sec Y*. *J. Gen. Microbiol.* **1992**, *138*, 1365–1370. [CrossRef]
100. Quirk, P.G.; Hicks, D.B.; Krulwich, T.A. Cloning of the *cta* operon from alkaliphilic *Bacillus firmus* OF4 and characterization of the pH-regulated cytochrome caa_3 oxidase it encodes. *J. Biol. Chem.* **1993**, *268*, 678–685.
101. Krulwich, T.A. Alkaliphiles: 'basic' molecular problems of pH tolerance and bioenergetics. *Mol. Microbiol.* **1995**, *15*, 403–410. [CrossRef]
102. McMillan, D.G.; Keis, S.; Dimroth, P.; Cook, G.M. A specific adaptation in the α subunit of thermoalkaliphilic F_1F_o-ATP synthase enables ATP synthesis at high pH but not at neutral pH values. *J. Biol. Chem.* **2007**, *282*, 17395–17404. [CrossRef]
103. Liu, J.; Fujisawa, M.; Hicks, D.B.; Krulwich, T.A. Characterization of the functionally critical AXAXAXA and PXXEXXP motifs of the ATP synthase c-subunit from an alkaliphilic *Bacillus*. *J. Biol. Chem.* **2009**, *284*, 8714–8725. [CrossRef]
104. Ivey, D.M.; Krulwich, T.A. Organization and nucleotide sequence of the *atp* genes encoding the ATP synthase from alkaliphilic *Bacillus firmus* OF4. *Mol. Gen. Genet.* **1991**, *229*, 292–300. [CrossRef]
105. Ivey, D.M.; Krulwich, T.A. Two unrelated alkaliphilic *Bacillus* species possess identical deviations in sequence from those of other prokaryotes in regions of F_o proposed to be involved in proton translocation through the ATP synthase. *Res. Microbiol.* **1992**, *143*, 467–470. [CrossRef]
106. Wang, Z.; Hicks, D.B.; Guffanti, A.A.; Baldwin, K.; Krulwich, T.A. Replacement of amino acid sequence features of a- and c-subunits of ATP synthases of alkaliphilic *Bacillus* with the *Bacillus* consensus sequence results in defective oxidative phosphorylation and nonfermentative growth at pH 10.5. *J. Biol. Chem.* **2004**, *279*, 26546–26554. [CrossRef] [PubMed]
107. Terahara, N.; Krulwich, T.A.; Ito, M. Mutations alter the sodium versus proton use of a *Bacillus clausii* flagellar motor and confer dual ion use on *Bacillus subtilis* motors. *Proc. Natl. Acad. Sci. USA* **2008**, *105*, 14359–14364. [CrossRef] [PubMed]

108. Fujinami, S.; Terahara, N.; Krulwich, T.A.; Ito, M. Motility and chemotaxis in alkaliphilic *Bacillus* species. *Future Microbiol.* **2009**, *4*, 1137–1149. [CrossRef] [PubMed]
109. Liu, R.; Ochman, H. Stepwise formation of the bacterial flagellar system. *Proc. Natl. Acad. Sci. USA* **2007**, *104*, 7116–7121. [CrossRef]

© 2020 by the authors. Licensee MDPI, Basel, Switzerland. This article is an open access article distributed under the terms and conditions of the Creative Commons Attribution (CC BY) license (http://creativecommons.org/licenses/by/4.0/).

 *microorganisms*

Article

Osmotic Adaptation and Compatible Solute Biosynthesis of Phototrophic Bacteria as Revealed from Genome Analyses

Johannes F. Imhoff [1,*], Tanja Rahn [1], Sven Künzel [2], Alexander Keller [3] and Sven C. Neulinger [4]

1. GEOMAR Helmholtz Centre for Ocean Research, 24105 Kiel, Germany; trahn@geomar.de
2. Max Planck Institute for Evolutionary Biology, 24306 Plön, Germany; kuenzel@evolbio.mpg.de
3. Center for Computational and Theoretical Biology, University Würzburg, 97074 Würzburg, Germany; a.keller@biozentrum.uni-wuerzburg.de
4. omics2view.consulting GbR, 24118 Kiel, Germany; s.neulinger@omics2view.consulting
* Correspondence: jimhoff@geomar.de

Abstract: Osmotic adaptation and accumulation of compatible solutes is a key process for life at high osmotic pressure and elevated salt concentrations. Most important solutes that can protect cell structures and metabolic processes at high salt concentrations are glycine betaine and ectoine. The genome analysis of more than 130 phototrophic bacteria shows that biosynthesis of glycine betaine is common among marine and halophilic phototrophic *Proteobacteria* and their chemotrophic relatives, as well as in representatives of *Pirellulaceae* and *Actinobacteria*, but are also found in halophilic *Cyanobacteria* and *Chloroherpeton thalassium*. This ability correlates well with the successful toleration of extreme salt concentrations. Freshwater bacteria in general lack the possibilities to synthesize and often also to take up these compounds. The biosynthesis of ectoine is found in the phylogenetic lines of phototrophic *Alpha*- and *Gammaproteobacteria*, most prominent in the *Halorhodospira* species and a number of *Rhodobacteraceae*. It is also common among *Streptomycetes* and *Bacilli*. The phylogeny of glycine-sarcosine methyltransferase (GMT) and diaminobutyrate-pyruvate aminotransferase (EctB) sequences correlate well with otherwise established phylogenetic groups. Most significantly, GMT sequences of cyanobacteria form two major phylogenetic branches and the branch of *Halorhodospira* species is distinct from all other *Ectothiorhodospiraceae*. A variety of transport systems for osmolytes are present in the studied bacteria.

Keywords: genomes of photosynthetic bacteria; glycine betaine biosynthesis; ectoine biosynthesis; osmotic adaptation; phylogeny of osmolyte biosynthesis

1. Introduction

Phototrophic bacteria are widely distributed at suitable habitats in the marine and hypersaline environment. They are exposed to sometimes dramatically changing salt concentrations and some are found in saturated brines of salt and soda lakes, where they regularly develop massive blooms, often forming patches and pinkish-red layers, even within deposits of crystalized salts [1,2]. One of the prerequisites to cope with high salt and solute concentrations is the ability to keep an osmotic balance, i.e., a positive turgor pressure inside the cells through the accumulation of solutes in the cytoplasm that are compatible with the metabolic processes, even at high concentrations, and preserve active structures of proteins and nucleic acids [3]. In consequence, these bacteria need proper mechanisms of osmotic adaptation and ways to accumulate osmotically active compatible solutes up to several molar concentrations at the extremes.

Limitation to protect cell structures and metabolism is given by the compatibility of the solutes and the ability to accumulate to high or extremely high, several molar concentrations inside the cell. Glycine betaine (in some cyanobacteria also glutamate betaine) and ectoine (also hydroxyectoine) are the top candidates for this function in bacteria. Glycine betaine (hereafter "betaine") accumulation is widespread among phototrophic and chemotrophic eubacteria [4–8]. Ectoine was first identified as a compatible solute in the extreme halophilic *Halorhodospira halochloris* [9] and was later shown to be widely distributed among marine and halophilic eubacteria [10]. Though a number of other solutes are accumulated in bacteria in response to osmotic stress, these can provide protection only at low to moderate osmotic stress. Such compounds include sugars such as trehalose and sucrose, amino acids in particular glutamate, glucosyl glycerol, N-acetyl-glutaminyl glutamine amide, N-carbamoyl-glutamine amide, and others [6,7,11–13]. Basically, the accumulation can be achieved by uptake from the environment or by biosynthesis.

Betaine biosynthesis can be achieved by three consecutive methylation steps from glycine and includes the formation of monomethylglycine (sarcosine) and dimethylglycine as intermediates. In most of the bacteria studied, these enzymes are encoded by two genes and have overlapping enzymatic activities. In *Halorhodospira halochloris*, the first enzyme (glycine and sarcosine methyltransferase GMT) catalyzes the formation of monomethylglycine and dimethylglycine and the second one (dimethylglycine methyltransferase DMT) catalyzes the methylations to dimethylglycine and betaine [14,15]. In the halophilic cyanobacterium *Aphanothece halophytica*, the second enzyme specifically catalyzes the methylation of dimethylglycine [16]. In *Actinopolyspora halophila*, the two genes are fused, showing corresponding sequence homologies to the two genes in *Halorhodospira halochloris* [14].

An alternative route of betaine biosynthesis starts from choline and oxidizes this compound in two steps to betaine, catalyzed by choline dehydrogenase (BetA) and betaine aldehyde dehydrogenase (BetB). This route is widely distributed among bacteria but requires the external presence and uptake of choline. One possible uptake system is the high-affinity secondary transporter BetT, considered to be a specific choline transporter in *E. coli* [17] and a betaine transporter in *Aphanothece halophytica* [18]. In *E. coli*, the *betT* gene, together with a regulatory *betI* gene, is included in the *bet* gene cluster.

The biosynthesis of ectoine, which was first identified in the extreme halophilic phototrophic bacterium *Halorhodospira halochloris* [9], starts from aspartate. Aspartate is activated to L-aspartate-phosphate (Ask_ect) and then reduced to L-aspartate-β-semialdehyde (Asd), followed by a transamination (with glutamate or alanine as donor of the amino group) to L-diaminobutyric acid (EctB), acetylation of the amino group to N-acetyldiaminobutyric acid (EctA), and finally ring closure (EctC) to form ectoine, as shown for *Halorhodospira halophila* and *Halomonas elongata* [19,20]. Genes responsible for ectoine biosynthesis and their osmotically regulated expression were first identified in *Marinococcus halophilus* [21]. The oxidation of ectoine to hydroxyectoine, which was first demonstrated in *Streptomyces parvulus* [22], turned out to be common to many Actinobacteria. The hydroxylation of ectoine (EctD) is strongly dependent on the presence of molecular oxygen and is accompanied by the oxidative decarboxylation of oxoglutarate forming CO_2 and succinate [23]. While the *ask_ect* gene, which encodes a specific aspartate kinase not underlying the feedback control of threonine [24,25], is often included in the *ect* gene cluster, *asd* usually is at a different locus in the genome. The coexpression of *ask_ect* together with the osmotically induced gene cluster *ectABC*, ensures optimal supply of the precursor L-aspartate-β-semialdehyde under osmotic stress conditions. Aspects of ectoine and hydroxyectoine biosynthesis were recently reviewed by Czech et al. [10].

Osmotic adaptation can also be achieved by uptake of osmolytes rather than biosynthesis. Provided that such solutes are available in the environment, uptake generally is the favored way, because it is far less energetically expensive than de novo synthesis. Laboratory culture media with complex carbon sources (e.g., proteose peptone and yeast extract), often contain these compounds. In such media, betaine was accumulated to high levels exceeding 1 M concentrations by a number of salt-tolerant chemoheterotrophic bacteria isolated from hypersaline soils [5]. Additionally, a number of phototrophic green and purple sulfur bacteria are able to take up and accumulate betaine but are incapable of its biosynthesis [13]. In nature, such situations can occur, whenever large decaying biomass is accumulated and betaine or choline becomes available, leaking from living cells or released upon cell lysis. Under such conditions, which are likely found in microbial mats, uptake might become an important strategy to accumulate compatible solutes, although dependent on biosynthesis as their primary producers. Uptake systems for betaine, choline, and ectoine are found in numerous bacteria, many of which are unable to biosynthesize these molecules. Best known is a widely distributed transport system first identified in the transport of proline ProU (ProVWX), which has low affinity to choline and also transports glycine betaine, proline betaine, carnitine, and ectoine [26,27]. Another betaine transport system is OpuA. Both of these systems are ABC type transport systems and include binding proteins for the substrate and for ATP and a permease (ProVWX, OpuAA,AB,AC). Several more other transport systems for osmolytes are known (see [27]). Of these, OpuD is a single-component secondary transporter for betaine. BetT might also be active in betaine transport as found in *Aphanothece halophytica* [18].

In the present study, we focused on the biosynthetic capability of phototrophic bacteria to produce betaine and ectoine, using genomic information available from new genome sequences and from databases. The genomic repertoire and the distribution of the studied biosynthesis pathways are related to salt responses of the bacteria to conclude on their requirements for environmental adaptation. In addition, the phylogeny of ectoine and betaine biosynthesis was studied by comparing sequences of glycine–sarcosine–methyltransferase (GMT) and diaminobutyrate–pyruvate aminotransferase (EctB) of phototrophic bacteria, together with selected chemotrophic bacteria.

2. Materials and Methods

2.1. Cultivation and DNA Extraction

Cells were grown in the appropriate media, as described for the purple sulfur bacteria [28,29] and several groups of phototrophic nonsulfur purple bacteria [30]. Extraction of DNA was done as described earlier [31]. DNA from 2 mL of a freshly grown culture was extracted with the DNeasy® Blood&Tissue Kit, according to the manufacturer's instructions (QIAgen, Hilden, Germany), including the pretreatment for Gram-positive bacteria (consisting of enzymatic lysis buffer, proteinase K, and RNAse) and then dissolved in the TE-buffer. The extracted DNA was checked for quantity and quality by agarose gel electrophoresis with linear and double-stranded Lambda DNA used as control (Thermo Fisher Scientific, Waltham, MA, USA, Cat.No. SD0011). DNA was purified using a gel extraction procedure with the MoBio Ultra DNA Purification Kit (Cat.No. 12100-300).

2.2. Sequencing and Assembly

Sequencing of DNA and the assembly of sequences were done, as described earlier [31]. Samples were prepared with the Nextera® XT DNA Sample Preparation kit from Illumina, following the manufacturer's protocol. Afterwards, the samples were pooled and sequenced on the MiSeq using the MiSeq® Reagent Kit v3 600 cycles sequencing chemistry. The library was clustered to a density of approximately 1200 K/mm^2.

Read quality filtering was performed with Trimmomatic v0.36 [32]. Reads were scanned for residues of Illumina Nextera XT adapters. Quality trimming was conducted with a 5-base pairs (bp) sliding window, trimming the read once the average Phred quality score within this window dropped below 30. Reads with a minimum length of 21 bp after quality trimming were retained. Single reads (i.e., reads with their mate deleted) were retained and included into downstream analysis. Reads were further checked for ambiguous base calls as well as for low complexity, employing the DUST algorithm [33]. They were filtered accordingly with an in-house R script in Microsoft R Open v3.3.2 (R Core Team 2016). Retained reads are referred to as 'filtered reads'. Filtered reads were pre-assembled with SPAdes v3.10.0 [34,35], using default k-mer lengths. Scaffolds ≥500 bp of this pre-assembly were subjected to extension and second-round scaffolding with SSPACE standard v3.0 [36].

2.3. Genome Annotation and Submission to GenBank

Genome sequences were annotated by the "Rapid Annotation using Subsystem Technology" (RAST) [37]. Sequences of EctB, GMT, and ProW were retrieved from the annotated genomes using the RAST and The SEED Viewer provided by this platform [38,39], which also offered the option to search with gene/protein sequences within the annotated genomes. In addition, standard protein BLAST of the NCBI database was used with EctB and GMT sequences to retrieve additional protein sequences. All genome sequences were deposited in the GenBank database (Supplementary Table S4). Accession numbers of gene and genome sequences, together with species and strain designations as well as the corresponding higher taxonomic ranks, are included in Supplementary Tables S1–S4.

2.4. Phylogenetic Sequence Analyses

For phylogenetic analysis, protein sequences of GMT, EctB, and ProW were aligned using ClustalX version 2.1 [40] and the trees were calculated by the neighbor-joining (NJ) method with correction for multiple substitutions, according to ClustalX [41]. NJ plot was used to draw the phylogenetic trees expressed in the Newick phylogenetic tree format [42]. The tree topologies were evaluated with bootstrap analyses, based on 1000 replicates and the values are indicated in the trees.

3. Results and Discussion

3.1. Osmotic Adaptation in Phototrophic Bacteria

According to genome analysis of representative phototrophic bacteria, the genetic repertoire and kind of possible responses to osmotic stress of these bacteria varies widely between the different groups and between freshwater, marine, and halophilic species. The genetic repertoire of all studied phototrophic bacteria is shown in Tables 1–3. While almost all marine and halophilic phototrophic bacteria can synthesize either betaine or ectoine or both, true freshwater bacteria lack the ability to synthesize betaine and ectoine. Often, they also lack the possibility of uptake of these osmolytes or their biosynthetic precursors, while marine and halophilic bacteria generally have this option. Obviously, biosynthesis or uptake of betaine more than that of ectoine is a prerequisite for their ability to thrive in marine and hypersaline habitats and to tolerate high salt concentrations.

Table 1. Genes and gene clusters of betaine and ectoine biosynthesis as well as relevant transport systems of phototrophic *Acidobacteria* (*Chloracidobacterium*), *Chlorobi*, *Chloroflexi*, *Cyanobacteria*, *Heliobacterium modesticaldum*, *Gemmatimonas phototrophica* and *Betaproteobacteria* are shown together with salt responses and their systematic affiliation [a,b,c].

	Gene Repertoire of Various Orders of Phototrophic Bacteria for Osmotic Adaptation								
				Betaine	biosynthesis from choline		Osmolyte transport		
Family	Species	Strain	Salt response	from glycine GMT-DMT	betAB	betT	opuA / opuC opuD	opuCB-proXV	proW1 proW2
Acidobacteria/Acidobacteriales									
Acidobacteriaceae	*Chloracidobacterium thermophilum*	B-G2	F	o	o	o	o		o
Chlorobi/Chlorobiales									
Chlorobiaceae	*Chlorobaculum thiosulfatophilum*	DSM 249	F	o	o	o	o		o
Chlorobiaceae	*Chlorobium limicola*	DSM 245	F	o	o	o	o		o
Chlorobiaceae	*Chlorobium phaeovibrioides*	DSM 265	F	o	o	o	o		o
Chlorobiaceae	*Chloroherpeton thalassium*	ATCC 35110	M	GMT-DMT-betT	o	o	o		o
Chlorobiaceae	*Prosthecochloris aestuarii*	DSM 271	M	o	o	o	o		o
Chlorobiaceae	*Prosthecochloris vibrioformis*	DSM 260	M	o	o	o	proVopuAB,AC		o
Chloroflexi/Chloroflexales									
Chloroflexaceae	*Chloroflexus aggregans*	DSM 9485	F		o	o		o	o
Chloroflexaceae	*Chloroflexus aurantiacus*	J-10-fl	F	o	o	o		o	o
Roseiflexaceae	*Roseiflexus castenholzii*	DSM 13941	F	o	o	o		opuCB-proXV	o
Roseiflexaceae	*Roseiflexus sp.*	RS-1	F	o	o	o		opuCB-proXV	o
Cyanobacteria/Synechococcales									
Prochloraceae	*Prochlorococcus marinus*	MIT 9313	M	GMT-DMTproVWX	o	o	o		o
Synechococcaceae	*Synechococcus species*	WH8102	M	GMT-DMTproVWX	o	o	o		o
Synechococcaceae	*Dactylococcopsis salina*	PCC 8305	H	GMT-DMT	o	o	o		W1

Table 1. Cont.

Gene Repertoire of Various Orders of Phototrophic Bacteria for Osmotic Adaptation

Family	Species	Strain						
Cyanobacteria/Chroococcales								
Aphanothecaceae	Halothece sp./Aphanothece halophytica	PCC 7418	H	GMT-DMT	o	/betT/	o	W1
Aphanothecaceae	Euhalothece natronophila	Z-M001	H	GMT-DMT	o	/betT/	o	W1
Firmicutes/Clostridiales								
Heliobacteriaceae	Heliobacterium modesticaldum	Ice1	F	o	o	o	o	o
Gemmatimonadetes/Gemmatimonadales								
Gemmatimonadaceae	Gemmatimonas phototrophica	AP64	F	o	o	o	o	o
Betaproteobacteria/Burkholderiales								
Burkholderiaceae	Polynucleobacter duraquae	MWH-MoK4	F	o	o	o	o	o
Comamonadaceae	Rhodoferax antarcticus	DSM 24876	F	o	o	o	o	W2
Comamonadaceae	Rhodoferax fermentans	DSM 10138	F	o	o	o	opuD	o
uncl. Burkholderiales	Rubrivivax gelatinosus	IL144	F	o	o	o	opuD	o
uncl. Burkholderiales	Rubrivivax gelatinosus	DSM 1709	F	o	o	o	opuD	o
uncl. Burkholderiales	Rubrivivax gelatinosus	IM 151	F	o	o	o	opuD	o
uncl. Burkholderiales	Rubrivivax gelatinosus 155	DSM 149	F	o	o	o	opuD	o
Betaproteobacteria/Rhodocyclales								
Rhodocyclaceae	Rhodocyclus purpureus	TEM	F	o	o	o	o	o
Rhodocyclaceae	Rhodocyclus tenuis	IM 230	F	o	o	o	o	o

[a] Variants of the transport system are abbreviated W1 = proVWX, W2 = proVWV, W3 = proXWV-bet, W4 = proXVWlarge for proU and opuA = opuAA,AB,AC. The code for salt responses (growth optimum/tolerance) is F = freshwater species (<1% NaCl), M = marine species (1–7%/<8–9%), M/H = marine species with elevated salt tolerance (1–8%/>10–20%), and H = moderate to extreme halophilic species (>6–25%/>15–>25%). As far as possible, the gene clusters are given and "/" denotes a separate locus of the genes in the genome. The genome accession numbers of GenBank are shown in the Supplementary Table S4. [b] Ectoine biosynthesis absent, [c] proW3, and ProW4 are absent. Color shades indicate different types of gene and gene associations: betaine synthesis from glycine (blue) and from choline (green), transport with betT (light-lila), opuA,C and D (shades of beige-brown), proU-W1 and W2 (shades of green); also marine (blue) and halophilic (rose-pink) growth response of the bacteria.

Table 2. Genes and gene clusters of betaine and ectoine biosynthesis as well as relevant transport systems of phototrophic *Alphaproteobacteria* [a].

Gene Repertoire of Phototrophic Alphaproteobacteria for Osmotic Adaptation

Family	Species	Strain	Salt response	Ectoine biosynthesis ectABC	Betaine from glycine GMT-DMT	biosynthesis from choline betAB	Osmolyte transport betT	opuA opuD	proW1	proW2	proW3	proW4
Rhizobiales												
Beijerinckiaceae	*Methylocella silvestris*	BL2	F	o	o	o	o	o	o	o	o	o
Beijerinckiaceae	*Rhodoblastus acidophilus*	DSM 137	F	o	o	o	o	o	o	o	o	o
Beijerinckiaceae	*Rhodoblastus sphagnicola*	DSM 16996	F	o	o	o	o	o	o	o	o	o
Bradyrhizobiaceae	*Bradyrhizobium oligotrophicum*	S58	F	o	o	o	o	o	o	o	o	o
Bradyrhizobiaceae	*Rhodopseudomonas palustris*	DSM 126	F	o	o	/betA/ 3x /betB/	o	o	o	o	o	o
Bradyrhizobiaceae	*Rhodopseudomonas pseudopalustris*	DSM 123T	F	o	o		o	o	o	o	o	o
Hyphomicrobiaceae	*Blastochloris tepida*	GI	F	o	o		o	o	o	o	o	o
Hyphomicrobiaceae	*Blastochloris viridis*	DSM 133	F	o	o		o	o	o	o	o	o
Hyphomicrobiaceae	*Rhodomicrobium vannielii*	ATCC 17100	F	o	o		o	o	o	o	o	o
Hyphomicrobiaceae	*Rhodoplanes elegans*	DSM 11907	F	o	o		o	o	o	o	o	o
Aurantimonadaceae	*Fulvimarina pelagi*	HTCC2506	M	ectB/	o	betABCIproXW	/betT/ 3x	o	W1	W2	W3	o
Phyllobacteriaceae	*Hoeflea phototrophica*	DFL-43	M	o	o	betABCI proXbetIA /betB/	/betT/	o	o	W2	o	W4
Rhodobiaceae	*Afifella marina* 125/2	IM 162	M	o	o	proXbetIA /betB/	/betT/	o	W1	o	o	o
Rhodobiaceae	*Afifella marina* 125/4	IM 163	M	o	o	proXbetIA /betB/	/betT/	o	W1	o	o	o
Rhodobiaceae	*Afifella marina* 985, 126 (166)	DSM 2698	M	o	o	proXbetIA /betB/ proWV	/betT/	o	W1	o	W3	o
Rhodobiaceae	*Afifella pfennigii*	DSM 17143	M	o	o	proXbetIA /betB/	/betT/	o	W1	o	o	o
Rhodobiaceae	*Rhodobium orientis*	DSM 11290	M	o	o	betABIbetCpro XWV	/betT/	o	o	W2	W3	o

Table 2. Cont.

Gene Repertoire of Phototrophic Alphaproteobacteria for Osmotic Adaptation												
Rhodobacterales												
Rhodobacteraceae	Rhodobacter capsulatus	SB 1003	F	o	o	o	o	o	o	o	o	o
Rhodobacteraceae	Rhodobacter sphaeroides	ATCC 17025	F	o	o	betABIproXWV	o	o	W2	W3	o	o
Rhodobacteraceae	Rhodobacter veldkampii	DSM 11550	F	o	o	betABIproXWV	o	o	W2	W3	o	W4
Rhodobacteraceae	Rhodobaca barguzinensis	DSM19920	M	ectRABCask-ect	o	betAB>proXWV /betT/2x	o	o	W2	W3	o	o
Rhodobacteraceae	Rhodobaculum claviforme	GOR B7-4	M	ectRABCask-ect	o	o /betT/2x	o	o	o	o	o	o
Rhodobacteraceae	Rhodovulum adriaticum	DSM2781	M	ectRABCask-ect	o	o /betT/	o	o	W2	o	o	o
Rhodobacteraceae	Rhodovulum imhoffii	DSM 18064	M	ectRABCask-ect	o	o /betT/2x	o	o	o	o	o	o
Rhodobacteraceae	Rhodovulum sulfidophilum	IM 196	M	ectRABCask-ect	o	o /betT/	o	o	W2	o	o	o
Rhodobacteraceae	Rhodovulum sulfidophilum	DSM 2351	M	ectRABCask-ect	o	o /betT/	o	o	W2	o	o	o
Rhodobacteraceae	Rhodovulum sulfidophilum	DSM 1374	M	ectRABCask-ect	o	o /betT/	o	o	W2	o	o	o
Rhodobacteraceae	Roseisalinus antarcticus	DSM 11466	M	ectR-X-ectABCask-ect	o	betABNCIproXWV/betAT /betT/2x	o	o	o	W3	o	o
Rhodobacteraceae	Roseivivax isoporae	LMG 25204	M/H	ectRABCask-ect	o	betABC/<bet>proWX /betT/4x	o	o	W2	o	o	o
Rhodobacteraceae	Roseivivax halotolerans	JCM 10271	M/H	ectRABCask-ect	o	betABC /betI>proWV /betT/3x	o	o	o	o	o	o
Rhodobacteraceae	Roseivivax halodurans	JCM 10272	M/H	ectRABCask-ect	fusedMT-MAT	betABTC /betAT/bet>proWX /betT/3x	o	o	W2	o	o	o
Rhodobacteraceae	Roseitivax roseus	DSM 23042	H	ectRABCask-ect	fusedMT-MTHFR-MS-MAT-SAHase	betABCIproX/proWV /betT/5x	o	W1	o	o	o	W4

Table 2. Cont.

Gene Repertoire of Phototrophic Alphaproteobacteria for Osmotic Adaptation											
Rhodobacteraceae	*Rhodosalinus sediminis*	KCTC52478	H	ectRABCask-ect	fusedMT-MS-MAT	betABIproXWV	betT/	o	W2	W3	o
Rhodobacteraceae	*Roseovarius nitratireducens*	KTCT52967	M/H	ectRABCask-ect	fusedMT-MTHFR-MS-MAT-SAHase	betACINproXWV	betT/	o	W2	W3	W4
Rhodobacteraceae	*Roseovarius halotolerans*	DSM 29507	M/H	ectRABCask-ect	o	betABCIN proXWV	/betT/ 2x	W1	o	W3	W4
Rhodobacteraceae	*Roseovarius mucosus*	SMR3	M	ectRABCask-ect	o	betACIproXWV	/betT/ 2x	o	o	W3	W4
Rhodobacteraceae	*Roseovarius litoreus*	DSM 28249	M	ectRABCask-ect	o	betABCIN proXWV	/betT/ 2x	o	o	W3	W4
Rhodospirillales											
Acetobacteraceae	*Acidiphilium cryptum*	JF-5	F	ectRABCDask-ect	o	proXWVbetBA	o	o	o	W3	o
Acetobacteraceae	*Acidiphilium multivorum*	AIU301	F	ectRABCDask-ect	o	proXWVbetBA	o	o	o	W3	o
Acetobacteraceae	*Paracraurococcus ruber*	DSM 15382	F	o	o	o	o	o	W2	o	o
Acetobacteraceae	*Rhodopila globiformis*	DSM 161	F	o	o	o	o	o	o	o	o
Rhodospirillaceae	*Pararhodospirillum photometricum*	DSM 122	F	o	o	o	o	o	o	o	o
Rhodospirillaceae	*Phaeospirillum fulvum*	MGU-K5	F	o	o	o	o	o	o	o	o
Rhodospirillaceae	*Phaeospirillum molischianum*	DSM 120	F	o	o	o	o	o	o	o	o
Rhodospirillaceae	*Rhodospirillum rubrum*	DSM 1068	F	o	o	proXbetAB/betI/	o	o	o	o	o
Rhodospirillaceae	*Rhodospirillum rubrum 220*	DSM 107	F	o	o	proXbetAB/betI/	o	o	o	o	o
Rhodospirillaceae	*Rhodospirillum rubrum*	ATCC 11170	F	o	o	o	o	W1	o	o	o
Rhodospirillaceae	*Rhodospirillum rubrum*	FR1Mutante-IV	F	o	o	o	o	W1	o	o	o

Table 2. Cont.

Gene Repertoire of Phototrophic Alphaproteobacteria for Osmotic Adaptation

Family	Species	Strain								
Rhodospirillaceae	*Rhodovibrio salinarum*	DSM 9154	H	ectBC/2x ectA	GMT-B-DMT-MAT-SAHase	betABIproX	betTproX	proVWopuAC	o	o	W4
Rhodospirillaceae	*Rhodovibrio sodomensis*	DSM 9895	H	ectBC/2x ectA	GMT-B-DMT-MAT-SAHase	betABIproX	/betT/6x	o	W1	o	W4
Rhodospirillaceae	*Caenispirillum salinarum*	AK4	M	ectABCD/A/	fusedMT-MAT	betABIproXWV	/betT/	opuD	o	W3	o
Rhodospirillaceae	*Rhodospira trueperi*	ATCC 700224	M	ectABC	fusedMT-MAT	o	/betT/2x	o	W1	W2	o
Rhodospirillaceae	*Roseospira marina*	DSM 15113	M	ectABC	MAT-fusedMT	betIBAproX	/betT/2x	o	W1	W2	o
Rhodospirillaceae	*Roseospira navarrensis*	DSM 15114	M	ectABC	MAT-fusedMT	betIBAproX	/betT/	o	W1	o	W4
Rhodospirillaceae	*Roseospirillum parvum*	DSM 12498	M	ectRABC	MAT-fusedMT	o	/betT/2x	o	W1	W2	o
Rhodothalassiales											
Rhodothalassiaceae	*Rhodothalassium salexigens*	IM 261	H	o	GMT-DMT	o	/betT/3x	o	W1	o	o
Rhodothalassiaceae	*Rhodothalassium salexigens*	IM 265	H	o	GMT-DMT	o	/betT/3x	o	W1	o	o
Rhodothalassiaceae	*Rhodothalassium salexigens*	DSM 2132	H	o	GMT-DMT	o	/betT/3x	o	o	o	o
Sphingomonadales											
Erythrobacteraceae	*Erythrobacter litoralis*	DSM 8509	M	o	GMT-DMT	o	o	o	o	o	o

[a] see footnote [a] in Table 1; Color shades indicate different types of gene and gene associations, also marine (blue) and halophilic (rose) growth response of the bacteria. Color shades indicate different types of gene and gene associations: ectoine synthesis in rose, betaine synthesis from glycine in blue and from choline in green (in association with proU-W3 in pink; transport with betT (light-lila), opuA and D (shades of beige-brown), proU-W1 and proU-W2 shades of green, proU-W3 pink, proU-W4 yellow; also marine (blue) and halophilic (rose-pink) growth response of the bacteria.

Table 3. Genes and gene clusters of betaine and ectoine biosynthesis as well as relevant transport systems of phototrophic *Gammaproteobacteria* [a,b].

Gene repertoire of Phototrophic Gammaproteobacteria for Osmotic Adaptation				Ectoine	Betaine biosynthesis			Osmolyte transport		
Family	Species	Strain	Salt response	biosynthesis ectABC	from glycine GMT-DMT	from choline betAB		betT	opuA opuD	proW1
Cellvibrionales										
Halieaceae	*Congregibacter litoralis*	KT71	M	o	o	no genes of osmotic stress synthesis and transport				
Chromatiales										
Chromatiaceae	*Allochromatium vinosum*	MT86	F	o	o	o		o	o	o
Chromatiaceae	*Allochromatium vinosum*	DSM 180	F	o	o	o		o	o	o
Chromatiaceae	*Allochromatium warmingii*	DSM 173	F	o	o	o		o	o	o
Chromatiaceae	*Chromatium okenii* 6010	DSM 169	F	o	o	o		o	o	o
Chromatiaceae	*Chromatium weissei* IM 5910	DSM 5161	F	o	o	o		o	o	o
Chromatiaceae	*Lamprocystis purpurea*	DSM 4197	F	o	o	o		o	o	o
Chromatiaceae	*Thiocystis minor*	DSM 178	F	o	o	o		o	o	o
Chromatiaceae	*Thiocystis violacea*	DSM 207	F	o	o	o		o	o	o
Chromatiaceae	*Thiocystis violascens*	DSM 198	F	o	GMTtruncated, DMT	o		o	o	o
Chromatiaceae	*Thiocapsa imhoffii*	DSM 21303	F	o	o	o		betT/	o	o
Chromatiaceae	*Thiocapsa roseopersicina*	DSM 217	F	o	o	o		betT/	o	o
Chromatiaceae	*Thiocapsa marina* 5811	DSM 5653	M	o	GMT-DMT-betT	o			opuAA,AB,AC-N-GMT,DMT,betT	o
Chromatiaceae	*Marichromatium gracile* 5210	DSM 203	M	o	GMT-DMT	betBAT		betT/2x	opuAA,AB,AC-N-opuAC-NN-betT	o

93

Table 3. Cont.

Gene repertoire of Phototrophic Gammaproteobacteria for Osmotic Adaptation									
Chromatiaceae	Marichromatium purpuratum 984	DSM 1591	M	o	GMT-DMT	o	betT/	opuAA,AB,AC,N,AC,AC,NN,betT	o
Chromatiaceae	Thiocystis violacea	DSM 208	M	o	GMT-DMT	o	betTopuAC	opuAA,AB,AC	o
Chromatiaceae	Imhoffiella purpurea AK35	AK35	M	o	GMT-DMT	o	betTopuAC	opuAA,AB,AC	o
Chromatiaceae	Thiorhodococcus drewsii	AZ1	M	o	GMT-DMT	betBAT/betAB	betT-opuAC/betT	opuAA,AB,AC	o
Chromatiaceae	Thiorhodococcus mannitoliphagus	DSM 18266	M	o	GMT-DMT	o	betT-opuAC/betT		W1
Chromatiaceae	Thiorhodococcus minor	DSM 11518	M	o	GMT-DMT	o	betT/ 4x	o	W1-N-betT
Chromatiaceae	Thioflavicoccus mobilis 8321	ATCC700959	M	o	GMT-DMT/GMT-B	o	betT/ 2x	o	W1
Chromatiaceae	Thiococcus pfennigii 4252	DSM 228	M	o	GMT-DMT	o	betT/ 2x	o	W1
Chromatiaceae	Thiococcus pfennigii 4254	Pfennig 8320	M	o	GMT-DMT	o	betT/ 3x	proVW1-opuAC	
Chromatiaceae	Thiorhodovibrio winogradskyi	06511	M	o	GMT-DMT	o	betT/ 3x	proVW1-opuAC	
Chromatiaceae	Rhabdochromatium marinum	DSM 5261	M	o	GMT-DMT/GMT-B	o	betT/ 3x	o	W1
Chromatiaceae	Lamprobacter modestohalophilus	DSM 25653	M	o	GMT-DMT	o	betT/ 3x	o	W1
Chromatiaceae	Halochromatium roseum	DSM 18859	M	o	GMT-DMT	o	betT/ 2x	o	
Chromatiaceae	Halochromatium glycolicum	DSM 11080	H	o	GMT-DMT/GMT-B	o	betT/ 4x	W1-NN-betTopuAC	
Chromatiaceae	Halochromatium salexigens IM6310	DSM 4395	H	o	GMT-DMT/GMT-B	o	betT/ 2x	W1-N-betTopuAC-NN-betT	
Chromatiaceae	Thiohalocapsa halophila IM4270	DSM 6210	H	o	GMT-DMT	o	betT/ 5x	o	W1
Ectothiorhodospiraceae	Ectothiorhodospira mobilis	DSM 237	M	o	GMT-DMT-MAT/GMT-B	o	betT/	betT-opuAA,AB,AC	o

Table 3. *Cont.*

Gene repertoire of Phototrophic Gammaproteobacteria for Osmotic Adaptation

Family	Species	Strain	M/H	ect	GMT pathway	betABIproX	betT	bet-opuAA,AB,AC	OpuD/W1
Ectothiorhodospiraceae	*Ectothiorhodospira marismortui*	DSM 4180T	M/H	o	GMT-DMT-MAT/GMT-B	o	betT/	betT-opuAA,AB,AC	o
Ectothiorhodospiraceae	*Ectothiorhodospira marina*	DSM 241T	M/H	o	GMT-DMT-MAT/GMT-B	betABIproX	betT	betT-opuAA,AB,AC	o
Ectothiorhodospiraceae	*Ectothiorhodospira haloalkaliphila*	ATCC 51935	M/H	o	GMT-DMT-MAT/GMT-B	betABIproX	betT/2x	betT-opuAA,AB,AC	OpuD
Ectothiorhodospiraceae	*Ectothiorhodospira* sp.	BSL-9	M	o	GMT-DMT-MAT/GMT-B	betABIproX	betT/	betT-opuAA,AB,AC	o
Ectothiorhodospiraceae	*Ectothiorhodospira shaposhnikovii*	DSM 243	M	ectB/	GMT-DMT-MAT	betABIproX		betT-opuAA,AB,AC	o
Ectothiorhodospiraceae	*Ectothiorhodospira magna* B7-7	DSM 22250	M	o		o	betT/	o	W1
Ectothiorhodospiraceae	*Ectothiorhodosinus mongolicus* M9	DSM 15479	M	o		o	betT/	o	o
Ectothiorhodospiraceae	*Thiorhodospira sibirica*	ATCC 700588	M	o		o	o	o	o
Ectothiorhodospiraceae	*Halorhodospira abdelmalekii*	DSM 2110	H	ectAB/C	GMT-DMT-MAT-SAHase-MTHFR	o	see proU	opuD/2x	W1-betT
Ectothiorhodospiraceae	*Halorhodospira halochloris*	DSM 1059	H	ectABC	GMT-DMT-MAT-SAHase-MTHFR	o	betT/2x	opuD	o
Ectothiorhodospiraceae	*Halorhodospira halophila*	IM 9626	H	ectABC	GMT-DMT-MAT-SAHase-MTHFR/GMT-B	o	betT/5x	opuD	W1
Ectothiorhodospiraceae	*Halorhodospira halophila* SL1	DSM 244	H	ectABC	GMT-DMT-MAT-SAHase-MTHFR/GMT-B	o	betT/4x	opuD	W1-N-betT

Table 3. Cont.

Gene repertoire of Phototrophic Gammaproteobacteria for Osmotic Adaptation									
Ectothiorhodospiraceae	Halorhodospira halophila D	IM 9620	H	ectABC	GMT-DMT-MAT-SAHase-MTHFR/GMT-B	o	betT/5x	opuD	W1-N-betT
Ectothiorhodospiraceae	Halorhodospira halophila 51/3	IM 9630	H	ectABC	GMT-DMT-MAT-SAHase-MTHFR/GMT-B	o	betT/5x	opuD	W1
Ectothiorhodospiraceae	Halorhodospira neutriphila	DSM 15116	H	ectABC	GMT-DMT-MAT-SAHase-MTHFR/GMT-B	o	betT/3x	o	W1

[a] footnote as [a] in Table 1; [b] proW2, proW3, and proW4 are absent from Gammaproteobacteria of this study; Color shades indicate different types of gene and gene associations: ectoine synthesis (rose), betaine synthesis from glycine (blue) and from choline (green); transport with betT (light-lila), opuA and D (shades of beige-brown), proU-W1 (green); also marine (blue) and halophilic (rose-pink) growth response of the bacteria.

The ability to produce betaine or ectoine from the currently known routes is absent from the freshwater bacteria examined here, which include *Heliobacteria*, *Chloracidobacterium*, *Chloroflexi*, and the majority of *Cyanobacteria* and *Chlorobiaceae*, as well as phototrophic *Betaproteobacteria* (Table 1). Most of these bacteria also lack corresponding transport systems. Both pathways are also absent from the phylogenetic groups of freshwater phototrophic species of *Rhizobiales*, *Acetobacteraceae*, *Rhodospirillaceae*, *Betaproteobacteria*, and *Chromatiaceae* (Tables 1–3). In all freshwater bacteria that lack any of the options to accumulate betaine or ectoine, a limited osmotic adaptation might be achieved by accumulation of sugars, in particular trehalose and sucrose or glucosylglycerol, but also N-acetyl-glutaminylglutamine amide and N-carbamoyl-L-glutamine amide (Severin et al., 1992). Even potassium glutamate to some extent might contribute to osmotic adaptation [1,6,7].

3.1.1. Cyanobacteria

It was demonstrated that the salt-tolerance of *Cyanobacteria* was clearly related to the compatible solutes accumulated, those of the lowest tolerance (freshwater strains growing below 0.7 M NaCl) accumulate sucrose and trehalose, those of moderate tolerance (marine strains growing up to 1.8 M NaCl) accumulate glucosylglycerol, and those with the highest tolerance (strains of marine and hypersaline origin, in great majority classified as *Synechococcus* strains) accumulate glycine betaine or glutamate betaine [11]. Apparently, the majority of cyanobacteria adapted to marine environments count on glucosylglycerol as osmoticum [43–45]. The formation of a fused molecule with glycerol (glucosyl glycerol) as a component appears to be a clever strategy to keep at least, in part, the excellent compatible nature of glycerol but reduce leakage through the cell membrane. Those *Cyanobacteria* originating from salt lakes and hypersaline ponds (tolerance of >20% NaCl) accumulate betaine, which apparently is essential to provide sufficient protection at moderately and extremely high salt concentrations [11].

Betaine biosynthesis is found in two major phylogenetic branches of *Cyanobacteria*, but ectoine biosynthesis is absent (Table 1, Figure 1). One branch is formed by *Aphanothece halophytica* (Halothece PCC7418), together with other *Chroococcales*. *Aphanothece halophytica* is a characteristic inhabitant of hypersaline environments and among the most halotolerant of *Cyanobacteria*. A gene cluster of the two methyltransferases (GMT and DMT) is present in these bacteria (Table 1). The BetT present in *Aphanothece halophytica*, which lacks choline-dependent betaine synthesis, is characterized as a specific transporter of betaine [18]. A second major branch of *Cyanobacteria* includes representatives of the heterogeneous groups of the *Synechococccus* and *Prochlorococcus* species and is phylogenetically quite distinct from the first branch (Figure 1). These Cyanobacteria have *proXWV* genes included in a cluster with GMT and DMT genes (Table 1).

The two groups are also distinguished by significant difference in the G + C content of the DNA. Representative strains of the *Prochlorococcus* group have a G + C content near 50 mol% for the *Prochlorococcus marinus* (strain MIT9313: 50.7 and strain MIT9303: 50.0%) and near 59% for *Synechococcus* (strain WH8102: 59.4% and strain WH8103: 59.5%). Much lower values are present in the *Halothece/Aphanothece* group: 42.9 mol% in *Aphanothece halophytica*, 41.1 mol% in *Euhalothece natronophila*, and 42.4 mol% in *Dactylococcopsis salina*. Although systematically assigned to the *Synechococcales*, *Dactylococcopsis salina* fits very well into the *Halothece* group, according to the gene repertoire and the phylogeny of the GMT sequence (Figure 1, Table 1) as well as the G + C content, which puts a question mark to its current taxonomic affiliation.

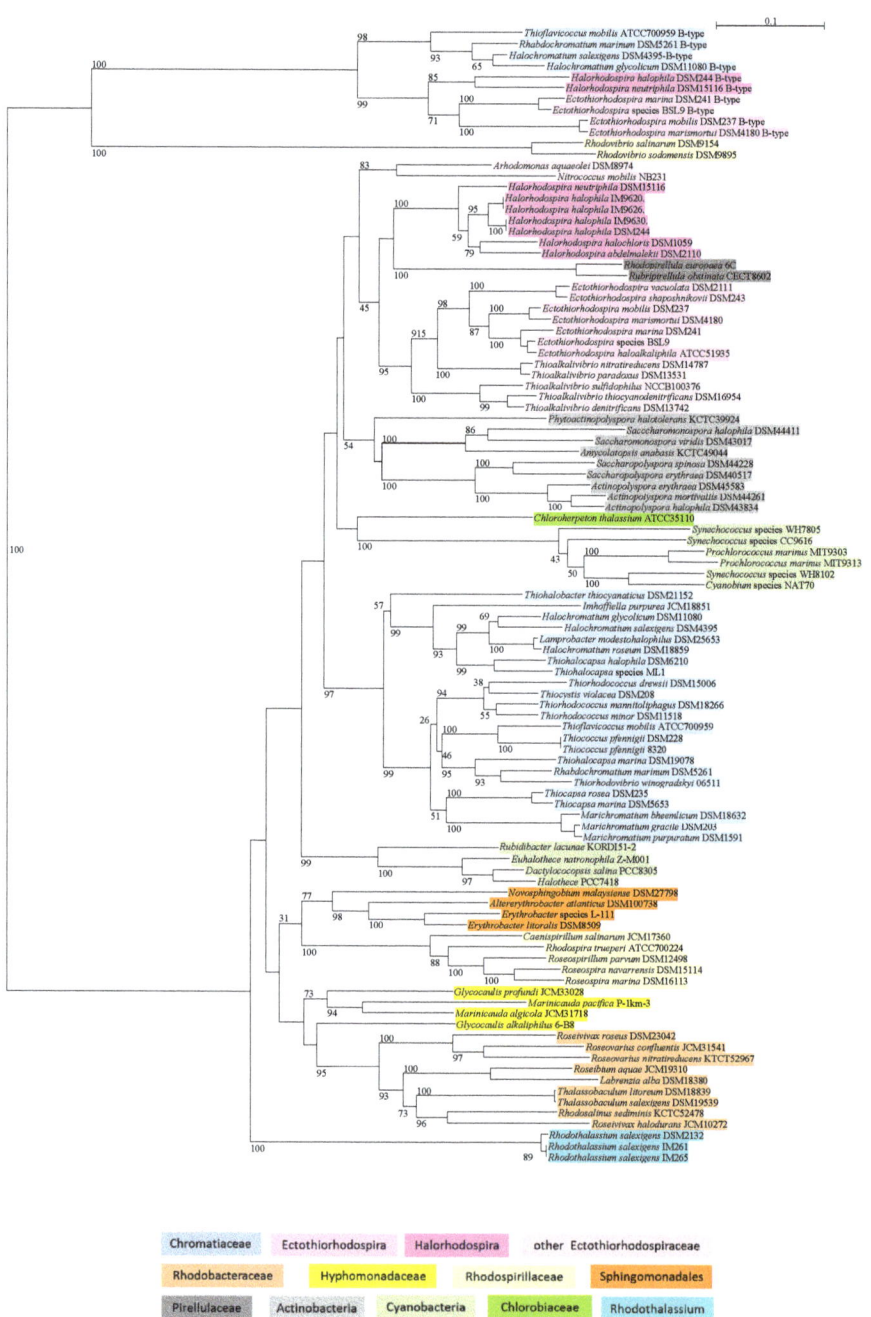

Figure 1. The phylogeny of betaine biosynthesis based on the sequences of glycine/sarcosine methyltransferase GMT is shown in a neighbor-joining tree. Sequences and gene bank accession numbers are shown in the Supplementary Table S1. Bootstrap values expressed as percentages of 1000 replications are given at the branches. The bar indicates an evolutionary distance of 0.1. The following color code highlights the different systematic groups.

3.1.2. Chlorobiaceae

As most of the green sulfur bacteria thrive in freshwater habitats, it is not surprising that they lack possibilities of synthesis of betaine and ectoine. The marine *Chloroherpeton thalassium* is the only one that can produce betaine from glycine (Table 1 and Figure 1). However, members of the genus *Prosthecochloris* and in particular *Prosthecochloris aestuarii* are also regularly found in brackish and marine coastal or saline habitats. They might cope with marine concentrations of salt by accumulation of trehalose, as shown to occur in *Prosthecochloris vibrioformis* DSM 260 (Pfennig 6030) and *Chlorobaculum thiosulfatophilum* DSM 249 (Pfennig 6230) [13]. If grown in marine media supplied with betaine, they can also accumulate the betaine [13]. Therefore, they have limited possibilities to cope with salt stress by accumulation of trehalose and uptake of betaine from the environment, to thrive at elevated salt concentrations. It is expected that they take advantage of betaine uptake when occurring in hypersaline habitats. A gene cluster annotated as *proVopuAB,AC* in *Prosthecochloris vibrioformis* DSM 260 (Table 1) presumably is a betaine transport system. The sequence of OpuAB is clearly distinct from those found in the *Chromatiaceae* and *Ectothiorhodospira* species, as well as from the ProW sequences found in other *Chromatiaceae* and *Halorhodospira* species and its ProV sequence is an outsider of the OpuAA branch (data not shown). This indicates that components of the transport system of *Prosthecochloris vibrioformis* might be related to an ancient ancestor of both the OpuA and the ProU transport systems, both of which might have evolved from a similar ancient ancestor. It would be interesting to study their catalytic properties and the evolutionary path of both systems.

3.1.3. Chloroflexi

Genes for the synthesis of ectoine and betaine were not found in the *Chloroflexus* and *Roseiflexus* species (Table 1). Additionally, transport systems for osmotica are absent from the *Chloroflexus* species, which characterizes them as strict freshwater bacteria. An ABC transport system (annotated as *opuCBproXV* and an ABC transport protein) found in the *Roseiflexus castenholzii* and *Roseiflexus* species RS-1 (Table 1) might as well represent an ancient evolutionary form of osmolyte transport and betaine uptake from the environment. Thereby, the *Roseiflexus* species might gain limited possibility for osmotic adaptation. As these bacteria are among the oldest mat-forming phototrophic bacteria and *Roseiflexus* might be able to take up betaine or other osmolytes, they or relatives thereof are expected to be found in marine microbial mats. It would be interesting to see, whether marine relatives have possibilities of compatible solute biosynthesis. The species known and characterized so far are expected to have lost such possibilities.

3.1.4. Marine Rhizobiales

Freshwater *Rhizobiales* (species of *Rhodoblastus*, *Rhodopseudomonas*, *Blastochloris*, *Rhodomicrobium*, and *Rhodoplanes*) lack both, biosynthesis genes for ectoine and betaine (from choline and glycine) as well as transport systems for these osmolytes (Table 2). In a group of marine *Rhizobiales* (*Fulvimarina pelagi*, *Hoeflea phototrophica*, *Rhodobium orientis*, *Afifella* species), the choline-dependent biosynthesis of betaine (BetABI) and a BetT transporter are present (Table 2). In addition, one or more copies/versions of the ProU system are found. Therefore, osmotic adaptation of these bacteria to the marine environment can be achieved by uptake of betaine or choline. In the absence of these compounds in the environment, other compatible solutes might be accumulated. For *Afifella marina* (formerly *Rhodopseudomonas marina*), the accumulation of trehalose was demonstrated [12].

3.1.5. Marine and Halotolerant Rhodobacteraceae

The *Rhodobacteraceae* include freshwater, marine and halophilic species. With the exception of the freshwater *Rhodobacter* species, all of them have a complete gene cluster for ectoine biosynthesis (*ectRectABCask-ect*). In addition, the extremely halotolerant *Roseivivax halodurans*, *Roseivivax roseus*, *Rhodosalinus sediminis*, and *Roseovarius nitratireducens* (not *Roseivivax halotolerans* and *Roseovarius halotolerans*) have genes of betaine biosynthesis from glycine, with gene clusters of varying composition (Table 2). In these bacteria, the GMT and DMT methyltransferase genes are fused, as in some of the halophilic *Rhodospirillaceae* (see below). Most of the *Rhodobacteraceae* have several transport systems for osmolytes. All have the BetT transport system, some strains have multiple copies. With the exception of *Rhodobaculum claviforme* and *Rhodovulum imhoffii*, they have several ProU systems, which according to the ProW sequences are phylogenetically distinct (Table 2 and see below under Section 3.5).

3.1.6. Acetobacteraceae

Phototrophic Acetobacteraceae are freshwater bacteria. *Rhodopila globiformis* lacks possibilities of synthesis and transport of betaine and ectoine, and *Paracraurococcus ruber* depends on external supply of such compounds and on transport via a ProU system (type W2) for osmotic adaptation (Table 2). *Acidiphilium* species, however, can produce ectoine and hydroxyectoine and in addition have a ProU (type W3) uptake system associated with the genes of betaine biosynthesis from choline (*proXWVbetBA*). *Acidiphilium* species are adapted to life in acidic freshwater habitats and their acidophilic nature should preclude their development in neutral/basic marine habitats. Therefore, the accumulation of ectoine and possibly also betaine might play a role in adaptation to highly acidic conditions rather than to saline habitats.

3.1.7. Marine and Halophilic Rhodospirillaceae

Freshwater species of *Rhodospirillaceae* including species of *Pararhodospirillum*, *Rhodospirillum*, and *Phaeospirillum* lack ectoine and betaine biosynthesis (from glycine) and only some strains of *Rhodospirillum rubrum* can synthesize betaine from choline or have a ProU transport system (Table 2). On the other hand, marine and halophilic *Rhodospirillaceae* (*Rhodospira*, *Rhodovibrio*, *Roseospira*, *Roseospirillum*, and *Caenispirillum* species) are genomically well equipped with possibilities of osmotic adaptation. These species are adapted to moderately and extremely high salt concentrations. In particular, the *Rhodovibrio* species tolerate more than 3 M (up to 20%) NaCl. All of them synthesize ectoine (*Caenispirillum salinarum* also hydroxyectoine) and betaine. With the exception of *Rhodospira trueperi* and *Roseospirillum parvum*, they can also transform choline to betaine (*betABI*). Different to other *Rhodospirillaceae*, in the *Rhodovibrio* species, the GMT and DMT genes are not fused and form a gene cluster with MAT and SHAase. Their GMT gene is the only example of a B-type GMT gene in *Alphaproteobacteria*, and the *ectA* gene is not included in an *ectABC* cluster, as in almost all other phototrophic bacteria producing ectoine. In addition to *betT* (multiple), marine *Rhodospirillaceae* have one or more ProU transport systems. These are related to the *proVWX* type W1 (*proU* of *E. coli*), *proXWV* (type W2), or *proXVW* (type W4) systems. In *Caenispirillum salinarum*, a type W3 ProW system is present (Table 2).

3.1.8. Halophilic *Rhodothalassium salexigens*

Rhodothalassium salexigens is a moderate halophilic and especially salt tolerant bacterium that grows at salt concentrations exceeding 3 M (20% NaCl). The outstanding properties of *Rhodothalassium salexigens* as distinct from all other phototrophic *Alphaproteobacteria* are demonstrated by sequences of the 16S rRNA gene and of the photosynthesis reaction center and bacteriochlorophyll biosynthesis genes [46]. These are in line with its recognition as a separate genus, family and order of the *Alphaproteobacteria* [47,48]. Ectoine biosynthesis is absent (Table 2). Betaine biosynthesis from glycine is possible and GMT sequences form a distinct lineage among those of the *Alphaproteobacteria* (Figure 1). In addition, BetT and a type W1 ProU transport systems are present.

3.1.9. Chromatiaceae

Freshwater Chromatiaceae including *Chromatium*, *Allochromatium*, and *Thiocystis* species lack betaine and ectoine biosynthesis and corresponding transport systems (Table 3). Marine and halophilic Chromatiaceae can synthesize betaine from glycine and have the

BetT transport system, but lack ectoine biosynthesis. With the exception of *Marichromatium gracile* and *Thiorhodococcus drewsii*, betaine biosynthesis from choline is absent. The group of marine representatives including the *Marichromatium, Thiorhodococcus,* and *Imhoffiella* species, *Thiocapsa marina* and *Thiocystis violacea* DSM 208 has an *opuA* gene cluster, but lacks the ProW permease protein of ProU. An exception is *Thiorhodococcus minor*, which has a ProU (type W1) instead of the OpuA system (Table 3) like all other marine and halophilic Chromatiaceae.

3.1.10. Ectothiorhodospiraceae

According to phylogeny of 16S rRNA and photosynthetic reaction center genes, the *Ectothiorhodospira* and *Halorhodospira* species form two clearly separated branches that might even require a separation at the family level [46]. This clear separation is also reflected in different options for osmotic adaptation and in different lineages of normal as well as B-type GMT sequences (Table 3, Figure 1, see Section 3.2.2).

The *Ectothiorhodospira* species lack ectoine biosynthesis, but can synthesize betaine from glycine (GMT-DMT-MAT), with the exception of *Ectothiorhodospira mobilis* and *Ectothiorhodospira marismortui* also from choline (*betABIproX*). They have a gene cluster including genes of the OpuA and BetT transport systems (*betT-opuAA,AB,AC*) and additional *betT* gene copies (except *Ectothiorhodospira vacuolata* and *Ectothiorhodospira shaposhnikovii*), but lack the otherwise common ProU transporter (Table 3). An exception is the *Ectothiorhodospira magna*, which obviously lacks the biosynthetic capabilities and entirely depends on the uptake of betaine and ectoine (*betT-opuAA,AB,AC*) (Table 3). *Ectothiorhodosinus mongolicus* and *Thiorhodospira sibirica* are exceptions among the Ectothiorhodospiraceae. While the first lacks genes for biosynthesis of betaine and ectoine and only has ProU (type W1) and BetT transport systems, the latter has no annotated osmotic stress genes at all. This limits the possible adaptation to elevated salt concentrations and suggests that alternative mechanisms/solutes are used to cope with the salt in the environment. Sucrose, N-carbamoyl-L-glutamine amide, or N-acetyl-glutaminylglutamine amide are possible candidates that were found to accumulate in other purple sulfur bacteria, though in marine and halophilic species, they are found only in addition to betaine [12].

The *Halorhodospira* species are the most halophilic and halotolerant phototrophic bacteria and can thrive even in saturated salt solutions [1,2]. All *Halorhodospira* species have complete gene clusters for ectoine (*ectABC*) and betaine biosynthesis from glycine, but lack genes for transformation of choline to glycine (Table 3). They are the only Gammaproteobacteria to include adenosylmethionine synthetase (MAT), adenosylhomocysteinase (SAHase) and 5,10-methylene tetrahydrofolate reductase (MTHFR) into a gene cluster, together with the two methyltransferase genes (GMT-DMT-MAT-SAHase-MTHFR). They are well equipped with the transport systems for osmolytes and have multiple copies of BetT, the secondary transporter OpuD (except *Halorhodospira neutriphila*) and the ProU (type W1) transport system (except *Halorhodospira halochloris*). These transport systems assure that osmolytes leaking out of the cells at very high, several molar cytoplasmic concentrations can be regained by the cells, and are not wasted to the environment. Therefore, it is assumed that the available transport systems are able to take up betaine and ectoine.

3.2. *Phylogeny of Glycine-Methyltransferase GMT*

The methyltransferases that transform glycine to betaine, glycine and sarcosine methyltransferase (GMT), and dimethylglycine methyltransferase (DMT) are present in a wide range of phototrophic bacteria, *Alpha-*, and *Gammaproteobacteria, Cyanobacteria* and green sulfur bacteria, as revealed by genome analysis using the SEED facility of the RAST platform [39]; shown in Tables 1–3. A phylogenetic tree of GMT methyltransferases, which in addition to phototrophic bacteria includes data from the genomes of chemotrophic bacteria as well as from BLAST searches, is shown in Figure 1. Though many of the deep branching points are poorly resolved and not supported by bootstrap values, it is obvious that the phylogeny of GMT is well depicted in a number of major phylogenetic branches. The phylogenetic grouping correlates well with differences in the gene clusters involved in betaine biosynthesis (Figure 1). The following major groups and distinct phylogenetic lineages are recognized.

3.2.1. Chromatiaceae

Among the phototrophic *Gammaproteobacteria*, marine and halophilic phototrophic *Chromatiaceae* and *Ectothiorhodospiraceae* species have the ability to produce betaine from glycine. Marine and halophilic species encoding this pathway of betaine biosynthesis are found in different phylogenetic lineages with (a) the halophilic species of the genus *Halochromatium, Lamprobacter modestohalophilus, Thiohalocapsa halophila* forming one branch and (b) the marine species of *Marichromatium, Thiocapsa, Thiorhodococcus, Thiorhodovibrio,* and *Thiocystis violacea* DSM 208, *Rhabdochromatium marinum*, together with *Thiococcus pfennigii* and *Thioflavicoccus mobilis*, forming a second one (Figure 1). In both of these branches of *Chromatiaceae*, including the separate lineages of *Thiohalobacter thiocyanaticus* and *Imhoffiella purpurea*, just the two methyltransferases (GMT and DMT) form a distinct gene cluster (Table 3).

3.2.2. Ectothiorhodospiraceae

According to GMT sequences, species of *Halorhodospira* and *Ectothiorhodospira* form two clearly separated groups (Figure 1). In *Ectothiorhodospira*, species including *Thioalkalivibrio nitratireducens* and related species, GMT and DMT genes form a cluster together with the gene encoding S-adenosylmethionine synthetase (methionine adenosyl transferase, MAT), which is essential for performance of the methylation by providing the methyl donor S-adenosylmethionine. In the *Halorhodospira* species, in addition, the genes encoding S-adenosyl homocysteinase (SAHase) and 5,10-methylene tetrahydrofolate reductase (MTHFR) are included in this gene cluster (Table 3). The methionine synthase (MS) that completes the methionine cycle is located at a different location within the genome. The coordinated action of these enzymes is expected to allow optimal performance of betaine biosynthesis by providing the essential methyl groups and removing the byproduct S-adenosylhomocysteine, which strongly inhibits the reaction [15]. The chemotrophic *Arhodomonas aquaeolei* and *Nitrococcus mobilis* (only GMT-DMT cluster) form a distinct subbranch within the *Ectothiorhodospiraceae*.

There is a curiosity with the presence of a second additional single GMT gene within a few species of *Chromatiales*, which is phylogenetically distinct from the genes commonly found in other phototrophic bacteria, except the two *Rhodovibrio* species (Figure 1). We refer to these genes as the B-type methyltransferases in betaine biosynthesis, compared to the "common" system. GMT sequences of this B-type group form three lineages of phototrophic Gammaproteobacteria, (i) marine and halophilic *Chromatiaceae* species (*Halochromatium salexigens, Halochromatium glycolicum, Rhabdochromatium marinum* and *Thioflavicoccus mobilis*), (ii) *Halorhodospira* species (*Halorhodospira halophila, Halorhodospira neutriphila*), and (iii) *Ectothiorhodospira* species (*Ectothiorhodospira mobilis, Ectothiorhodospira marismortui, Ectothiorhodospira marina*) (Figure 1). In contrast, the B-type GMT genes of *Rhodovibrio sodomensis* and *Rhodovibrio salinarum* are the only ones for biosynthesis of betaine in these bacteria that are unique among *Alphaproteobacteria*. While this gene is included in a functional gene cluster and is quite likely active in the betaine synthesis of the *Rhodovibrio* species (GMT-DMT-MAT-SAHase), its role in the *Chromatiales* is unclear and it might represent an evolutionary relict or a backup.

3.2.3. Cyanobacteria

Clearly two distinct branches of common GMT sequences of betaine biosynthesis are found in *Cyanobacteria* (Figure 1). The most divergent branch in the tree is represented by the *Prochlorococcus* group, including *Synechoccoccus* WH8102 and *Prochlorococcus marinus*. The second major phylogenetic branch of *Cyanobacteria*, the *Halothece/Aphanothece* group, is represented by the most prominent member of halophilic *Cyanobacteria, Aphanothece halophytica* (*Halothece* PCC7418), and several *Chroococcales*, including *Euhalothece natronophila* and *Rubidibacter lacunae*, but also *Dactylococcopsis salina*, which might be misclassified as a member of the *Synechococcales*.

3.2.4. Alphaproteobacteria

The GMT sequences of *Alphaproteobacteria* represent a diverse major branch, which includes the species of *Sphingomonadales, Rhodothalassiales, Rhodobacterales,* and *Rhodospirillales,* but no *Rhizobiales* (Figure 1). The following distinct subbranches of *Alphaproteobacteria* are formed:

- *Rhodospirillaceae*, including *Caenispirillum salinarum, Rhodospira trueperi, Roseospira navarrensis, Roseospira marina,* and *Roseospirillum parvum* (but not the *Rhodovibrio* species) have a fused GMT/DMT gene of the methyltransferases included in a small cluster with S-adenosylmethionine synthase (MAT).
- *Rhodothalassium salexigens*, which represents the most distant line to all other *Alphaproteobacteria*, has a small gene cluster with just the two methyltransferases.
- *Sphingomonadales* include *Novosphingobium malayensis, Erythrobacter litoralis, Altererythrobacter atlanticus*; in *Erythrobacter litoralis* just the two methyltransferases form a small gene cluster.
- *Hyphomonadaceae* include the *Glycocaulis* and *Marinicauda* species with a small gene cluster of the two methyltransferases only.
- *Rhodobacteraceae* included in the study, as indicated in Figure 1, form a subbranch together with *Thalassobaculum litoreum* and *Thalassobaculum salexigens* (according to 16S rRNA phylogeny forming a branch with the *Oceanobaculum* species at an almost equal distance to the *Rhodobacteraceae* and *Rhodospirillaceae* species; data not shown). A fused GMT/DMT gene is associated with the MAT gene in *Roseivivax halodurans*, with the MS-MAT genes in *Rhodosalinus sediminis*, and with the MTHFR-MS-MAT-SAHase genes in *Roseivivax roseus* and *Roseivarius nitratireducens* (Table 2).

3.2.5. Actinobacteria

The Actinobacteria are the only major group of chemotrophic bacteria that show betaine biosynthesis from glycine. They form a distinct branch distantly related to *Chromatiales* and *Chloroherpeton* (Figure 1). In *Actinopolyspora halophila*, for example, a fused GMT/DMT gene is included in a gene cluster with MAT, MTHFR, and SAHase (Table 4).

3.2.6. Chlorobiaceae

Among the green sulfur bacteria, betaine biosynthesis is found only in *Chloroherpeton thalassium*. The GMT sequence forms a separate line that associates distantly with those of the halophilic *Chromatiales* and *Actinobacteria* (Figure 1). Both methyltransferase genes are found in a small cluster together with the BetT transport system (Table 1).

3.2.7. Pirellulaceae

The two methyltransferases of betaine biosynthesis (GMT, DMT) are also found in the chemotrophic *Rhodopirellula europaea* and *Rubripirellula obstinata* (*Pirellulaceae*). Phylogenetically, these GMT sequences form a distinct branch distantly associated with *Ectothiorhodospiraceae*, though with low confidence. In *Rhodopirellula europaea*, the *betT* gene forms a cluster with the two methyltransferases (GMT-DMT-betT).

3.3. Phylogeny of EctB

The ability for ectoine biosynthesis is found in several phylogenetic distant lineages of anoxygenic phototrophic bacteria. Two major phylogenetic branches (type-1 and type-2 EctB sequences) can be distinguished, which also show differences in the *ect* gene cluster structure (Figure 2).

Table 4. Ectoine and betaine biosynthesis of selected chemotrophic bacteria [a].

Family	Species	Strain	Ectoine biosynthesis ectABC	Betaine from glycine GMT-DMT	biosynthesis from choline betAB
Actinobacteria					
Actinopolysporaceae	Actinopolyspora halophila	DSM 43834	ectABC	fusedMT-MAT-o-MS-SAHase-MTHFR	o
Gordoniaceae	Gordonia alkanivorans	NBRC16433	ectABC	o	betABT
Mycobacteriaceae	Mycolicibacterium thermoresistibile	DSM 44167	ectABCD	o	betA/betB
Streptomycetaceae	Streptomyces clavuligerus	ATCC 27064	ectABCD/AB	o	betAB/betI/betT
Bacilli/Bacillales					
Bacillaceae	Halobacillus halophilus	DSM 2266	ectABC/D	o	betABbetIopuAC
Proteobacteria/Alphaproteobacteria					
Rhodobacteraceae	Jannaschia seosinensis	CECT7799	ectRectABCask-ect	o	proVWXbetICBA
Rhodobacteraceae	Maritimibacter alkaliphilus	HTCC2654	ectRectABCask-ect	o	proVWooXbetICBA
Rhodobacteraceae	Paracoccus halophilus	JCM 14014	ectRectABCask-ect	o	proVWXbetIBA
Rhodobacteraceae	Salipiger bermudensis	HTCC2601	ectRectABCask-ect	o	proVWXbetICBA
Rhodobacteraceae	Thioclava pacifica	DSM 10166	ectRectABCask-ect	o	proVWoXbetIBA
Rhodospirillaceae	Ferruginivarius sediminum	WD2A32	ectRectABC/B/C/D	o	proXbetIBA
Proteobacteria/Betaproteobacteria					
Alcaligenaceae	Achromobacter xylosoxidans	SOLR10	ectRABCD	o	betA/betB
Alcaligenaceae	Bordetella avium	197N	ectRABCD	o	betA/
Burkholderiaceae	Paucimonas limoigeni	DSM 7445	ectRABCD	o	betA/
Oxalobacteraceae	Herminiimonas arsenicooxidans	DSM 17148	ectRABCD	o	betA/
Proteobacteria/Gammaproteobacteria					
Halomonadaceae	Halomonas elongata	DSM 2581	ectABC/D	o	proXbetIBA
Halomonadaceae	Chromohalobacter salexigens	DSM 3043	ectABC/D	o	proXbetIBA
Halomonadaceae	Chromohalobacter marismortui	DSM 6770	ectABC/D	o	proXbetIBA
Haliaceae	Haliea salexigens	DSM 19537	ectRectABCask-ect	o	betA/
Chromatiaceae	Nitrosococcus oceani	ATCC 19707	ectRCask-ect/ectABD	o	betA/
Chromatiaceae	Nitrosococcus halophilus	Nc4	ectRCask-ect/ectABD	o	o
Hahellaceae	Hahella chejuensis	KCTC2396	ectABCD	o	betIBAproXWV
Oceanospirillaceae	Marinomonas mediterranea	MMB1	ectABC/R/ask-ect	o	betAB / betIproXWV
Ectothiorhodospiraceae	Nitrococcus mobilis	NB-231	ectABC	GMT-DMT	betA/
Chromatiaceae	Acidihalobacter prosperus	DSM 14174	ectABC	o	o

[a] Gene clusters are given and "/" denotes a separate locus of the genes in the genome. Color shades indicate different types of gene and gene associations, also marine (blue) and halophilic (rose) growth response of the bacteria.

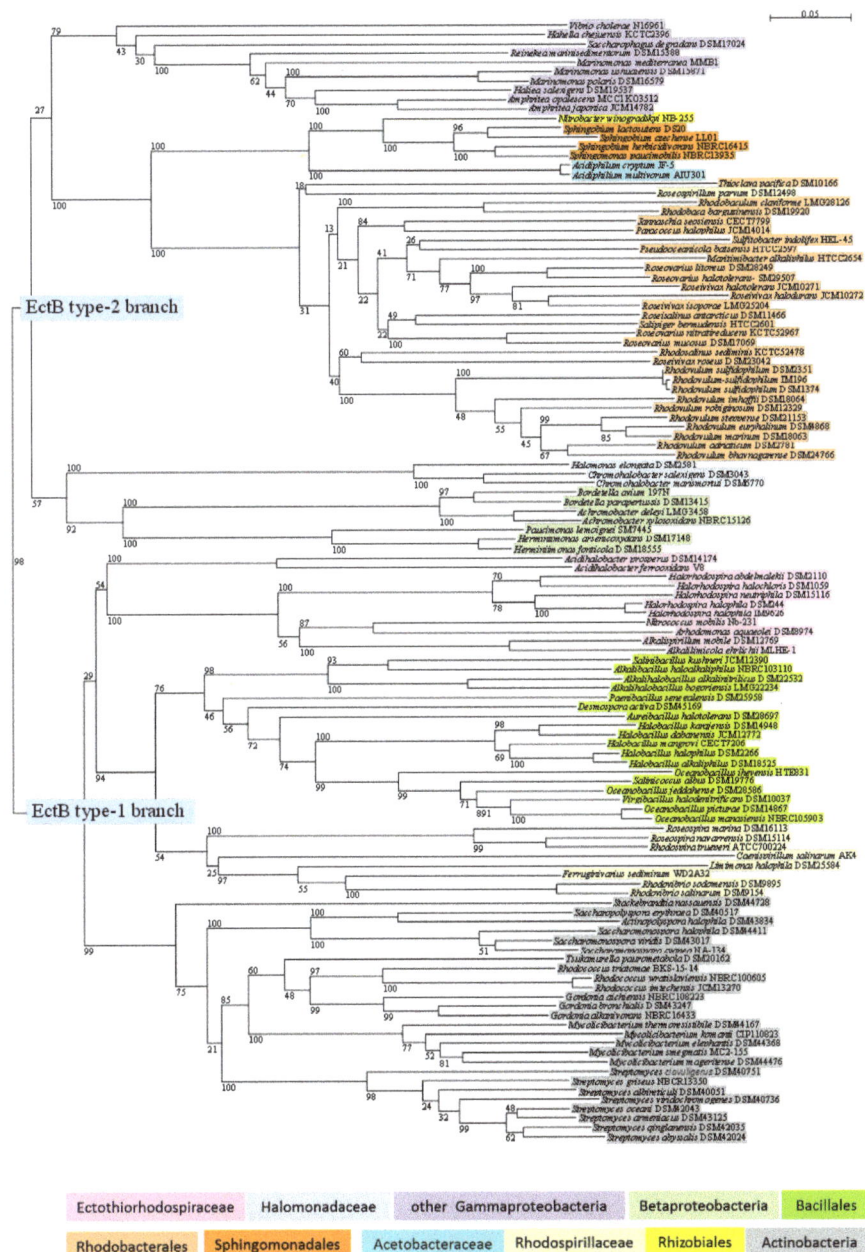

Figure 2. The phylogeny of ectoine biosynthesis on the basis of sequences of diaminobutyrate-pyruvate aminotransferase EctB is shown in a neighbor-joining tree. Sequences and gene bank accession numbers are shown in the Supplementary Table S2. Bootstrap values expressed as percentages of 1000 replications are given at the branches. The bar indicates an evolutionary distance of 0.05. The following color code highlights different systematic groups.

Gammaproteobacteria are found in both branches—representatives of *Ectothiorhodospiraceae* in the EctB type-1 branch and representatives of *Oceanospirillales, Cellvibrionales*, and *Vibrionales* in the EctB type-2 branch. According to genome analysis of the selected species and also BLAST search with representative EctB sequences, ectoine biosynthesis is absent from all tested *Cyanobacteria*, from *Heliobacteria, Chlorobi, Chloroflexi, Chloracidobacterium*, as well as from all studied phototrophic *Betaproteobacteria, Rhizobiales, Chromatiaceae*, and the *Ectothiorhodospira* species (Tables 1–3).

As shown in Figure 2, a first major phylogenetic branch (EctB type-1) contains four clearly separated groups: (1) the *Ectothiorhodospiraceae*, (2) *Rhodospirillaceae* including the non-phototrophic relatives, (3) *Bacillales*, and (4) *Actinobacteria*. This branch is characterized by a gene cluster lacking the regulatory *ectR* gene and the *ask_ect* gene.

- The *Ectothiorhodospiraceae* form a branch with representatives of the genera *Halorhodospira, Alkalilimnicola, Alkalispirillum, Nitrococcus, Arhodomonas*, and *Acidihalobacter*, but lack species of *Ectothiorhodospira, Thiorhodospira*, and *Ectothiorhodosinus*, in which ectoine biosynthesis is absent. A cluster of the *ectABC* genes that lacks the regulatory gene is found in the extremely halophilic *Halorhodospira* species.
- The marine and halophilic *Rhodospirillaceae* also have an *ectABC* gene cluster and form deeply branching separate lineages with *Rhodospira trueperi* and the *Roseospira* species in one, *Rhodovibrio sodomensis* and *Rhodovibrio salinarum* and the non-phototrophic *Ferruginivarius sediminum* and *Limimonas halophila* in another, and *Caenispirillum salinarum* in a third lineage (Figure 2).
- Different lineages of the *Bacillales* branch are represented by species of the genera *Halobacillus*, of *Oceanobacillus* (including *Salinicoccus albus* (*Staphylococcaceae*) and *Virgibacillus halodenitrificans*), of *Salinibacillus, Alkalibacillus* and *Alkalihalobacillus*. Distinct separate lines of *Aureibacillus halotolerans* (*Bacillaceae*), *Paenibacillus senegaliensis* (*Paenibacillaceae*) and *Desmospora activa* (*Thermoactinomycetaceae*) are found (Figure 2).
- The branch of *Actinobacteria* shows distinct subbranches of *Streptomyces, Mycolicibacterium, Gordonia*, and *Rhodococcus* (associated with the *Tsukamurella paurometabola*) species, and of *Saccharomonospora, Actinopolyspora*, and *Saccharopolyspora* species (Figure 2). Distinct from these and as an outsider of the group is *Stackebrandtia nassauensis*. Common among many of the *Actinobacteria* is the ability to form hydroxyectoine, as demonstrated here by genome analysis of *Mycolicibacterium thermoresistibile*, and BLAST search with the EctD sequences from these bacteria (data not shown).

A second major phylogenetic branch (EctB type-2) is characterized by the presence of an extended gene cluster for ectoine biosynthesis, often including the regulatory gene (*ectR*) and a specific isoenzyme of aspartokinase (*ask_ect*). The EctB type-2 group shows considerable deeper branching points and is phylogenetically even more diverse compared to the EctB type-1 group (Figure 2).

The *Alphaproteobacteria* form two related subbranches. One of these is represented by *Rhodobacterales*, with numerous anaerobic and aerobic phototrophic bacteria, as well as chemotrophic relatives that are unable to perform photosynthesis and has an extended *ask_ect_ectABCR* gene cluster. Species of *Rhodovulum* and the related genera *Roseivivax, Roseovarius, Rhodosalinus*, and *Roseisalinus* are included (Table 2, Figure 2). However, in the related *Rhodobacter* species, ectoine biosynthesis and the *ectABCR* genes are absent. One distinct separate lineage is formed by *Rhodobaca barguzinensis* and *Rhodobaculum claviforme*. The marine *Roseospirillum parvum* is the only representative of *Rhodospirillaceae* within the type-2 group. Its gene cluster includes the regulatory gene (*ectABCR*) but lacks the *ask_ect* gene. However, this species and *Thioclava pacifica* (*Rhodobacteraceae*) are outsider of this subbranch.

A second subbranch of *Alphaproteobacteria* is distantly associated with the *Rhodobacterales* branch. It includes the aerobic phototrophic *Acidiphilium multivorum* (*Acetobacteraceae, Rhodospirillales*) and *Acidiphilium cryptum* (non-phototrophic), which also encode the formation of hydroxyectoine and have *ectD* included in the gene cluster *ectRABCDask_ect* (Table 2). The related phototrophic *Acetobacteraceae Rhodopila globiformis* and *Paracrauro-*

coccus ruber are unable to produce ectoine (Table 2). A second lineage of this subbranch contains *Nitrobacter winogradskyi* (*Rhizobiales*) and representatives of the *Sphingomonadales*.

Although none of the phototrophic *Betaproteobacteria* included in this study was able to synthesize ectoine, in several chemotrophic *Betaproteobacteria* (*Burkholderiales*) EctB is present and a distinct branch of type-2 EctB sequences is formed. Two separate lineages include representatives of *Alcaligenaceae* (*Achromobacter* and *Bordetella* species) in one, and the *Paucimonas* (*Burkholderiaceae*) and *Herminiimonas* (*Oxalobacteraceae*) species in another lineage (Figure 2). An *ectRABCD* gene cluster was found in the genomes of *Achromobacter xylosoxidans*, *Bordetella avium*, *Herminiimonas arsenicoxydans*, and *Paucimonas limoigeni* (Table 4).

Two branches of *Gammaproteobacteria* are found among the EctB type-2 sequences. One is represented by *Halomonadaceae* with *Halomonas elongata* and *Chromohalobacter salexigens* and related species (*ectABC*/*ectD*), and associates distantly with the sequences of *Betaproteobacteria*. The second one is poorly associated and is quite distant to the branch of *Alphaproteobacteria*. It includes representatives of *Oceanospirillaceae* (*Marinomonas* and *Amphritea* species) and deeply branching lines with representatives of other families (Table 4, Figure 2). There is a considerable variation of the composition of the *ect* gene cluster in this branch (*Marinomonas mediterranea*: *ectABC*/*ectR*/*ask_ect*; *Hahella chejuvensis*: *ectABCD*; *Haliea salexigens*: *ectRABCask_ect*), as shown in Table 4.

3.4. Betaine Synthesis from Choline—Distribution of Bet Genes in Phototrophic Bacteria

An alternative and independent pathway of betaine synthesis that starts from choline exists in a number of bacteria [49]. It depends on an external source of choline that needs to be taken up by the cells and is then converted into betaine. Thus, the availability of choline is a crucial factor that eliminates this pathway from consideration, as an important and independent option to adapt to high salt concentrations. Though this pathway offers a good chance for bacteria living in eutrophic locations rich in biomass and provide choline as the source, such habitats with extremes of salt concentrations are almost devoid of higher developed eukaryotes that could produce and release choline. Here, mass developments of halophilic microorganisms that would be the primary colonizers could be a possible source under such conditions. If the presence of this pathway is compared with the ability to cope with even low (marine) salt concentrations, it is then obvious that it does not play a primary role in conquering marine and hypersaline habitats. Most marine species that can perform the choline pathway, are not dependent on this pathway, but have alternative options of compatible solute biosynthesis. Examples are *Rhodobaca barguzinensis* (produces also ectoine), *Rhodovibrio* species (also have the ectoine and glycine-dependent pathway), *Marichromatium gracile* (also has the glycine-dependent pathway), and *Ectothiorhodospira* species (have the glycine-dependent pathway) (see Tables 2 and 3). The only marine species that exclusively rely on the choline-dependent pathway of biosynthesis for betaine synthesis (in addition to several transport systems) are members of the *Rhizobiales* including *Fulvimarina pelagi*, *Hoeflea phototrophica*, *Rhodobium orientis*, and the *Afifella* species (Table 2). It is also found in some freshwater bacteria, such as *Rhodopseudomonas palustris*, *Rhodospirillum rubrum*, and *Rhodobacter* species.

Quite interestingly, the choline-dependent pathway of betaine synthesis is also an additional option in a few extremely halotolerant species, which are capable of betaine synthesis from glycine and of ectoine synthesis. These include a few *Rhodobacteraceae* (species of *Roseivivax*, *Rhodosalinus*, and *Roseovarius*), as well as the *Rhodovibrio* species (Table 2).

3.5. Transport System for Uptake of Glycine Betaine and Choline

All marine and halophilic phototrophic bacteria have one or more transport systems for betaine or choline. This underlines the importance of transport to gain or regain osmolytes for successful adaptation to elevated salt concentrations. In contrast, freshwater phototrophic bacteria not only lack possibilities of biosynthesis but also of common transport systems for betaine, ectoine, and related osmolytes. While there are no obvious correlations between these transport systems and the biosynthetic pathways present, various phylogenetic groups have characteristic sets of transport systems for betaine, choline, and possibly ectoine and related osmolytes.

The most common and highly variable system is the ABC transporter ProU, which is present in different distinct variants and in different association with other genes. Often multiple forms are found in one and the same bacterium (Figure 3, Tables 1–3).

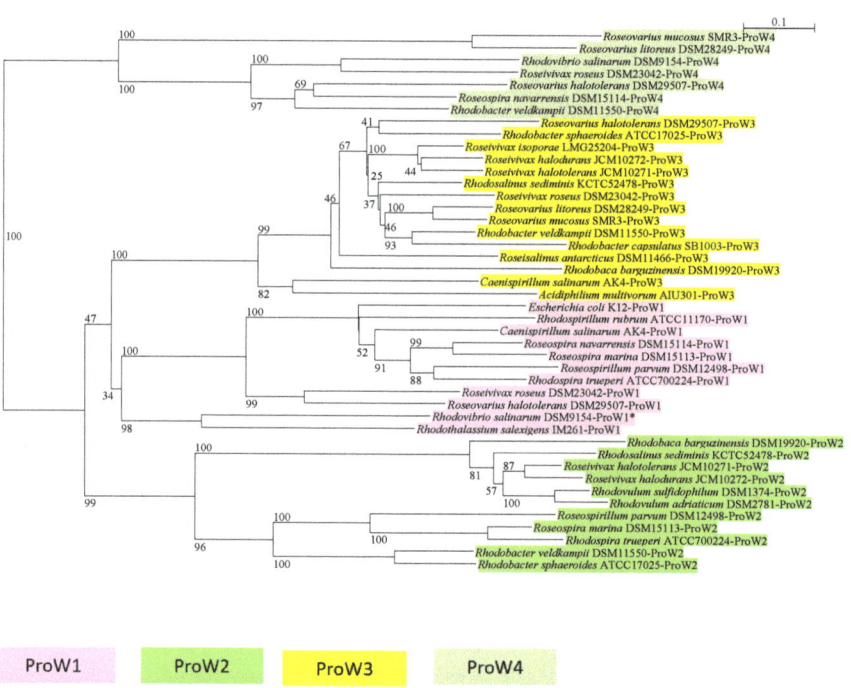

Figure 3. The phylogeny of the permease protein ProW of the L-proline/glycine betaine transport system ProU is shown in a neighbor-joining tree. The ProW1-ProW4 gene clusters are defined with Table 1. Sequences and gene bank accession numbers are shown in Supplementary Table S3. Bootstrap values expressed as percentages of 1000 replications are given at the branches. The bar indicates an evolutionary distance of 0.1. The following color code is used for the different ProW sequence types.

According to the ProW sequences, at least four distinct types of this transporter can be distinguished (Figure 3). One separate *proVWX* gene cluster is related to the *E. coli proU* cluster (type ProW1). Another separate *proXWV* cluster represents a second phylogenetic group (type ProW2). Phylogenetically distinct is a third group (type ProW3), which has *proXWV* genes associated with a cluster of genes for choline-dependent betaine biosynthesis (*betABI-proXWV*), e.g., in *Rhodosalinus sediminis*. A fourth group with a particularly large ProW sequences (type ProW4) is clearly distant to the first three groups. Finally, a gene cluster including a YehZ betaine binding protein is especially found in some species restricted to freshwater habitats (data not shown). In fact, YehZYXW mediated activity in *E. coli* was inhibited at increased salinity and it was concluded that this transport system is not relevant to osmotic protection [50]. A branch of sequences of ProW of this type of transport system is distantly related to the four others and includes sequences of the *E. coli* YehW (data not shown). This was not considered in the phylogenetic tree of ProW sequences (Figure 3). In this tree of selected ProW sequences, the presence of several sequence types is demonstrated in *Rhodobacter veldkampii* (types W2, W3, W4), *Rhodobaca barguzinensis* (types W2, W3), *Roseovarius halotolerans* (types W1, W3, W4), and others. Though sequence variations might suggest possible differences in catalytic properties, gene regulation or substrate specificities between the 4 types of ProU transport systems, these are yet to be demonstrated in future studies.

Common to most marine and halophilic phototrophic bacteria, no matter whether they synthesize betaine or ectoine, is the BetT secondary transporter, which is known as a specific choline transport system in *E. coli*, but is shown to specifically transport betaine in *Aphanothece halophytica* [18].

Another secondary transporter for glycine betaine, OpuD, is found in the freshwater Betaproteobacteria *Rhodoferax fermentans* and *Rubrivivax gelatinosus*, as only a transport system for betaine. Whereas, it is present in the marine *Caenispirillum salinarum* as well as in the extreme halophilic *Halorhodospira* species, both of which synthesize betaine and ectoine, (not in *Hlr. neutriphila*) together with BetT and ProU (type W1) (not in *Hlr. halochloris*). The only species of *Ectothiorhodospira* that encodes OpuD is *Ectothiorhodospira haloalkaliphila*.

The ABC transporter OpuA was found together with BetT in a small group of marine Chromatiaceae producing betaine (*Thiocapsa*, *Thiocystis*, *Marichromatium*, *Imhoffiella*, and *Thiorhodococcus* species), and also in all *Ectothiorhodospira* species. In all of these latter species, *betT* clusters with the *opuA* genes (*betT-opuAA,AB,AC*) and in a few Chromatiaceae, a copy of *betT* also specifically associates with the *opuAC* gene (Table 3). In a few cases, a chimera of the *proU* and *opuA* genes occur. In marine *Prosthecochloris vibrioformis*, this is *proVopuAB,AC*, in *Rhodovibrio salinarum* and also in two Chromatiaceae (*Rhabdochromatium marinum* and *Thiorhodovibrio winogradskyi*) this is *proVWopuAC*. Furthermore, a chimera between *opuC* and *proU* (*opuCB-proXV*) is annotated in the *Roseiflexus* species.

3.6. Evolutionary Considerations

From an evolutionary point of view, the primary adaptation to high solute concentrations clearly requires a biosynthesis of compatible solutes. Already in the early prokaryote era the ability for biosynthesis and accumulation of organic solutes was a prerequisite for archaic eubacteria to conquer saline and hypersaline environments. Inorganic ions and common simple organic metabolites might have allowed a basic osmotic adaptation in freshwater and marine bacteria. Potassium very likely is a suitable candidate and is responsible for a kind of basic osmotic adaptation in many bacteria in which it accumulates together with glutamic acid, and thereby contributes to the overall osmotic balance [7]. In the extremely halophilic Archaea, the *Halobacteria*, it is even the primary osmolyte and accumulates to several molar concentrations [51,52]. In freshwater bacteria such as *E. coli*, its uptake or release makes possible rapid responses to small changes in osmotic conditions. For osmotic adaptation of marine and halophilic eubacteria, however, it is not important.

In marine eubacteria, a number of non-charged neutral organic molecules contribute to different degrees in achieving osmotic balance. As was shown for *Cyanobacteria*, the degree of salt tolerance depends on the kind of osmolytes that are accumulated. Only limited osmotic adaptation to lower ranges of salt concentrations is possible by carbon compounds such as sugars (sucrose, trehalose), glucosylglycerol, and others [1,6,7,11–13]. The advantage of accumulating these compounds is the comparable cheap biosynthesis and the absence of nitrogen as one of the most severe limiting elements in the environment. The disadvantage is their comparatively low compatibility or solubility. For example, trehalose has a solubility of approximately 2 M, ectoine of 3.87 M, and glycine betaine of 13.6 M. The extreme compatibility of betaine, which goes hand in hand with its excellent solubility, and the almost perfect ability to protect macromolecules against denaturation even at very high concentrations, make betaine the top compatible solute. The excellent osmotic protection of betaine is also reflected in the finding, that the phototrophic bacteria that have most successfully adapted to extremely high salt concentrations accumulate betaine as compatible solute through de novo biosynthesis from glycine. In fact, a striking dependence on betaine synthesis for adaption, to live in extreme and even moderate salt concentrations as found in this study, points out that the kind of osmolytes accumulated in a critical way determine the success of salt adaptation. Therefore, the establishment of betaine biosynthesis is expected to represent an evolutionary breakthrough in osmotic adaptation of eubacteria to high salt concentrations, and is considered to be a prerequisite to conquer highly saline environments.

Nonetheless, the synthesis and accumulation of high concentrations of betaine is a costly process. Though glycine, the immediate biosynthetic precursor of betaine, is a common intermediate of central metabolic processes and can be transformed to betaine with a few enzymatic steps, the three methylation steps have a high energy demand. Additionally, the accumulation of several molar concentrations can be easily limited by the availability of nitrogen. Therefore, bacteria that have no shortage in energy and nitrogen supply, such as phototrophic bacteria, which use sunlight as an energy source and are capable of fixing dinitrogen, are likely to be the first and also to be the most successful in conquering highly saline habitats. In fact, mass developments of phototrophic bacteria regularly occur in salt and soda lakes, as well as in coastal lagoons, and cause colored blooms [1,7,53–55].

Compared to betaine, the accumulation of ectoine has two major and significant disadvantages. The physiological disadvantage of ectoine is its low solubility compared to betaine, which limits its accumulation and protective action at hypersaline salt concentrations. Its ecological disadvantage is the content of 2 nitrogen atoms in the molecule, which doubles the requirement for nitrogen compared to betaine. This restricts accumulation to environments with a high content of combined nitrogen compounds, unless nitrogen can be supplied by nitrogen fixation. In this case, the energy requirement for osmotic adaptation is further increased, because twice as much nitrogen has to be fixed, compared to betaine synthesis. Phylogenetically, it appears as a late event compared to betaine biosynthesis. The most deeply branching points within both type-1 and type-2 EctB sequences are those of the Gammaproteobacteria, forming one branch in the type-1 group and two branches in the type-2 group. Quite remarkably, also the structure of the *ect* gene cluster is much more variable in Gammaproteobacteria compared to other groups. In addition to the *ectABC* genes, either *ectD* or *ask_ect* or *ectR*, or combinations thereof might be present, but might not be part of the gene cluster (Table 3). Therefore, it is assumed that the two types of EctB sequences have their origin in ancestors of the *Gammaproteobacteria*, which might also represent the most ancient ectoine producers. Both phylogenetic lineages might have separated early and given rise to the two independent major branches. Today, the two most prominent and best studied representatives of the two branches that have evolved are found in the Gammaproteobacteria: *Halorhodospira halochloris* (type-1) and *Halomonas elongata* (type-2). Among phototrophic bacteria, ectoine biosynthesis is restricted to a few distinct groups, with representatives of *Gammaproteobacteria* (*Halorhodospira* species) and *Alphaproteobacteria* (*Rhodobacterales* and *Rhodospirillales*).

4. Conclusions

The comprehensive analysis of more than 130 genomes of phototrophic bacteria gives insight into their ability to synthesize the compatible solutes betaine and ectoine, and to take up these compounds. The potential to accumulate these compatible solutes and the kind of solute accumulated, clearly define the range of salt concentrations and the habitats where these bacteria can develop. The data suggest that betaine is the primary compatible solute at high salt concentrations and the most ancient one in evolutionary terms. All halophilic phototrophic bacteria rely on betaine synthesis, and only few of them have additional options of ectoine biosynthesis, or betaine synthesis from choline. Ectoine synthesis as a sole compatible solute is only found in some marine bacteria, in particular *Rhodobacteraceae* and the acidophilic *Acidiphilum* species. This is in accord with its potential to achieve osmotic protection in moderately halotolerant marine bacteria. As information on the presence of choline in marine and hypersaline habitats is missing, it is unclear whether the transformation of choline to betaine is of relevance for osmotic adaptation in these environments. Only few marine species of the *Rhizobiales*, exclusively rely on this pathway, but in addition have several transport systems (betT, ProU) to take up betaine and ectoine from the environment. Therefore, these bacteria are more limited than others to adapt to marine salt concentrations, and their occurrence is restricted to habitats where appropriate solutes are available.

The possibility to take up and accumulate compatible solutes or their precursors from the environment and also from the culture media, was been largely neglected in previous studies on salt relations of marine and halophilic species. In order to consider osmotic adaptation in an ecological context and to characterize salt relations of individual species, the potential of uptake needs to be considered. In complex microbial communities such a microbial mats, the study of synthesis, release and uptake at the community level would be rewarding and would shed light on possible complex syntrophic interactions based on the exchange of compatible solutes between members of the community. In view of the variety and the presence of multiple transport systems in several species, characterization of their catalytic properties and regulation is necessary to identify their functional roles. Quite rewarding from both an evolutionary point of view and from a functional context, should be the analysis of chimeric transport systems as they occur in *Prosthecochloris vibrioformis* (*proVopuAB,AC*) and in the *Roseiflexus* species (*opuCBproXV*).

Despite the fact that synthesis and accumulation of betaine is common to all known halophilic phototrophic bacteria, there is a considerable variation in gene arrangement and formation of gene clusters. In addition, fused GMT-DMT genes occur in some groups of the studied bacteria and a second type called the B-type of GMT sequences is found in several species. The phylogeny of the biosynthesis pathway (of GMT), suggests that the roots are manifested early in bacterial evolution and are most likely before diversification of bacteria as we know today. The recognition of two major phylogenetic branches of *Cyanobacteria* and their relations to others suggests that they represent one of the most ancient betaine-producing bacterial phyla and betaine biosynthesis and might have originated in one of their early ancestor. In another very ancient phylum of phototrophic bacteria, the *Chlorobi* (green sulfur bacteria), betaine biosynthesis was found only in *Chloroherpeton thalassium*. This is one of the most ancient representatives of green sulfur bacteria that is known and the deeply rooted branch of its GMT sequence points out that it might represent one of the most ancient betaine producers as well. Other green sulfur bacteria such as the *Chlorobium* and *Chlorobaculum* species and their relatives might have lost the capability of betaine synthesis during adaptation to freshwater habitats. Most remarkable is the occurrence of B-type GMT sequences, which are phylogenetically distant to all other GMT sequences, and might represent a much older system of betaine biosynthesis. Such a GMT gene is included in a gene cluster for betaine biosynthesis only in the *Rhodovibrio* species. Therefore, studies of betaine biosynthesis of the *Rhodovibrio* species should be quite rewarding.

In addition to these few examples, data are presented form a comprehensive basis for more detailed studies on osmotic adaptation of phototrophic bacteria.

Supplementary Materials: The Supplementary Material for this article is available online at https://www.mdpi.com/2076-2607/9/1/46/s1.

Author Contributions: Cultivation of bacterial cultures, DNA extraction and purification were performed by T.R.; genomic sequencing and quality assurance by S.K. and A.K.; sequence assembly by S.C.N.; sequence annotation, retrieval of sequences from annotated genomes and databases, phylogenetic calculations and design of the study, as well as writing of the manuscript was done by J.F.I. All authors contributed to and approved the work for publication. All authors have read and agreed to the published version of the manuscript.

Funding: This research received no external funding.

Institutional Review Board Statement: Not applicable.

Informed Consent Statement: Not applicable.

Data Availability Statement: Data on gene and genome sequences and accession numbers are contained within supplementary material of the article. All sequence data are available in a publicly accessible repository of gene bank.

Conflicts of Interest: The authors declare no conflict of interest.

References

1. Imhoff, J.F. Minireview—True marine and halophilic anoxygenic phototrophic bacteria. *Arch. Microbiol.* **2001**, *176*, 243–254. [CrossRef] [PubMed]
2. Imhoff, J.F. Diversity of anaerobic anoxygenic phototrophic purple bacteria. In *Modern Topics in the Phototrophic Prokaryotes: Environmental and Applied Aspects*; Hallenbeck, P.C., Ed.; Springer: Cham, Switzerland, 2017; pp. 47–85.
3. Brown, A.D. Microbial water stress. *Bacteriol. Rev.* **1976**, *40*, 803–846. [CrossRef] [PubMed]
4. Galinski, E.A.; Trüper, H.G. Betaine, a compatible solute in the extremely halophilic phototrophic bacterium Ectothiorhodospira halochloris. *FEMS Microbiol. Lett.* **1982**, *13*, 357–360. [CrossRef]
5. Imhoff, J.F.; Rodriguez-Valera, F. Betaine is the main compatible solute of halophilic eubacteria. *J. Bacteriol.* **1984**, *160*, 478–479. [CrossRef]
6. Imhoff, J.F. Osmoregulation and compatible solutes in eubacteria. *FEMS Microbiol. Rev.* **1986**, *39*, 57–66. [CrossRef]
7. Imhoff, J.F. Osmotic adaptation in halophilic and halotolerant microorganisms. In *The Biology of Halophilic Bacteria*; Vreeland, R.W., Hochstein, L.J., Eds.; The CRC Press: Boca Raton, FL, USA, 1993; pp. 211–253.
8. Trüper, H.G.; Galinski, E.A. Biosynthesis and fate of compatible solutes in extremely halophilic phototrophic eubacteria. *FEMS Microbiol. Rev.* **1990**, *75*, 247–254. [CrossRef]
9. Galinski, E.A.; Pfeiffer, H.-P.; Trüper, H.G. 1,4,5,6-Tetra-hydro-2-methyl-4-pyrimidinecarboxylic acid: A novel cyclic amino acid from halophilic phototrophic bacteria of the genus *Ectothiorhodospira*. *Eur. J. Biochem.* **1985**, *149*, 135–139. [CrossRef]
10. Czech, L.; Hermann, L.; Stöveken, N.; Richter, A.A.; Höeppner, A.; Smits, S.H.J.; Heider, J.; Bremer, E. Role of the Extremolytes Ectoine and Hydroxyectoine as Stress Protectants and Nutrients: Genetics, Phylogenomics, Biochemistry, and Structural Analysis. *Genes* **2018**, *9*, 177. [CrossRef]
11. Mackay, M.A.; Norton, R.S.; Borowitzka, L.J. Organic Osmoregulatory Solutes in Cyanobacteria. *J. Gen. Microbiol.* **1984**, *130*, 2177–2191. [CrossRef]
12. Severin, J.; Wohlfarth, A.; Galinski, E.A. The predominant role of recently discovered tetrahydropyrimidines for the osmoadaptation of halophilic eubacteria. *J. Gen. Microbiol.* **1992**, *138*, 1629–1638. [CrossRef]
13. Welsh, D.T.; Herbert, R.A. Identification of organic solutes accumulated by purple and green sulfur bacteria during osmotic stress using natural abundance 13C nuclear magnetic resonance spectroscopy. *FEMS Micobiol. Ecol.* **1993**, *13*, 145–150. [CrossRef]
14. Nyyssölä, A.; Kerovuo, J.; Kaukinen, P.; Von Weymarn, N.; Reinikainen, T. Extreme Halophiles Synthesize Betaine from Glycine by Methylation. *J. Biol. Chem.* **2000**, *275*, 22196–22201. [CrossRef] [PubMed]
15. Nyyssölä, A.; Reinikainen, T.; Leisola, M. Characterization of Glycine Sarcosine *N*-Methyltransferase and Sarcosine Dimethylglycine *N*-Methyltransferase. *Appl. Environ. Microbiol.* **2001**, *67*, 2044–2050. [CrossRef] [PubMed]
16. Waditee, R.; Tanaka, Y.; Aoki, K.; Hibino, T.; Jikuya, H.; Takano, J.; Takabe, T.; Takabe, T. Isolation and functional characterization of *N*-methyltransferases that catalyze betaine synthesis from glycine in a halotolerant photosynthetic organism *Aphanothece halophytica*. *J. Biol. Chem.* **2003**, *278*, 4932–4942. [CrossRef]
17. Andresen, P.A.; Kaasen, I.; Styrvold, O.B.; Boulnois, G.; Strøm, A.R. Molecular cloning, physical mapping and expression of the bet genes governing the osmoregulatory choline-glycine betaine pathway of *Escherichia coli*. *J. Gen. Microbiol.* **1988**, *134*, 1737–1746. [CrossRef]
18. Laloknam, S.; Tanaka, K.; Buaboocha, T.; Waditee, R.; Incharoensakdi, A.; Hibino, Y.; Tanaka, Y.; Takabe, T. Halotolerant cyanobacterium *Aphanothece halophytica* contains a betaine transporter active at alkaline pH and high salinity. *Appl. Environ. Microbiol.* **2006**, *72*, 6018–6026. [CrossRef]
19. Peters, P.; Galinski, E.A.; Trüper, H.G. The biosynthesis of ectoine. *FEMS Microbiol. Lett.* **1990**, *71*, 157–162. [CrossRef]
20. Ono, H.; Sawada, K.; Khunajakr, N.; Tao, T.; Yamamoto, M.; Hiramoto, M.; Shinmyo, A.; Takano, M.; Murooka, Y. Characterization of biosynthetic enzymes for ectoine as a compatible solute in a moderately halophilic eubacterium *Halomonas elongata*. *J. Bacteriol.* **1999**, *181*, 91–99. [CrossRef]
21. Louis, P.; Galinski, E.A. Characterization of genes for the biosynthesis of the compatible solute ectoine from *Marinococcus halophilus* and osmoregulated expression in *Escherichia coli*. *Microbiology* **1997**, *143*, 1141–1149. [CrossRef]
22. Inbar, L.; Labidot, A. The structure and biosynthesis of new tetrahydropyrimidine derivatives in actinomycin D producer *Streptomyces parvulus*. Use of 13C- and 15N-labeled L-glutamate and 13C and 15N NMR spectroscopy. *J. Biol. Chem.* **1988**, *263*, 16014–16022.
23. Widderich, N.; Höppner, A.; Pittelkow, M.; Heider, J.; Smits, S.H.; Bremer, E. Biochemical properties of ectoine hydroxylase from extremophiles and their wide taxonomic distribution among microorganisms. *PLoS ONE* **2014**, *9*, e93809. [CrossRef] [PubMed]
24. Stöveken, N.; Pittelkow, M.; Sinner, T.; Jensen, R.A.; Heider, J.; Bremer, E. A specialized aspartokinase enhances the biosynthesis of the osmoprotectants ectoine and hydroxyectoine in *Pseudomonas stutzeri* A1501. *J. Bacteriol.* **2011**, *193*, 4456–4468. [CrossRef] [PubMed]

25. Reshetnikov, A.S.; Khmelenina, V.N.; Trotsenko, Y.A. Characterization of the ectoine biosynthesis genes of haloalkalotolerant obligate methanotroph "Methylomicrobium alcaliphilum 20Z". *Arch. Microbiol.* **2006**, *184*, 286–297. [CrossRef] [PubMed]
26. Sleator, R.D.; Hill, C. Bacterial osmoadaptation: The role of osmolytes in bacterial stress and virulence. *FEMS Microbiol. Rev.* **2002**, *26*, 49–71. [CrossRef]
27. Kempf, B.; Bremer, E. Uptake and synthesis of compatible solutes as microbial stress response to high-osmolality environments. *Arch. Microbiol.* **1998**, *170*, 319–330. [CrossRef]
28. Pfennig, N.; Trüper, H.G. Isolation of members of the families Chromatiaceae and Chlorobiaceae. In *The Prokaryotes, a Handbook on Habitats, Isolation and Identification of Bacteria*; Starr, M.P., Stolp, H., Trüper, H.G., Balows, A., Schlegel, H.G., Eds.; Springer: New York, NY, USA, 1981; pp. 279–289.
29. Imhoff, J.F. The Chromatiaceae. In *The Prokaryotes. A Handbook on the Biology of Bacteria*, 3rd ed.; Dworkin, M., Falkow, S., Rosenberg, E., Schleifer, K.-H., Stackebrandt, E., Eds.; Springer: New York, NY, USA, 2006; Volume 6, pp. 846–873.
30. Imhoff, J.F. Anoxygenic phototrophic bacteria. In *Methods in Aquatic Bacteriology*; Austin, B., Ed.; John Wiley & Sons: Chichester, UK, 1988; pp. 207–240.
31. Imhoff, J.F.; Rahn, T.; Künzel, S.; Neulinger, S.C. New insights into the metabolic potential of the phototrophic purple bacterium *Rhodopila globiformis* DSM 161^T from its draft genome sequence and evidence for a vanadium-dependent nitrogenase. *Arch. Microbiol.* **2018**, *200*, 847–857. [CrossRef]
32. Bolger, A.M.; Lohse, M.; Usadel, B. Trimmomatic: A flexible trimmer for Illumina sequence data. *Bioinformatics* **2014**, *30*, 2114–2120. [CrossRef]
33. Morgulis, A.; Gertz, E.M.; Schäffer, A.A.; Agarwala, R. A Fast and Symmetric DUST Implementation to Mask Low-Complexity DNA Sequences. *J. Comput. Biol.* **2006**, *13*, 1028–1040. [CrossRef]
34. Nurk, S.; Bankevich, A.; Antipov, D.; Gurevich, A.; Korobeynikov, A.; Lapidus, A. Assembling genomes and mini-metagenomes from highly chimeric reads. In *Research in Computational Molecular Biology*; Deng, M., Jiang, R., Sun, F., Zhang, X., Eds.; Springer: Heidelberg, Germany, 2013; pp. 158–170.
35. Bankevich, A.; Nurk, S.; Antipov, D.; Gurevich, A.A.; Dvorkin, M.; Kulikov, A.S.; Lesin, V.M.; Nikolenko, S.I.; Pham, S.; Prjibelski, A.D.; et al. SPAdes: A new genome assembly algorithm and its applications to single-cell sequencing. *J. Comput. Biol.* **2012**, *19*, 455–477. [CrossRef]
36. Boetzer, M.; Henkel, C.V.; Jansen, H.J.; Butler, D.; Pirovano, W. Scaffolding pre-assembled contigs using SSPACE. *Bioinformatics* **2011**, *27*, 578–579. [CrossRef]
37. Aziz, R.K.; Bartels, D.; Best, A.A.; DeJongh, M.; Disz, T.; Edwards, R.A.; Formsma, K.; Gerdes, S.; Glass, E.M.; Kubal, M.; et al. The RAST Server: Rapid Annotations using Subsystems Technology. *BMC Genom.* **2008**, *9*, 75. [CrossRef] [PubMed]
38. Altschul, S.F.; Madden, T.L.; Schäffer, A.A.; Zhang, J.; Zhang, Z.; Miller, W.; Lipman, D. Gapped BLAST and PSI-BLAST: A new generation of protein database search programs. *Nucleic Acids Res.* **1997**, *25*, 3389–3402. [CrossRef] [PubMed]
39. Overbeek, R.; Olson, R.; Pusch, G.D.; Olsen, G.J.; Davis, J.J.; Disz, T.; Edwards, R.A.; Gerdes, S.; Parrello, B.; Shukla, M.; et al. The SEED and the Rapid Annotation of microbial genomes using Subsystems Technology (RAST). *Nucleic Acids Res.* **2014**, *42*, D206–D214. [CrossRef] [PubMed]
40. Larkin, M.A.; Blackshields, G.; Brown, N.P.; Chenna, R.; Mcgettigan, P.A.; McWilliam, H.; Valentin, F.; Wallace, I.M.; Wilm, A.; Lopez, R.; et al. Clustal W and Clustal X version 2.0. *Bioinformatics* **2007**, *23*, 2947–2948. [CrossRef]
41. Saitou, N.; Nei, M. The neighbor-joining method: A new method for reconstructing phylogenetic trees. *Mol. Biol. Evol.* **1987**, *4*, 406–425. [CrossRef]
42. Perrière, G.; Gouy, M. WWW-query: An on-line retrieval system for biological sequence banks. *Biochimie* **1996**, *78*, 364–369. [CrossRef]
43. Reed, R.; Stewart, W.D.P. Osmotic adjustment and organic solute accumulation in unicellular cyanobacteria from freshwater and marine habitats. *Mar. Biol.* **1985**, *88*, 1–9. [CrossRef]
44. Reed, R.; Borowitzka, L.J.; Mackay, M.A.; Chudek, J.A.; Foster, R.; Warr, S.C.; Moort, D.J.; Stewart, W.D.P. Organic solute accumulation in osmotically stressed cyanobacteria. *FEMS Microbiol. Rev.* **1986**, *39*, 51–56. [CrossRef]
45. Klähn, S.; Hagemann, M. Compatible solute biosynthesis in cyanobacteria. *Environ. Microbiol.* **2011**, *13*, 551–562. [CrossRef]
46. Imhoff, J.F.; Rahn, T.; Künzel, S.; Neulinger, S.C. Phylogeny of anoxygenic photosynthesis based on protein sequences of the photosynthetic reaction center and of bacteriochlorophyll biosynthesis. *Microorganisms* **2019**, *7*, 576. [CrossRef]
47. Imhoff, J.F.; Petri, R.; Süling, J. Reclassification of species of the spiral-shaped phototrophic purple nonsulfur bacteria of the alpha-proteobacteria: Description of the new genera *Phaeospirillum* gen. nov., *Rhodovibrio* gen. nov., *Rhodothalassium* gen. nov. and *Roseospira* gen. nov. as well as transfer of *Rhodospirillum fulvum* to *Phaeospirillum fulvum* comb. nov., of *Rhodospirillum molischianum* to *Phaeospirillum molischianum* comb. nov., of *Rhodospirillum salinarum* to *Rhodovibrio salinarum* comb. nov., of *Rhodospirillum sodomense* to *Rhodovibrio sodomensis* comb. nov., of *Rhodospirillum salexigens* to *Rhodothalassium salexigens* comb. nov., and of *Rhodospirillum mediosalinum* to *Roseospira mediosalina* comb. nov. *Int. J. Syst. Bacteriol.* **1998**, *48*, 793–798. [PubMed]
48. Ramana, V.V.; Chakravarthy, S.K.; Ramaprasad, E.V.V.; Thiel, V.; Imhoff, J.F.; Sasikala, C.; Ramana, C.V. Emended description of the genus *Rhodothalassium* Imhoff et al., 1998 and proposal of *Rhodothalassiaceae* fam. nov. and *Rhodothalassiales* ord. nov. *Syst. Appl. Microbiol.* **2013**, *36*, 28–32. [CrossRef] [PubMed]
49. Cánovas, D.; Vargas, C.; Csonka, L.N.; Ventosa, A.; Nieto, J.J. Synthesis of Glycine Betaine from Exogenous Choline in the Moderately Halophilic Bacterium *Halomonas elongata*. *Appl. Environ. Microbiol.* **1998**, *64*, 4095–4097. [CrossRef] [PubMed]

50. Lang, S.; Cressatti, M.; Mendoza, K.E.; Coumoundouros, C.N.; Plater, S.M.; Culham, D.E.; Kimber, M.S.; Wood, J.M. YehZYXW of *Escherichia coli* is a low-affinity, non-osmoregulatory betaine-specific ABC transporter. *Biochemistry* **2015**, *54*, 5735–5747. [CrossRef] [PubMed]
51. Christian, J.H.G.; Whalto, J.A. Solute concentrations within cells of halophilic and non-halophilic bacteria. *Biochim. Biophys. Acta* **1962**, *65*, 506–508. [CrossRef]
52. Kushner, D.J. (Ed.) Life in high salt and solute concentrations: Halophilic bacteria. In *Microbial Life in Extreme Environments*; Academic Press: New York, NY, USA; London, UK, 1978; pp. 317–368.
53. Imhoff, J.F.; Hashwa, F.; Trüper, H.G. Isolation of extremely halophilic phototrophic bacteria from the alkaline Wadi Natrun, Egypt. *Arch. Hydrobiol.* **1978**, *84*, 381–388.
54. Imhoff, J.F.; Sahl, H.G.; Soliman, G.S.H.; Trüper, H.G. The Wadi Natrun: Chemical composition and microbial mass developments in alkaline brines of Eutrophic Desert Lakes. *Geomicrobiol. J.* **1979**, *1*, 219–234. [CrossRef]
55. Imhoff, J.F. Anoxygenic phototrophic bacteria from extreme environments. In *Modern Topics in the Phototrophic Prokaryotes: Environmental and Applied Aspects*; Hallenbeck, P.C., Ed.; Springer: Cham, Switzerland, 2017; pp. 427–480.

Article

Discovery of Siderophore and Metallophore Production in the Aerobic Anoxygenic Phototrophs

Steven B. Kuzyk, Elizabeth Hughes and Vladimir Yurkov *

Department of Microbiology, University of Manitoba, Winnipeg, MB R3T 2N2, Canada; umkuzyks@myumanitoba.ca (S.B.K.); elizabeth.hughes@umanitoba.ca (E.H.)
* Correspondence: Vladimir.Yurkov@umanitoba.ca

Abstract: Aerobic anoxygenic phototrophs have been isolated from a rich variety of environments including marine ecosystems, freshwater and meromictic lakes, hypersaline springs, and biological soil crusts, all in the hopes of understanding their ecological niche. Over 100 isolates were chosen for this study, representing 44 species from 27 genera. Interactions with Fe^{3+} and other metal(loid) cations such as Mg^{2+}, V^{3+}, Mn^{2+}, Co^{2+}, Ni^{2+}, Cu^{2+}, Zn^{2+}, Se^{4+} and Te^{2+} were tested using a chromeazurol S assay to detect siderophore or metallophore production, respectively. Representatives from 20 species in 14 genera of α-*Proteobacteria*, or 30% of strains, produced highly diffusible siderophores that could bind one or more metal(loid)s, with activity strength as follows: Fe > Zn > V > Te > Cu > Mn > Mg > Se > Ni > Co. In addition, γ-proteobacterial *Chromocurvus halotolerans*, strain EG19 excreted a brown compound into growth medium, which was purified and confirmed to act as a siderophore. It had an approximate size of ~341 Da and drew similarities to the siderophore rhodotorulic acid, a member of the hydroxamate group, previously found only among yeasts. This study is the first to discover siderophore production to be widespread among the aerobic anoxygenic phototrophs, which may be another key method of metal(loid) chelation and potential detoxification within their environments.

Keywords: aerobic anoxygenic phototrophs; siderophore; metallophore; CAS assay; *Chromocurvus halotolerans* strain EG19

Citation: Kuzyk, S.B.; Hughes, E.; Yurkov, V. Discovery of Siderophore and Metallophore Production in the Aerobic Anoxygenic Phototrophs. *Microorg* **2021**, *9*, 959. https://doi.org/10.3390/microorganisms9050959

Academic Editor: Johannes F. Imhoff

Received: 22 March 2021
Accepted: 28 April 2021
Published: 29 April 2021

Publisher's Note: MDPI stays neutral with regard to jurisdictional claims in published maps and institutional affiliations.

Copyright: © 2021 by the authors. Licensee MDPI, Basel, Switzerland. This article is an open access article distributed under the terms and conditions of the Creative Commons Attribution (CC BY) license (https://creativecommons.org/licenses/by/4.0/).

1. Introduction

Iron (Fe) is an essential element for life. In microorganisms, it is used as a co-factor for enzymatic processes, such as in electron transfer during respiration and photosynthesis, nucleic acid or chlorophyll synthesis, nitrate reduction, nitrogen fixation and detoxification of oxygen radicals [1–3]. The use of Fe in bacteriochlorophyll (BChl) synthesis and the process of photophosphorylation makes it particularly important to phototrophic organisms. Aerobic anoxygenic phototrophs (AAP) are one such physiological group that uses photosynthesis in oxic conditions as an additional energy source to respiration [4]. They make up a significant proportion of many bacterial communities from a host of environments [5], and therefore likely require a substantial Fe uptake. In addition, Fe may be crucial to protect the cells from oxidative stress due to singlet oxygen formation during BChl *a* synthesis. While necessary for metabolism, biologically active Fe is typically quite sparse in nature as its soluble level is very low at soil and water surfaces [2]. In response to this limitation, both bacteria and fungi have developed siderophores to compete for the available Fe [6]. Siderophores, aptly named from the Greek root representing "iron-bearing" [7], are low weight molecules (no more than 1500–2000 Da and generally lower than 1000 Da) with a high-affinity for Fe [2,3,8–10]. The molecules are often short polypeptides with modified or D-amino acids [2,9,11]. They can also be made from dicarboxylic acids and diamine or amino alcohols, linked by amide and ester bonds. These building blocks retain some characteristics of amino acids [2]. Siderophores can be classified into two main functional groups. The first is the hydroxamate group, which involves hydroxamic acid and is produced by both fungi and bacteria. Second is the catechol group, compounds of which contain catechol rings and are only produced by bacteria [6,9,10]. Smaller groups such as the hydroxyacids and the α-hydroxy carboxylates are only rarely used by bacteria [9,10]. Usually, siderophores are synthesized and secreted by cells under iron-deplete growth conditions [8]. Once bound to iron, the complex is taken up by the cells in a substrate specific process [2].

While the main purpose of siderophores is Fe acquisition, they may also play some additional roles. In *Pseudomonas aeruginosa*, pyoverdine controls virulence factor production [12]. *Escherichia coli* can be protected from oxidative stress by the catechol type siderophore enterobactin [13]. Pyochelin in *P. aeruginosa* has a toxic effect on eukaryotic cells, possibly aiding in the bacterium's virulence [13]. Watasemycins, sideromycins, oxachelin and fusigen have antibacterial activity, which may aid in community competition by preventing other populations from growing [13]. Additionally, siderophores have been shown to bind more than one metal [3,14], including some that have higher affinity for Cu or Zn rather than Fe [15,16]. This broad-spectrum activity has required the classification "metallophore" [17,18], which is a term used for secondary metabolites capable of binding a range of metal(loid) cations. When a metallophore would have a specific metal affinity, it would have a sub-categorical name, where siderophore is for Fe-binding, chalkophore for Cu, or zincophore for Zn, all named when discovered.

This concept, of metallophores capable of capturing multiple metal cations has implicated usefulness in toxic heavy metal tolerances. Particularly in extreme environments, where metals can be present at elevated concentrations that inhibit a variety of life [19,20]. Many metal ions can diffuse freely through the cellular membrane, which is inhibited if the metal is bound to a siderophore that is too large to move without active transport. Furthermore, membrane receptors specific to siderophore-iron complexes can differentiate between those containing substitute cations, causing the cell to reject the alternative metal-containing siderophore. The reduction of free metal concentrations in proximity of the bacterium and decreased passive diffusion of unwanted metals into the cell will lower their overall toxic effect [3]. This could be a compelling concept, as AAP possess very high levels of resistance to toxic heavy metal(loid) oxides [21]. While internal enzymatic reduction takes place [22], the external production of siderophores may provide an additional layer of defense. Additionally, as mentioned above, the siderophores' ability to reduce reactive oxygen species could help AAP as they need protection against oxidative stress due to their aerobic production of BChl *a*.

Chromocurvus halotolerans EG19, is a γ-proteobacterial AAP that was isolated during the spring of 2002 from floating microbial mats within the East German Creek System, Manitoba, Canada [23]. As this is a landlocked hypersaline spring system, it likely contains highly endemic communities of microorganisms that have not been mixed with or affected by allochthonous populations [23]. EG19 forms motile, short rod or longer curved rod-shaped, orange-pink bacteria. When grown with complex carbon sources, EG19 produces a brown pigmented hydrophilic compound, which is excreted into the growth medium. While a similar phenomenon had never been reported in other AAP, it was hypothesized that the compound could be a siderophore [24], as ferric bound siderophores can be visually yellow-brown or red-brown [25]. Our study confirms the identity of this extracellular product and describes it as the first siderophore discovered in an AAP. Other AAP from a vast array of environments that do not pigment growth medium were also tested for their production of siderophores, as most of these metal chelating small molecules are colourless, and synthesis is therefore possible.

2. Materials and Methods

2.1. Bacterial Strains and Growth Conditions

For this study, 101 strains of AAP originating from an assortment of environments, as well as phylogenetically diverse throughout numerous proteobacterial clades were selected from Dr. Vladimir Yurkov's vast collection. A complete list of chosen strains, original source of isolation, relatedness to type species, and 16S rRNA partial gene sequence accession numbers are listed in Tables A1 and A2.

Freshwater AAP were cultivated on rich organic (RO) medium as described [26], with one minor modification. Bactopeptone was reduced to 0.5 g/L and casamino acids were supplemented at 0.5 g/L, which provided a larger variety of complex nutrients. Marine AAP requiring salt were propagated on RO medium described above supplemented with 2% NaCl. AAP originally isolated from the East German Creek System, Manitoba, Canada, were grown using medium A (MA) [23]. Those isolated from biological soil crusts of Sandy Lands Forest and Spruce Woods National Park were cultured on Biological Soil Crust Medium A or B (BSCA or BSCB) [27]. Strains from meromictic Mahoney Lake in British Columbia, Canada were grown on N1 medium [28].

In addition to bacterial isolates formerly described elsewhere, some of those chosen for siderophore testing had not been previously published. AAP isolated from the meromictic Blue Lake in British Columbia, Canada, were cultured using a Blue Lake medium (BLM) containing (g/L): $MgSO_4$, 0.5; NH_4Cl, 0.3; KH_2PO_4, 0.3; KCl, 0.3; $CaCl_2$, 0.05; NaCl, 12.0; $NaHCO_3$, 0.5; Na-acetate, 1.0; malic acid, 1.0; yeast extract, 1.0; bactopeptone, 0.5; with vitamins and trace elements solutions, 2.0 mL each; autoclaved at pH 5.9 and then adjusted after autoclaving to pH 7.5 with 0.5 N NaOH.

Furthermore, several AAP had been recovered from the freshwaters of Lake Winnipeg whilst the habitat was under study [29,30]. Specifically, in the spring of 2017, strain AJ72 was collected at Grand Beach from littoral water, AM19, AM27 and AM91 were from littoral sediment of Victoria Beach. During that summer, BA23 and BE100 were isolated at limnetic sites S1 and S5, respectively, while BL67 and BK61 originated within the littoral water and sediment of Grand Beach and Victoria Beach, respectively. Fall samples of Grand Beach sediment contained CK155 and CK182, while Victoria Beach littoral waters revealed isolate CL63. All were purified on the slightly modified freshwater RO medium as described.

2.2. Iron Chelating Chromeazurol S Assay

Every chosen strain was grown on 2% agar plates containing their specific growth medium supplemented with the dye chromeazurol S (CAS), which turns blue when bound to Fe and reverts to yellow/orange being released [31]. To make media, 60.5 mg CAS were dissolved in 50 mL ddH_2O, then mixed with 10 mL of an iron solution containing 1 mM $FeCl_3$ and 10 mM HCl. HDTMA (72.9 mg) was dissolved in 40 mL ddH_2O prior to mixing with the CAS/iron solution, bringing the total volume to 100 mL. Separately, 900 mL of each growth medium (MA, RO, RO 2% NaCl, BSCA, or BLM) were prepared without the addition of iron, but with the correct amount of components for 1 L, as that would be the final volume of medium after combining with CAS/iron solution. Media were then autoclaved at pH 5.9, separate from the CAS mixture. After autoclaving, each medium was adjusted to pH 6.8 and the CAS solution was added. Agar plates would turn blue when solidified. If measurements of ingredients were not exact or added in an incorrect order, CAS would precipitate out of solution and plates with pH higher than 6.8 would appear green instead of blue. CAS plates were heavily inoculated with each strain, and streaked to achieve isolated colonies after 5 days of growth at 28 °C in the dark.

2.3. Variant Cation Chromeazurol S Assay

Due to the principle chemistry behind the siderophore assay [31], which used negatively charged CAS dye that would weakly bind the cation Fe^{3+}, we decided to replace Fe^{3+} with different metal(loid) cations, including Mg^{2+}, V^{3+}, Mn^{2+}, Co^{2+}, Ni^{2+}, Cu^{2+}, Zn^{2+}, Se^{4+} and Te^{2+}. Microbial growth with an aura of colour change in medium from blue to yellow hue would signify that microbially produced siderophore could bind the substituted cation, causing the released dye to revert to its yellow colour. These variant cation CAS plates were successfully created as explained in the previous section, cured to diverse shades of blue dependent on the metal(loid), and were screened for metallophore production identically as in the Fe-chelation CAS assay. All cations were purchased from Sigma Aldrich, USA as chloride salts, which were soluble once acidified as described.

2.4. Phylogeny

To assess the phylogenetic position, 16S rRNA gene sequences were either acquired from previously published sources, or newly sequenced in this study. A list of accession numbers to partial 16S rRNA gene sequences attained from public repositories was provided in Tables A1 and A2. From unsequenced strains, DNA was extracted via the phenol chloroform method [32], and sequenced using Sanger technique [33]. Primer set 27F and 1492R, 5′-AGAGTTTGATCCTGGCTCAG-3′ and 5′-GGTTACCTTGTTACGACTT-3′, respectively, was used to achieve a total contiguous 16S rRNA gene sequence length of >1400 bp per strain. New sequences of 16S rRNA genes from previously unidentified AAP strains were deposited to GenBank under accession numbers (MW970346–MW970408) as listed in Tables A1 and A2. Genetic relation of 16S rRNA gene sequences acquired from each AAP isolate were compared to the archived sequences of type species using the web-based software Basic Local Alignment Search Tool, BLAST [34]. Phylogenetic trees were constructed via MEGA X software [35] with 1000 bootstraps [36], using Maximum Likelihood method to align all AAP 16S rRNA gene sequences to one another based on the General Time Reversible model [37]. Initial tree(s) for the heuristic search were obtained automatically by applying Neighbor-Join and BioNJ algorithms to a matrix of pairwise distances estimated using the Maximum Composite Likelihood (MCL) approach, and then selecting the topology with superior log likelihood value.

2.5. Siderophore Isolation and Concentration from C. halotolerans

Strain EG19 was grown at 28 °C on MA agarized (2%) plates in the dark for 5 days. Pink-purple colonies developed with brown hue dispersed in agarized medium (Figure 1A), which corresponded to zone of clearing on Fe-Chromeazurol S Assay (Figure 1B). To obtain liquid cultures, EG19 was inoculated at 5% in Fe-free MA and grown for 5 days at 28 °C shaking at 200 rpm in the dark. Fe-free MA was prepared as described in Chromeazurol S Assay, excluding Fe from the trace element solution. In addition, after all components were added, 5 g/L of Chelex resin enclosed in dializing tubing was placed in the medium for 1 h prior to autoclaving. This step allowed the Chelex resin to remove traces of Fe introduced through the complex nutrients such as yeast extract, bactopeptone and casamino acids [38].

After the cultivation of *C. halotolerans*, Fe-free medium became dark brown. The pink-purple cells were pelleted in ~450 mL bottles at 10,000 rpm for 30 min using a Beckman J2HS centrifuge and a JA-10 rotor (Figure 1C). Siderophore containing supernatant was collected and 1.25 L was transferred into 2 L Erlenmeyer flasks (Figure 1D) to freeze overnight at −20 °C. Highly concentrated dark brown-pigmented high-salt slurry was formed predominantly on top of ice, while the lower part of ice was close to colourless. The flask with frozen material was removed from freezer and allowed to defrost at room temperature inserted upside down in a new collection beaker. Thawed supernatant was fractioned into 250 mL batches, with concentrated siderophore thawed and collected first, leaving frozen medium behind (Figure 1D), both observable visually (Figure 1E), and from the absorbance spectrum (Figure 1F).

A batch-type method of siderophore purification was used in this study [39]. Here, the combined freeze-concentrated samples were adjusted to pH 6.0 and XAD 7-HP resin was added (20 g resin per L of supernatant). This slurry was shaken at 200 rpm for 1 h on a rotary shaker and then filtered through a glass Millipore filter funnel that would collect resin, but allow supernatant to easily flow through. Concentrated siderophore required several extractions with resin to remove all pigment from supernatant. Siderophore-bound-resin (now brown in colour) was thoroughly washed with ddH_2O to remove all residual salts and other soluble contaminants. Vacuum assisted drying using Millipore system was performed, ddH_2O discarded, and the cleaned siderophore-bound-resin was soaked in methanol for 30 min to release the siderophore. Pigmented methanol was collected, and resin was soaked once more using the same batch of solvent for additional extraction. Methanol extracts were combined and then evaporated to dryness. The resulting dark brown powder was the presumed highly concentrated siderophore, and kept at −20 °C until further testing. Unbound-resin was washed with methanol, then soaked with ddH$_2$O prior to reuse.

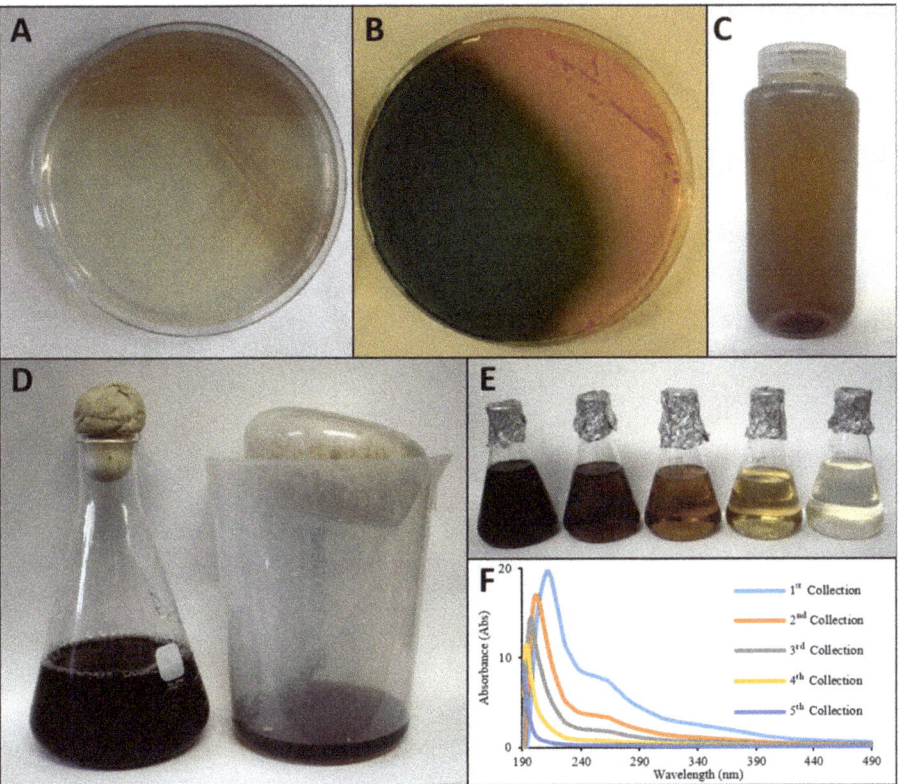

Figure 1. Extraction and concentration of brown pigment excreted by *C. halotolerans*. (**A**) Strain EG19 released a brown pigment into MA medium. (**B**) Fe-CAS plate with siderophore production by EG19. (**C**) Pelleted cells under supernatant. (**D**) Pigment concentrated with freezing-out technique. (**E**) Collected fractions during thawing with concentrated pigment released first. (**F**) Absorbance spectrum of each fraction.

2.6. Purification and Fe-Chelation of Siderophore from C. halotolerans

Concentrated siderophore powder from resin purification was resuspended in ddH$_2$O (10 mg in 50 µL), and decimally diluted up to 10^{-3}. In addition, a solution containing 5 mg of concentrate in 100 µL ddH$_2$O was filtered through an Amicon® Ultra 0.5 mL 3000 Da spin column manufactured by Millipore Ireland, which removed any contaminating proteins over that size. This filtered concentrate was also decimally diluted up to 10^{-2}. Once the dilutions were made, 10 µL of each sample, and an aliquot of freeze–thawed concentrate, were individually mixed with 10 µL of loading buffer, prior to filling into the wells of a Mini-Protean tris-tricine gel, 16.5% from Bio-Rad, Hercules, CA, USA. Loading buffer contained 200 mM tris-HCl, pH 6.8, 2% SDS, 40% glycerol and 0.04% Coomassie Brilliant Blue G-250 (CBB) from Bio-Rad, USA. Running buffer was made up of 1 M tris, 1 M tricine and 1% SDS at pH 8.3. Wells were 30 µL, filled with 20 µL of sample (10 µL of sample, 10 µL of loading buffer). Protein ladder was a C6210—color marker ultra-low range (M.W.

1060–26,600) from Sigma-Aldritch, St. Louis, MO, USA. Gel electrophoresis was run at 100 volts for 1 h prior to being stained in a CBB solution (1 g of CBB dissolved in 1 L of [50% MeOH, 10% glacial acetic acid, 40% H_2O]). The gel was then destained with a mixture containing 5% acetic acid and 20% MeOH overnight.

Fe chelation was tested with the concentrated siderophore powder from the resin concentration step, the brown pigment filter-purified below 3 kDa, as well as with a sample from the remaining proteins that were >3 kDa from the spin filtration procedure. Each fraction was solubilized, where 10 mg of dried pigment was dissolved in 500 µL of 60% MeOH, prior to applying each solution into a blank diffusion disk, manufactured by Oxoid in the UK, allowing it to become dry and concentrated within the disk. These were then placed on a Fe-CAS plate and left to react with chromeazurol S overnight.

3. Results

3.1. Bacterial Growth and Fe-CAS-Plate Reactions

Comparative phenotypic analysis was achieved by allowing each strain to develop for 5 days on their respective agarized media under conditions that promoted the best growth. This elapsed time ensured stationary phase was reached for each representative, which resulted in the formation of sufficient colonies for analysis. Triplicate CAS-supplemented and CAS-free controls were simultaneously plated to identify the viability of inoculum. In all cases, growth occurred on CAS plates, but was marginally reduced in comparison to controls. Both after 3 and 5 days, the average zone of colour change (blue to yellow) was recorded, revealing several phenotypic attributes. As some strains grew slower than others, 5 days' period was chosen to analyze and compare all simultaneously. We identified siderophore production based on zone of diffusion/colour change around colonies as follows (Figure A1): no zone (−), a zone <1 mm around colonies (+), a moderate diffusion <10 mm (++), and considerable diffusion >10 mm (+++). Fe-chelating CAS reactions after 5 days are listed in Tables 1 and 2, as well as shown in Figure 2 beside each strain name.

3.2. Substitute Cation CAS Assays

Once Fe-chelating siderophore production was confirmed using the standard Fe-CAS assay, all strains were tested on CAS plates that contained one of 9 other metal(loid) cations: Mg^{2+}, V^{3+}, Mn^{2+}, Co^{2+}, Ni^{2+}, Cu^{2+}, Zn^{2+}, Se^{4+} and Te^{2+}. Results for varied CAS assays are listed along-side Fe^{2+} data in Tables 1 and 2 using the same zone distinctions as described. While some strains only produced siderophores that reacted with Fe^{2+}, others had secondary metabolites capable of chelating additional metal(loid) cations. Most strains capable of chelating metals other than Fe could also chelate iron, with the exception of two strains P4 and SS56 (Figure A2), which were found to produce metallophore capable of acting predominantly on other tested metal(loid)s, rather than Fe.

3.3. Phylogenetic Diversity of Siderophore Producing AAP

Isolates were chosen to represent AAP from a variety of environments as well as embody a host of phylogenetically diverse species. In this way, the 101 representatives listed in Tables 1 and 2 were cultivated, and 16S rRNA gene sequences acquired either from repositories, or decoded in this work. Phylogenetic relation to sequences of known type strains was determined by BLAST search (Tables A1 and A2). In addition, these sequences were used to create a phylogenetic tree (Figure 2), that also included some previously described type species not tested for siderophore production, but were included as key placeholders of phylogenetic groups. The evolutionary analysis performed on Mega X using Maximum Likelihood method involved 132 nucleotide sequences and had a total of 1717 positions in the final data set. The chosen AAP diversely spread throughout *Erythrobacteraceae* and *Sphingomonadaceae* relating to known AAP type species. While some aligned to reported AAP in *Acetobacteraceae*, and *Rhodobacteraceae*, many others aligned to organisms previously undescribed as AAP within these clades, as well as to some species within *Hyphomonadaceae* and *Methylobacteraceae*.

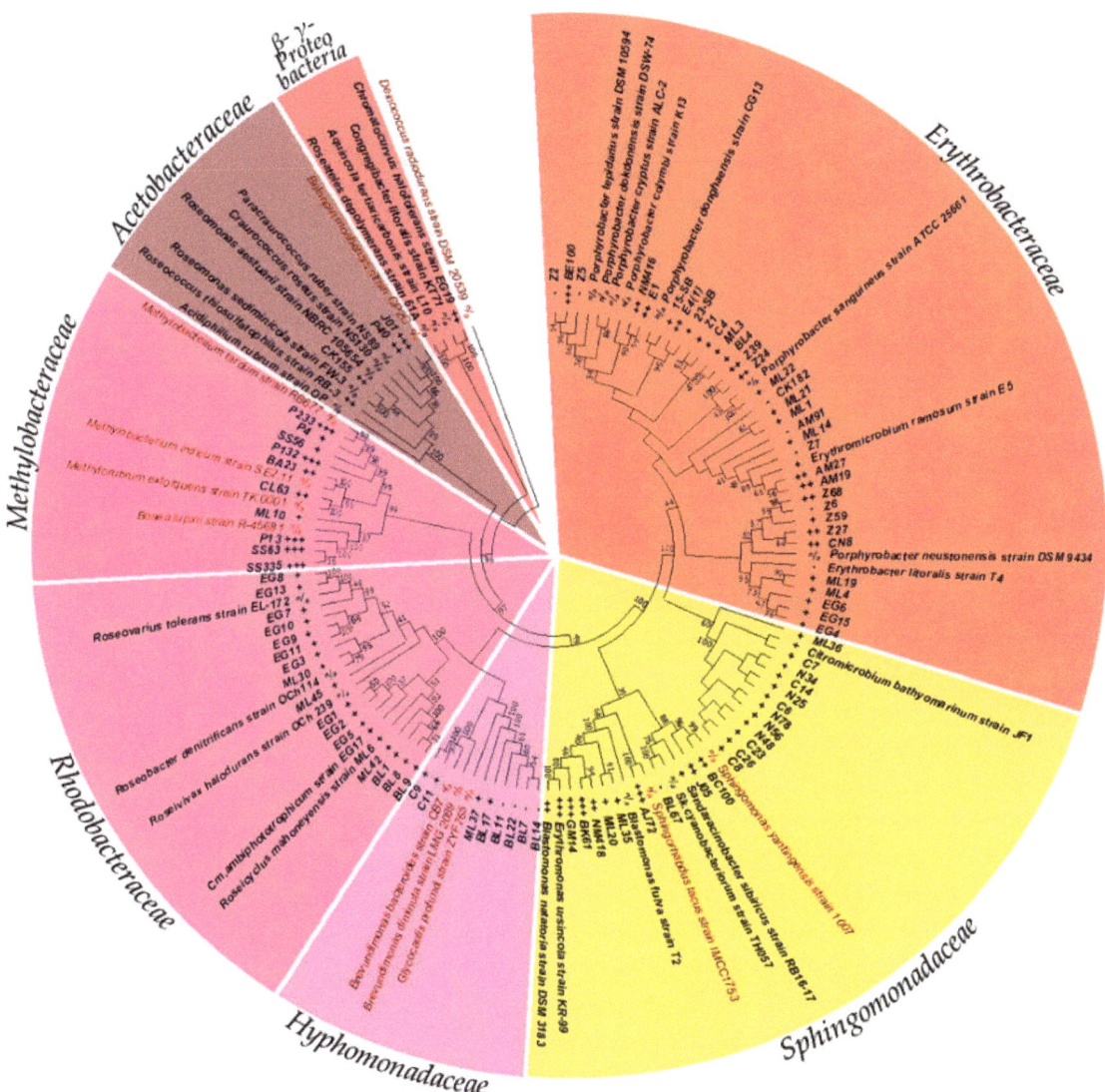

Figure 2. Phylogenetic tree of AAP tested for siderophore production. Isolates hailed from α-proteobacterial families *Erythrobacteraceae*, *Sphingomonadaceae*, *Acetobacteraceae*, *Rhodobacteraceae*, *Hyphomonadaceae* and *Methylobacteraceae*, as well as a representative within the γ- subclass of *Proteobacteria* (titled sections around circumference of circle). Fe-chelation siderophore activities are listed between strain names and phylogenetic position. Strain names in bold are confirmed AAP, red highlighted are not.

Table 1. Freshwater and saline AAP analyzed for metal(loid)-chelation via CAS assays with varied cations.

Environment	Strain	Medium	Mg	V	Mn	Fe	Co	Ni	Cu	Zn	Se	Te
Hot spring, Kamchatka Isl.	KR99[T]	RO	++	+++	++	+++	++	++	++	+++	++	+++
Warm temperature spring, Baikal Lake, Russia	E1	RO	−	−	+	++	+	+	−	−	−	−
	E4(1)	RO	−	−	−	++	−	−	−	−	−	+
	E5[T]	RO	−	−	−	+	−	−	−	−	−	+
	RB3[T]	RO	−	−	++	++	−	++	+	+	+	+
	RB16-17[T]	RO	−	−	−	+	−	−	+	−	−	−
	T4[T]	RO-NaCl	+	+	+	−	+	+	−	−	+	+
Deep Ocean, Juan De Fuca Ridge, Pacific Ocean	JF1[T]	RO-NaCl	+	+	+	+	+	+	+	+	+	+
	C6	RO-NaCl	+	+	+	+	+	+	+	+	+	+
	C7	RO-NaCl	+	+	+	+	−	+	+	+	+	+
	C8	RO-NaCl	+	+	+	+	−	+	+	+	+	+
	C14	RO-NaCl	+	+	+	+	−	+	+	+	+	+
	C23	RO-NaCl	−	+	+	+	+	+	+	+	+	+
	C26	RO-NaCl	+	+	+	+	+	+	+	+	+	+
	N25	RO-NaCl	+	+	+	+	+	+	+	+	+	+
	N34	RO-NaCl	+	+	+	+	+	+	+	+	+	+
	N48	RO-NaCl	+	+	+	+	+	+	+	+	+	+
	N56	RO-NaCl	+	+	+	+	+	+	+	+	+	+
	N78	RO-NaCl	+	+	+	+	+	+	+	+	+	+
Rag Beach sediment, BC, Canada	15-SB	RO-NaCl	+	−	+	+	−	−	+	+	+	+
	23-SB	RO-NaCl	+	−	+	+	+	+	−	+	−	−
Central Gold Mine, MB, Canada	C4	RO	−	+	+	+	+	+	+	+	−	+
	C9	RO	−	+	−	+	+	−	+	+	−	+
	C11	RO	−	+	−	+	−	−	+	+	−	−
	NM4.16	RO	−	+	+	++	+	+	+	+	−	+
	NM4.18	RO	++	+	+	++	++	++	++	+	++	++
Lake Winnipeg, MB, Canada	AJ 72	RO	−	+	+	+++	+	+	+	++	−	+
	AM 19	RO	−	+	−	+++	−	−	−	++	−	−
	AM 27	RO	−	−	−	++	−	−	−	+	−	−
	AM 91	RO	−	−	−	-	−	−	−	−	−	−
	BA 23	RO	+	++	++	++	++	+	++	++	++	++
	BC 100	RO	−	−	−	++	+	−	−	−	−	−
	BE 100	RO	++	++	++	+++	++	++	++	++	++	+
	BK 61	RO	+	+	+	+++	+	−	+	++	+	−
	BL 67	RO	−	−	−	−	+	−	−	−	−	−
	CK 155	RO	−	−	−	++	−	−	−	−	−	−
	CK 182	RO	−	−	−	−	−	−	−	−	−	−
	CL 63	RO	+	++	++	++	++	++	++	++	++	++
	CN8	RO	+	++	+	++	+	+	−	++	++	+
	GM14	RO	++	++	++	+++	++	+	−	++	++	+
Zebra mussels, Lake Winnipeg, MB, Canada	Z1	RO	−	−	−	+	−	−	−	−	−	−
	Z2	RO	−	−	−	−	−	−	−	−	−	−
	Z5	RO	−	−	−	−	−	−	−	−	−	−
	Z6	RO	+	−	+	−	+	+	−	−	+	−
	Z7	RO	+	−	−	−	−	−	−	−	−	−
	Z24	RO	−	−	−	++	−	−	−	+	-	−
	Z27	RO	+	+	+	++	−	−	−	++	+	++
	Z39	RO	−	−	+	++	+	−	−	+	−	−
	Z59	RO	+	−	+	+	−	+	−	−	+	−
	Z68	RO	−	++	−	++	−	+	−	++	−	+

Table 2. Meromictic lake, saline spring and biological soil crust AAP analyzed for metal(oid)-chelation via CAS assays with varied cations.

Environment	Strain	Medium	Mg	V	Mn	Fe	Co	Ni	Cu	Zn	Se	Te
Mahoney Lake, BC, Canada	ML1	N1	−	+	−	+	−	−	+	+	−	+
	ML3	N1	−	+	−	+	−	−	+	+	−	+
	ML4^T	N1	−	+	−	+	−	−	+	+	−	+
	ML6^T	N1	+	+	+	+	−	−	+	+	−	+
	ML10	N1	−	+	−	+	−	−	+	+	−	+
	ML14	N1	−	+	−	+	−	−	+	+	−	+
	ML19	N1	+	+	+	+	+	−	+	+	+	+
	ML20	N1	−	+	−	+	−	−	+	+	−	+
	ML21	N1	−	+	−	+	−	−	+	+	−	+
	ML22	N1	−	+	+	+	−	−	+	+	−	+
	ML30	N1	+	+	−	+	−	−	+	+	+	+
	ML35	N1	−	+	−	+	−	−	+	+	−	+
	ML36	N1	+	+	+	+	+	−	+	+	+	+
	ML37	N1	−	+	+	+	−	−	+	+	−	+
	ML42	N1	+	+	+	+	+	+	+	+	+	+
	ML45	N1	+	+	+	+	−	+	+	+	+	+
Blue Lake, BC, Canada	BL1	BLM	+	+	+	+	+	+	+	+	+	+
	BL4	BLM	−	+	−	+	−	−	+	+	−	+
	BL5	BLM	+	+	+	+	+	+	+	+	+	+
	BL7	BLM	+	+	+	−	+	+	+	−	+	+
	BL8	BLM	+	+	+	+	+	+	+	+	+	+
	BL9	BLM	+	+	+	+	+	+	+	+	+	+
	BL11	BLM	+	+	+	−	+	+	+	+	+	+
	BL14	BLM	+	+	+	−	+	+	+	+	+	+
	BL17	BLM	+	+	+	+	+	+	+	+	+	+
	BL22	BLM	+	−	+	−	+	+	+	−	+	+
East German Creek System, MB, Canada	EG1	MA	−	+	−	+	−	−	+	+	−	+
	EG2	MA	+	+	+	+	−	−	+	+	−	+
	EG3	MA	−	-	−	+	−	−	−	−	−	−
	EG4	MA	−	+	−	+	−	−	+	+	−	+
	EG5	MA	−	+	+	+	−	−	+	+	−	+
	EG6	MA	−	+	−	+	−	−	+	+	−	+
	EG7	MA	−	+	+	+	−	−	+	+	−	+
	EG8	MA	−	+	−	+	+	−	+	+	+	+
	EG9	MA	−	+	−	+	−	−	+	+	−	+
	EG10	MA	−	+	−	+	−	−	+	+	+	+
	EG11	MA	−	+	−	+	−	−	+	+	−	+
	EG13	MA	−	+	−	+	−	−	+	+	−	+
	EG15	MA	−	+	−	+	+	−	+	+	−	+
	EG17^T	MA	+	+	+	+	+	+	+	+	−	+
	EG19^T	MA	+	+	+	++	−	+	+	++	+	+
Sandy Lands Forest soil crust, MB, Canada	SS56	BSCA	++	++	++	+	++	++	++	++	++	++
	SS63	BSCA	++	+++	++	+++	++	+++	+++	+++	+++	+++
	SS335	BSCA	+++	+++	+++	+++	++	++	+++	++	++	+++
Spruce Woods National Park soil crust, MB, Canada	J01	BSCA	+++	++	+++	+++	+	++	++	++	+	++
	J05	BSCA	+	+	+	++	−	−	+	+	−	+
	P4	BSCB	++	++	++	+	+	+	++	++	++	++
	P13	BSCB	+++	+++	+++	+++	++	+++	+++	+++	++	++
	P40	BSCB	++	++	++	++	+	++	++	++	+	+
	P132	BSCB	++	++	++	+++	++	++	++	++	++	++
	P233	BSCB	−	++	++	+++	++	++	++	++	++	++

3.4. C. halotolerans Pigment Purification and Identification

When purified via 16.5% tris-tricine gel electrophoresis, the brown pigment migrated further than the loading buffer's running dye, CBB, after 1 h (Figure 3A). This gel-shift revealed the pigment under study to be smaller than CBB, which has a known molecular weight of 856.03 g/mol. Since well 1 contained a standard ladder, the measurement of migration distance for the siderophores' brown band, and of each protein in the ladder allowed for the rough estimation of pigment size to be near ~341 Da (Figure S1). Gel staining and destaining revealed that the siderophore sample collected after resin concentration and diluted in methanol at 200, 20, 2, and 0.2 µg/mL loaded into wells 2 through 5, respectively, had some contaminating small proteins as expected (Figure 3B). In addition, samples that received the subsequent removal of proteins larger than 3 kDa via spin-column were also run on the same gel. Wells 6 through 8 contained brown pigment, which passed through the <3 kDa spin column and then diluted in 60% methanol to 100, 10, and 1 µg/mL, respectively. This step purified the brown pigment of any contaminating proteins (Figure 3A,B). Often, small molecules below 1 kDa are lost from the gel during destaining step [40], which likely caused the small brown pigment to escape similarly to CBB, (Figure 3B). Regardless, testing the crude dried pigment (Disk 1), as well as fractions <3 kDa and >3 kDa (Disks 2 and 3), confirmed the smaller fraction containing brown pigment acted as a siderophore, while the larger proteins did not (Figure 3C). Since the <3 kDa fraction had no contaminants (Figure 3B), but contained the brown pigment prior to destaining (Figure 3A), the small ~341 Da molecule produced by *C. halotolerans* acted as a siderophore.

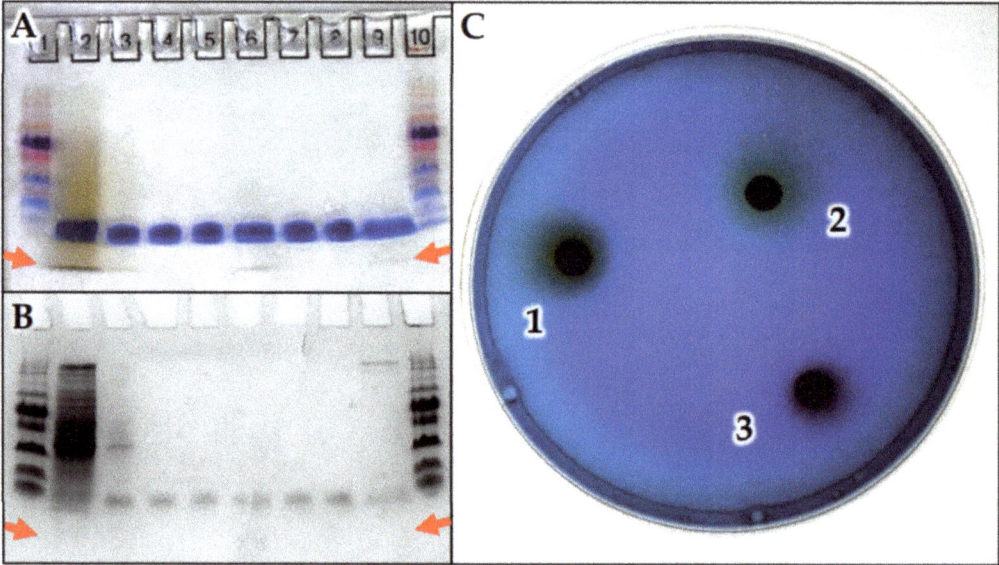

Figure 3. Gel purification of *C. halotolerans* brown pigment, and confirmation of siderophore activity. (**A**) Unstained and (**B**) stained tris-tricine gel electrophoresis performed on: siderophore from resin concentration, wells 2–5; siderophore sample smaller than 3 kDa, wells 6–8; proteins larger than 3 kDa and remaining in solution well 9; standard ladder, wells 1 and 10. (**C**) Siderophore activity observed for: (1) Crude resin extract positive reaction visible as yellowing area; (2) <3 kDa dark brown fraction produced positive yellowing reaction; (3) Proteins >3 kDa negative results observed as darkening of blue due to alkaline pH, without siderophore activity present.

4. Discussion
4.1. Siderophore Production Revealed by CAS-Assay

While effectively employed for the identification of numerous siderophore producing bacteria, the CAS assay has a notable limitation; microbial growth may be hindered due to a few factors [31]. Since the metal cation of interest weakly bound to the dye, it was less freely diffusible into the bacteria, resulting in lower availability. In addition, the medium needed to be at pH 6.8 for the indicative colour change to occur successfully. Finally, HDTMA has been known for its slight bacterial toxicity. If an organism had weak Fe transport, required a basic or acidic optimal pH, or was sensitive to HDTMA, it would of had reduced growth. In our experiments, we found that while a range of media compositions could be used, all AAP did have reduced growth on CAS plates, when compared to control. However, since growth did indeed occur, siderophore production could therefore be analyzed.

Other, more high-throughput, alternative methods were considered, but the chosen agarized CAS-assay was most ideal for determining siderophore and metallophore production. Recently, a bulk screening assay for siderophore detection was proposed [41]. However, both the "traditional" and "modified microplate" qualitative techniques described could not be used for our application, since there was an assumption that siderophores were always constitutively expressed, whereas cultures were grown in complex media without manipulating the concentration of any metal of interest. While this may be the case for some strains, the expression/production of most siderophores or other secondary metabolites usually requires induction from an external factor, which can be either the presence or absence of a stimuli [18]. For biologically significant metals, including Fe, Zn, and Cu, the related cation-specific metallophores are expected to be produced only under limiting conditions. In opposite, metallophores that act on V, Te, Se, or other more toxic metal(loid)s are presumably only synthesized when such toxins are present at higher concentrations. Therefore, growth on agarized plates that contained each metal of interest pre-bound to CAS-reagent as a stimulant for metallophore production was our chosen method. Future works are required to test varying concentrations of metal cations to determine which yields more/less production of specific metallophores.

Regarding the Fe-chelation, a range of phenotypes was observed, when detecting siderophore activity after the 5-day incubation (Figure A1). In particular, 4 phenotypes were distinguished among all isolates based on size of clearing/colour change zone. Negative results (−) had to have bacterial growth, but without a change in medium opacity. The smallest zone of clearing (+), seen previously after prolonged growth [42], was likely not due to siderophore production [43]. Rather, this small aura could be due to high rates of metal uptake from the surrounding medium. Prolonged bacterial metabolism allowed for increased simple metal diffusion into cells, which reduced its amount in the near-by medium, rendering the dye in that narrow area void of metal, turning yellow. Hence, there is a very small <1 mm zone. In comparison, moderate or highly diffusible siderophore release was quite evident, and was segregated into two phenotypic groups. A zone <10 mm compared to a zone >10 mm, where each represented less or more diffusible secondary metabolite, respectfully, and where lower or higher concentrations were produced and captured additional metal cations. In this test, all 4 zonal varieties for Fe-chelation were discovered (Tables 1 and 2, and Figure 2).

Modifying the CAS assay to monitor chelation of elements other than Fe^{2+} included Mg^{2+}, V^{3+}, Mn^{2+}, Co^{2+}, Ni^{2+}, Cu^{2+}, Zn^{2+}, Se^{4+} and Te^{2+} (Figure 4A). The method could be successfully adapted for all selected metal(loid)s, where only Mn^{2+}, Co^{2+}, Ni^{2+}, Cu^{2+}, Zn^{2+} had been proposed previously [44]. In addition, the assay could be used with variable nutrient and organic carbon concentrations that did not inhibit the activity. Furthermore, we discovered that a wide range of AAP produced metallophores, which bound a variety of metal(loid)s in addition to Fe (Tables 1 and 2). Some AAP had siderophores that specifically bound Fe only, including strains E1, E4(1), RB3, NM416, AM27, CK155, BC100, Z24, Z39, and J05. Most AAP with highly diffusible siderophores, and >10 mm zones on Fe-CAS plates, had activity towards a large variety of metal(loid)s. Few strains, including SS56 and

P4 preferentially bound metal(loid)s other than Fe. Since all 9 additional cations Mg^{2+}, V^{3+}, Mn^{2+}, Co^{2+}, Ni^{2+}, Cu^{2+}, Zn^{2+}, Se^{4+} and Te^{2+} could be bound by metallophores, future work may consider a broader range of cations and the extent of metals that can be exogenously chelated. To determine if specific metal(loid)s were more readily chelated than others, all positive CAS assay results were tallied for each, Figure 4B. Here, multiple +++ represented strong production or significant interaction, and could be compared to weaker reactions such as +, or ++. In this way, Fe was the highest cumulative acquired cation, with the activity ranked as Fe > Zn > V > Te > Cu > Mn > Mg > Se > Ni > Co. It would appear slight variation in cation size, from Fe (55.85) to Co (58.93), had a strong impact on activity, where Fe was most frequently captured, and Co the rarest. Indeed, since strongest reactions existed for Fe and Zn, with less reactivity found for metals in between the sizes of these two within the periodic table, specific mechanisms likely existed to capture either metal, inferring cation specificity for each metallophore. Further analysis will be required to see if the broad range metal(loid) acquisition is due to the production of a single siderophore that reacts with a variety of metals, or the bacteria produced specific metallophores for each. AAP metallophore production could be explained by requirement of trace elements and the need of toxicity prevention. Both Fe and Zn are biologically necessary and commonly limited or unavailable in dissolved forms, therefore acquisition via specific siderophores would be an asset. The 3rd and 4th highest captured metal(loid)s were V and Te, which have known toxic properties, and could have been sequestered as a result of a protection mechanism alone. Bound to a metallophore, these cations would be too large to freely diffuse through outer membranes and be restricted from entering the cell, to prevent any toxic influence. The remaining 6 had reduced activity, likely because they are less toxic and easier available in microbial environment, and therefore less metallophores could be expected.

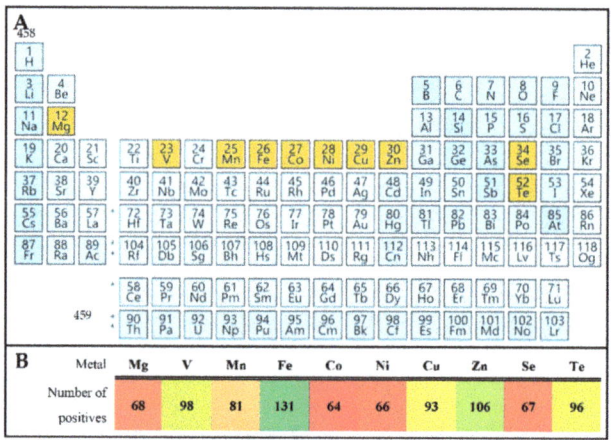

Figure 4. Variant metal(loid)s tested in CAS assay. (**A**) Chosen elements highlighted in yellow. (**B**) Tallied number of positive results from Tables 1 and 2.

4.2. Environmental Distribution of Siderophore Producers

In relation to the origin of isolation a few patterns were observed. Collections of AAP strains that originated from hot springs, freshwater lakes, and biological soil crusts all had a high proportion of siderophore producers. In opposition, the isolates from marine, meromictic lake, and saline spring habitats produced less, or no siderophores. Rather, they seem to rely on sufficient metal uptake directly from the local microenvironment by diffusion. The main differentiating factor in this case is the requirement of NaCl. It appears that AAP capable of halotolerance or halophilic growth do not produce siderophores of equal activity or quantity as bacteria that do not depend on NaCl for growth. This correlation may be due to each strains' reaction to osmotic pressure. Cells that are adapted to tolerate higher levels of NaCl will likely have additional cation pumps to survive naturally and resist the high levels of solutes in a saline environment. As seen elsewhere, cation pumps can be non-specific, where many are capable of removing several toxic cations from the cytoplasm of microorganisms [45]. In comparison, freshwater AAP are comparably less osmophilic, and would therefore have less need of copious cation pumps in membranes. Therefore, they are more likely to evolve defensive mechanisms that modify the local environment to suit their needs, including the production of external small molecules that would render cations less diffusible.

4.3. Phylogenetic Diversity of Fe-Chelating AAP

Comparing phylogenetic diversity and the production of siderophores on CAS plates by AAP delivered a few notable trends (Figure 2). Broadly, no studied representatives of the order *Rhodobacterales* or *Hyphomonadaceae* had siderophores, as no or <1 mm zone of colour change was present. The *Acetobacteraceae* that were in closest relation to known type species of AAP, strains RB-3T and CK155 had siderophores that only chelated Fe as discussed above. In comparison, strains P40 and J01 were more genetically distant from known type species of AAP, and also produced significant zones of clearing, ≥10 mm for all metal(loid)s tested, signifying their own group. The predominantly strong expression of siderophores by AAP among the *Methylobacteraceae* warrant further study as most type species in this clade have not been previously recognized as phototrophs. Of note, previous research had found that isolates relating to *Methylobacterium mesophilicum* and *M. extorquens* produced siderophores, but the activity had not been linked to aerobic anoxygenic phototrophy [46]. One clade of AAP which related by 98.7–99.1% 16S rRNA sequence similarity to *Bosea lupini*, including strains P13, SS335, and SS63 were all capable of strong production of metallophores, >10 mm zones for multiple metal(loid)s. In comparison, strains P4 and SS56, which have 99.6% relation to *M. phylloshaerae* and 99.6% to *M. branchiatum*, respectively, both had stronger reactions against metal(loid)s that were not Fe (Figure A2). This activity may be explained through the findings of related works, where methanotrophs produced methanobactin, a chalkophore, which is a Cu specific metallophore [15]. The isolates we have tested may indeed possess similar mechanisms as they sequester Cu strongly, as well as Mg, V, Mn, Zn, and Te more favorably than Fe.

Sphingomonadaceae could be separated into 3 groups, as those related to *Citromicrobium* did not produce siderophores, *Sphingomonas* relatives produced some siderophores that were predominantly Fe specific, while relatives of *Blastomonas* could produce siderophores that acted on all metal(loid)s tested. A few *Sphingomonas* relatives had been previously found to produce siderophores against Fe [47], but not other metal(loid)s. The *Blastomonas/Erythromonas* grouping was of particular interest as most representatives revealed strong metallophore production against all 10 cations tested. Our results corresponded well with previous analysis of *E. ursincola*, strain KR99, which had very high resistance to V, Te, and Se oxides, internally reducing them to elemental states [48]. Since KR99 can both acquire Se, Te, and V via metallophore activity (Figure A2), and internally reduce metal(loid) oxides, it appears to require them in reduced form for some reason. Future study of such associations will determine if *E. ursincola* sequesters these cations as a protective measure, or uses them for some metabolic purpose. In comparison, the *Erythrobacteraceae* were not as concisely divided as other families, where those closest to *Erythromicrobium* had siderophores, but most *Porphyrobacter* had very small <1 mm or negligible zones. An exception was strain BE100, which branched distantly from its nearest relative *P. colymbi* (Figure 2), and showed a significant presence of metallophores, ~10 mm zones for all metal(loid)s except Te. Finally, strain EG19 that hailed from γ- rather than α-*Proteobacteria*, had moderate siderophore production for both Fe and Zn, <10 mm. While many siderophores have been discovered as products of bacteria within the γ -proteobacterial clade (Table S1), none have been documented as highly pigmented.

4.4. Analysis of the Brown-Coloured Siderophore

Gel purification of brown pigment produced by *C. halotolerans* (Figure 3), revealed that while CBB bound to proteins remained in the gel, both unbound CBB and brown pigment were released during destaining process. Comparing lanes 2–5, the sample prepared by resin concentration clearly contained brown pigment, but also accumulated proteins smaller than 26 kDa. The use of the 3 kDa cut off spin column did indeed remove these contaminants, as shown in lanes 6–8. Since the small brown pigment passed through the spin column, was purified on gel electrophoresis, and maintained activity on CAS plates, these tests confirmed that it acted as a siderophore. Furthermore, these procedures established the brown compound to clearly be under 800 Da, and approximated to be around ~341 Da when correlated to the ladder during TRIS-tricine gel electrophoresis (Figure S1). Both the estimated small size and the brown appearance of the siderophore synthesized by *C. halotolerans* were useful for its tentative identification. The most comparable small molecule described in literature was rhodotorulic acid (Table S1), a 344.4 Da siderophore that was pigmented red when bound to Fe [49]. However, this acid has only been naturally found in yeasts including *Rhodotorula pilimanae*, with no known bacterial producers [50]. With that in mind, hydroxamic acids are produced by both bacteria and fungi [51,52], and therefore similar secondary metabolites can be expected in other species. In addition, the colour disparity, red compared to orange-brown, may indicate an altered structure among siderophores produced by *R. pilimanae* and *C. halotolerans*, respectively.

Since *C. halotolerans* hails from the γ-*Proteobacteria*, comparisons must be drawn between its siderophore and those produced by other species in the γ-subclass. The most similar in size was acinetobactin, a 346.4 Da molecule from *Acinetobacter baumannii* expressed using the operon containing *basABCDEFGHIJ*, *bauABCDEF* and *barAB* genes [53]. *C. halotolerans* genome, published within the One Thousand Microbial Genomes Phase 4 (KMG IV) project by the DOE Joint Genome Institute, submitted online in 2019 with accession number PRJNA520330 [54], contained neither similar genes nor was the operon present, while using very low homology search. Due to the divergence between *C. halotolerans*' pigmented siderophore and *A. baumannii*'s lack of colour, and the absence of similar genes, we assume that acinetobactin was not the siderophore of *C. halotolerans*. Further structural analysis will be necessary to confirm the structural identity of this novel compound.

5. Conclusions

We have discovered that many AAP produce siderophores or metallophores as diffusible secondary metabolites. Production could be related to acquisition of metal(loid)s including magnesium, vanadium, manganese, iron, cobalt, nickel, copper, zinc, selenium and tellurium, or to provide resistance to toxic metal(loid)s in environments with elevated concentrations. A correlation existed between site of isolation and production of siderophores, whereas tested freshwater AAP produced siderophores, and AAP that required NaCl predominantly did not. Furthermore, there could be connection between phylogeny of isolates and their ability to form siderophores, but as with many phenotypes to genotype comparisons, it did not appear as a strictly followed rule. With such considerations, siderophore production cannot be recommended as a taxonomic marker for AAP identification, as variable production types occurred. However, a potential application exists to use this phenotypic feature during taxonomic differentiation between species. Future work will hopefully identify the siderophores, and potential metallophores, produced by each AAP, and determine the total list of metal cations that can be targeted.

Supplementary Materials: The following are available online at https://www.mdpi.com/article/10.3390/microorganisms9050959/s1, Figure S1: Determination of siderophore size, Table S1: Siderophores examples listed from largest to smallest.

Author Contributions: Conceptualization, S.B.K., E.H. and V.Y.; methodology, S.B.K., E.H. and V.Y.; software, S.B.K.; validation, S.B.K., V.Y.; formal analysis, S.B.K.; investigation, S.B.K., E.H.; resources,

V.Y.; data curation, S.B.K.; writing—original draft preparation, S.B.K., E.H.; writing—review and editing, S.B.K., V.Y.; visualization, S.B.K.; supervision, V.Y.; project administration, V.Y.; funding acquisition, V.Y. All authors have read and agreed to the published version of the manuscript.

Funding: This research was funded by an NSERC Discovery Grant and a University of Manitoba GETS grant, both held by V. Yurkov.

Institutional Review Board Statement: Not applicable.

Informed Consent Statement: Not applicable.

Data Availability Statement: Data is contained within the article.

Conflicts of Interest: The authors declare no conflict of interest.

Appendix A

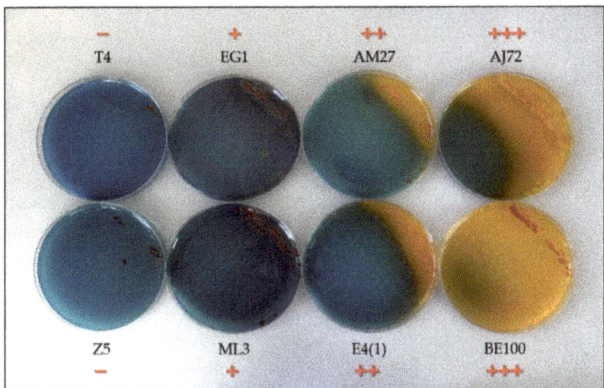

Figure A1. Iron chelating CAS assay after 5 days' growth. Zones of activity are separated into 4 groups based on range of colour change, as indicated in red by each strain name.

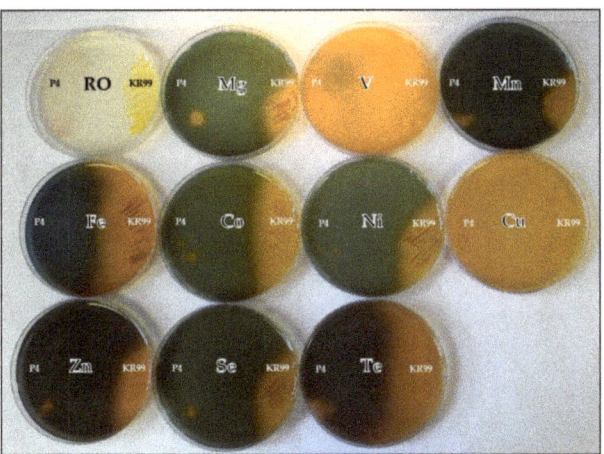

Figure A2. Variant metal(loid) CAS assay after 5 days' growth for strains P4 and KR99. Growth on RO without CAS was compared to those CAS plates supplemented with Mg, V, Mn, Fe, Co, Ni, Cu, Zn, Se, or Te.

Table A1. Freshwater and saline AAP chosen for siderophore testing.

Environment	Strain	Medium	Most Related Type Species	Accession #
Hot spring, Kamchatka island	KR99^T	RO	Erythromonas ursincola	NR_119243.1
Warm temperature spring, Bikal Lake, Russia	E1	RO	99.6% Porphyrobacter colymbi	MW970346
	E4(1)	RO	99.7% Porphyrobacter donghaensis	MW970347
	E5^T	RO	Erythromicrobium ramosum	NR_041891.1
	RB3^T	RO	Roseococcus thiosulfatophilus	NR_026114.1
	RB16-17^T	RO	Sandaracinobacter sibiricus	NR_026382.1
	T4^T	RO-NaCl	Erythrobacter litoralis	NR_119016.1
Deep Ocean, Juan DeFuco Ridge, Pacific Ocean	JF1^T	RO-NaCl	Citromicrobium bathyomarinum	Y16267.1
	C6	RO-NaCl	98.1% Citromicrobium bathyomarinum	MW970348
	C7	RO-NaCl	99.8% Citromicrobium bathyomarinum	MW970349
	C8	RO-NaCl	99.9% Citromicrobium bathyomarinum	MW970350
	C14	RO-NaCl	99.6% Citromicrobium bathyomarinum	MW970351
	C23	RO-NaCl	99.7% Citromicrobium bathyomarinum	MW970352
	C26	RO-NaCl	99.8% Citromicrobium bathyomarinum	MW970353
	N25	RO-NaCl	99.7% Citromicrobium bathyomarinum	MW970354
	N34	RO-NaCl	99.7% Citromicrobium bathyomarinum	MW970355
	N48	RO-NaCl	99.8% Citromicrobium bathyomarinum	MW970356
	N56	RO-NaCl	99.9% Citromicrobium bathyomarinum	MW970357
	N78	RO-NaCl	99.6% Citromicrobium bathyomarinum	MW970358
Rag Beach sediment, BC, Canada	23-SB	RO-NaCl	99.7% Porphyrobacter donghaensis	MW970359
	15-SB	RO-NaCl	99.8% Porphyrobacter donghaensis	MW970360
Central Gold Mine, MB, Canada	C4	RO	99.4% Porphyrobacter colymbi	KX148515
	C9	RO	99.1% Brevundimonas variabilis	KX148516
	C11	RO	98.6% Brevundimonas bacteroides	KX148517
	NM4.16	RO	99.7% Porphyrobacter colymbi	KX148518
	NM4.18	RO	99.3% Blastomonas fulva	KX148519
Lake Winnipeg, MB, Canada	AJ 72	RO	99.7% Sphingorhabdus lacus	MW970361
	AM 19	RO	99.7% Erythromicrobium ramosum	MW970362
	AM 27	RO	99.6% Erythromicrobium ramosum	MW970363
	CK 182	RO	93.4% Porphyrobacter donghaensis	MW970364
	BL 67	RO	98.9% Sandarakinorhabdus cyanobacteriorum	MW970365
	BK 61	RO	99.1% Blastomonas fulva	MW970366
	BE 100	RO	99.4% Porphyrobacter colymbi	MW970367
	AM 91	RO	98.5% Porphyrobacter sanguineus	MW970368
	CK 155	RO	98.1% Roseomonas sediminicola	MW970369
	CL 63	RO	100% Methylorubrum extorquens	MW970370
	BA 23	RO	96.2% Methylobacterium indicum	MW970371
	BC100	RO	96.6% Sphingomonas yantingensis	MW970372
	CN8	RO	99.2% Porphyrobacter neustonensis	MW970373
	GM14	RO	99.2% Erythromonas ursincola	MW970374
Zebra mussels, Lake Winnipeg, MB, Canada	Z1	RO	99.1% Porphyrobacter colymbi	MN987006
	Z2	RO	99.4% Porphyrobacter tepidarius	MN987007
	Z5	RO	99.6% Porphyrobacter tepidarius	MN987008
	Z6	RO	99.3% Porphyrobacter neustonensis	MN987009
	Z7	RO	99.6% Erythromicrobium ramosum	MN987010
	Z24	RO	98.7% Porphyrobacter sanguineus	MN987011
	Z27	RO	99.4% Porphyrobacter neustonensis	MN987012
	Z39	RO	98.9% Porphyrobacter sanguineus	MN987013
	Z59	RO	99.5% Porphyrobacter neustonensis	MN987014
	Z68	RO	99.4% Porphyrobacter neustonensis	MN987015

Table A2. Meromictic lake and biological soil crust AAP chosen for siderophore testing.

Environment	Strain	Medium	Most Related Type Species	Accession #
Mahoney Lake, BC, Canada	ML1	N1	98.4% *Erythromicrobium ramosum*	MW970375
	ML3	N1	99.4% *Porphyrobacter sanguineus*	MW970376
	ML4^T	N1	*Porphyrobacter meromictius*	NR_115007.1
	ML6^T	N1	*Roseicyclus mahoneyensis*	NR_042080.1
	ML10	N1	97.9% *Salinarimonas ramus*	MW970377
	ML14	N1	98.9% *Erythromicrobium ramosum*	MW970378
	ML19	N1	96.9% *Porphyrobacter meromiticus*	MW970379
	ML20	N1	99.3% *Blastomonas fulva*	MW970380
	ML21	N1	98.5% *Erythromicrobium ramosum*	MW970381
	ML22	N1	99.8% *Porphyrobacter sanguineus*	MW970382
	ML30	N1	98.8% *Seohaeicola saemankumensis*	MW970383
	ML35	N1	99.4% *Blastomonas fulva*	MW970384
	ML36	N1	97.1% *Porphyrobacter sanguineus*	MW970385
	ML37	N1	95.4% *Glycocaulis profundi*	MW970386
	ML42	N1	98.0% *Roseicyclus marinus*	MW970387
	ML45	N1	94.7% *Ruegeria intermedia*	MW970388
Blue Lake, BC, Canada	BL1	BLM	98.1% *Roseicyclus marinus*	MW970389
	BL4	BLM	99.3% *Porphyrobacter sanguineus*	MW970390
	BL5	BLM	99.2% *Seohaeicola saemankumensis*	MW970391
	BL7	BLM	95.6% *Glycocaulis profundi*	MW970392
	BL8	BLM	98.1% *Roseicyclus marinus*	MW970393
	BL9	BLM	98.0% *Roseicyclus marinus*	MW970394
	BL11	BLM	95.8% *Glycocaulis profundi*	MW970395
	BL14	BLM	95.6% *Glycocaulis profundi*	MW970396
	BL17	BLM	95.1% *Glycocaulis profundi*	MW970397
	BL22	BLM	95.0% *Glycocaulis profundi*	MW970398
East German Creek System, MB, Canada	EG1	MA	95.2% *Roseovarius pacificus*	AM691094
	EG2	MA	94.4% *Roseovarius bejariae*	AM691093
	EG3	MA	97.3% *Yoonia vestfoldensis*	AM691092
	EG4	MA	98.8% *Erythrobacter longus*	AM691105
	EG5	MA	95.7% *Roseovarius pacificus*	AM691095
	EG6	MA	98.4% *Porphyrobacter meromictius*	AM691106
	EG7	MA	97.3% *Roseovarius nitratireducens*	AM691097
	EG8	MA	99.0% *Roseovarius tolerans*	AM691101
	EG9	MA	97.3% *Roseovarius nitratireducens*	AM691098
	EG10	MA	98.2% *Roseovarius tibetensis*	AM691100
	EG11	MA	97.7% *Roseovarius nitratireducens*	AM691099
	EG13	MA	99.0% *Roseovarius tolerans*	AM691102
	EG15	MA	98.9% *Erythrobacter aquimaris*	AM691107
	EG17^T	MA	*Charonomicrobium ambiphototrophicum*	AM691091
	EG19^T	MA	*Chromocurvus halotolerans*	AM691088
Sandy Lands Forest Soil Crust, MB, Canada	SS56	BSCA	99.6% *Methylobacterium brachiatum*	MW970399
	SS63	BSCA	99.1% *Bosea lupini*	MW970400
	SS335	BSCA	98.3% *Bosea lupini*	MW970401
Spruce Woods National Park Soil Crust, MB, Canada	J01	BSCA	98.4% *Belnapia moabensis*	MW970402
	J05	BSCA	96.6% *Sphingomonas pruni*	MW970403
	P4	BSCB	99.6% *Methylobacterium phyllosphaerae*	MW970404
	P13	BSCB	98.7% *Bosea lupini*	MW970405
	P40	BSCB	99.9% *Belnapia soli*	MW970406
	P132	BSCB	99.6% *Methylobacterium brachiatum*	MW970407
	P233	BSCB	99.8% *Methylobacterium tardum*	MW970408

References

1. Granger, J.; Price, N.M. The importance of siderophores in iron nutrition of heterotrophic marine bacteria. *Limnol. Oceanogr.* **1999**, *44*, 541–555. [CrossRef]
2. Challis, G.L. A widely distributed bacterial pathway for siderophore biosynthesis independent of nonribosomal peptide synthetases. *ChemBioChem* **2005**, *6*, 601–611. [CrossRef] [PubMed]
3. Schalk, I.J.; Hannauer, M.; Braud, A. New roles for bacterial siderophores in metal transport and tolerance. *Environ. Microbiol.* **2011**, *13*, 2844–2854. [CrossRef] [PubMed]
4. Yurkov, V.V.; van Gemerden, H. Impact of light/dark regimen on growth rate, biomass formation and bacteriochlorophyll synthesis in Erythromicrobium hydrolyticum. *Arch. Microbiol.* **1993**, *159*, 84–89. [CrossRef]
5. Yurkov, V.; Hughes, E. Aerobic Anoxygenic Phototrophs: Four Decades of Mystery. In *Modern Topics in the Phototrophic Prokaryotes: Environmental and Applied Aspects*; Hallenbeck, P.C., Ed.; Springer: Cham, Switzerland, 2017; pp. 193–217.
6. Faraldo-Gómez, J.D.; Sansom, M.S.P. Acquisition of siderophores in gram-negative bacteria. *Nat. Rev. Mol. Cell Biol.* **2003**, *4*, 105–116. [CrossRef]
7. Lankford, C.E. Bacterial assimilation of iron. *Crit. Rev. Microbiol.* **1973**, *2*, 273–331. [CrossRef]
8. Reid, R.T.; Livet, D.H.; Faulkner, D.J.; Butler, A. A siderophore from a marine bacterium with an exceptional ferric ion affinity constant. *Nature* **1993**, *366*, 455–458. [CrossRef]
9. Wandersman, C.; Delepelaire, P. Bacterial iron sources: From siderophores to hemophores. *Annu. Rev. Microbiol.* **2004**, *58*, 611–647. [CrossRef]
10. Grobelak, A.; Hiller, J. Bacterial siderophores promote plant growth: Screening of catechol and hydroxamate siderophores. *Int. J. Phytoremed.* **2017**, *19*, 825–833. [CrossRef]
11. Crosa, J.H.; Walsh, C.T. Genetics and Assembly Line Enzymology of Siderophore Biosynthesis in Bacteria. *Microbiol. Mol. Biol. Rev.* **2002**, *66*, 223–249. [CrossRef]
12. Lamont, I.L.; Beare, P.A.; Ochsner, U.; Vasil, A.I.; Vasil, M.L. Siderophore-mediated signaling regulates virulence factor production in Pseudomonas aeruginosa. *Proc. Natl. Acad. Sci. USA* **2002**, *99*, 7072–7077. [CrossRef]
13. Adler, C.; Corbalán, N.S.; Seyedsayamdost, M.R.; Pomares, M.F.; de Cristóbal, R.E.; Clardy, J.; Kolter, R.; Vincent, P.A. Catecholate Siderophores Protect Bacteria from Pyochelin Toxicity. *PLoS ONE* **2012**, *7*, e46754. [CrossRef]
14. Ghssein, G.; Brutesco, C.; Ouerdane, L.; Fojcik, C.; Izaute, A.; Wang, S.; Hajjar, C.; Lobinski, R.; Lemaire, D.; Richaud, P.; et al. Biosynthesis of a broad-spectrum nicotianamine-like metallophore in Staphylococcus aureus. *Science* **2016**, *352*, 1105–1109. [CrossRef]
15. Kim, H.J.; Graham, D.W.; DiSpirito, A.A.; Alterman, M.A.; Galeva, N.; Larive, C.K.; Asunskis, D.; Sherwood, P.M.A. Methanobactin, a copper-acquisition compound from methane-oxidizing bacteria. *Science* **2004**, *305*, 1612–1615. [CrossRef]
16. Morey, J.R.; Kehl-Fie, T.E. Bioinformatic Mapping of Opine-Like Zincophore Biosynthesis in Bacteria. *mSystems* **2020**, *5*, 1–16. [CrossRef]
17. Welch, R.M.; Shuman, L. Micronutrient Nutrition of Plants. *CRC Crit. Rev. Plant Sci.* **1995**, *14*, 49–82. [CrossRef]
18. Pedler, J.F.; Parker, D.R.; Crowley, D.E. Zinc deficiency-induced phytosiderophore release by the Triticaceae is not consistently expressed in solution culture. *Planta* **2000**, *211*, 120–126. [CrossRef]
19. Nies, D.H. Heavy metal-resistant bacteria as extremophiles: Molecular physiology and biotechnological use of Ralstonia sp. CH34. *Extremophiles* **2000**, *4*, 77–82. [CrossRef]
20. Rothschild, L.J.; Mancinelli, R.L. Life in extreme environments. *Nature* **2001**, *409*, 1092–1101. [CrossRef]
21. Csotonyi, J.T.; Maltman, C.; Yurkov, V. Influence of tellurite on synthesis of bacteriochlorophyll and carotenoids in aerobic anoxygenic phototrophic bacteria. *Res. Trends Photochem. Photobiol.* **2014**, *16*, 1–17.
22. Maltman, C.; Yurkov, V. The Effect of Tellurite on Highly Resistant Freshwater Aerobic Anoxygenic Phototrophs and Their Strategies for Reduction. *Microorganisms* **2015**, *3*, 826–838. [CrossRef]
23. Csotonyi, J.T.; Swiderski, J.; Stackebrandt, E.; Yurkov, V.V. Novel halophilic aerobic anoxygenic phototrophs from a Canadian hypersaline spring system. *Extremophiles* **2008**, *12*, 529–539. [CrossRef]
24. Csotonyi, J.T.; Stackebrandt, E.; Swiderski, J.; Schumann, P.; Yurkov, V. Chromocurvus halotolerans gen. nov., sp. nov., a gammaproteobacterial obligately aerobic anoxygenic phototroph, isolated from a Canadian hypersaline spring. *Arch. Microbiol.* **2011**, *193*, 573–582. [CrossRef]
25. Drechsel, H.; Jung, G. Peptide siderophores. *J. Pept. Sci.* **1998**, *4*, 147–181. [CrossRef]
26. Yurkov, V. Aerobic Phototrophic Proteobacteria. In *The Prokaryotes*; Dworkin, M., Falkow, S., Rosenberg, E., Schleifer, K.-H., Stackebrandt, E., Eds.; Springer: New York, NY, USA, 2006; pp. 562–584.
27. Csotonyi, J.T.; Swiderski, J.; Stackebrandt, E.; Yurkov, V. A new environment for aerobic anoxygenic phototrophic bacteria: Biological soil crusts. *Environ. Microbiol. Rep.* **2010**, *2*, 651–656. [CrossRef]
28. Yurkova, N.; Rathgeber, C.; Swiderski, J.; Stackebrandt, E.; Beatty, J.T.; Hall, K.J.; Yurkov, V. Diversity, distribution and physiology of the aerobic phototrophic bacteria in the mixolimnion of a meromictic lake. *FEMS Microbiol. Ecol.* **2002**, *40*, 191–204. [CrossRef]
29. Kuzyk, S.B.; Pritchard, A.O.; Plouffe, J.; Sorensen, J.L.; Yurkov, V. Psychrotrophic violacein-producing bacteria isolated from Lake Winnipeg, Canada. *J. Great Lakes Res.* **2020**. [CrossRef]
30. Kuzyk, S.B.; Wiens, K.; Ma, X.; Yurkov, V. Association of aerobic anoxygenic phototrophs and zebra mussels, Dreissena polymorpha, within the littoral zone of Lake Winnipeg. *J. Great Lakes Res.* **2020**. [CrossRef]

31. Schwyn, B.; Neilands, J.B. Universal chemical assay for the detection and determination of siderophores. *Anal. Biochem.* **1987**, *160*, 47–56. [CrossRef]
32. Rainey, F.A.; Ward-Rainey, N.; Kroppenstedt, R.M.; Stackebrandt, E. The genus Nocardiopsis represents a phylogenetically coherent taxon and a distinct actinomycete lineage: Proposal of *Nocardiopsaceae* fam. nov. *Int. J. Syst. Bacteriol.* **1996**, *46*, 1088–1092. [CrossRef]
33. Sanger, F.; Coulson, A.R. A rapid method for determining sequences in DNA by primed synthesis with DNA polymerase. *J. Mol. Biol.* **1975**, *94*, 441–448. [CrossRef]
34. Madden, T. The BLAST Sequence Analysis Tool. In *The NCBI Handbook [Internet]*; McEntyre, J., Ostell, J., Eds.; National Center for Biotechnology Information (US): Bethesda, MD, USA, 2002; pp. 1–15.
35. Kumar, S.; Stecher, G.; Li, M.; Knyaz, C.; Tamura, K. MEGA X: Molecular evolutionary genetics analysis across computing platforms. *Mol. Biol. Evol.* **2018**, *35*, 1547–1549. [CrossRef] [PubMed]
36. Felsenstein, J. Confidence Limits on Phylogenies: An Approach Using the Bootstrap. *Evolution* **1985**, *39*, 783–791. [CrossRef] [PubMed]
37. Nei, M.; Kumar, S. *Molecular Evolution and Phylogenetics*; Oxford University Press: New York, NY, USA, 2000.
38. Barker, R.; Boden, N.; Cayley, G.; Charlton, S.C.; Henson, R.; Holmes, M.C.; Kelly, I.D.; Knowles, P.F. Properties of cupric ions in benzylamine oxidase from pig plasma as studied by magnetic-resonance and kinetic methods. *Biochem. J.* **1979**, *177*, 289–302. [CrossRef]
39. Yamamoto, S.; Okujo, N.; Sakakibara, Y. Isolation and structure elucidation of acinetobactin., a novel siderophore from *Acinetobacter baumannii*. *Arch. Microbiol.* **1994**, *162*, 249–254.
40. Schägger, H.; von Jagow, G. Tricine-sodium dodecyl sulfate-polyacrylamide gel electrophoresis for the separation of proteins in the range from 1 to 100 kDa. *Anal. Biochem.* **1987**, *166*, 368–379. [CrossRef]
41. Arora, N.K.; Verma, M. Modified microplate method for rapid and efficient estimation of siderophore produced by bacteria. *3 Biotech* **2017**, *7*, 1–9. [CrossRef]
42. Ames-Gottfred, N.P.; Christie, B.R.; Jordan, D.C. Use of the Chrome Azurol S Agar Plate Technique to Differentiate Strains and Field Isolates of Rhizobium leguminosarum biovar trifolii. *Appl. Environ. Microbiol.* **1989**, *55*, 707–710. [CrossRef]
43. Amaro, C.; Aznar, R.; Alcaide, E.; Lemos, M.L. Iron-binding compounds and related outer membrane proteins in Vibrio cholerae non-O1 strains from aquatic environments. *Appl. Environ. Microbiol.* **1990**, *56*, 2410–2416. [CrossRef]
44. Patel, P.R.; Shaikh, S.S.; Sayyed, R.Z. Modified chrome azurol S method for detection and estimation of siderophores having affinity for metal ions other than iron. *Environ. Sustain.* **2018**, *1*, 81–87. [CrossRef]
45. Rensing, C.; Ghosh, M.; Rosen, B.P. Families of Soft-Metal-Ion-Transporting ATPases. *J. Bacteriol.* **1999**, *181*, 5891–5897. [CrossRef]
46. Idris, R.; Trifonova, R.; Puschenreiter, M.; Wenzel, W.W.; Sessitsch, A. Bacterial communities associated with flowering plants of the Ni hyperaccumulator Thlaspi goesingense. *Appl. Environ. Microbiol.* **2004**, *70*, 2667–2677. [CrossRef]
47. Sun, L.N.; Zhang, Y.F.; He, L.Y.; Chen, Z.J.; Wang, Q.Y.; Qian, M.; Sheng, X.F. Genetic diversity and characterization of heavy metal-resistant-endophytic bacteria from two copper-tolerant plant species on copper mine wasteland. *Bioresour. Technol.* **2010**, *101*, 501–509. [CrossRef]
48. Maltman, C.; Donald, L.; Yurkov, V. Tellurite and Tellurate Reduction by the Aerobic Anoxygenic Phototroph Erythromonas ursincola, Strain KR99 Is Carried out by a Novel Membrane Associated Enzyme. *Microorganisms* **2017**, *5*, 20. [CrossRef]
49. Atkin, C.L.; Neilands, J.B. Rhodotorulic Acid, a Diketopiperazine Dihydroxamic Acid with Growth-Factor Activity. I. Isolation and Characterization. *Biochemistry* **1968**, *7*, 3734–3739. [CrossRef]
50. Andersen, D.; Renshaw, J.C.; Wiebe, M.G. Rhodotorulic acid production by *Rhodotorula mucilaginosa*. *Mycol. Res.* **2003**, *107*, 949–956. [CrossRef]
51. Carson, K.C.; Meyer, J.M.; Dilworth, M.J. Hydroxamate siderophores of root nodule bacteria. *Soil Biol. Biochem.* **2000**, *32*, 11–21. [CrossRef]
52. Holinsworth, B.; Martin, J.D. Siderophore production by marine-derived fungi. *BioMetals* **2009**, *22*, 625–632. [CrossRef]
53. Mihara, K.; Tanabe, T.; Yamakawa, Y.; Funahashi, T.; Nakao, H.; Narimatsu, S.; Yamamoto, S. Identification and transcriptional organization of a gene cluster involved in biosynthesis and transport of acinetobactin, a siderophore produced by *Acinetobacter baumannii* ATCC 19606T. *Microbiology* **2004**, *150*, 2587–2597. [CrossRef]
54. Goeker, M. *The One Thousand Microbial Genomes Phase 4 Project (KMG-4) Sequencing the Most Valuable Type-Strain Genomes for Metagenomic Binning, Comparative Biology and Taxonomic Classification*; DOE Joint Genome Institute: Oak Ridge, TN, USA, 2016. [CrossRef]

Article

Succession and Colonization Dynamics of Endolithic Phototrophs within Intertidal Carbonates

Daniel Roush [1,2] and Ferran Garcia-Pichel [1,2,*]

[1] School of Life Sciences, Arizona State University, Tempe, AZ 85282, USA; dwroush@asu.edu
[2] Center for Fundamental and Applied Microbiomics, Biodesign Institute, Arizona State University, Tempe, AZ 85282, USA
* Correspondence: ferran@asu.edu

Received: 20 December 2019; Accepted: 4 February 2020; Published: 5 February 2020

Abstract: Photosynthetic endolithic communities are common in shallow marine carbonates, contributing significantly to their bioerosion. Cyanobacteria are well known from these settings, where a few are euendoliths, actively boring into the virgin substrate. Recently, anoxygenic phototrophs were reported as significant inhabitants of endolithic communities, but it is unknown if they are euendoliths or simply colonize available pore spaces secondarily. To answer this and to establish the dynamics of colonization, nonporous travertine tiles were anchored onto intertidal beach rock in Isla de Mona, Puerto Rico, and developing endolithic communities were examined with time, both molecularly and with photopigment biomarkers. By 9 months, while cyanobacterial biomass and diversity reached levels indistinguishable from those of nearby climax communities, anoxygenic phototrophs remained marginal, suggesting that they are secondary colonizers. Early in the colonization, a novel group of cyanobacteria (unknown boring cluster, UBC) without cultivated representatives, emerged as the most common euendolith, but by 6 months, canonical euendoliths such as *Plectonema* (*Leptolyngbya*) sp., *Mastigocoleus* sp., and Pleurocapsalean clades displaced UBC in dominance. Later, the proportion of euendolithic cyanobacterial biomass decreased, as nonboring endoliths outcompeted pioneers within the already excavated substrate. Our findings demonstrate that endolithic cyanobacterial succession within hard carbonates is complex but can attain maturity within a year's time.

Keywords: bioerosion; anoxygenic phototroph; microbiome; euendolith

1. Introduction

The endolithic microbiome of intertidal carbonate rocks has been the subject of intensive study since the 1800s [1,2], with a main focus on the characterization of bioerosive agents within these communities. The agents, boring organisms referred to as euendoliths, excavate the rock substrate and create pore spaces for their own growth. Le Campion-Alsumard and colleagues [3,4] first examined succession and colonization by microscopic inspection in order to better understand the ecological principles that drive euendolith community formation. As concern for coral destruction rose in the 1990s, others [5–11] applied the same procedures to understand these dynamics and mitigate bioerosion in reef ecosystems. These studies on porous, biogenic carbonates from coral skeletons showed swift initial colonization by euendolithic algae, with successional changes occurring within months and communities reaching maturity after a year. Kiene [10], Gektidis [9] and Chacón et al. [12] examined hard mineral carbonates as well, finding that euendolithic cyanobacteria, not algae, were the dominant boring organisms there and that hard substrates led to more diverse cyanobacterial populations than those of corals.

Although early research was informative in identifying and characterizing major euendolithic players, the use of morphological characterization alone has been found to underestimate microbial diversity in endolithic cyanobacterial communities [12,13]. Indeed, in the case of marine carbonate communities, high-throughput amplicon sequencing has demonstrated that morphology-based studies can underrepresent cyanobacterial diversity estimates by factors of 10 to 100 [12,14,15]. Early research identified three major morphotypical groups of euendolithic cyanobacteria. One is represented by the thin, filamentous, *Leptolyngbya*-like organisms most commonly assigned to *Plectonema terebrans* (or *Leptolyngbya terebrans*), which are typically one of the most abundant euendolith morphotypes, at times exceeding 80% of total euendolithic biovolume [6]. Unfortunately, no 16S rRNA gene sequence of *P. terebrans* has been obtained from cultures, making it impossible to identify it with certainty in molecular surveys. Environmental sequences best matching *Halomicronema* and *Leptolyngbya* species have been tentatively suggested to represent the elusive *P. terebrans* [14]. The second group corresponds to the species *Mastigocoleus testarum*, which is characterized by a complex, true-branching filamentous morphology, making it easily identifiable from microscopic examination. It has been recently redescribed on the basis of a polyphasic approach based on strain BC008, showing congruency between molecular and traditional approaches [16], has served as a model to elucidate the physiological mechanism of boring [16–18], and is the first euendolith whose genome has been fully sequenced [19]. *Mastigocoleus testarum* is one of the earliest colonizers in soft carbonates, being found as early as one week after initial exposure [3,11,20]. A third, diverse group includes several members of the order Pleurocapsales in the genera *Hyella, Solentia, Hormathonema*, and the recently described *Candidatus* Pleuronema. Members of the Pleurocapsales typically act as pioneer borers but can bore only to shallow depths and are easily preyed upon by grazers, leading to low abundance in mature communities [6,7,11,20].

Through comprehensive, high-throughput molecular surveys, we recently found a diverse phototrophic community in the endolithic habitat of coastal hard carbonates, which included four distinct anoxygenic phototrophic bacterial (APB) groups. The most dominant APBs were members of the Chloroflexales (green nonsulfur bacteria) [21–23] and *Erythrobacter* (aerobic anoxygenic phototrophs) [24,25]. APBs could comprise upwards of 80% of the total phototroph community [15] in some samples. Our findings broadened the known habitats for APBs and suggested that some microscopic characterizations of endolithic thin filamentous organisms (*Plectonema*-like) may have in fact been APBs. Thus, APBs could be euendolithic in nature, potentially upending the long established understanding of endolith ecology by broadening the pool of possible pioneer organisms and boring mechanisms.

Therefore, to provide new molecular insights into euendolith colonization and succession and to attempt to answer questions that arose from our prior work, we set up a colonization experiment in the intertidal zone of Playa Ulvero, Isla de Mona, Puerto Rico. We anchored nonporous travertine (a dense, compact form of calcium carbonate) tiles onto beach rock 5 m from shore and collected samples every 3 months over a 9-month period. Our study had four specific aims: (1) to elucidate APB colonization timing to identify if APBs are pioneer organisms with the ability to bore; (2) to examine cyanobacterial euendolith colonization and succession using molecular methods; (3) to measure the colonization dynamics of the *Leptolyngbya*-like (*Plectonema*), *Mastigocoleus*-like, and Pleurocapsalean euendolithic cyanobacterial groups; and (4) to compare colonization progress to previously described steady-state climax communities of similar geological composition and geographic location in order to gauge community maturity

2. Materials and Methods

2.1. Tile Placement and Sample Collection

Commercial, 4 inches wide, 1.5 inches thick travertine square tiles, were anchored onto intertidal beach rock some 5 m from the high-tide shoreline at Playa Uvero on Isla de Mona, Puerto Rico, (18°03′36.2″ N 67°54′21.8″ W) (Figure 1) after having received permits from the Departamento de Recursos Naturales y Ambientales (Commonwealth of Puerto Rico). Tiles were fastened to the beach rock using a combination of Red Head 5″ × 3/8″ 316 Stainless Steel Wedge Anchors and JB Weld

Waterweld putty. Three tiles were sacrificially collected every three months, air-dried and shipped, reaching the laboratory in less than a week, and then stored on arrival at −80 °C until analysis.

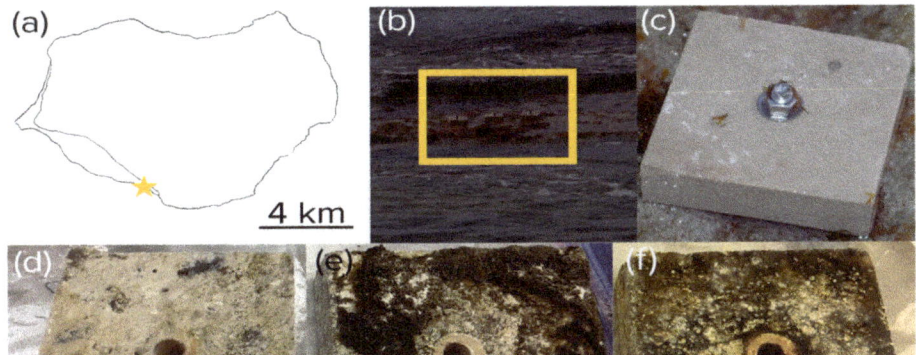

Figure 1. Experimental tile placement. (**a**) Location near Playa Uvero (yellow star) on Isla de Mona, Puerto Rico. (**b**) Anchoring on a stretch of intertidal beach rock (yellow box) as seen at low tide. (**c**–**f**) Aspect of virgin (**c**) and exposed tiles harvested after harvested after 3 (**d**), 6 (**e**), and 9 months (**f**). Part of the growth observable in the pictures was epilithic in nature.

2.2. Endolithic Community DNA Extraction

Tiles were vigorously brushed with sterile toothbrushes and sterilized seawater to remove epilithic biomass. To ensure a consistent sampling effort, each tile was sampled four times in 2 cm by 2 cm squares, 1 cm from the edge of the tile (sampling is shown in Figure 2d–f). Sampled fragments were ground in sterile mortars following the protocol described in Wade and Garcia-Pichel 2003 [26], and 0.5 g of powered rock was used as the input material for a MoBio PowerPlant Pro DNA extraction kit (Mo Bio Laboratories, Inc., Carlsbad, CA, USA) following the protocol provided, except that, before the first lysis step, the contents of the bead tubes were homogenized horizontally at 2200 rpm for 10 min, and, additionally, subjected to seven freeze–thaw cycles using liquid nitrogen to ensure full disruption of bacterial membranes.

2.3. Quantitative PCR of 16S rRNA Gene Content

In order to quantify the number of 16S rRNA gene copies in the extracts, quantitative real-time PCR was conducted using universal V3 16S rRNA gene primers 338F (5′- ACTCCTACGGGAGGCAGCAG-3′) and 518R (5′-GTATTACCG CGGCTGCTGG-3′). PCRs were performed in triplicate using Sso Fast mix (Bio-Rad, Hercules, CA, USA) following Couradeau et al. [27]. Following quantification, triplicate 16S rRNA gene counts were averaged and then converted to counts per square meter using the surface area of the tile analyzed. The total counts per square meter were then multiplied by the associated proportional abundance of any clade of interest in order to obtain absolute population size for that clade. Separate biological replicates (i.e., tiles) were then averaged.

2.4. 16S rRNA Gene Library Preparation and Illumina Sequencing

Amplicon sequencing of the V3–V4 variable region of the 16S rRNA gene was performed using the universal bacterial PCR primers 341F (5′-CCTACGGGNGGCWGCAG) [28] and 806R (5′-GGACTACVSGGGTATCTAAT) [29]. PCR amplifications were done in triplicate, then pooled and quantified using Quant-iT™ PicoGreen®dsDNA Assay Kit (Invitrogen). Two hundred forty nanograms of DNA per sample were pooled and then cleaned using QIA quick PCR purification kit (QIAGEN). The PCR pool was quantified by Illumina library Quantification Kit ABI Prism®(Kapa Biosystems). DNA pool was determined and diluted to a final concentration of 4 nM then denatured diluted to a final concentration of 4 pM with a 30% of PhiX. Finally, the DNA library was loaded in the MiSeq Illumina sequencer using the chemistry version 3 (2 × 300 paired-end) and following the guidelines of the manufacturer.

2.5. Data Analysis Pipeline

Raw sequences were processed using the QIIME2 2018.2 analysis pipeline [30]. Demultiplexed sequences were imported into QIIME2 and processed using the DADA2 [31] denoised-paired plugin with the following parameters: trunc_len_f:280, trunc_len_r:235, trim_left_f:20, trim_left_r:25, and max_ee:8, so as to obtain amplicon sequence variants (ASVs). After resolving ASVs, any sequences found in the control tile extracts (uncolonized tiles) were filtered from the final feature table. Sequencing depth of the experimental tiles ranged from 21,897 to 180,168 (post filtering), and alpha-rarefaction analysis indicated that all samples had reached convergence (Figure S1). In order to conduct diversity analysis, representative sequences were aligned using MAFFT7 [32], and a phylogenetic tree was generated using FastTree [33]. Diversity metrics were calculated using the core-metrics-phylogenetic plugin, including Weighted and Unweighted UniFrac metrics [34]. ASVs were initially classified using the classify-sklearn plugin, (Available online: https://github.com/qiime2/q2-feature-classifier) with a Green Genes 13_8 [35] based classifier. The feature table was then exported, and differential abundance analysis was conducted using the QIIME1 [36] plugin *differential_abundance.py* and the DESeq2 algorithm [37]. PCoAs were generated using the vegan package [38], and graphics were created using R [39] and the ggplot2 package [40]. Statistical analyses were conducted either using R (Student's *t*-test) or within Qiime2 (Kruskal–Wallis, PERMANOVA).

2.6. Cyanobacterial ASV Classification

To identify key euendolithic cyanobacterial clades, the representative sequence output from QIIME2 was filtered to only include cyanobacterial sequences (plastids were removed). These comprised at least 95% of the total number of reads within each sample. Next, the sequences were aligned to the Cydrasil reference alignment [41] using PaPaRa [42], and placed into the Cydrasil reference tree using the Evolutionary Placement Algorithm (based on the maximum-likelihood model) feature of RAxML8 [43]. The output was visualized using the ITOL3 website [44]. An ASV was considered a likely euendolith if it was placed on a branch containing only known euendolithic cyanobacteria with a >70% certainty. Biomass was calculated for each tile by multiplying the total relative abundance of the cluster by the total areal concentration of 16S rRNA genes in that sample. Then, biological replicates for each time point were averaged and graphed using R and ggplot2.

2.7. Steady-State Climax Community Comparisons

Three natural substrate samples from Couradeau et al. [14] and Roush et al. [15] (samples denoted as H001-H003 in SRA) were used as proxies for steady-state climax communities for comparison of colonization progress. The samples were chosen based upon their geographic proximity to the tile placement location and their similar geological composition (calcite). The raw sequencing data was processed using the same parameters and pipeline as described above. Pigment analysis was conducted in the same manner as for the tiles.

2.8. Unknown Boring Cluster (UBC) Phylogenetic Tree

In order to assess the nearest neighbors of the unknown boring cluster, a multiple sequence alignment (MSA) was generated using SSU-Align [45]. The MSA was comprised of the three most differentially abundant ASVs identified from DESeq2 and EPA placement analysis, the nearest sequences from Cydrasil, and the top seven most similar NCBI nr database sequences identified using BLAST [46]. The resulting alignment of 398 sequences was then used as input into RAxML8 [47] to generate a phylogenetic tree using the rapid-bootstrap algorithm with 1000 bootstraps and the GTR GAMMA model. The remaining ASVs were then checked using BLAST and the nr database for proximity to the resulting clade.

2.9. Pigment Extraction and Analysis

In order to extract lipid-soluble pigments, the remaining powdered sample (the same samples used for DNA extraction) was suspended in 7:2 acetone:methanol solvent and sonicated twice for 30 s in an ice bath in the dark. Extracts were centrifuged at 2100× g for 10 min, decanted, and the supernatant filtered through a 0.22 μm nylon filter. These steps were repeated and the supernatants were pooled until the extract was devoid of color. The resulting extract was then evaporated under a N_2 stream in the dark and resuspended in 200 μL of HPLC-grade acetone. HPLC analysis was conducted on a Waters Alliance e2695 HPLC with an inline Waters 2998 photodiode array detector, using a protocol adapted from Frigaard et al. [48] for use on a CORTECS C18 4.6 mm × 150 mm (90 Å pore size, 2.7 μm particles) column. Separation was performed as follows: the initial solvent gradient composed of ethyl-acetate:methanol:acetonitrile:water in a 21:23.9:47.6:7.5 ratio by volume and linearly changed to 30:20:50:0 ratio by volume in 13.43 min, held for 3.87 min, and then immediately returned to the initial ratio (21:23.9:47.6:7.5 by volume) and held for 5.7 min. Total runtime per sample was 23 min, at a flow rate of 2 mL min^{-1} and column temperature of 30 °C. Pigment identification was done by comparison of retention time and spectrum against standards of Chl a and BChl a obtained from Sigma Aldrich. All other pigments were identified from known spectra [49]. Injected pigment mass was calculated from the chromatogram using the equation m = FA $(e_m d)^{-1}$, where m is the mass of BChl or Chl in milligrams, F is the solvent flow rate (1 mL min^{-1}), A is the peak area (in Au), e_m is the extinction coefficient in L mg^{-1} cm^{-1}, and d is the path length of the PDA detector (1 cm). Extinction coefficients were taken from Ley et al. 2006 [50].

2.10. Data Availability

Isla de Mona steady-state climax community raw sequencing data is deposited under NCBI BioProject PRJNA603780. Raw sequencing data from the experimental tiles is deposited under NCBI BioProject PRJNA596277.

3. Results

3.1. Endolithic Bacterial and Phototrophic Growth

Visual inspection of the colonized tiles showed a marked increase in both pigmentation and erosion with time (Figure 2c–f). The tiles sustained both nonphototrophic and phototrophic bacterial growth over the 9-month exposure period. Bacterial biomass increased at an average rate of 3×10^{10} 16S rRNA gene copies per m^{-2} $month^{-1}$, reaching a mean value of 1.1×10^{11} 16S rRNA gene copies per m^{-2} at 9 months (Figure 3a). As expected, phototroph colonization followed a similar trend with photopigment content increasing at a rate of 2.3 mg m^{-2} $month^{-1}$, reaching an average value of 7.22 mg m^{-2} by 9 months. (Figure 3b), at which point 16S rRNA gene counts were not significantly different from those found in steady-state climax communities described by Couradeau et al. [14] and Roush et al. [15] (Student's t-test, $p < 0.05$). While total chlorophyll pigment concentrations were not significantly different between 9 months and steady-state climax communities either, cyanobacteria-specific counts were actually higher at 9 months than in steady-state climax communities.

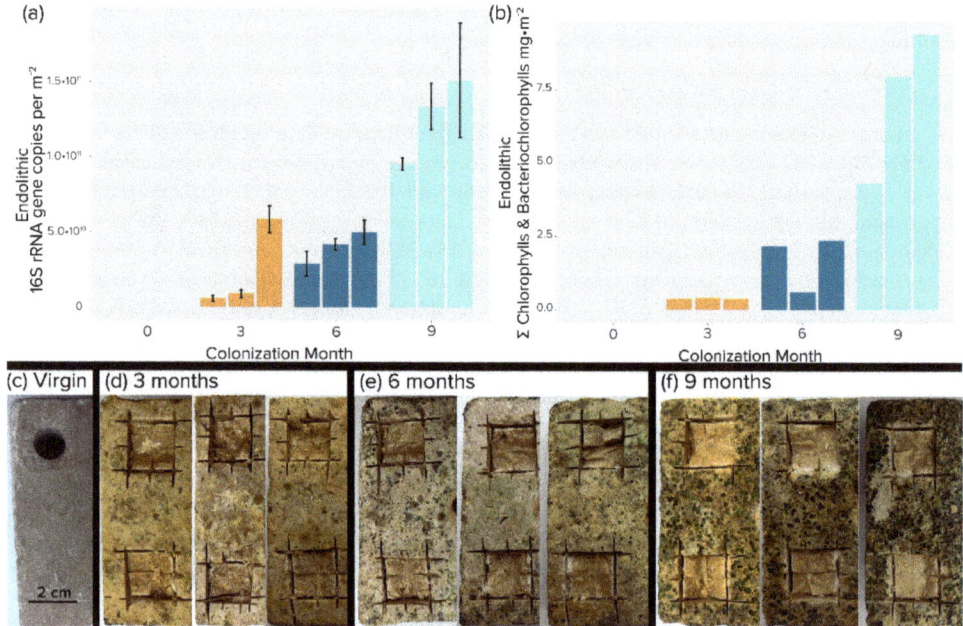

Figure 2. Endolithic colonization of travertine tiles. (**a**) Areal concentration of 16S rRNA gene copies. Each bar is an independent replicate. Error bars are from biological replicates. (**b**) Areal concentration of total photosynthetic chlorins (chlorophylls plus bacteriochlorophylls). Single determinations were carried out for each replicate tile. (**c–f**) Photographic evidence of colonization after removal of epilithic biomass by brushing. (**c**) Initial, virgin tile. Excising squares were samples used for analyses.

3.2. Incidence of Anoxygenic Phototrophs

APB abundance measured by bacteriochlorophylls increased with time but trailed in concentration by some two orders of magnitude to cyanobacterial abundance measured by chlorophylls during the colonization period. This situation obviously changed significantly later during succession, as bacteriochlorophylls were statistically as abundant as chlorophylls when compared to steady-state climax communities (Figure 3b). The magnitude of the difference between APB and cyanobacteria was less marked, but still very significant, when measured by 16S rRNA gene abundance (Figure 3a). By using this metric it was obvious that although APB trailed cyanobacteria during the colonization period, they eventually matched and even exceeded cyanobacteria in steady-state climax communities. We found very differing dynamics between populations of relevant APB groups: while Chloroflexales were only present in very small quantities during early phases (Figure 3c) and reached only 6.2×10^6 16s rRNA gene copies per m^{-2} after 9 months, *Erythrobacter* abundance was stable throughout the colonization, with an average of 1.2×10^9 16S rRNA gene copies per m^{-2} at 9 months. In comparison, the situation was reversed in steady-state mature communities, where *Erythrobacter* sp. decreased to some 1.6×10^8 16s rRNA gene copies per m^{-2} in steady-state climax communities, but Chloroflexales increased to populations close to those of cyanobacteria (Figure 3c). The apparent differences in trends between bacteriochlorophyll and 16S rRNA genes as proxies for population size can be explained by the relatively low bacteriochlorophyll content of *Erythrobacter* spp. compared to members of the Chloroflexales [51,52], which essentially made the total content of bacteriochlorophyll be very sensitive to the population size of the latter.

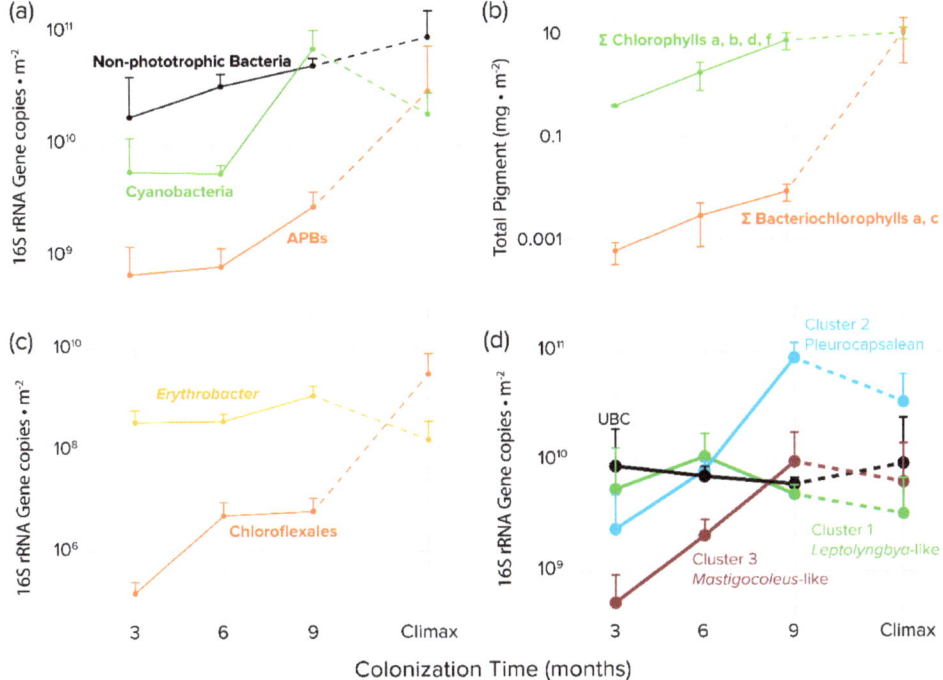

Figure 3. Time series of bacterial biomass proxies detected in colonized tiles and steady-state climax communities by guild or taxon. (**a**) Areal concentrations of 16S rRNA gene copies based on quantitative PCR and high-throughput sequencing phylogenetic assignments (**b**) Areal photosynthetic chlorins as biomarkers for oxygenic phototrophs (total chlorophylls) or APB (total bacteriochlorophylls) (**c**) areal population size of APB clades *Erythrobacter* spp. and Chloroflexales based on quantitative PCR and high-throughput sequencing phylogenetic assignments. (**d**) Endolithic colonization dynamics of specific microboring cyanobacterial clades, based on qPCR and bioinformatic placement of high-throughput environmental sequences using the Cydrasil cyanobacterial reference tree and database. Error bars are for biological sample triplicates.

3.3. Cyanobacterial Succession: Diversity and Composition

Unexpectedly, cyanobacterial richness gauged by the number of observed amplicon sequence variants (ASVs) was not significantly different across time points and when compared to steady-state climax communities (Kruskal–Wallis, $p = 0.33$; Table 1), whereas ASV evenness (measured as Pielou's Evenness) decreased significantly (Kruskal–Wallis, $p = 0.04$) with time. Pairwise Kruskal–Wallis comparisons indicated that the difference was driven by a drop in evenness between early (3 and 6 months) and late succession communities (9 month and steady-state climax) (adjusted $p = 0.07$ for all four comparisons). Shannon's diversity also followed the evenness trend, with significant differences with time (Kruskal–Wallis, $p = 0.02$) where late succession samples were less diverse than early succession samples (adjusted $p = 0.06$ for all four comparisons). Regarding cyanobacterial community composition (beta-diversity), all time points and steady-state climax communities were significantly different from each other (PERMANOVA, $p < 0.05$, pairwise Kruskal–Wallis $p < 0.05$), a result also supported statistically by a PCOA (principal coordinates ordination analysis; Weighted UniFrac metric; Figure S2).

Table 1. Alpha diversity metrics of cyanobacterial endolithic communities in tiles placed in the intertidal zone of Isla de Mona and metrics from geographically similar natural substrate communities on Isla de Mona described by Roush et al. [15].

Timepoint	n	Observed ASVs	Pielou's Evenness	Shannon's Diversity
3 months	3	78 ± 4	0.74 ± 0.06 [a]	4.55 ± 0.45 [a]
6 months	3	98 ± 6	0.79 ± 0.01 [a]	5.22 ± 0.07 [b]
9 months	3	73 ± 3	0.60 ± 0.06 [b]	3.67 ± 0.50 [c]
Climax	3	69 ± 3	0.62 ± 0.02 [b]	3.67 ± 0.45 [c]

Community composition of steady-state climax communities was taken from calcite samples published in Roush 2018. Lower-case letters denote samples not significantly different ($\alpha = 0.1$). ASVs, amplicon sequence variants.

3.4. Identification of Endolithic Cyanobacteria Clades

In nonporous virgin substrates, only euendolithic organisms can colonize and grow to large abundance. Since we removed all epilithic biomass before sequencing, those organisms found to be abundant early on can be deemed to be bona fide euendoliths since they must have been able to excavate the substrate. Therefore, to identify pioneer euendolithic cyanobacterial clades, the most abundant cyanobacterial ASVs from the 3-month-old tiles were placed using the RAxML Evolutionary Placement Algorithm into the Cydrasil reference cyanobacterial 16S rRNA gene tree containing 980 curated cyanobacterial sequences, which includes all full-length 16S rRNA gene sequences traceable to known euendolithic cyanobacteria (Figure 3d and Figure S3). Euendolithic sequences that were not full length were included in the query sequence list and checked for correlation with known clades. In order to pare down the dataset for placement, we ranked each sample's cyanobacterial ASVs in order of abundance until cumulative counts reached 95% of the total abundance in each sample, yielding 213 unique ASVs across all tile samples and steady-state climax communities. We then placed the resulting pared ASV dataset into the Cydrasil reference tree. Of the 213 initial ASVs, 139 were placed with high confidence and clustered onto four distinct tree nodes. Two of the nodes contained known euendolithic species: Cluster 2 (containing 37 unique ASVs) encompassed endolithic members in the Pleurocapsales, and Cluster 3 (27 ASVs) contained *Mastigocoleus testarum*. The other two did not align with known euendoliths: one contained *Leptolyngbya* species (Cluster 1; 60 ASVs) and the other was a novel clade that contained only environmental sequences lacking taxonomic assignment and only distantly related (<95.2% similarity) to *Stanieria cyanosphaera*. We named this clade UBC (15 ASVs), for "unknown boring cluster".

3.5. Colonization Dynamics of Euendolithic Cyanobacterial Clades

To quantify colonization dynamics, qPCR-normalized abundances of the euendolithic clusters were plotted over time (Figure 3d and Figure S3). Members of the UBC were double to an order of magnitude more abundant than the other groups after 3 months of exposure, with an average biomass of 9.1×10^9 16S rRNA gene copies per m^{-2}. UBC abundance remained stable throughout the experiment and was not significantly different when compared to steady-state climax communities. Cluster 1 (*Leptolyngbya*-like) population size lagged that of UBC, reaching a maximum after 6 months (1.1×10^9 16s rRNA gene copies per m^{-2}). Clusters 2 (Pleurocapsalean) and 3 (*Mastigocoleus*-like) colonized substrate at the slowest rate, reaching maximum populations after 9 months (8.7×10^{10} and 9.9×10^9 16S rRNA gene copies per m^{-2}, respectively). Clusters 2 (Pleurocapsalean) and 3 (*Mastigocoleus*-like) also decreased in abundance in steady-state climax communities.

3.6. Differential Abundance Analysis

In order to identify which cyanobacterial colonizers were driving compositional differences between early (3-month) and late (9-month) tiles, we conducted a differential abundance analysis. The most abundant and significant ASVs at both time points were members of the four clades delineated above (Figure S4). At 3 months, representatives of the UBC were three of the four most abundant cyanobacteria ($p < 0.05$), both in total sequence count and in differential relative abundance (fold change) with respect to 9-month communities. The fourth ASV was a member of Cluster 1, allied to *Leptolyngbya*. At 9 months, Cluster 2 (Pleurocapsalean) and Cluster 3 (*Mastigocoleus*-like) sequences were found to be the most differentially abundant with respect to 3-month communities.

3.7. New Pioneer Euendolith Clade

Both qPCR-adjusted relative abundance and differential abundance analysis revealed that the previously unknown UBC clade played a significant role in early colonization of hard intertidal carbonates. In order to better constrain its identification, we conducted a maximum-likelihood phylogenetic reconstruction of 395 sequences, largely from cyanobacterial isolates (Figure 4), but including those of the most differentially abundant UBC and the seven sequences most similar to UBC that we could find by BLAST analyses. As before (i.e., Figure 3d), UBC members were only distantly related (<5.2% similar) to cultured cyanobacteria, the nearest being *Stanieria cyanosphaera* (formerly *Chroococcidiopsis cyanosphaera*), an epilithic freshwater unicellular cyanobacterium [53]. UBC was distant from the canonical euendolithic groups, with the Cluster 2 (Pleurocapsalean) being the closest. However, UBC members were phylogenetically close to environmental sequences obtained from marine carbonate microbialites, a habitat not dissimilar from the interior of hard carbonates and containing known euendoliths [54].

4. Discussion

We recently reported that APBs can be major components of endolithic intertidal ecosystems and could potentially be euendolithic in nature [15], for which no precedent existed. Alternatively, these APBs may constitute secondary colonizers of opened pore space that rely on metabolic interactions with cyanobacteria, as they commonly do in other benthic environments like microbial mats or microbialites [50,54,55]. We hypothesized that examining colonization using molecular techniques and photopigment analysis specifically targeting APBs could help solve this question, in that early colonizers of bare substrates can be logically assumed to be active borers, while a dependency on cyanobacteria should result on delayed colonization by APBs. The temporal dynamics of endolithic population of Chloroflexales indeed suggest that this group of APBs are not euendoliths but instead act as secondary colonizers whose populations do not attain significance until communities of cyanobacteria are mature and the substrate has significantly eroded. The case of the proteobacterium *Erythrobacter* sp. was clearly different, since significant populations of *Erythrobacter* were present early in the colonization process and were sustained through the experimental period. *Erythrobacter* are aerobic anoxygenic phototrophs that conduct photoheterotrophy, have a low BChl *a* content, and require a source of organic carbon [24,56]. Our endolithic sequences were most similar to those in Group I *Erythrobacter* genomes [57]. Under our hypothesis, these organisms could still be euendoliths, even though their populations remained low throughout the experiment. Alternatively, since these small unicellular bacteria are abundant in coastal marine waters [24,56], they could have easily washed into fresh pits made by cyanobacteria in exposed tiles. Our current data cannot fully solve these alternatives. In fact, the metabolic action of photoheterotrophs can increase pH levels around cells, leading the precipitation, not dissolution, of calcium carbonate [15], which would make a boring activity more difficult [58]. By contrast, the lack of the more complex photosynthetic Chloroflexales and low total bacteriochlorophylls suggests that, during colonization, euendolithic cyanobacteria dominate the photosynthetic niche due to their ability to excavate habitable space and utilize the mineral carbon for autotrophy [18,59]. Only once sufficient habitable space has been created by cyanobacteria can significant populations of APBs develop.

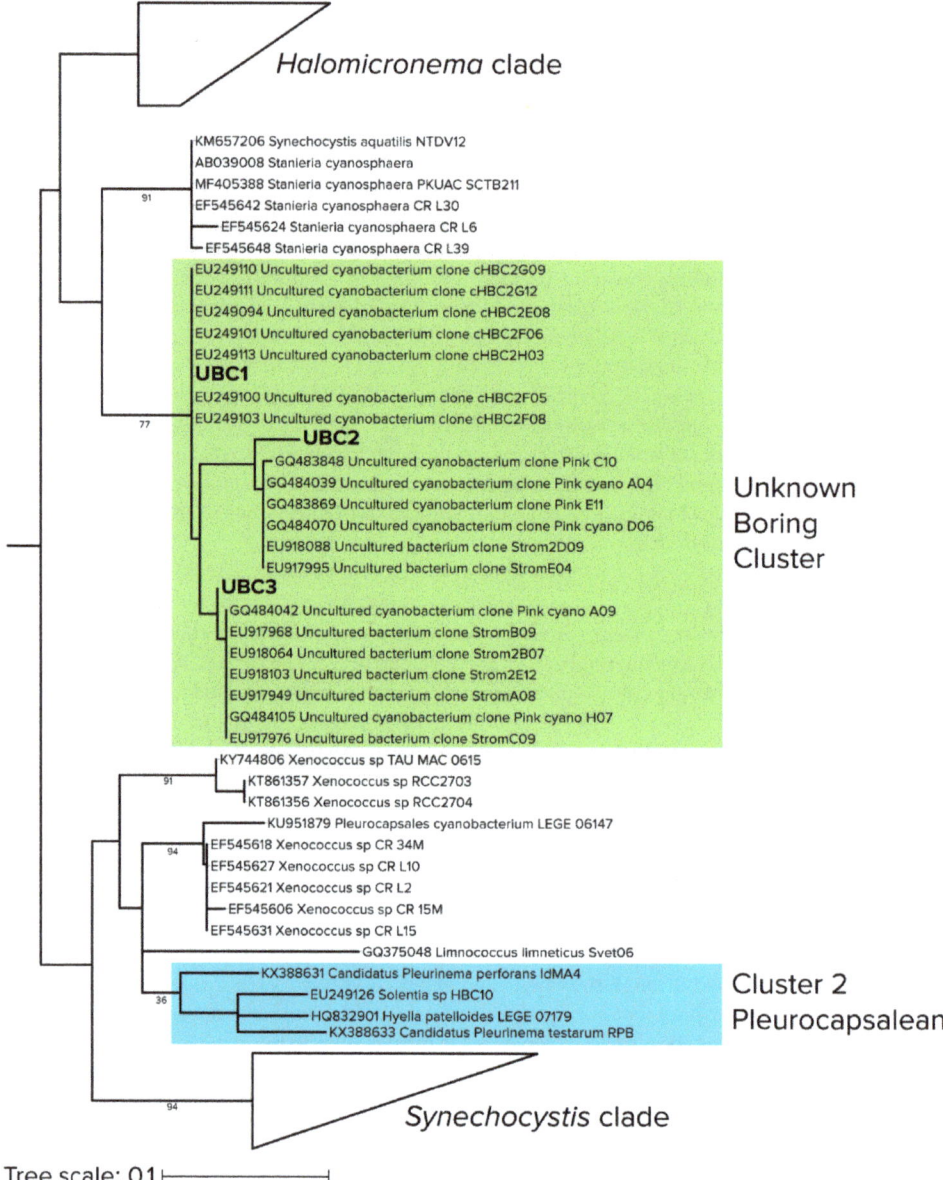

Figure 4. Detailed phylogenetic relationships of sequences in the "unknown boring cluster (UBC)", with environmental (uncultured) cyanobacterial sequences from stromatolites (shaded in green) and the closest known euendolith cluster (shaded in blue). Branch lengths are substitutions per site and node labels indicate bootstrap values.

We found that the patterns of endolithic cyanobacterial succession within hard intertidal carbonates sustain three distinct phases (early, late succession, and steady-state climax). In our habitat, early colonization is predominantly conducted by a previously undescribed group of euendolithic cyanobacteria (UBC) that rapidly colonizes rock to maximal levels within 3 months. This clade could exceed 40% of endolithic cyanobacterial populations early on. Cluster 1 (*Leptolyngbya*-like) organisms also contribute to early colonization but only reach 60% of the biomass of UBC. By 9 months of incubation, the three canonical groups of euendolithic cyanobacteria, *Leptolyngbya* (which we tentatively equate to the *Plectonema terebrans* morphotype), boring members of the Pleurocapsales, and *Mastigocoleus testarum* gain a foothold. Finally, as the community reaches a steady-state climax composition, euendolithic cyanobacteria are displaced in relative importance by other cyanobacteria and by significant populations of Chloroflexalean APBs. The initial large abundance of the UBC could be explained by the presence of fast-growing propagules in natural seawater that quickly attach and bore into the substrate. Since boring microorganisms are fixed in place in their boreholes, competition for space, which can influence patterns of distribution in benthic cyanobacterial communities [60] is likely not a relevant factor until significant proportions of the rock surface become colonized. Hence, having an early foothold on the substrate may have ensured their persistence through time, as we observed. However, UBC did not continue to increase in population size through the colonization, unlike the total cyanobacterial population, which did. The dynamics of the Cluster 1 (*Leptolyngbya*-like) members were not very different from those of UBC, although they seemed to sustain net population losses in late stages of colonization. The net gains in later stages can be attributed to Cluster 3 (*Mastigocoleus*) and, even more so, Cluster 2 (Pleurocapsalean) cyanobacteria (Figure 3d). As these slow colonizers begin to excavate more carbonate, they could reach a threshold where individual pore spaces become connected and pioneer organisms are no longer fully insulated from competition for space. Chlorophyll and qPCR data suggest that this carrying capacity is reached by 9 months of incubation. This density-dependent competition would also explain the overall decline in cyanobacterial evenness/Shannon diversity with successional progress. Finally, at maturity, as endolithic space has been colonized and the rock has become porous, nonboring endoliths can begin to colonize. One can imagine a scenario where nonboring endoliths, which need not spend energy for excavation, can outcompete borers in the outermost sections of the rock. Euendoliths would still have a competitive advantage deeper within the rock. This would be consistent with the relative decline of boring cyanobacterial ASVs in steady-state climax communities, as they are better adapted to diffusion-limited conditions. Interestingly, we did not see a difference in cyanobacterial pigment concentrations between the 9-month samples and steady-state climax communities, which suggests that nonboring phototrophs may colonize the upper interior of the rock, shading the deeper euendoliths and contributing to their decline.

A comparison of our results with prior colonization studies shows that there exist similarities, as well as marked differences, with the dynamics of porous, biogenic coral skeletons. For example, early work [4,61] demonstrated the divergence in euendolith composition between shells and inorganic calcites. However, careful consideration must be taken as both substrate composition [9,10,12,14] and water depth [9,10] influence community structure, and, as discussed above, there are substantial differences in methodology. Even bearing those caveats in mind, the fact that all four major euendolithic clades are present after 3 months of colonization corroborates the prior conclusions that cyanobacterial colonization happens swiftly, in as little as 4 weeks, with *Plectonema*, *Mastigocoleus*, *Solentia*, and *Hyella* species all present [3,8–10]. Interestingly, there are no reports of any *Chroococcidiopsis*-like organism that could potentially represent our UBC. We also found that though *Mastigocoleus* does colonize quickly, it does not reach large abundances until the community approaches a steady-state climax composition, in contrast to the findings from corals where it is one of the first and most abundant pioneer organisms. Our observations on Cluster 1 *Leptolyngbya*-like euendoliths agree with the patterns of *P. terebrans* described by Grange et al. [11]. We find that this cluster peaks in abundance after 6 months, which was also found for coral systems. However, when comparing 9-month Cluster 1 *Leptolyngbya*-like populations to those of steady-state climax communities, we found that Cluster 1 *Leptolyngbya*-like populations were less than 10% of the 9-month totals, whereas in corals, *P. terebrans* remains very abundant through maturity [6]. Cluster 2 Pleurocapsalean euendoliths were not very abundant (sometimes < 1%) in previous colonization experiments and surveys, which was attributed to their alleged susceptibility to grazing by fish and chitons due to their shallow mode of boring [11]. This was clearly not the case in our system, with Cluster 2 Pleurocapsalean organisms being the most abundant euendoliths after 9 months. Perhaps grazing pressure was unusually low in our setting, even though we did see abundant, actively grazing chitons during sampling. Though the abundance of eukaryotic euendoliths are widely reported in coral systems [6,11], we did not find a significant contribution of plastid 16S rRNA genes in our samples, and those that were there were not phylogenetically related to known euendoliths.

In summary, by applying molecular approaches to euendolithic systems we were able to confirm that Chloroflexalean APBs act as secondary colonizers of marine carbonates, illustrate the complex dynamics of cyanobacterial colonization, and define a new clade of likely euendolithic cyanobacteria, highlighting the differences and similarities in succession dynamics between mineral and biogenic carbonates. Our work provides a first look at the complex colonization dynamics that drive bioerosion on these substrates.

Supplementary Materials: The following are available online at http://www.mdpi.com/2076-2607/8/2/214/s1, Figure S1: Rarefaction curves of tile communities, Figure S2: PCoA analysis of bacterial and cyanobacterial community composition, Figure S3: Phylogenetic placement of euendolith clusters and colonization dynamics, Figure S4: ASV differential abundance analysis of 3-month and 9-month colonized tiles. Table S1: BIOM tables containing sequences counts for each euendolith cluster at all 3 time points and steady-state climax communities.

Author Contributions: Conceptualization, D.R. and F.G.-P.; methodology, D.R.; validation, D.R., and F.G.-P.; formal analysis, D.R.; investigation, D.R.; resources, D.R. and F.G.-P.; data curation, D.R.; writing—original draft preparation, D.R.; writing—review and editing, D.R. and F.G.-P.; visualization, D.R.; supervision, F.G.-P.; project administration, D.R.; funding acquisition, F.G.-P. All authors have read and agree to the published version of the manuscript.

Funding: This research was partly funded by the National Science Foundation, grant number EAR 1224939 to F.G.-P.

Acknowledgments: We would like to thank A. Garrástazu, S. Velasco Ayuso, and B. Guida for field support.

Conflicts of Interest: The authors declare no conflict of interest.

References

1. Kölliker, A. On the frequent occurrence of vegetable parasites in the hard structures of animals. *Proc. R. Soc. Lond.* **1859**, *10*, 95–99. [CrossRef]
2. Duerden, J.E. Boring algae as agents in the disintegration of corals. *Bull. Am. Museum Nat. Hist.* **1902**, *889*, 323–332.
3. Le Campion-Alsumard, T. Étude Expérimentale De La Colonisation D'Éclats De Calcite Par Les Cyanophycées Endolithes Marines. *Cah. Biol. Mar.* **1975**, *16*, 177–185.
4. Le Campion-Alsumard, T. Les cyanophycées endolithes marines–Systématique, ultrastructure, écologie et biodestruction. *Ocean. Acta* **1979**, *2*, 143–156.
5. Tribollet, A.; Golubic, S. Cross-shelf differences in the pattern and pace of bioerosion of experimental carbonate substrates exposed for 3 years on the northern Great Barrier Reef, Australia. *Coral Reefs* **2005**, *24*, 422–434. [CrossRef]
6. Tribollet, A. Dissolution of dead corals by euendolithic microorganisms across the northern Great Barrier Reef (Australia). *Microb. Ecol.* **2008**, *55*, 569–580. [CrossRef]
7. Chazottes, V.; Le Campion-Alsumard, T.; Peyrot-Clausade, M. Bioerosion rates on coral reefs: interactions between macroborers, microborers and grazers (Moorea, French Polynesia). *Palaeogeogr. Palaeoclimatol. Palaeoecol.* **1995**, *113*, 189–198. [CrossRef]
8. Vogel, K.; Gektidis, M.; Golubic, S.; Kiene, W.E.; Radtke, G. Experimental studies on microbial bioerosion at Lee Stocking Island, Bahamas and One Tree Island, Great Barrier Reef, Australia: Implications for paleoecological reconstructions. *Lethaia* **2000**, *33*, 190–204. [CrossRef]
9. Gektidis, M. Development of microbial euendolithic communities: The influence of light and time. *Bull. Geol. Soc. Denmark* **1999**, *45*, 147–150.
10. Kiene, W.; Radtke, G.; Gektidis, M.; Golubić, S.; Vogel, K. Factors controlling the distribution of microborers in Bahamian Reef environments. *Facies* **1995**, *32*, 174–188.

11. Grange, J.S.; Rybarczyk, H.; Tribollet, A. The three steps of the carbonate biogenic dissolution process by microborers in coral reefs (New Caledonia). *Environ. Sci. Pollut. Res.* **2015**, *22*, 13625–13637. [CrossRef]
12. Chacón, E.; Berrendero, E.; Garcia Pichel, F. Biogeological signatures of microboring cyanobacterial communities in marine carbonates from Cabo Rojo, Puerto Rico. *Sediment. Geol.* **2006**, *185*, 215–228. [CrossRef]
13. Ramírez-Reinat, E.L.; Garcia-Pichel, F. Prevalence of Ca2+-ATPase-mediated carbonate dissolution among cyanobacterial euendoliths. *Appl. Environ. Microbiol.* **2012**, *78*, 7–13. [CrossRef]
14. Couradeau, E.; Roush, D.; Guida, B.S.; Garcia-Pichel, F. Diversity and mineral substrate preference in endolithic microbial communities from marine intertidal outcrops (Isla de Mona, Puerto Rico). *Biogeosciences* **2017**, *14*, 311–324. [CrossRef]
15. Roush, D.; Couradeau, E.; Guida, B.; Neuer, S.; Garcia-Pichel, F. A new niche for anoxygenic phototrophs as endoliths. *Appl. Environ. Microbiol.* **2018**, *84*, AEM.02055-17. [CrossRef]
16. Ramírez-Reinat, E.L.; Garcia-Pichel, F. Characterization of a marine cyanobacterium that bores into carbonates and the redescription of the genus *Mastigocoleus*. *J. Phycol.* **2012**, *48*, 740–749. [CrossRef]
17. Garcia-Pichel, F.; Ramirez-Reinat, E.; Gao, Q. Microbial excavation of solid carbonates powered by P-type ATPase-mediated transcellular Ca2+ transport. *Proc. Natl. Acad. Sci. USA* **2010**, *107*, 21749–21754. [CrossRef]
18. Guida, B.S.; Garcia-Pichel, F. Extreme cellular adaptations and cell differentiation required by a cyanobacterium for carbonate excavation. *Proc. Natl. Acad. Sci. USA* **2016**, *113*, 5712–5717. [CrossRef]
19. Guida, B.S.; Garcia-Pichel, F. Draft Genome Assembly of a Filamentous Euendolithic (True Boring) Cyanobacterium, Mastigocoleus testarum Strain BC008. *Genome Announc.* **2016**, *4*, 1–2. [CrossRef]
20. Tribollet, A.; Langdon, C.; Golubic, S.; Atkinson, M. Endolithic microflora are major primary producers in dead carbonate substrates of Hawaiian coral reefs. *J. Phycol.* **2006**, *42*, 292–303. [CrossRef]
21. Pierson, B.K.; Valdez, D.; Larsen, M.; Morgan, E.; Mack, E.E. Chloroflexus-like organisms from marine and hypersaline environments: Distribution and diversity. *Photosynth. Res.* **1994**, *41*, 35–52. [CrossRef]
22. Klappenbach, J.A.; Pierson, B.K. Phylogenetic and physiological characterization of a filamentous anoxygenic photoautotrophic bacterium "Candidatus Chlorothrix halophila" gen. nov., sp. nov., recovered from hypersaline microbial mats. *Arch. Microbiol.* **2004**, *181*, 17–25. [CrossRef]
23. Hanada, S.; Takaichi, S.; Matsuura, K.; Nakamura, K. *Roseiflexus castenholzii* gen. nov., sp. nov., a thermophilic, filamentous, photosynthetic bacterium that lacks chlorosomes. *Int. J. Syst. Evol. Microbiol.* **2002**, *52*, 187–193. [CrossRef]
24. Koblížek, M.; Béjà, O.; Bidigare, R.R.; Christensen, S.; Benitez-Nelson, B.; Vetriani, C.; Kolber, M.K.; Falkowski, P.G.; Kolber, Z.S. Isolation and characterization of Erythrobacter sp. strains from the upper ocean. *Arch. Microbiol.* **2003**, *180*, 327–338. [CrossRef]
25. Koblížek, M.; Janouškovec, J.; Oborník, M.; Johnson, J.H.; Ferriera, S.; Falkowski, P.G. Genome sequence of the marine photoheterotrophic bacterium *Erythrobacter* sp. Strain NAP1. *J. Bacteriol.* **2011**, *193*, 5881–5882. [CrossRef]
26. Wade, B.; Garcia-Pichel, F. Evaluation of DNA Extraction Methods for Molecular Analyses of Microbial Communities in Modern Calcareous Microbialites. *Geomicrobiol. J.* **2003**, *20*, 549–561. [CrossRef]
27. Couradeau, E.; Karaoz, U.; Lim, H.C.; Nunes da Rocha, U.; Northen, T.; Brodie, E.; Garcia-Pichel, F. Bacteria increase arid-land soil surface temperature through the production of sunscreens. *Nat. Commun.* **2016**, *7*, 1–7. [CrossRef]
28. Muyzer, G.; De Waal, E.; Uitterlinden, A. Profiling of complex microbial populations by denaturing gradient gel electrophoresis analysis of polymerase chain Reaction-Amplified Genes Coding for 16S rRNA. *Appl. Environ. Microbiol.* **1993**, *59*, 695–700. [CrossRef]
29. Caporaso, J.G.; Lauber, C.L.; Walters, W.A.; Berg-Lyons, D.; Lozupone, C.A.; Turnbaugh, P.J.; Fierer, N.; Knight, R. Global patterns of 16S rRNA diversity at a depth of millions of sequences per sample. *Proc. Natl. Acad. Sci. USA* **2011**, *108*, 4516–4522. [CrossRef]
30. Bolyen, E.; Rideout, J.R.; Dillon, M.R.; Bokulich, N.A.; Abnet, C.C.; Al-Ghalith, G.A.; Alexander, H.; Alm, E.J.; Arumugam, M.; Asnicar, F.; et al. Reproducible, interactive, scalable and extensible microbiome data science using QIIME 2. *Nat. Biotechnol.* **2019**, *37*, 852–857. [CrossRef]
31. Callahan, B.J.; McMurdie, P.J.; Rosen, M.J.; Han, A.W.; Johnson, A.J.A.; Holmes, S.P. DADA2: High-resolution sample inference from Illumina amplicon data. *Nat. Methods* **2016**, *13*, 581–583. [CrossRef]

32. Katoh, K.; Standley, D.M. MAFFT multiple sequence alignment software version 7: Improvements in performance and usability. *Mol. Biol. Evol.* **2013**, *30*, 772–780. [CrossRef] [PubMed]
33. Price, M.N.; Dehal, P.S.; Arkin, A.P. FastTree 2—Approximately maximum-likelihood trees for large alignments. *PLoS ONE* **2010**, *5*. [CrossRef] [PubMed]
34. Lozupone, C.; Knight, R. UniFrac: A New Phylogenetic Method for Comparing Microbial Communities UniFrac: A New Phylogenetic Method for Comparing Microbial Communities. *Appl. Environ. Microbiol.* **2005**, *71*, 8228–8235. [CrossRef]
35. DeSantis, T.Z.; Hugenholtz, P.; Larsen, N.; Rojas, M.; Brodie, E.L.; Keller, K.; Huber, T.; Dalevi, D.; Hu, P.; Andersen, G.L. Greengenes, a chimera-checked 16S rRNA gene database and workbench compatible with ARB. *Appl. Environ. Microbiol.* **2006**, *72*, 5069–5072. [CrossRef]
36. Caporaso, J.G.; Kuczynski, J.; Stombaugh, J.; Bittinger, K.; Bushman, F.D.; Costello, E.K.; Fierer, N.; Peña, A.G.; Goodrich, J.K.; Gordon, J.I.; et al. QIIME allows analysis of high-throughput community sequencing data. *Nat. Methods* **2010**, *7*, 335–336. [CrossRef]
37. Love, M.I.; Huber, W.; Anders, S. Moderated estimation of fold change and dispersion for RNA-seq data with DESeq2. *Genome Biol.* **2014**, *15*, 1–21. [CrossRef]
38. Oksanen, A.J.; Blanchet, F.G.; Friendly, M.; Kindt, R.; Legendre, P.; Mcglinn, D.; Minchin, P.R.; Hara, R.B.O.; Simpson, G.L.; Solymos, P.; et al. Vegan: Community Ecology Package. 2018. Available online: https://github.com/vegandevs/vegan (accessed on 19 December 2019). [CrossRef]
39. R Development Core Team. R: A Language and Environment for Statistical Computing. Available online: https://repo.bppt.go.id/cran/web/packages/dplR/vignettes/intro-dplR.pdf (accessed on 15 July 2019).
40. Wickham, H. *ggplot2: Elegant Graphics for Data Analysis*; Springer: New York, NY, USA, 2016; ISBN 978-0-387-98140-6.
41. Roush, D.; Giraldo-Silva, A.; Fernandes, V.M.C.; Maria Machado de Lima, N.; McClintock, S.; Velasco Ayuso, S.; Klicki, K.; Dirks, B.; Arantes Gama, W.; Sorochkina, K.; et al. Cydrasil: A Comprehensive Phylogenetic Tree of Cyanobacterial 16s rRNA Gene Sequences. Available online: https://github.com/FGPLab/cydrasil (accessed on 8 August 2018).
42. Berger, S.A.; Stamatakis, A. Aligning short reads to reference alignments and trees. *Bioinformatics* **2011**, *27*, 2068–2075. [CrossRef]
43. Berger, S.A.; Krompass, D.; Stamatakis, A. Performance, accuracy, and web server for evolutionary placement of short sequence reads under maximum likelihood. *Syst. Biol.* **2011**, *60*, 291–302. [CrossRef]
44. Letunic, I.; Bork, P. Interactive tree of life (iTOL) v3: An online tool for the display and annotation of phylogenetic and other trees. *Nucleic Acids Res.* **2016**, *44*, W242–W245. [CrossRef]
45. Madden, T.L.; Camacho, C.; Ma, N.; Coulouris, G.; Avagyan, V.; Bealer, K.; Papadopoulos, J. BLAST+: architecture and applications. *BMC Bioinform.* **2009**, *10*, 421. [CrossRef]
46. Nawrocki, E. Structural RNA Homology Search and Alignment Using Covariance Models. Ph. D. Thesis, Washington University School of Medicine, St. Louis, MO, USA, 20 December 2009.
47. Stamatakis, A. RAxML version 8: A tool for phylogenetic analysis and post-analysis of large phylogenies. *Bioinformatics* **2014**, *30*, 1312–1313. [CrossRef] [PubMed]
48. Frigaard, N.U.; Takaichi, S.; Hirota, M.; Shimada, K.; Matsuura, K. Quinones in chlorosomes of green sulfur bacteria and their role in the redox-dependent fluorescence studied in chlorosome-like bacteriochlorophyll *c* aggregates. *Arch. Microbiol.* **1997**, *167*, 343–349. [CrossRef]
49. Frigaard, N.U.; Larsen, K.L.; Cox, R.P. Spectrochromatography of photosynthetic pigments as a fingerprinting technique for microbial phototrophs. *FEMS Microbiol. Ecol.* **1996**, *20*, 69–77. [CrossRef]
50. Ley, R.E.; Harris, J.K.; Wilcox, J.; Spear, J.R.; Miller, S.R.; Bebout, B.M.; Maresca, J.A.; Bryant, D.A.; Sogin, M.L.; Pace, N.R. Unexpected Diversity and Complexity of the Guerrero Negro Hypersaline Microbial Mat Unexpected Diversity and Complexity of the Guerrero Negro Hypersaline Microbial Mat. *Appl. Environ. Microbiol.* **2006**, *72*, 3685–3695. [CrossRef] [PubMed]
51. Koblížek, M. Ecology of aerobic anoxygenic phototrophs in aquatic environments. *FEMS Microbiol. Rev.* **2015**, *39*, 854–870. [CrossRef]
52. Golecki, J.R.; Oelze, J. Quantitative relationship between bacteriochlorophyll content, cytoplasmic membrane structure and chlorosome size in *Chloroflexus aurantiacus*. *Arch. Microbiol.* **1987**, *148*, 236–241. [CrossRef]
53. Komarek, J.; Hindak, F. Taxonomy of the new isolated strains of Chroococcidiopsis (Cyanophyceae). *Arch. Hydrobiol.* **1975**, *13*, 311–329.

54. Couradeau, E.; Benzerara, K.; Moreira, D.; Gérard, E.; Kaźmierczak, J.; Tavera, R.; López-García, P. Prokaryotic and eukaryotic community structure in field and cultured microbialites from the alkaline Lake Alchichica (Mexico). *PLoS ONE* **2011**, *6*. [CrossRef]
55. Lee, J.Z.; Burow, L.C.; Woebken, D.; Craig Everroad, R.; Kubo, M.D.; Spormann, A.M.; Weber, P.K.; Pett-Ridge, J.; Bebout, B.M.; Hoehler, T.M. Fermentation couples Chloroflexi and sulfate-reducing bacteria to Cyanobacteria in hypersaline microbial mats. *Front. Microbiol.* **2014**, *5*, 1–17. [CrossRef]
56. Shiba, T.; Simidu, U. Erythrobacter longus gen. nov., sp. nov., an aerobic bacterium which contains bacteriochlorophyll a. *Int. J. Syst. Bacteriol.* **1982**, *32*, 211–217. [CrossRef]
57. Zheng, Q.; Lin, W.; Liu, Y.; Chen, C.; Jiao, N. A comparison of 14 Erythrobacter genomes provides insights into the genomic divergence and scattered distribution of phototrophs. *Front. Microbiol.* **2016**, *7*. [CrossRef]
58. Garcia-Pichel, F. Plausible mechanisms for the boring on carbonates by microbial phototrophs. *Sediment. Geol.* **2006**, *185*, 205–213. [CrossRef]
59. Guida, B.S.; Bose, M.; Garcia-Pichel, F. Carbon fixation from mineral carbonates. *Nat. Commun.* **2017**, *8*, 1–6. [CrossRef] [PubMed]
60. Nübel, U.; Garcia-Pichel, F.; Kühl, M.; Muyzer, G. Spatial scale and the diversity of benthic cyanobacteria and diatoms in a salina. In *Molecular Ecology of Aquatic Communities*; Zehr, J.P., Voytek, M.A., Eds.; Springer: Dordrecht, The Netherlands, 1999; pp. 199–206. ISBN 978-94-011-4201-4.
61. Perkins, R.D.; Tsentas, C.I. Microbial infestation of carbonate substrates planted on the St. Croix shelf, West Indies. *Bull. Geol. Soc. Am.* **1976**, *87*, 1615–1628. [CrossRef]

© 2020 by the authors. Licensee MDPI, Basel, Switzerland. This article is an open access article distributed under the terms and conditions of the Creative Commons Attribution (CC BY) license (http://creativecommons.org/licenses/by/4.0/).

Article

Distribution of Phototrophic Purple Nonsulfur Bacteria in Massive Blooms in Coastal and Wastewater Ditch Environments

Akira Hiraishi [1,*], **Nobuyoshi Nagao** [1], **Chinatsu Yonekawa** [1], **So Umekage** [1], **Yo Kikuchi** [1], **Toshihiko Eki** [1,2] **and Yuu Hirose** [1,2,*]

1. Department of Environmental and Life Sciences, Toyohashi University of Technology, Toyohashi 441-8580, Japan; velvetschild@gmail.com (N.N.); hgm.cnt@gmail.com (C.Y.); soumekage@gmail.com (S.U.); kikuchi@tut.jp (Y.K.); eki@chem.tut.ac.jp (T.E.)
2. Department of Applied Chemistry and Life Science, Toyohashi University of Technology, Toyohashi 441-8580, Japan
* Correspondence: hiraishi@tut.jp (A.H.); hirose@chem.tut.ac.jp (Y.H.)

Received: 20 December 2019; Accepted: 20 January 2020; Published: 22 January 2020

Abstract: The biodiversity of phototrophic purple nonsulfur bacteria (PNSB) in comparison with purple sulfur bacteria (PSB) in colored blooms and microbial mats that developed in coastal mudflats and pools and wastewater ditches was investigated. For this, a combination of photopigment and quinone profiling, *pufM* gene-targeted quantitative PCR, and *pufM* gene clone library analysis was used in addition to conventional microscopic and cultivation methods. Red and pink blooms in the coastal environments contained PSB as the major populations, and smaller but significant densities of PNSB, with members of *Rhodovulum* predominating. On the other hand, red-pink blooms and mats in the wastewater ditches exclusively yielded PNSB, with *Rhodobacter*, *Rhodopseudomonas*, and/or *Pararhodospirillum* as the major constituents. The important environmental factors affecting PNSB populations were organic matter and sulfide concentrations and oxidation–reduction potential (ORP). Namely, light-exposed, sulfide-deficient water bodies with high-strength organic matter and in a limited range of ORP provide favorable conditions for the massive growth of PNSB over co-existing PSB. We also report high-quality genome sequences of *Rhodovulum* sp. strain MB263, previously isolated from a pink mudflat, and *Rhodovulum sulfidophilum* DSM 1374[T], which would enhance our understanding of how PNSB respond to various environmental factors in the natural ecosystem.

Keywords: anoxygenic phototrophic bacteria; purple nonsulfur bacteria; massive blooms; *pufM* gene; *Rhodovulum*; phylogenomics

1. Introduction

The phototrophic purple bacteria are widely distributed in nature and play important roles in global carbon, nitrogen, and sulfur cycles. Sunlight-exposed stagnant water bodies in natural and engineered environments that contain sulfide and/or high-strength organic matter are rich sources of the phototrophic purple bacteria, which often exhibit massive growth, as seen as red, pink, and brown blooms and microbial mats. The massive development of the phototrophic purple bacteria can be seen in specific habitats [1], including meromictic lakes [2–6], lagoons in intertidal zones [7,8], salt marsh and lakes [9,10], shallow soda pans [11], and wastewater stabilization ponds [12–18]. Naturally occurring massive blooms are mainly caused by the proliferation of the purple sulfur bacteria (PSB), which comprise the family *Chromatiaceae* within the class *Gammaproteobacteria*. On the other hand, it is believed that the purple nonsulfur bacteria (PNSB), belonging to *Alpha*- and *Betaproteobacteria*, rarely form colored blooms in the environment; in fact, there has been only scattered information about the involvement of PNSB in blooming phenomena.

As a rare case, Okubo et al. [19] reported that a swine wastewater ditch allowed PNSB to exclusively develop into red microbial mats. This massive development was possibly achieved under specific conditions characterized by the exposure to light and air, the presence of high-strength organic acids, and the absence of sulfide. Also, *Rhodovulum* (*Rdv.*) *strictum* and some other marine PNSB related to *Rhv. sulfidophilum* were isolated from colored blooms in coastal mudflats and tide pools [20,21]. Although these previous reports suggest the potential of PNSB to naturally develop massive populations under particular conditions, it is still uncertain how many PNSB populations co-exist with PSB in natural blooming communities and what the factors are that allow for the massive growth of PNSB in the environment.

This study was undertaken to determine how many populations of PNSB occur in colored blooms and microbial mats developing in coastal and wastewater environments. In order to obtain quantitative and qualitative information on PNSB populations, we used a polyphasic approach by photopigment and quinone profiling, *pufM* gene-targeted quantitative PCR (qPCR), and *pufM* gene clone library analysis, in addition to conventional cultivation-dependent approaches. Quinone profiling is a chemotaxonomic biomarker method useful for roughly determining microbial populations in terms of quantity and quality [22,23]. The clone library analysis of *pufLM* genes, encoding the L and M subunits of photochemical reaction center proteins, is useful to classify PSB and PNSB phylotypes in mixed populations for which 16S rRNA genes are not suitable as phylogenetic markers [8,10,19,24–26]. We also determined the genome sequence of *Rhodovulum* sp. strain MB263, which was previously isolated from a pink-blooming mudflat [21], and the type strain of *Rdv. sulfidophilum*. Based on the results of these analyses, we discuss the environmental conditions under which PNSB can overgrow PSB and co-existing chemoheterotrophic bacteria and the importance of *Rhodovulum* members as the major PNSB populations in coastal environments.

2. Materials and Methods

2.1. Studied Sites and Samples

Colored bloom and mat samples were collected from different environments in Japan from 2002 to 2014 (Table 1). The studied sites, geographical coordinates of the locations, and sample designations were as follows: mudflat (Yatsuhigata, Narashino (35°40'32" N, 140°00'17" E); samples Y1–Y3), tide pools (Jogashima, Yokosuka (35°07'52" N, 139°37'06" E); samples J1–J3), and wastewater ditches in Kosai (34°41'23" N, 137°29'27" E; sample D1), Shizuoka (35°01'00" N, 138°26'57" E, sample D2), and Matsudo (35°48'06" N, 139°58'58" E, sample D3). For comparison, a red microbial mat sample was taken from Nikko Yumoto hot spring (36°48'30" N, 139°25'29" E, sample H1). All samples were stored in polyethylene bottles, transported in an insulated cooler, and used for analysis immediately upon return to the laboratory. Samples used for direct cell counting were fixed in situ with ethanol (final concentration, 50% (*v/v*)).

Table 1. Physicochemical characteristics of colored bloom/mat samples studied.

Sample	Sampling Month/Year	Color	Temp. (°C)	pH	COD (mg L^{-1})	Salinity (‰)	S^{2-} (mg L^{-1})	ORP (mV)
Hot spring								
H1	October/2000	Red	35.0	7.5	nd *	nd	nd	nd
Tidal flat								
Y1	May/2007	Pink	24.5	8.1	62	21.8	6.8	−170
Y2	May/2007	Pink	24.5	8.1	44	21.8	6.2	−180
Y3	August/2009	Pink	28.7	8.0	54	19.8	8.4	−220
Tide pool								
J1	September/2002	Yellow-green	28.9	7.8	19	28.8	9.4	−280
J2	September/2004	Red-brown	29.5	8.2	98	27.3	7.3	−240
J3	September/2014	Yellow-green	29.4	8.5	37	27.5	12	−320
Ditch								
D1	May/2004	Red	22.3	8.9	890	3.8	0	23
D2	August/2012	Pink-brown	28.6	8.2	280	1.1	0.5	−56
D3	August/2002	Pink-brown	28.9	8.1	210	2.6	0.6	−93

* nd, not determined.

2.2. Physicochemical Analysis

The temperature, pH, salinity, and oxidation–reduction potential (ORP) as Eh were measured in situ using Horiba potable water quality meters (Horiba, Kyoto, Japan). Sulfide was measured colorimetrically by the methylene blue method using a PACKTEST WAK-S kit (Kyoritsu Chemical-Check Lab, Tokyo, Japan). The color intensity of the test papers was analyzed using the ImageJ 1.47v program (http://imagej.nih.gov/ij) to quantify the S^{2-} concentration. Chemical oxygen demand (COD), as the index of organic matter concentration, was determined by the standard method [27].

2.3. Photopigment Analysis

Microbial biomass from 20 to 30 mL of samples was harvested by centrifugation at 12,600× g for 10 min, washed with 50 mM phosphate buffer (pH 6.8), and re-suspended in a 60% sucrose solution. In vivo absorption spectra of cells in the sucrose solutions were measured with a BioSpec-1600 spectrophotometer (Shimadzu, Kyoto, Japan) at 300–1000 nm. Bacteriochlorophylls (BChl) from centrifuged biomass were extracted with an acetone–methanol mixture (7:2, v/v) and subjected to spectroscopy. The concentration of BChl a was calculated by the absorption peak at 770 nm and a molar extinction coefficient of 75 mM cm^{-1} [28].

2.4. Quinone Profiling

Quinones were extracted according to the method of Minnikin et al. [29] with slight modifications. Microbial biomass from 100 to 200 mL of samples was harvested by centrifugation at 12,600× g for 10 min, washed twice with 50 mM phosphate buffer supplemented with 1 mM ferricyanide (pH 6.8), and re-suspended in 10 mL of a methanol–0.3% saline mixture (9:1, v/v). Then, the suspensions were mixed with 10 mL of n-hexane and extracted twice by agitating for 30 min each. The hexane extracts were combined, fractionated into the menaquinone and ubiquinone fractions, and analyzed by reverse-phase HPLC and photodiode array detection to identify quinone components with external standards, as described previously [22,23]. In some cases, quinones in the hexane extract were directly separated by HPLC or purified by thin-layer chromatography before HPLC analysis [30,31]. In this study, ubiquinones, rhodoquinones, and menaquinones with n isoprene units were abbreviated as Q-n, RQ-n, and MK-n, respectively. Partially saturated menaquinones and chlorobiumquinone were abbreviated as MK-n(H$_x$) and CK, respectively.

2.5. Phase-Contrast and Epifluorescence Microscopy

Phase-contrast and epifluorescence microscopy was performed using an Olympus BX50 microscope equipped with an Olympus DP70 camera (Olympus, Tokyo, Japan). Direct total cell counts were determined by SYBR Green staining as described previously [32].

2.6. Enumeration of Viable Phototrophs

Bloom and mat samples (1 mL each) were mixed with 9 mL of autoclaved phosphate-buffered saline (PBS, pH 7.2) supplemented with 0.1% yeast extract and 2 mM sodium ascorbate (filter-sterilized). For marine samples, PBS was supplemented with 3% NaCl in addition. These samples were sonicated weakly on ice for 1 min (20 kHz; output power 50 W) to disperse cells. For enumerating PNSB, samples were serially diluted with the buffered solution, and appropriate dilutions were plated by the pour-plating method with RPL2 agar medium (pH 6.8) [33] supplemented with 0.2 mM Na$_2$S × 9H$_2$O. For coastal seawater samples, the medium was modified by adding 30 g of NaCl and 0.4 g of MgCl$_2$ × 6H$_2$O per liter. Inoculated plates were introduced into an AnaeroPak system (Mitsubishi Gas Chemical Co., Niigata, Japan) before incubation. PSB were enumerated by serial dilution in agar tubes (10 mL of medium in 20 mL capacity screw-capped test tubes) using a previously described medium for PSB [34] with slight modifications. The modified agar medium (designated PSB2 agar, pH 7.0) contained (per liter) 0.25 g ammonium acetate, 0.5 g NH$_4$Cl, 1.0 g KH$_2$PO$_4$, 0.2 g NaCl, 0.4

g MgSO$_4$ × 7H$_2$O, 0.05 g CaCl$_2$ × 2H$_2$O, 2.0 g NaHCO$_3$ (filter-sterilized), 0.3 g Na$_2$S × 9H$_2$O, 1 mL each of trace element solution SL8 [35] and a vitamin B$_{12}$ solution (10 mg 100 mL^{-1}), and 1% agar. The NaCl and MgSO$_4$ × 7H$_2$O concentrations were elevated to 30 g and 0.8 g per liter, respectively, for seawater samples. The test tubes were further overlaid with 2 mL of 1% agar containing 1 mM sulfide solution (pH 7.0) before incubation. All plates and test tubes were incubated at 30 °C under incandescent illumination at 2000 lux. The number of colony-forming units (CFU) was recorded after 10–14 days of incubation.

2.7. Isolation and Phylogenetic Identification of PNSB

Single-colored colonies on RPL2 agar plates used for enumeration were picked at random and subjected to a standard purification procedure by streaking of plates. Purified isolates were preserved in RPL2 agar medium as stub cultures. The 16S rRNA genes from the cell lysate [36] were PCR-amplified with bacterial universal primers 27f and 1492r (see Table S1) [37] and sequenced by the Sanger method using a cycle sequencing kit and an automated DNA sequencer [38]. Sequence data were compiled using the GENETYX-MAC program (GENETYX, Tokyo, Japan) and subjected to EzBioCloud [39] and BLAST [40] homology searches for phylogenetic identification.

2.8. DNA Extraction from Bloom Samples

Microbial biomass from samples was harvested by centrifugation as noted above and washed twice with PBS (pH 7.2). Bulk DNA from the biomass was extracted according to the protocol previously described [41]. The crude DNA extracted was purified according to a standard protocol consisting of RNase digestion, chloroform-isoamylalcohol treatment, and ethanol precipitation [42]. The purified DNA was dissolved in TE buffer, diluted as needed, and used as the PCR template.

2.9. Real-Time Quantitative PCR (RT-qPCR)

RT-qPCR assays were performed to target at the 16S rRNA and *pufM* genes, for which pair primer sets of 357f/517r and pufM.557mF/pufM.750mR was used, respectively (Table S1). The primers for *pufM* gene amplification were modifications of pufM.557F and pufM.750R [24]. RT-qPCR was performed using a LightCycler FAStStart DNA MAstr SYBR GREEN kit (Roche Molecular Biochemicals, Indianapolis, IN, USA) according to the protocol previously described [43], where the *Rhodobacter sphaeroides* ATCC 17023^T DNA was used as the control. The copy number of the amplicons was calculated using LightCycler software version 3.5 (Roche Diagnosis, Mannheim, Germany). The available information on bacterial 16S rRNA genes shows that the average gene copy number of *Alpha-*, *Beta-*, and *Gammaproteobacteria* and *Chlorobi*, to which the anoxygenic phototrophs are classified, is 3.1 [44]. Thus, the *pufM* gene copy number obtained was corrected by multiplying the direct total count by a 3.1-fold ratio of *pufM* to 16S rRNA genes.

2.10. pufM Gene Clone Library Analysis

The *pufM* genes from the biomass DNA extracted were amplified by nested PCR using an r*Taq* DNA polymerase kit (Takara, Otsu, Japan) and a Takara Thermal Cycler. The first PCR was performed using a primer set of M150f [19] and pufM.750mR. The thermocycling conditions consisted of pre-heating at 95 °C for 2 min, denaturation at 94 °C for 1 min, and annealing at 53 °C for 1 min, with a total of 20 cycles. Then the second amplification procedure was performed by touchdown PCR with a primer set of pufM.557F and pufM.750mR under the thermocycling conditions as previously described [19]. The PCR products were purified using a GENECLEAN Spin kit (Bio 101, Vista, CA, USA) and subcloned using a pMosBlue blunt-ended vector kit (Amersham Bioscience, Amersham, UK). Ligation and transformation into *Escherichia coli*-competent cells were performed according to the manufacturer's instructions. Plasmid DNA was extracted and purified using a Wizard Plus Minipreps DNA Purification System (Promega Inc., Madison, WI, USA) following the manufacturer's instructions. Sequencing was performed by the Sanger method as described above. The identity of the nucleotide and amino acid sequences were examined using the BLAST search system. Multiple alignment of sequences, calculation of the nucleotide substitution rate with Kimura's two-parameter model, and reconstruction of phylogenetic trees by the neighbor-joining and maximum likelihood algorithms were performed using the MEGA7 program [45]. The topology of phylogenetic trees was evaluated by bootstrapping with 1000 resamplings [46].

2.11. Genome Analysis

The whole genome sequence of *Rhodovulum* sp. strain MB263, which was isolated previously from a pink-blooming pool [21] in the tidal flat area where we found blooms Y1 and Y3 in this study, was determined. An axenic culture of this strain has been deposited with the Biological Resource Center, National Institute of Technology and Evaluation, Kisarazu, Japan with accession number NBRC 112775. For comparison, *Rdv. sulfidophilum* strain DSM 1374^T obtained from DSMZ-German Collection of Microorganisms and Cell Cultures GmbH, Braunschweig, Germany, was subjected to whole-genome sequencing. Genomic DNA was extracted from phototrophically grown cultures using the CTAB method [47]. Genome sequencing and gap closing of the *Rhodovulum* strains were performed using a previously established pipeline [48]. Briefly, a PCR-free paired-end library was prepared with a KAPA Hyper prep kit (Roche Sequencing and Life Science KAPA Biosystems, Wilmington, MA, USA) after shearing of genomic DNA into ~550 bp using an M-220 focused-ultrasonicator (Covaris, Woburn, MA, USA). A mate-pair library of ~8 kbp insert length was prepared with a Nextera mate-pair sample preparation kit (Illumina, San Diego, CA, USA). Both libraries were sequenced on an Illumina MiSeq system with a MiSeq reagent kit version 3 (600 cycles) for *Rhodovulum* sp. MB263 and a MiSeq reagent kit version 2 (500 cycles) for *Rdv. sulfidophilum* DSM 1374^T. Removal of junction adapter sequence and conversion of RF to FR orientation of the mate pair reads were performed by ShortReadManager, an accessory tool of GenoFinisher (http://www.ige.tohoku.ac.jp/joho/genoFinisher/) [49]. The paired-end and mate pair reads were assembled with newbler version 2.9 [50]. Sequence gaps between the scaffolds and contigs were determined in silico using GenoFinisher and AceFileViewer [49], followed by PCR and Sanger sequencing. The finished sequence was validated by FinishChecker, an accessory tool of GenoFinisher. Annotation was performed using the NCBI Prokaryotic Genome Annotation Pipeline (PGAP, https://www.ncbi.nlm.nih.gov/genome/annotation_prok/) [51].

2.12. Phylogenomic Analysis

Average nucleotide identity (ANI) values [52] between the genome sequence of *Rhodovulum* sp. MB263 and other *Rhodovulum* strains were estimated using an ANI calculator (http://enve-omics.ce.gatech.edu/ani/index). A phylogenetic tree of 25 strains of *Rhodovulum* species was reconstructed by the maximum likelihood method based on concatenated sequences of 92 up-to-date bacterial core genes (UBCGs), which were prepared using the UBCG pipeline [53]. The tree reconstruction and gene support index and bootstrapping (100 replications) tests were performed using RAxML version 8.2.11 with the -m GTRCAT -f a -# 100 options [54]. The UBCG tree was visualized with the MEGA7 program [45].

2.13. Statistical and Numerical Analysis

Correlation analysis between different parameters was performed using Microsoft Excel. Differences in *pufM* gene sequence-based community structure among environmental samples were evaluated using the dissimilarity (D) index [22], which is a modification of city-block distance between two samples with k dimensions. In this study, k corresponded to the number of *pufM* phylotypes (22 phylotypes) as described below. Multi-dimensional scaling (MDS) of D matrix data was performed using the XLSTAT program (Addinsoft, New York, NY, USA).

2.14. Accession Numbers

The *pufM* gene sequences determined in this study have been deposited under DDBJ accession numbers LC512373–LC512431. The complete genome sequence of *Rhodovulum* sp. MB263 was deposited with GenBank with accession numbers CP020384.1 for chromosome, CP020385.1 for plasmid pRSMBA, and CP020386.1 for plasmid pRSMBB. The BioSample and BioProject IDs are SAMN06610252 and PRJNA379495, respectively. The complete genome sequence of *Rdv. sulfidophilum* DSM 1374^T was deposited with GenBank with accession numbers CP015418.1 for the chromosome, CP015419.1 for plasmid unamed1, and CP015420.1 for plasmid unamed2. The BioSample and BioProject IDs are SAMN04903811 and PRJNA319729, respectively.

3. Results

3.1. Appearance and General Characteristics of Colored Blooms

The colored blooms in the coastal mudflats and tide pools studied were pink, red, brown, or yellow-green (Figure S1) and had the odor of sulfide. The sulfide concentration and ORP (Eh) in these coastal environments varied between 6.2 and 12 mg-S^{2-} L^{-1}, and −170 and −320 mV, respectively (Table 1), suggesting that the studied sites were under strongly anaerobic anoxic conditions. On the other hand, the colored blooms and mats in the wastewater ditches were red and pink-brown and had much lower concentrations of sulfide (≤0.6 mg-S^{2-} L^{-1}) and higher Eh (−93 to 23 mV). These data evidenced marked physicochemical differences between the blooms/mats in the coastal environments and wastewater ditches.

3.2. Microscopic Observations

Phase-contrast microscopy showed that the morphotypes of the phototrophic bacteria were quite different from sample to sample. Representatives of the phase-contrast micrographs are shown in Figure 1.

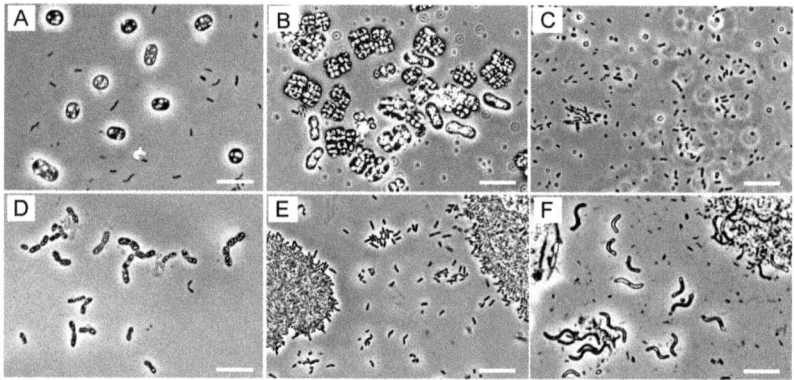

Figure 1. Phase-contrast micrographs of microorganisms in the colored blooms and mats. (**A**) Hot spring mat H1; (**B**) mudflat Y1; (**C**) tide pool J1; (**D**) tide pool J2; (**E**) swine wastewater ditch D1; (**F**) sewage ditch D3. Scale bars = 10 µm.

In the red samples, H1 and J1, PSB cells containing elemental sulfur granules predominated (Figure 1A,D), whereas pink mud samples Y1–3 contained abundant PSB with gas vacuoles that resembled members of the genera *Thiolamprovum* and *Thiodictyon* (Figure 1B). In yellow-green tide pools J1 and J3, PSB cells were scarce, while smaller rod-shaped and oval cells predominated (Figure 1C). Also, none of the ditch samples seemed to contain PSB cells as the major populations (Figure 1E,F). It is noticeable that large spiral cells resembling members of the genera *Pararhodospirillum*, *Phaeospirillum*, and *Rhodospirilum* predominated in sewage ditches D2 and D3 (Figure 1F).

3.3. Photopigment and Quinone Profiles

The biomass collected from the red and pink samples (H1, Y1–3, J2, and D1-3) showed in vivo absorption maxima at 800 nm and 850–861 nm in the near infrared region (Figure S2), indicating that the presence of BChl *a* incorporated into the photosynthetic reaction center and peripheral pigment–protein complexes. On the other hand, the biomass from the yellow-green samples J1 and J3 showed an in vivo absorption maximum at 745 nm, indicating the presence of BChl *c*, typical of the green sulfur bacteria (GSB). The absorption spectra of acetone–methanol extract from all these samples, except samples J1 and J3, showed an extinctive peak at 770–771 nm, which is typical of BChl *a*. The amount of BChl *a* in the samples ranged from 0.4 to 4.6 µmol mL^{-1}.

Quinone profiling studies showed that the red and pink samples H1, Y1–3, and J1 contained Q-8 as the predominant quinone and MK-8 or Q-10 as the second most abundant components (Figure 2 and Figure S3). On the other hand, the yellow-green samples J1 and J3 contained MK-7 and CK as the major quinone species. The ditch samples produced Q-10 (D1) or Q-8 (D2 and D3) as the most abundant quinones. Also, significant proportions of RQ-8, Q-9, and MK-9 were found in samples D2 and D3.

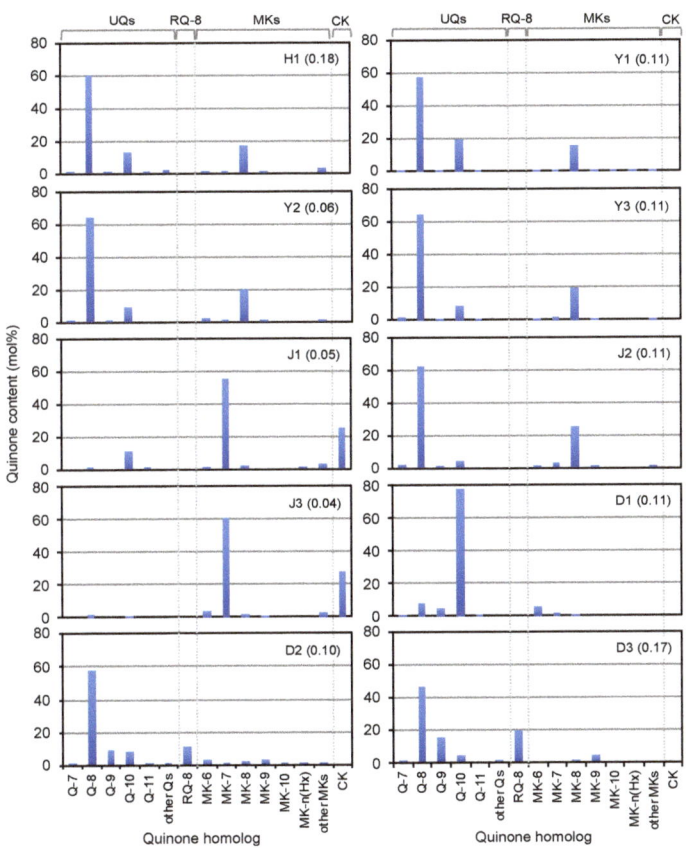

Figure 2. Quinone profiles as biomarkers of microorganisms in colored blooms and mats in hot spring H1, mudflats (Y1–3), tide pools (J1–3), and wastewater ditches (D1–3). The figures in parentheses shows the concentrations of quinones detected (nmol L^{-1}).

The results of quinone profiling together with microscopic observations demonstrate that PSB with Q-8 and MK-8 as the major quinones, i.e., members of *Chromatiaceae*, predominated in the coastal red-pink blooms. Although the co-existence of chemoorganotrophic bacteria should be taken into consideration in quinone composition, smaller amounts of Q-10 in these environments suggest the presence of marine phototrophic alphaproteobacteria, possibly those of the genus *Rhodovulum*, as described below. The predominant phototrophs in the wastewater ditches might be PNSB with Q-10 in D1 and those with Q-8 + RQ-8 or Q-9 + MK-9 in D2 and D3. These quinone systems can be assigned to those of *Rhodobacter* (Q-10), *Rhodopseudomonas* (Q-10), *Pararhodospirillum* (Q-8 + RQ-8), and *Phaeospirillum* (Q-9 + MK-9) [30,31,55].

3.4. PNSB and PSB Populations

All test samples yielded viable PNSB and PSB as well as *pufM* genes, except sample D1, which produced no PSB in detectable counts (Figure 3). In coastal samples Y2, Y3, J2, and J3, the PSB counts were 3–10-fold higher (10^3–10^6 CFU mL^{-1}) than the PNSB counts (10^2–10^6 CFU mL^{-1}). An exceptional case was yellow-green tide pool J1, which yielded slightly higher PNSB counts than PSB counts. The low PSB populations in yellow-green tide pool J1 were in accordance with the failure to detect PSB cells by phase-contrast microscopy (Figure 1) and the low content of Q-8 and MK-8 (Figure 2). All ditch samples (D1–D3) yielded much higher populations of PNSB (10^6–10^7 CFU mL^{-1}) than PSB ($\leq 10^2$ CFU mL^{-1}), and these findings agreed well with the results of microscopic observations and quinone profiling.

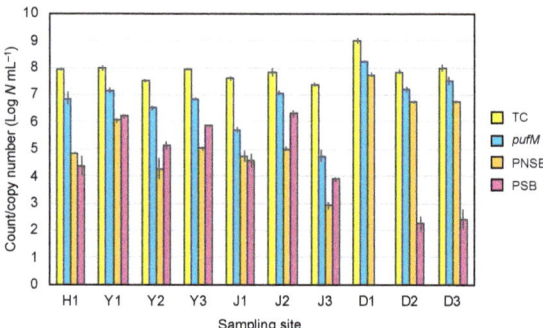

Figure 3. Direct total counts, *pufM* gene copy number, and viable counts of PNSB and PSB in colored blooms and mats in hot spring H1, mudflats (Y1–3), tide pools (J1–3), and wastewater ditches (D1–3). Color of histograms: yellow, direct total count; blue, *pufM* copy number; orange, PNSB viable count; purple, PSB viable count. The vertical bars show standard deviation ($n = 3$).

In all of the pink-red blooms and mats studied, high copy numbers of *pufM* genes (10^6–10^8 copies mL^{-1}) were detected, whereas they were low (10^4–10^5 copies mL^{-1}) in the GSB-predominating blooms J1 and J3 (Figure 3). The *pufM* gene copy numbers, as well as the total bacterial counts in the bloom/mat samples, were much higher than expected from the sum of the viable PNSB and PSB counts. One of the possible reasons for this is that the selective media used under anaerobic light conditions for the enumeration could not fully recover PNSB and PSB present in these environments. Also, we can presume that the qPCR counts might partly include *pufM* amplicons from dead or metabolically low cells and/or other microorganisms than PSB and PNSB. It has been shown that the *pufLM* genes are widely distributed among members of *Alpha-*, *Beta-*, and *Gammaproteobacteria* [56] and in phototrophic members of *Gemmatimonadetes* [57,58].

3.5. Environmental Factors Affecting PNSB

To know about environmental factors affecting PNSB populations in blooming phenomena, we investigated the relationships between viable PNSB counts and sulfide concentrations, ORP, or organic matter concentrations expressed as COD. As shown in Figure 4, the PNSB count had a significant negative correlation with sulfide-S ($p < 0.001$). Also, the PNSB count had significant positive correlations with ORP ($p < 0.001$) and COD ($p < 0.005$). No significant correlations at $p < 0.05$ levels were noted between the PSB count and these physicochemical parameters.

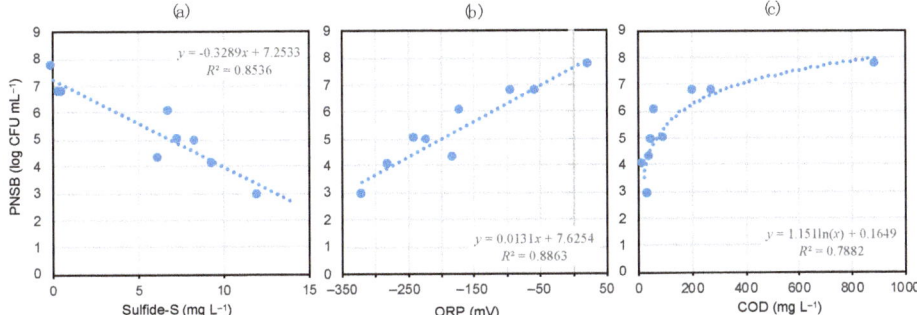

Figure 4. Correlations between the viable PNSB count and sulfide concentration (**a**), ORP (**b**), or COD (**c**) in the colored mat- and bloom-developing environments. The shaded areas show data on coastal samples. Deduced regression equations and correlation coefficients are given in the plots. The correlations in (**a–c**) are significant at $p < 0.001$, 0.001, and 0.005, respectively.

In interpreting the data in Figure 4, not only the marked differences in environmental and biotic conditions between the coastal environments and wastewater ditches (e.g., salinity, effects of co-existing microorganisms), but also the cultivation biases in counting PNSB should be taken into account. Despite these possible biases and factors, there was a highly positive correlation ($p < 0.001$) between the sum of PNSB and PSB viable counts and the number of *pufM* gene copies detected (Figure S4), suggesting the significance of the data shown in Figure 4.

3.6. pufM Gene Clone Phylotyping

The phylogenetic composition of PNSB and PSB in six selected samples, Y1, Y3, J1, J2, D1, and D3, was studied on the basis of *pufM* gene clone library analysis. A total of 354 clones (approximately 60 each for one sample) were sequenced, among which 218 clones, having 233 nt each, were identified as *pufM* genes without uncertainty by BLAST homology search and translation to amino acid sequences. This clone library produced 64 unique sequences that were grouped into 22 phylotypes at a ≥97% identity level.

A neighbor-joining phylogenetic tree of the *pufM* gene sequences representing the 22 phylotypes showed that they were classified mostly into two major clades, *Gammaproteobacteria* and *Alphaproteobacteria*, as expected (Figure 5). A similar topography of the phylogenetic tree was obtained by the maximum likelihood algorithm (Figure S5). Smaller numbers of the phylotypes were classified as members of *Betaproteobacteria*, *Gemmatimonadetes*, and unidentified phylogenetic groups. Phylotypes 12 and 16, assigned to *Gemmatimonadetes* and *Betaproteobacteria*, respectively, nested into the clade of *Alphaproteobacteria*, which can be explained by ancestral lateral transfer of a photosynthetic gene set with *pufM* from phototrophic alphaproteobacteria [57,59,60]. Except for the *Gemmatimonadetes* (phylotype 12), *Dinoroseobacter* (phylotype 15), and unidentified clones, all of the *pufM* clones detected could be assigned to members of PSB and PNSB. Because of the relatively small size of the clone library set, however, there remained the possibility of many of the major clones present (e.g., those of anoxygenic aerobic phototrophic bacteria) being overlooked.

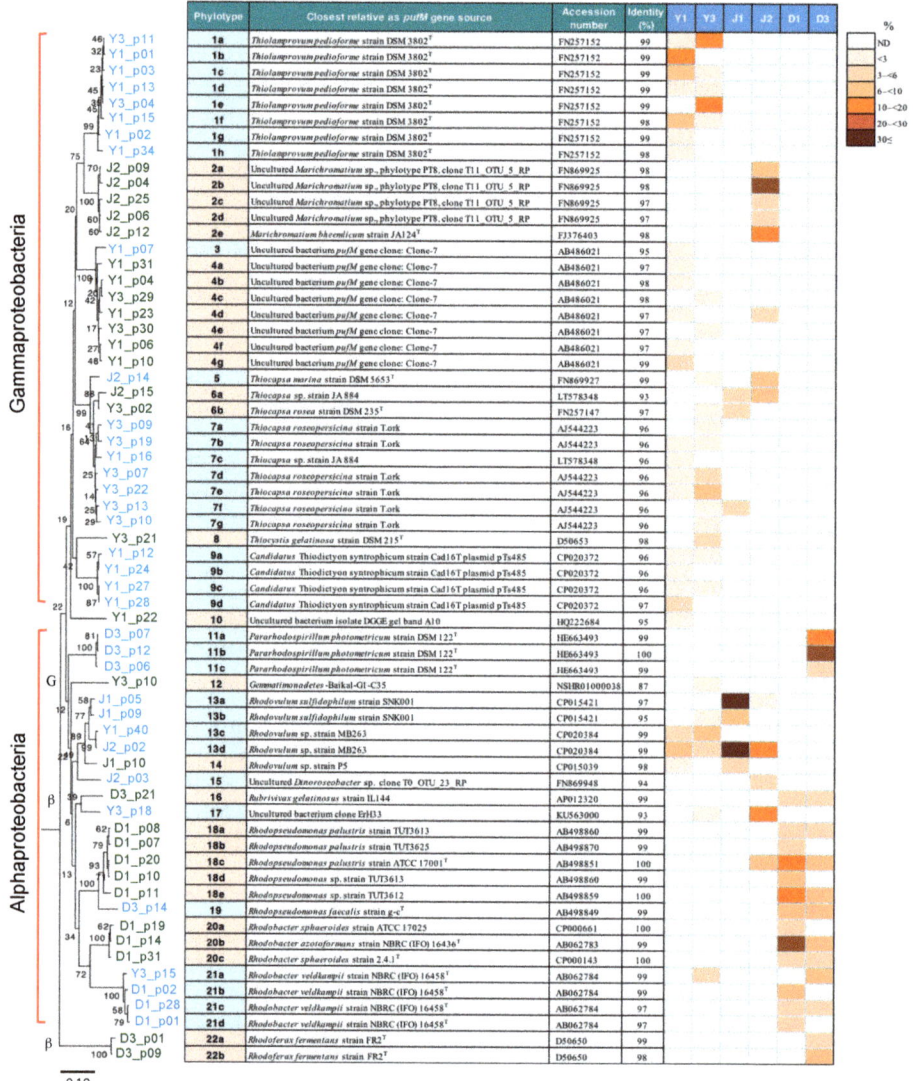

Figure 5. Neighbor-joining phylogenetic tree of the 64 unique *pufM* gene clones, divided into 22 major phylotypes, the assignment to their closest relatives, and a heat map showing their percentage distribution in six selected blooms/mat samples, Y1, Y3, J1, J2, D1, and D3. *Chloroflexus aggregans* DSM 9485^T (accession number, CP001337) was used as an outgroup to root the tree. The percentages of bootstrap confidence values by 1000 replications are given at the nodes of the tree. The letters G and β with shaded parts in the phylogenetic tree shows the lineages of *Gemmatimonadetes* and *Betaproteobacteria*, respectively. Scale bar = 0.1 substitution per position.

Most of the phylotypes detected in mudflat blooms Y1 and Y3 were identified as PSB genera including *Thiolamprovum* (phylotype 1), *Thiocapsa* (phylotypes 5–7), *Thiocystis* (phylotype 8), and *Thiodictyon* (phylotype 9). In particular, the *Thiolamprovum* clones predominated with the *Thiocapsa* clones as the second most abundant constituents in these environments. In addition, mudflats Y1 and/or Y3 yielded clones of unidentified *Gammaproteobacteria* (phylotypes 3 and 4) and PNSB of the genera *Rhodovulum* (phylotype 13) and *Rhodobacter* (phylotype 21). It was of special interest that the yellow-green bloom J1 as well as the mudflat blooms harbored phylotypes 13 as the major PNSB clones, the majority of which were assigned to *Rhodovulum* sp. MB263. Another tide pool investigated, J2, contained *Marichromatium* (phylotype 2) as the major clones and *Thiocapsa* and some PNSB as minor constituents. In contrast, wastewater ditches D1 and D3 did not yield any PSB clones; instead, PNSB phylotypes identified as members of the genera *Pararhodospirillum* (phylotype 11), *Rhodopseudomonas* (phylotypes 18 and 19), or *Rhodobacter* (phylotypes 20 and 21), were the major constituents. Recently, the *Rhodobacter* species to which phylotypes 20 and 21 were assigned have been proposed to be transferred to the new genera "*Luteovulum*" and "*Phaeovulum*", respectively [61]. These results suggest that different taxa of PNSB and PSB rely upon differently suitable environmental conditions to form massive blooms.

Differences in the *pufM* gene-based phylotype composition among the six selected samples were evaluated using the D index. An MDS analysis of the D values calculated showed that the massive blooms/mats in mudflats, tide pools, and wastewater ditches had respective unique phototrophic community structures that separated from each other at a D level of >70% (Figure 6). These results suggest that the formation and community structure of massive blooms of the phototrophs are strongly affected by geographical and environmental conditions of their habitats.

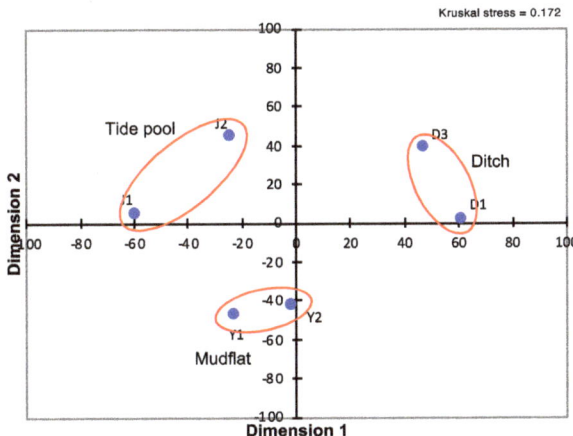

Figure 6. Multidimensional scaling of D matrix data showing differences in *pufM* phylotype composition among the six selected samples taken from mudflat (Y1 and Y2), tide pools (J1 and J2), and wastewater ditches (D1 and D3). The distance between samples are shown by blue dots.

3.7. Phylogenetic Identification of Isolates

A total of 78 strains of PNSB were isolated from mudflat Y1, tide pools J1 and J2, and wastewater ditch D1 through plate counting and cultivation. These isolates were phylogenetically identified by 16S rRNA gene sequencing and EZbioCloud homology search targeting the type strains of established species. The isolates from Y1, J1, and J2 were identified as being members of the genus *Rhodovulum* with either *Rdv. sulfidophilum* DSM 1374^T or *Rdv. adriaticum* DSM 2581^T as their closest relatives (99.6–100% similarities) (Figure S6). Actually, however, a BLAST search showed that most of the marine isolates completely matched *Rhodovulum* sp. MB263, which has similarity levels of 99.6% to *Rdv. sulfidophilum* DSM 1374^T and *Rdv. algae* JA877^T as its nearest phylogenetic neighbors (Table S2). The isolates from D1 were members of the genera *Rhodopseudomonas*, *Rhodobacter* ("*Luteovulum*" and "*Phaeovulum*"), and *Rhodoferax*.

Although the aforementioned cultivation-based studies provided limited information and might produce some biased results, the phylogenetic composition of the PNSB isolates is consistent with that obtained with the *pufM* gene clone library analysis.

3.8. Genomic Analysis of Rhodovulum *Strains*

Our clone library and cultivation-based phylogenetic studies provide circumstantial evidence that, in the coastal massive blooms, members of *Rhodovulum*, especially those corresponding to *Rhodovulum* sp. MB263, constitute the major PNSB population and play important ecological roles. To address this issue and taxonomic problem of *Rhodovulum* sp. MB263, a genome-wide approach is useful. However, the available genomic and phylogenomic information on *Rhodovulum* strains was limited before this study. Also, since most of the available genome assemblies of *Rhodovulum* strains were at the contig level, it is difficult to analyze the whole structure or repeat regions. Therefore, we performed whole-genome sequencing of *Rhodovulum* sp. MB263, previously isolated from a pink-blooming pool in the tidal flat area [21] where we found blooms Y1 and Y3 in this study. For comparison, we also determined the genome sequence of *Rdv. sulfidophilum* DSM 1374^T, which was previously determined by another project but included sequence gaps [62].

We succeeded at determining the complete genome sequence of *Rhodovulum* sp. MB263 and *Rdv. sulfidophilum* DSM 1374^T. The genome size, G+C content, and the number of CDS, rRNA operon, tRNA genes, and plasmids in these strains are shown in Table 2 in comparison with those of *Rhv. sulfidophilum* DSM 2351 [63]. The chromosome size of *Rhodovulum* sp. MB263 (3.86 Mbp) is smaller by 6.6% and 13.3% than those of *Rdv. sulfidophilum* DSM 1374^T (4.13 Mbp) and *Rdv. sulfidophilum* DSM 2351 (4.45 Mbp). Members of the genus *Rhodovulum* are known to utilize sulfide as an electron donor for photolithotrophic growth [64,65]. In this context, it has been shown that *Rdv. sulfidophilum* has the 12 genes of the *sox* operon, which encode cytochrome *c* and other redox proteins involved in sulfide oxidation (Sox) [66,67]. Our genome analysis confirmed that the *sox* operons are completely conserved in *Rhodovulum* sp. MB263 as well as in *Rdv. sulfidophilum* strains DSM 1374^T and DSM 2351. The nucleotide identity level in the *sox* locus (approximately 11 kb stretch) between strains MB263 and DSM 1374^T is 92.22%.

Table 2. Comparison of genome assemblies of *Rhodovulum* sp. strain MB263 and the two authentic strains of *Rdv. sulfidophilum*.

Feature	*Rhodovulum* sp. Strain MB263	*Rdv. sulfidophilum* Strain DSM 1374^T	*Rdv. sulfidophilum* Strain DSM 2351
Reference	This study	This study	Nagao et al. [63]
Genome size (bp)	4,162,560	4,347,929	4,732,772
G+C content (%)	67.03	66.92	66.89
Number of:			
CDS	3725	3899	4146
rRNA operons	3	3	3
tRNA genes	49	50	50
Plasmids	2	2	3

3.9. Phylogenomics

We compared the whole chromosome structure of *Rhodovulum* sp. MB263 and the two authentic *Rdv. sulfidophilum* strains by dot plotting. The arrangement of chromosomes is less similar between strains MB263 and DSM 1374^T than between strains DSM 1374^T and DSM 2351 (Figure S7). In agreement with this, the pair-wise ANI scores for whole genomes calculated were 91.21% between strains MB263 and DSM 1374^T and 97.35% between strains DSM 1374^T and DSM 2351. *Rhodovulum* sp. MB263 had lower ANI values to the type strains of other *Rhodovulum* species (Table S2). Since the ANI value between strain MB263 and its closest relative *Rdv. sulfidophilum* DSM 1374^T is lower than the threshold value (95–96%) recommended for the boundary of prokaryotic species delineation [52,68], the former strain may represent a novel genospecies of the genus *Rhodovulum*.

A phylogenomic analysis of 25 strains of *Rhodovulum* species on which whole-genome information is available (Table S3) [63,69–71] was performed on the basis of 92 protein-coding gene sequences. The reconstructed phylogenomic tree revealed that *Rhodovulum* sp. MB263 is closely related to *Rdv. sulfidophilum* but clearly branched off the clade of the latter species. This suggests that strain MB263 has a distinct phylogenetic position at the species level within the genus *Rhodovulum* (Figure 7).

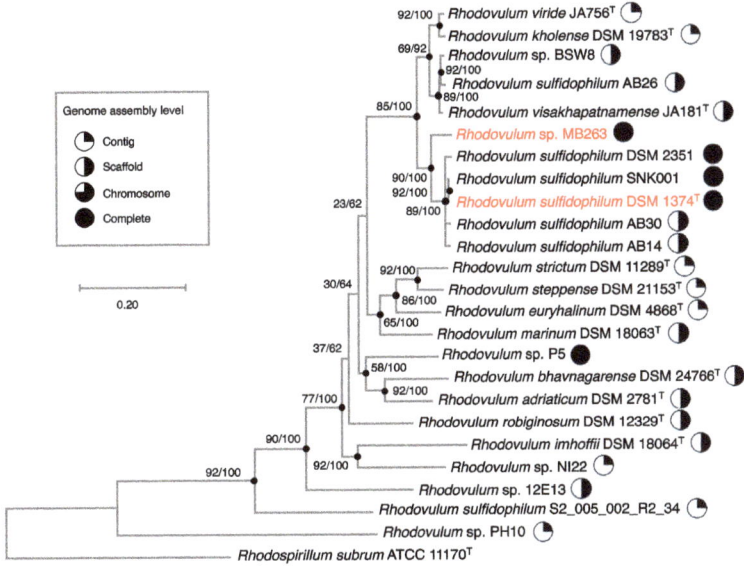

Figure 7. Maximum likelihood phylogenomic tree based on UBCGs (concatenated alignment of 92 core gene sequences) showing genealogical relationships between *Rhodovulum* sp. MB263 and *Rdv. sulfidophilum* DSM 1374^T (shown by red letters) or other members of the genus *Rhodovulum*. The database accession numbers for the sequences incorporated are shown in Table S2. The genome sequence of *Rhodospirillum rubrum* ATCC 11170^T was used as an outgroup to root the tree. Gene support indices and percentage bootstrap values are given at the nodes of the tree. Scale bar = 0.2 substitution per position.

4. Discussion

The available information on the occurrence of PNSB in colored microbial mats and blooms in the environment has so far been only scattered and fragmentary [8,19–21]. As reported herein, our polyphasic approach to address this issue by culture-independent techniques as well as by conventional cultivation methods has improved information on the distribution of PNSB in colored blooms/mats in terms of quantity and quality. We need to carefully consider that there might be cultivation and PCR biases in the used approach and that the numbers of the *pufM* gene clones sequenced and the isolates phylogenetically identified in this study are not sufficient to describe the phototrophic community structures at the studied sites. However, the results of direct phase-contrast microscopy, quinone profiling, and the clone library analysis match and complement each other relatively well, thereby increasing the reliability of our data.

One of the most important findings in the present study is that there were significant differences in the contribution of PNSB to blooming phenomena and their biodiversity between the coastal environments and wastewater ditches. In the coastal red-pink blooms, PNSB constituted significant proportions of the phototrophic bacterial populations, but usually occurred in smaller CFU numbers than PSB. Actually, direct microscopic observations and quinone profiling data have shown that the overwhelming majority as the biomass in the coastal red-pink blooms was represented by elemental sulfur globe- and/or gas vacuole-containing PSB, whose main quinones are Q-8 and MK-8. In addition to these findings, the *pufM* gene amplicon analysis has clearly shown that the major phylotypes detected were those assigned to the genera *Thiolamprovum* and *Thiocapsa* in the pink mudflat and to *Marichromatium* and *Thiocapsa* in the red tide pools. Similar PSB members accompanied by smaller numbers of PNSB and aerobic anoxygenic phototrophic bacteria have been found in massive blooms in a brackish lagoon [8]. To our knowledge, therefore, the formation of red-pink blooms in coastal environments is attributable to massive development of PSB, and the contribution of PNSB as the colorants to the bloom formation is less significant.

In contrast, the red-pink blooms/mats in the swine wastewater and sewage ditches exclusively yielded PNSB, as demonstrated by a combination of phase-contrast microscopy, quinone profiling, *pufM* gene clone sequencing, and cultivation-based phylogenetic analysis. Our data have shown that PNSB assigned to the genera *Rhodobacter* ("*Luteovulum*" and "*Phaeovulum*" [61]), *Rhodopseudomonas*, and *Pararhodospirillum* predominated in the ditches, but the PNSB community structure differed from sample to sample. These results fully support the previous observation that visible massive development by PNSB themselves takes place in the environment under specific conditions [19]. Obviously, while coastal blooms commonly harbor *Rhodovulum* species as the major PNSB populations, wastewater blooms/mats may contain different major taxa of PNSB depending on the environmental conditions.

One of the most important environmental factors affecting the proliferation of PNSB is the concentration of sulfide. Since most of the PNSB species are unable to tolerate high concentrations of sulfide, there are few chances for them to grow massively in such sulfide-rich environments as coastal environments. However, marine species belonging to the genus *Rhodovulum* can use relatively high concentrations of sulfide as the electron donor for photolithotrophic growth [64,65], and *Rdv. sulfidophilum* has the genes of the Sox pathway involved in the complete eight-electron oxidation of sulfide to sulfate [66,67]. Our genomic studies have confirmed that *Rhodovulum* sp. strain MB263 as well as *Rdv. sulfidophilum* strains DSM 1374^T and DSM 2351 have the complete gene set of the *sox* operon. Also, the previous study has shown that *Rhodovulum* sp. MB263 is capable of photolithotrophic growth with 2 mM sulfide as the electron donor [21]. These facts provide a plausible reason why the group of *Rhodovulum* sp. MB263 and *Rdv. sulfidophilum* occurred in significant phototrophic bacterial populations in the coastal blooms with high concentrations of sulfide. Nevertheless, the biomass density of the *Rhodovulum* members would not become so high as exceeding those of PSB and GSB, both of which have growth advantages, with higher affinities to sulfide as the electron donor for photolithotrophy.

The second important factor controlling PNSB populations is ORP, which is also related to sulfide concentrations. The mudflat and tide pools we studied exhibited an Eh level of −320 to −170 mV with high concentrations of sulfide. Such low Eh levels as well as high sulfide concentrations are apparently more favorable for growth and survival of PSB or GSB than PNSB. On the other hand, a limited higher range of Eh (−93 to 23 mV), as seen in the ditch blooms/mats, may be effective for stimulating the growth of PNSB while suppressing that of PSB and GSB.

The concentration of organic matter, which PNSB can use as energy and carbon sources for photoorganotrophy as their best life mode, should be noted as the third important factor. The coastal blooming mudflat and tide pools studied had low concentrations of organic matter, expressed as COD, whereas the sewage and wastewater ditches were at much higher COD levels. This provides an additional explanation for why PNSB could overgrow PSB in the wastewater ditches but not in the coastal environments. Taken together, it may be logical to conclude that light-exposed, sulfide-deficient water bodies with high-strength simple organic matter and in the limited range of ORP provide the best conditions for the massive growth of PNSB. This also explains the basis for wastewater treatment systems using PNSB for purifying highly concentrated organic matter, where they can compete with co-existing PSB and chemoorganotrophic bacteria under limited aerobic conditions [72,73].

As mentioned above, members of the genus *Rhodovulum*, especially the *Rhodovulum* sp. MB263 genospecies and *Rdv. sulfidophilum*, occurred as the major PNSB populations in coastal blooms, as revealed by *pufM* clone library analysis and cultivation-based phylogenetic studies. Apparently, the ability of the *Rhodovulum* species to utilize sulfide as the electron donor for photolithotrophic growth is an advantage in such sulfide-rich environments as blooming seawater pools and mudflats. In addition, it is noteworthy that *Rhodovulum* sp. MB263 and *Rdv. sulfidophilum* are capable of floc formation depending on the culture conditions [74]. This capacity might be another advantage for the *Rhodovulum* members to protect themselves against environmental stress. Further study with the genomic information obtained in this study should give a more comprehensive understanding of how *Rhodovulum* species respond to various environmental factors in the ecosystem.

Rhodovulum sp. strain MB263 was isolated previously from a pink pool in a mudflat [21] in the same area where we found mud blooms Y1 and Y3 in this study. Genomic DNA–DNA hybridization assays in previous work showed that strain MB263 had a similarity level of 57% to *Rdv. sulfidophilum* DSM 1374^T as its closest relative, suggesting that the strains are closely related to each other but differ at the species level [21]. In the present study, this suggestion is fully supported by genomic and phylogenomic information. An ANI value of 91.21% between the genomes of strains MB263 and DSM 1374^T is lower than the recommended threshold value (95–96%) for bacterial genospecies circumscription [52,68]. Genome-wide comparisons of *Rhodovulum* sp. MB263 with other closely related *Rhodovulum* species, e.g., *Rdv. algae*, should provide more definitive information on whether strain MB263 represents a novel species of the genus *Rhodovulum*.

5. Conclusions

In polluted freshwater environments like wastewater ditches, PNSB can form visible dense populations over PSB and GSB under specific conditions. These conditions are represented by light-exposed, sulfide-deficient water bodies with high-strength organic matter and in a limited range of ORP (−93 to 23 mV). On the other hand, coastal environments provide more favorable conditions for the massive growth of PSB or GSB because of the high availability of sulfide and lower concentrations of organic matter. In coastal colored blooms, nevertheless, PNSB with *Rhodovulum* members predominating constitute a significant proportion of the phototrophic bacterial population. These results expand our knowledge of the phototrophic community structure of marine and wastewater massive blooms and the ecological significance of PNSB in these environments. Also, the high-quality genomic information on *Rhodovulum* sp. strain MB263 and *Rdv. sulfidophilum* strain DSM 1374^T obtained in this study enhances our understanding of how PNSB respond to various environmental factors in the ecosystem.

Supplementary Materials: The following are available online at http://www.mdpi.com/2076-2607/8/2/150/s1, Table S1: Appearance of colored blooms and microbial mats studied, Figure S2: In vivo absorption spectra of the biomass collected from pink mudflat Y1 (A), yellow-green tide pool J3 (B), and red ditch mat D3 (C), Figure S3: HPLC elution profiles of isoprenoid quinones extracted from pink mudflat Y1 (A), yellow-green tide pool J3 (B), and red ditch mat D3 (C), Figure S4: Relationship between *pufM* gene copy numbers and viable counts of PNSB + PSB in colored blooms and mats investigated, Figure S5: Maximum-likelihood phylogenetic tree of the *pufM* gene clones detected and their closest relatives, Figure S6: 16S rRNA gene sequence-based phylogenetic composition of the PSNB isolates from red mudflat J1, red tide pool J2, yellow-green tide pool J3, and red ditch mat D1, Figure S7: Harr plots showing genomic similarities between *Rhodovulum sulfidophilum* strains DSM 1374^T and DSM 2351 (A) and between *Rhodovulum sulfidophilum* DSM 1374^T and *Rhodovulum* sp. MB263 (B), Table S1: PCR primers used for specific gene amplification in this study, Table S2: Similarity levels of 16S rRNA gene and whole genome sequences between *Rhodovulum* sp. MB263 and authentic strains of established *Rhodovulum* species, Table S3: List of *Rhodovulum* species and strains and the accession numbers for genome sequences used for reconstruction of the phylogenomic tree based on 92 core protein-coding genes.

Author Contributions: Conceptualization, project administration, methodology, investigation, and writing—original draft preparation, A.H.; investigation, N.N. and C.Y.; investigation, data curation, and manuscript writing, Y.H.; writing—review, editing, and supervision, S.U., Y.K., and T.E. All authors have read and agreed to the published version of the manuscript.

Funding: This research received no external funding.

Acknowledgments: We are grateful to Yoko Okubo, Toyohashi University of Technology, for her early contributions to this study. We also thank Takashi Kurogi for his technical assistance in qPCR and sequencing experiments.

Conflicts of Interest: The authors declare no conflict of interest.

References

1. Imhoff, J.F. Diversity of anaerobic anoxygenic phototrophic purple bacteria. In *Modern Topics in the Phototrophic Prokaryotes, Environmental and Applied Aspects*; Hallenbeck, P.C., Ed.; Springer: New York, NY, USA, 2017; pp. 47–85.
2. Takahashi, M.; Ichimura, S. Vertical distribution and organic matter production of photosynthetic sulfur bacteria in Japanese lakes. *Limnol. Oceanogr.* **1968**, *13*, 644–655. [CrossRef]
3. Matsuyama, M.; Shirouzu, E. Importance of photosynthetic sulfur bacteria, *Chromatium* sp. as an organic matter producer in Lake Kaiike. *Jpn. J. Limnol.* **1978**, *39*, 103–111. [CrossRef]
4. Overmann, J.; Beauty, J.T.; Hall, K.J.N.; Pfennig, N.; Northcote, T.G. Characterization of a dense, purple sulfur bacterial layer in a meromictic salt lake. *Limnol. Oceanogr.* **1991**, *36*, 846–859. [CrossRef]
5. Schanz, F.; Fischer-Romero, C.; Bachofen, R. Photosynthetic production and photoadaptation of phototrophic sulfur bacteria in Lake Cadagno (Switzerland). *Limnol. Oceanogr.* **1998**, *43*, 1262–1269. [CrossRef]
6. Nakajima, Y.; Okada, H.; Oguri, K.; Suga, H.; Kitazato, H.; Koizumi, Y.; Fukui, M.; Ohkouchi, N. Distribution of chloropigments in suspended particulate matter and benthic microbial mat of a meromictic lake, Lake Kaiike, Japan. *Environ. Microbiol.* **2003**, *5*, 1103–1110. [CrossRef] [PubMed]
7. Caumette, P. Phototrophic sulfur bacteria and sulfate-reducing bacteria causing red waters in a shallow brackish coastal lagoon (Prévost Lagoon, France). *FEMS Microbiol. Lett.* **1986**, *38*, 113–124. [CrossRef]
8. Tank, M.; Blümel, M.; Imhoff, J.F. Communities of purple sulfur bacteria in a Baltic Sea coastal lagoon analyzed by *pufLM* gene libraries and the impact of temperature and NaCl concentration in experimental enrichment cultures. *FEMS Microbiol. Ecol.* **2011**, *78*, 428–438. [CrossRef]
9. Nicholson, J.A.M.; Stolz, J.F.; Pierson, B.K. Structure of a microbial mat at Great Sippewissett Marsh, Cape Cod, Massachusetts. *FEMS Microbiol. Ecol.* **1987**, *45*, 343–364. [CrossRef]
10. Thiel, V.; Tank, M.; Neulinger, S.C.; Gehrmann, L.; Dorador, C.; Imhoff, J.F. Unique communities of anoxygenic phototrophic bacteria in saline lakes of Salar de Atacama (Chile): Evidence for a new phylogenetic lineage of phototrophic *Gammaproteobacteria* from *pufLM* gene analyses. *FEMS Microbiol. Ecol.* **2010**, *74*, 510–522. [CrossRef]
11. Korponai, K.; Szabó, A.; Somogyi, B.; Boros, E.; Borsodi, A.K.; Jurecska, L.; Vörös, L.; Felföldi, T. Dual bloom of green algae and purple bacteria in an extremely shallow soda pan. *Extremophiles* **2019**, *23*, 467–477. [CrossRef]
12. Sletten, O.; Singer, R.H. Sulfur bacteria in red lagoons. *J. Water Pollut. Control Fed.* **1971**, *43*, 2118–2122.
13. Cooper, D.E.; Rands, M.B.; Woo, C.P. Sulfide reduction in fellmongery effluent by red sulfur bacteria. *J. Water Pollut. Control Fed.* **1975**, *47*, 2088–2100. [PubMed]
14. Wenke, T.L.; Vogt, J.C. Temporal changes in a pink feedlot lagoon. *Appl. Environ. Microbiol.* **1981**, *41*, 381–385. [CrossRef] [PubMed]
15. McGarvey, J.A.; Miller, W.G.; Lathrop, J.R.; Silva, C.J.; Bullard, G.L. Induction of purple sulfur bacterial growth in dairy wastewater lagoons by circulation. *Lett. Appl. Microbiol.* **2009**, *49*, 427–433. [CrossRef] [PubMed]
16. Belila, A.; Abbas, B.; Fazaa, I.; Saidi, N.; Snoussi, M.; Hassen, A.; Muyzer, G. Sulfur bacteria in wastewater stabilization ponds periodically affected by the red-water phenomenon. *Appl. Microbiol. Biotechnol.* **2013**, *97*, 379–394. [CrossRef]
17. Belila, A.; Fazaa, I.; Hassen, A.; Ghrabi, A. Anoxygenic phototrophic bacterial diversity within wastewater stabilization plant during red water phenomenon. *Int. J. Environ. Sci. Technol.* **2013**, *10*, 837–846. [CrossRef]
18. Dungan, R.S.; Leytem, A.B. Detection of purple sulfur bacteria in purple and non-purple dairy wastewaters. *J. Environ. Qual.* **2015**, *44*, 1550–1555. [CrossRef]
19. Okubo, Y.; Futamata, H.; Hiraishi, A. Characterization of phototrophic purple nonsulfur bacteria forming colored microbial mats in a swine wastewater ditch. *Appl. Environ. Microbiol.* **2006**, *72*, 6225–6233. [CrossRef]

20. Hiraishi, A.; Kitamura, H. Distribution of phototrophic purple nonsulfur bacteria in activated sludge systems and other aquatic environments. *Bull. Jpn. Soc. Sci. Fish.* **1984**, *50*, 1929–1937. [CrossRef]
21. Hiraishi, A.; Ueda, Y. Isolation and characterization of *Rhodovulum strictum* sp. nov. and some other members of purple nonsulfur bacteria from colored blooms in tidal and seawater pools. *Int. J. Syst. Bacteriol.* **1995**, *45*, 319–326. [CrossRef]
22. Hiraishi, A. Isoprenoid quinones as biomarkers of microbial populations in the environment. *J. Biosci. Bioeng.* **1999**, *88*, 449–460. [CrossRef]
23. Hiraishi, A.; Ueda, Y.; Ishihara, J.; Mori, T. Comparative lipoquinone analysis of influent sewage and activated sludge by high-performance liquid chromatography and photodiode array detection. *J. Gen. Appl. Microbiol.* **1996**, *42*, 457–469. [CrossRef]
24. Achenbach, L.A.; Carey, J.; Madigan, M.T. Photosynthetic and phylogenetic primers for detection of anoxygenic phototrophs in natural environments. *Appl. Environ. Microbiol.* **2001**, *67*, 2922–2926. [CrossRef] [PubMed]
25. Hirose, S.; Nagashima, K.V.P.; Matsuura, K.; Haruta, S. Diversity of purple phototrophic bacteria, Inferred from pufM gene, within epilithic Biofilm in Tama River, Japan. *Microbes Environ.* **2012**, *27*, 327–329. [CrossRef] [PubMed]
26. Imhoff, J.F. New dimensions in microbial ecology-functional genes in studies to unravel the biodiversity and role of functional microbial groups in the environment. *Microorganisms* **2016**, *4*, 19. [CrossRef] [PubMed]
27. American Public Health Association; American Water Works Association; Water Pollution Control Federation. *Standard Methods for the Examination of Water and Wastewater*, 19th ed.; American Public Health Association: Washington, DC, USA, 1995.
28. Clayton, R.K. Toward the isolation of a photochemical reaction center in *Rhodopseudomonas sphaeroides*. *Biochim. Biophys. Acta* **1963**, *75*, 312–323. [CrossRef]
29. Minnikin, D.E.; O'Donnell, A.G.; Goodfellow, M.; Alderson, G.; Athalye, M.; Schaal, A.; Parlett, J.H. An integrated procedure for the extraction of bacterial isoprenoid quinones and polar lipids. *J. Microbiol. Methods* **1984**, *2*, 233–241. [CrossRef]
30. Hiraishi, A.; Hoshino, Y.; Kitamura, K. Isoprenoid quinone composition in the classification of *Rhodospirillaceae*. *J. Gen. Appl. Microbiol.* **1984**, *30*, 197–210. [CrossRef]
31. Hiraishi, A.; Hoshino, Y. Distribution of rhodoquinone in *Rhodospirillaceae* and its taxonomic implications. *J. Gen. Appl. Microbiol.* **1984**, *30*, 435–448. [CrossRef]
32. Yoshida, N.; Hiraishi, A. An improved redox dye-staining method using 5-cyano-2, 3-ditoryl tetrazolium chloride for detection of metabolically active bacteria in activated sludge. *Microbes Environ.* **2004**, *19*, 61–70. [CrossRef]
33. Hiraishi, A.; Okamura, K. *Rhodopseudomonas telluris* sp. nov., a phototrophic alphaproteobacterium isolated from paddy soil. *Int. J. Syst. Evol. Microbiol.* **2017**, *67*, 3369–3374. [CrossRef] [PubMed]
34. Imhoff, J.F. Anoxygenic phototrophic bacteria. In *Methods in Aquatic Bacteriology*; Austin, B., Ed.; John Wiley & Sons: New York, NY, USA, 1988; pp. 207–240.
35. Biebl, H.; Pfennig, N. Growth yields of green sulfur bacteria in mixed cultures with sulfur and sulfate reducing bacteria. *Arch. Microbiol.* **1978**, *117*, 9–16. [CrossRef]
36. Hiraishi, A.; Shin, Y.K.; Ueda, Y.; Sugiyama, J. Automated sequencing of PCR-amplified 16S rDNA on "Hydrolink" gels. *J. Microbiol. Methods* **1994**, *19*, 145–154. [CrossRef]
37. Lane, D.J. 16S/23S rRNA sequencing. In *Nucleic Acid Techniques in Bacterial Systematics*; Stackebrandt, E., Goodfellow, M., Eds.; Wiley: New York, NY, USA, 1991; pp. 115–175.
38. Hanada, A.; Kurogi, T.; Giang, N.M.; Yamada, T.; Kamimoto, Y.; Kiso, Y.; Hiraishi, A. Bacteria of the candidate phylum TM7 are prevalent in acidophilic nitrifying sequencing-batch reactors. *Microbes Environ.* **2014**, *29*, 353–362. [CrossRef] [PubMed]
39. Yoon, S.H.; Ha, S.M.; Kwon, S.; Lim, J.; Kim, Y.; Seo, H.; Chun, J. Introducing EzBioCloud: A taxonomically united database of 16S rRNA gene sequences and whole-genome assemblies. *Int. J. Syst. Evol. Microbiol.* **2017**, *67*, 1613–1617. [CrossRef] [PubMed]
40. Altschul, S.F.; Madden, T.L.; Schäffer, A.A.; Zhang, J.; Zhang, Z.; Miller, W.; Lipman, D.J. Gapped BLAST and PSI-BLAST: A new generation of protein database search programs. *Nucleic Acids Res.* **1997**, *25*, 3389–3402. [CrossRef]

41. Hiraishi, A.; Iwasaki, M.; Shinjo, H. Terminal restriction pattern analysis of 16S rRNA genes for the characterization of bacterial communities of activated sludge. *J. Biosci. Bioeng.* **2000**, *90*, 148–156. [CrossRef]
42. Marmur, J. A procedure for the isolation of deoxyribonucleic acid from microorganisms. *Methods Enzymol.* **1963**, *6*, 726–738.
43. Hiraishi, A.; Sakamaki, N.; Miyakoda, H.; Maruyama, T.; Kato, K.; Futamata, H. Estimation of "*Dehalococcoides*" populations in lake sediment contaminated with low levels of polychlorinated dioxins. *Microbes Environ.* **2005**, *20*, 216–226. [CrossRef]
44. Větrovský, T.; Baldrian, P. The Variability of the 16S rRNA gene in bacterial genomes and its consequences for bacterial community analyses. *PLoS ONE* **2013**, *8*, e57923. [CrossRef]
45. Kumar, S.; Stecher, G.; Tamura, K. MEGA7: Molecular evolutionary genetics analysis version 7.0 for bigger datasets. *Mol. Biol. Evol.* **2016**, *33*, 1870–1874. [CrossRef] [PubMed]
46. Felsenstein, J. Confidence limits on phylogenies: An approach using the bootstrap. *Evolution* **1985**, *39*, 783–791. [CrossRef] [PubMed]
47. Wilson, K. Preparation of genomic DNA from bacteria. *Curr. Protoc. Mol. Biol.* **2001**, *56*, 2–4. [CrossRef] [PubMed]
48. Hirose, Y.; Katayama, M.; Ohtsubo, Y.; Misawa, N.; Iioka, E.; Suda, W.; Oshima, K.; Hanaoka, M.; Tanaka, K.; Eki, T.; et al. Complete genome sequence of cyanobacterium *Geminocystis* sp. strain NIES-3709, which harbors a phycoerythrin-rich phycobilisome. *Genome Announc.* **2015**, *3*, e00385-15. [CrossRef]
49. Ohtsubo, Y.; Maruyama, F.; Mitsui, H.; Nagata, Y.; Tsuda, M. Complete genome sequence of *Acidovorax* sp. strain KKS102, a polychlorinated-biphenyl degrader. *J. Bacteriol.* **2012**, *194*, 6970–6971. [CrossRef]
50. Margulies, M.; Egholm, M.; Altman, W.E.; Attiya, S.; Bader, J.S.; Bemben, L.A.; Berka, J.; Braverman, M.S.; Chen, Y.-J.; Chen, Z.; et al. Genome sequencing in microfabricated high-density picolitre reactors. *Nature* **2005**, *437*, 376–380. [CrossRef]
51. Tatusova, T.; DiCuccio, M.; Badretdin, A.; Chetvernin, V.; Nawrocki, E.P.; Zaslavsky, L.; Lomsadze, A.; Pruitt, K.D.; Borodovsky, M.; Ostell, J. NCBI prokaryotic genome annotation pipeline. *Nucleic Acids Res.* **2016**, *44*, 6614–6624. [CrossRef]
52. Goris, J.; Konstantinidis, K.T.; Coenye, T.; Vandamme, P.; Tiedje, J.M. DNA-DNA hybridization values and their relation to whole genome sequence. *Int. J. Syst. Evol. Microbiol.* **2007**, *57*, 81–91. [CrossRef]
53. Na, S.-I.; Kim, Y.O.; Yoon, S.-H.; Ha, S.; Baek, I.; Chun, J. UBCG: Up-to-date bacterial core gene set and pipeline for phylogenomic tree reconstruction. *J. Microbiol.* **2018**, *56*, 280–285. [CrossRef]
54. Stamatakis, A. RAxML version 8: A tool for phylogenetic analysis and post-analysis of large phylogenies. *Bioinformatics* **2014**, *30*, 1312–1313. [CrossRef]
55. Imhoff, J.F. Quinones of phototrophic purple bacteria. *FEMS Microbiol. Lett.* **1984**, *25*, 85–89. [CrossRef]
56. Imhoff, J.F.; Rahn, T.; Künzel, S.; Neulinger, S.C. Photosynthesis is widely distributed among *Proteobacteria* as demonstrated by the phylogeny of PufLM reaction center proteins. *Front. Microbiol.* **2018**, *8*, 2679. [CrossRef] [PubMed]
57. Zeng, Y.; Feng, F.Y.; Medová, H.; Dean, J.; Koblížek, M. Functional type 2 photosynthetic reaction centers found in the rare bacterial phylum Gemmatimonadates. *Proc. Natl. Acad. Sci. USA* **2014**, *111*, 7795–7800. [CrossRef]
58. Zeng, Y.; Baumbach, J.; Barbosa, E.G.; Azevedo, V.; Zhang, C.; Koblížek, M. Metagenomic evidence for the presence of phototrophic Gemmatimonadetes bacteria in diverse environments. *Environ. Microbiol. Rep.* **2016**, *8*, 139–149. [CrossRef]
59. Nagashima, K.V.; Hiraishi, A.; Shimada, K.; Matsuura, K. Horizontal transfer of genes coding for the photosynthetic reaction centers of purple bacteria. *J. Mol. Evol.* **1997**, *45*, 131–136. [CrossRef]
60. Imhoff, J.F.; Rahn, T.; Künzel, S.; Neulinger, S.C. Phylogeny of anoxygenic photosynthesis based on sequences of photosynthetic reaction center proteins and a key enzyme in bacteriochlorophyll biosynthesis, the chlorophyllide reductase. *Microorganisms* **2019**, *7*, 576. [CrossRef]
61. Suresh, G.; Lodha, T.D.; Indu, B.; Sasikala, C.; Ramana, C.V. Taxogenomics resolves conflict in the genus *Rhodobacter*: A two and half decades pending thought to reclassify the genus *Rhodobacter*. *Front. Microbiol.* **2019**, *10*, 2840. [CrossRef]
62. Masuda, S.; Hori, K.; Maruyama, F.; Ren, S.; Sugimoto, S.; Yamamoto, N.; Mori, H.; Yamada, T.; Sato, S.; Tabata, S.; et al. Whole-genome sequence of the purple photosynthetic bacterium *Rhodovulum sulfidophilum* strain W4. *Genome Announc.* **2013**, *1*, e00577-13. [CrossRef]

63. Nagao, N.; Hirose, Y.; Misawa, N.; Ohtsubo, Y.; Umekage, S.; Kikuchi, Y. Complete genome sequence of *Rhodovulum sulfidophilum* DSM 2351, an extracellular nucleic acid-producing bacterium. *Genome Announc.* **2015**, *3*, e00388-15. [CrossRef]
64. Hansen, T.A.; Veldkamp, H. *Rhodopseudomonas sulfidophila*, nov. spec., a new species of the purple nonsulfur bacteria. *Arch. Mikrobiol.* **1973**, *92*, 45–58. [CrossRef]
65. Hiraishi, A.; Ueda, Y. Intrageneric structure of the genus *Rhodobacter*: Transfer of *Rhodobacter sulfidophilus* and related marine species to the genus *Rhodovulum* gen. nov. *Int. J. Syst. Bacteriol.* **1994**, *44*, 15–23. [CrossRef]
66. Appia-Ayme, C.; Little, P.J.; Matsumoto, Y.; Leech, A.P.; Berks, B.C. Cytochrome complex essential for photosynthetic oxidation of both thiosulfate and sulfide in *Rhodovulum sulfidophilum*. *J. Bacteriol.* **2001**, *183*, 6107–6118. [CrossRef] [PubMed]
67. Ghosh, W.; Dam, B. Biochemistry and molecular biology of lithotrophic sulfur oxidation by taxonomically and ecologically diverse bacteria and archaea. *FEMS Microbiol. Rev.* **2009**, *33*, 999–1043. [CrossRef] [PubMed]
68. Richter, M.; Rosselló-Móra, R. Shifting the genomic gold standard for the prokaryotic species definition. *Proc. Natl. Acad. Sci. USA* **2009**, *106*, 19126–19131. [CrossRef]
69. Brooks, B.; Olm, M.R.; Firek, B.A.; Baker, R.; Thomas, B.C.; Morowitz, M.J.; Banfield, J.F. Strain-resolved analysis of hospital rooms and infants reveals overlap between the human and room microbiome. *Nat. Commun.* **2017**, *8*, 1814. [CrossRef]
70. Brown, L.M.; Gunasekera, T.S.; Bowen, L.L.; Ruiz, O.N. Draft genome sequence of *Rhodovulum* sp. strain NI22, a naphthalene-degrading marine bacterium. *Genome Announc.* **2015**, *3*, e01475-14. [CrossRef]
71. Khatri, I.N.; Korpole, S.; Subramanian, S.; Pinnaka, A.K. Draft genome sequence of *Rhodovulum* sp. strain PH10, a phototrophic alphaproteobacterium isolated from a soil sample of mangrove of Namkhana, India. *J. Bacteriol.* **2012**, *194*, 6363. [CrossRef]
72. Kobayashi, M.; Kobayashi, M. Waste remediation and treatment using anoxygenic phototrophic bacteria. In *Anoxygenic Photosynthetic Bacteria*; Blankenship, R.E., Madigan, M.T., Bauer, C.E., Eds.; Kluwer Academic Publishers: Dordrecht, The Netherlands, 1995; pp. 1269–1282.
73. Hiraishi, A.; Shi, J.L.; Kitamura, H. Effects of organic nutrient strength on the purple nonsulfur bacterial content and metabolic activity of photosynthetic sludge for wastewater treatment. *J. Ferment. Bioeng.* **1989**, *68*, 269–276. [CrossRef]
74. Suzuki, H.; Daimon, M.; Awano, T.; Umekage, S.; Tanaka, T.; Kikuchi, Y. Characterization of extracellular DNA production and flocculation of the marine photosynthetic bacterium *Rhodovulum sulfidophilum*. *Appl. Microbiol. Biotechnol.* **2009**, *84*, 349–356. [CrossRef]

© 2020 by the authors. Licensee MDPI, Basel, Switzerland. This article is an open access article distributed under the terms and conditions of the Creative Commons Attribution (CC BY) license (http://creativecommons.org/licenses/by/4.0/).

 *microorganisms*

Article

Adaptation to Photooxidative Stress: Common and Special Strategies of the Alphaproteobacteria *Rhodobacter sphaeroides* and *Rhodobacter capsulatus*

Mathieu K. Licht [1], Aaron M. Nuss [2], Marcel Volk [1,3], Anne Konzer [4], Michael Beckstette [5], Bork A. Berghoff [1,*] and Gabriele Klug [1,*]

[1] Institute of Microbiology and Molecular Biology, University of Giessen, 35392 Giessen, Germany; Mathieu.Licht@mikro.bio.uni-giessen.de (M.K.L.); Marcel.Volk@ukmuenster.de (M.V.)
[2] Department of Molecular Infection Biology, Helmholtz Centre for Infection Research, 38124 Braunschweig, Germany; aaron.nuss@microsynth.seqlab.de
[3] Institute for Infectiology, Center for Molecular Biology of Inflammation, University of Münster, 48149 Münster, Germany
[4] Biomolecular Mass Spectrometry, Max Planck Institute for Heart and Lung Research, 61231 Bad Nauheim, Germany; anne.konzer@gmail.com
[5] Department of Computational Biology for Individualized Medicine, Centre for Individualized Infection Medicine, 30625 Hannover, Germany; Michael.Beckstette@helmholtz-hzi.de
* Correspondence: Bork.A.Berghoff@mikro.bio.uni-giessen.de (B.A.B.); Gabriele.Klug@mikro.bio.uni-giessen.de (G.K.); Tel.: +49-641-99-35558 (B.A.B.); +49-641-99-35542 (G.K.)

Received: 28 January 2020; Accepted: 13 February 2020; Published: 19 February 2020

Abstract: Photosynthetic bacteria have to deal with the risk of photooxidative stress that occurs in presence of light and oxygen due to the photosensitizing activity of (bacterio-) chlorophylls. Facultative phototrophs of the genus *Rhodobacter* adapt the formation of photosynthetic complexes to oxygen and light conditions, but cannot completely avoid this stress if environmental conditions suddenly change. *R. capsulatus* has a stronger pigmentation and faster switches to phototrophic growth than *R. sphaeroides*. However, its photooxidative stress response has not been investigated. Here, we compare both species by transcriptomics and proteomics, revealing that proteins involved in oxidation–reduction processes, DNA, and protein damage repair play pivotal roles. These functions are likely universal to many phototrophs. Furthermore, the alternative sigma factors RpoE and RpoH$_{II}$ are induced in both species, even though the genetic localization of the *rpoE* gene, the RpoE protein itself, and probably its regulon, are different. Despite sharing the same habitats, our findings also suggest individual strategies. The *crtIB-tspO* operon, encoding proteins for biosynthesis of carotenoid precursors and a regulator of photosynthesis, and *cbiX*, encoding a putative ferrochelatase, are induced in *R. capsulatus*. This specific response might support adaptation by maintaining high carotenoid-to-bacteriochlorophyll ratios and preventing the accumulation of porphyrin-derived photosensitizers.

Keywords: *Rhodobacter capsulatus*; *Rhodobacter sphaeroides*; photooxidative stress; transcriptomics; proteomics; stress defense

1. Introduction

Microbes in aquatic habitats need to adapt to frequent changes in environmental parameters like temperature, O_2-saturation, or light conditions. While phototrophic bacteria can take advantage of pigment-protein complexes to use light energy for ATP production, they face the special challenge of photooxidative stress: (bacterio-) chlorophylls can act as photosensitizers and transfer energy to the ground state triplet oxygen ($^3O_2$), causing a spin conversion in the π*2p orbital that generates highly reactive singlet oxygen ($^1O_2$). While other photosensitizers like humic acids also contribute to photooxidative stress, (bacterio-) chlorophyll *a* is regarded as the main cause of $^1O_2$-generation in photosynthetic bacteria. Independently of light, processes like lipid peroxide decomposition or hypochloric acid reacting with hydrogen peroxide can generate $^1O_2$ [1].

Facultative anoxygenic phototrophic bacteria of the genus *Rhodobacter* adjust their lifestyle to the light and oxygen conditions. Due to a high metabolic versatility, they do not rely on photosynthesis for ATP production, but can also perform aerobic or anaerobic respiration or fermentation. They do not form photosynthetic complexes at high oxygen concentrations, and at intermediate oxygen concentration, light inhibits the accumulation of pigment-protein complexes [2,3], which reduces the risk of photooxidative stress. Several protein regulators, including redox-responsive factors, photoreceptors, and even proteins with dual-sensing function like AppA [2], but also RNA regulators [4,5], contribute to the regulated formation of photosynthetic complexes.

Nevertheless, situations that cause photooxidative stress cannot be completely avoided, and consequently, mechanisms to defend this stress are important for survival. As seen across all kingdoms, $^1O_2$ damages a wide variety of biomolecules, including nucleic acids, amino acids, fatty acid lipids or thiols, and glutathione [1,6,7]. Singlet oxygen can directly oxidize its targets or generate other reactive oxygen species (ROS) like endo- or hydroperoxides via (4 + 2) cycloaddition or the ene reaction [8]. Without a proper cellular response, $^1O_2$-stress can be cytotoxic [9,10]. In the case of DNA, the mutagenic potential of $^1O_2$ in *Escherichia coli* can be assigned to the oxidation of guanine sites to 8-oxo-7,8-dihydro-2′-deoxyguanosine (8-OHdG), which is susceptible to single strand breaks [9,11]. Regardless of the occurrence of 8-OHdG, $^1O_2$ can also affect RNA, specifically viral RNA, by mediating RNA-protein-crosslinking, as shown for photoinactivation of HIV-1 [12]. As $^1O_2$ can form peroxides, it also targets unsaturated fatty acids, causing lipid peroxidation which impairs membranes in their potential, integrity, or transport activities [13,14]. However, due to their high abundance in the cell, proteins are the primary targets of $^1O_2$ [15]. Protein damage by $^1O_2$ may often be traced back to the oxidization of amino acids containing sulfur or aromatic compounds [7,16], but other ROS generated by $^1O_2$ might target the proteome as well. Unfolded or aggregated proteins and the loss of enzyme activities are likely consequences of $^1O_2$ [17].

For more than a decade, *R. sphaeroides* have served as the bacterial model organism to elucidate the photooxidative stress response. Increased expression of certain genes in response to $^1O_2$ was demonstrated, and an important role of the alternative sigma factors RpoE, RpoH$_I$, and RpoH$_{II}$ in this response was revealed [10,18–21]. An early step in $^1O_2$-dependent gene activation is proteolytic degradation of the antisigma factor ChrR [22,23]. The released RpoE sigma factor directly activates a small number of genes like the DNA photolyase gene *phrA* or *cfaS* (cyclopropane fatty acyl-phospholipid synthase) [24,25], but also the *rpoH$_{II}$* gene. RpoH$_{II}$, together with RpoH$_I$, activates a high number of genes upon photooxidative stress, but also in response to other stresses [20,21]. Genes that are activated upon photooxidative stress have functions, e.g., in the detoxification of toxic molecules like peroxides or methylglyoxal, in protein quality control and turnover, in $^1O_2$ quenching, DNA repair, and transport [1,26]. Although carotenoids provide protection against $^1O_2$ in *R. sphaeroides*, genes for carotenoid synthesis are not activated by $^1O_2$ in this bacterium [27,28].

Regarding the photoprotective function of carotenoids, both *R. capsulatus* and *R. sphaeroides* accumulate mainly spheroidene (SE) under anaerobic conditions and spheroidenone (SO) under (semi-) aerobic conditions [27,29–33]. An oxygen-activated spheroidene monooxygenase (CrtA) causes this shift and incorporates a keto-group into SE to form SO [31,32]. The shift from SE to SO helps *Rhodobacter* to counteract $^1O_2$ [34]. The introduced keto-group stabilizes the intramolecular charge transfer state of excited carotenoids by binding to the reaction center of the light-harvesting complex I. Due to its low energy, the intramolecular charge transfer state of carotenoids enables the quenching of $^1O_2$ without sacrificing light-harvest. This could explain why different carotenoid-deficient mutants of *R. sphaeroides* showed decreased survival rates under photooxidative stress when (hydroxy-) SO amounts were very low [27].

Many small RNAs (sRNAs) are induced by $^1O_2$ [35], and for some, the regulatory function could be elucidated. The sRNA-mRNA interactions are often stabilized by the RNA chaperone Hfq, which is another crucial element of the photooxidative stress response in *R. sphaeroides* [36]. Some of these sRNAs (CcsR1-4, Pos19) are involved in balancing the glutathione pool and in the downregulation of the pyruvate dehydrogenase complex and aerobic electron transport, a primary source of ROS [37,38]. Other sRNAs (SorY, SorX) reduce the metabolic flux into the tricarboxylic acid (TCA) cycle or affect polyamine transport [39,40]. a switch from glycolysis to the pentose phosphate cycle and reduced activity of the TCA cycle upon oxidative stress reduce the production of the pro-oxidant NADH and increase production of the protective NADPH. An integrative "omics" approach supports the importance of posttranscriptional regulation in the $^1O_2$ response of *R. sphaeroides* [28].

R. capsulatus, another member of the *Rhodobacteraceae*, shares a very similar life style with *R. sphaeroides*, and was also intensely studied with regard to its adaptation to different oxygen- and light conditions [41,42]. Under high oxygen tension, *R. capsulatus* cultures show more pigmentation than *R. sphaeroides*, implying a faster adaptation to phototrophic conditions but a higher risk of $^1O_2$ production. However, the response to $^1O_2$ has not been elucidated in *R. capsulatus*. In this study, we applied omics approaches to analyze and compare the response of the two *Rhodobacter* species to photooxidative stress. Although both species share the same habitats, our findings suggest individual strategies to defend against photooxidative stress in addition to a common core response.

2. Materials and Methods

2.1. Bacterial Strains and Growth Conditions

Rhodobacter strains (Table S1) were cultivated at 32 °C in minimal medium containing malate as a carbon source [43]. For microaerobic conditions (~25 µM O_2), cultures were incubated in Erlenmeyer flasks with a culture volume of 80% and shaking at 140 rpm. To cultivate *Rhodobacter* under aerobic conditions (160–180 µM O_2), cultures were grown either in baffled flasks with shaking at 140 rpm and a culture volume of 20%, or in flat glass bottles gassed with air. To establish phototrophic growth, airtight flat glass bottles were completely filled with medium and cultures were illuminated continuously with white light (60 W·m^{-2}; fluorescent tube: Omnilux 18W). In order to shift *Rhodobacter* between two different growth conditions, exponentially growing cultures (OD_{660} of ~0.4) were diluted to an OD_{660} of 0.2.

2.2. Photooxidative Stress Experiments

Photooxidative stress experiments were carried out as previously described in Glaeser and Klug, 2005 [27]. In short, pigmented cultures from microaerobic cultivation were shifted to aerobic conditions in air-gassed flat glass bottles in the dark. Methylene blue was added at a final concentration of 0.2 µM. After an OD_{660} of ~0.4 was reached, cultures were exposed to 800 W·m^{-2} white light to generate $^1O_2$ (photooxidative stress).

For zone of inhibition assays, exponentially growing cultures were diluted into soft agar (0.8%, w/v) and poured onto malate minimal salt medium agar (1.6%, w/v). Five microliters of methylene blue (10 µM) were spotted onto a filter paper disk, which was placed in the center of the agar plate. Cultures were incubated for 48 h at 32 °C under illumination with 20 W·m^{-2} white light.

2.3. Analysis of Pigmentation

BChl *a* and carotenoids were extracted and measured as described in Glaeser and Klug, 2005 [27]. Briefly, 1 mL samples of *Rhodobacter* cultures were harvested at 17,000× *g* for 5 min. Pellets were resuspended in 50 µL ddH_2O and mixed with 500 µL of acetone/methanol (7/2, v/v) by vortexing for 30 s. Samples were centrifuged at 17,000× *g* for 5 min, and the absorption of the supernatant was measured in a Specord 50 Plus spectrometer (Analytik Jena, Jena, Germany), using acetone/methanol (7/2, v/v) as a reference. The carotenoid and BChl *a* concentrations were calculated from the absorptions at 484 and 770 nm, respectively, with extinction coefficients of 128 mM^{-1}·cm^{-1} for carotenoids [44] and 76 mM^{-1}·cm^{-1} for BChl *a* [45]. Concentrations were normalized to the OD$_{660}$.

2.4. Measurement of Reactive Oxygen Species

Singlet oxygen levels were measured using the fluorescent probe Singlet Oxygen Sensor Green (SOSG, Molecular Probes, Eugene, OR, USA). a SOSG stock solution of 100 µM was prepared in HEPES buffer (40 mM, pH 7, 1% methanol). Six microliters of the SOSG stock solution were added to 114 µL culture samples (final SOSG concentration of 5 µM). Technical duplicates were incubated for 30 min at 32 °C and 450 rpm in a Vibramax 100 shaker (Heidolph Instruments, Schwabach, Germany). Samples were either kept in the dark or illuminated with 800 W·m^{-2} red light. Samples without SOSG served as background controls. Cells were centrifuged at 8,000 rpm for 5 min and resuspended in 100 µL HEPES buffer (40 mM, pH 7, 1% methanol). Fluorescence intensities (excitation 500 nm, emission 532 nm) were measured in an Infinite M200 microplate reader (Tecan, Crailsheim, Germany). After subtraction of the background control, fluorescence intensities were normalized to BChl *a* levels. Ratios between illuminated samples and dark controls were subsequently calculated.

General ROS levels were measured as previously described [43], using the oxidation-sensitive fluorescent probe 2,7-dihydrodichlorofluorescein diacetate (H$_2$DCFDA, Molecular Probes, Eugene, OR, USA). Culture samples of 100 µL were incubated at 32 °C with H$_2$DCFDA (final concentration of 10 µM) for 30 min and shaking at 140 rpm in technical triplicates. a culture sample without H$_2$DCFDA served as background control. Fluorescence intensities (excitation 492 nm, emission 525 nm) were measured in an Infinite M200 microplate reader (Tecan, Crailsheim, Germany). After subtraction of the background control, fluorescence intensities were normalized to the OD$_{660}$.

2.5. Transcriptome Analysis by RNA-Sequencing

2.5.1. Sample Preparation for RNA-seq

Cultures of *R. capsulatus* were shifted from microaerobic to aerobic growth in the dark followed by photooxidative stress as described by Berghoff and colleagues [28]. Samples of 20 mL before (0 min) and after stress (10 min) were collected, cooled on ice, and centrifuged at 10,000× *g* for 10 min at 4 °C. Cell pellets were resuspended in 1 mL minimal medium and centrifuged at 10,000× *g* for 10 min at 4 °C. RNA was extracted via the hot phenol protocol [46]. The RNA was resolved in RNase-free water (Roth) and treated with DNaseI (Invitrogen, Carlsbad, CA, USA) to remove traces of DNA. a test PCR (rpoZ-for: 5′-GAT GAT CTG CGC GAG CGT CT-3′; rpoZ-rev: 5′-CCT TGC GCG TCC ATC AAT GC-3′) was performed to ensure that the RNA was free of DNA. RNA integrity was assessed using the Agilent RNA 6000 Nano Kit on the Agilent 2100 Bioanalyzer (Agilent Technologies, Santa clara, CA, USA) to ensure high quality RNA (RIN ≥ 9) for downstream processing. rRNA was depleted from 5 µg of total RNA using the Ribo-Zero rRNA Removal Kit (Gram-Negative Bacteria, Epicentre Biotechnologies, Madison, WI, USA) as recommended by the manufacturer. One microliter of either 1:10 diluted ERCC ExFold RNA Spike-in Mix 1 or Mix 2 (Ambion, Austin, TX, USA) was added to 1 µg of rRNA-depleted RNA. To create 5′-monophosphorylated RNA, rRNA-depleted RNA (including ERCC Spike-in Mixes) was treated with RNA 5′ polyphosphatase as recommended by the manufacturer.

2.5.2. Strand-Specific Library Preparation and Illumina Sequencing

Strand-specific RNA-seq cDNA library preparation and barcode introduction was based on RNA adapter ligation as described earlier [47]. The quality of the libraries was validated using an Agilent 2100 Bioanalyzer (Agilent Technologies) following the manufacturer's instruction. Cluster generation was performed using the Illumina cluster station. Single-end sequencing on the HiSeq2500 followed a standard protocol. The fluorescent images were processed to sequences and transformed to FastQ format using the Genome Analyzer Pipeline Analysis software 1.8.2 (Illumina, San Diego, CA,

USA). The sequence output was controlled for general quality features, sequencing adapter clipping, and demultiplexing using the fastq-mcf and fastq-multx tool of ea-utils [48].

2.5.3. Read Mapping, Bioinformatics and Statistics

The quality of the sequencing output and potential contamination was analyzed using FastQC (Babraham Bioinformatics, http://www.bioinformatics.babraham.ac.uk/projects/fastqc/). Identified adapter contamination and remaining artificial sequence (barcode) were removed using program fastx_trimmer from the FASTX-210 toolkit version 0.0.13 (http://hannonlab.cshl.edu/fastx_toolkit/). On the 3′-end, reads were trimmed if the per base Phred score fell short of 20. Trimmed reads with a remaining length < 20 nucleotides were discarded. All sequenced libraries were mapped to the *R. capsulatus* genome (accession no. NC_014034) and the pRCB133 plasmid (accession no. NC_014035.1) using Bowtie2 (version 2.1.0) in end-to-end alignment mode [49]. After read mapping, the resulting bam files were filtered for uniquely mapped reads using SAMtools (both strands) [50]. The determined uniquely mapped read counts served as inputs to DESeq2 [51] for the pairwise detection and quantification of differential gene expression. For DESeq2 parametrization, we used a beta prior and disabled Cook distance cut off filtering. All other parameters remained unchanged. In addition, RPKM (reads per kilobase max. transcript length per million mapped reads) values were computed for each library from the raw gene counts. The list of DESeq2 determined differentially expressed genes (DEGs) was filtered with a conservative absolute log2 fold change cutoff of at least 1 and a cutoff for a multiple testing corrected *p*-value of at most 0.05.

2.5.4. ERCC Spike-in Control Analysis

To assess the platform dynamic range and the accuracy of fold-change responses, ERCC RNA Spike-in controls were used. Spike-in control sequences were added to the *R. capsulatus* reference genome/annotation prior to read alignment and read counts for Spike-in controls were determined, along with normal gene counts with program htseq-count. Further data analyses and the generation of dose- and fold-change response plots were performed as described by the manufacturer (Ambion, Carlsbad, CA, USA).

2.5.5. RNA-seq Data Accessibility

The RNA-seq analysis can be found in Table S2. Raw RNA-seq data have been deposited in NCBI's Gene Expression Omnibus, and are accessible through GEO Series accession number GSE134200.

2.6. Quantitative RT-PCR

For quantitative RT-PCR, total RNA was isolated after 10 min of photooxidative stress as described for RNA-seq. Samples were treated with TURBO DNA-free™ Kit (Invitrogen, Thermo Fisher Scientific, Schwerte, Germany) to remove DNA contaminations. The Brilliant III Ultra-Fast SYBR Green QRT-PCR Master Mix (Agilent Technologies, Waldbronn, Germany) was applied using 4 ng·μL^{-1} of total RNA per reaction. RT-PCR was performed in a CFX Connect™ Real-Time System (Bio-Rad). Cycle threshold (Ct) values were determined using the CFX Maestro™ Software (Bio-Rad, Feldkirchen, Germany), and relative transcript levels calculated according to Pfaffl (2001) [52]. The *rpoZ* gene was used for normalization. Primers and their amplification efficiencies are listed in Table S3.

2.7. Protein Sample Preparation and Mass Spectrometry

Protein sample preparation was performed as previously described [53]. For mass spectrometry (MS) analysis, peptides were eluted from STAGE tips by solvent B (80% acetonitrile, 0.1% formic acid), dried down in a SpeedVac Concentrator (Thermo Fisher Scientific, Schwerte, Germany) and dissolved in solvent a (0.1% formic acid). Peptides were separated using an UHPLC system (EASY-nLC 1000, ThermoFisher Scientific, Waltham, MA, USA) and 20 cm, in-house packed C18 silica columns

(1.9 µm C18 beads, Dr. Maisch GmbH) coupled in line to a Q-Exactive HF orbitrap mass spectrometer (ThermoFisher Scientific) using an electrospray ionization source. a gradient of 240 min was applied using a linearly increasing concentration of solvent B (80% acetonitrile, 0.1% formic acid) over solvent a (0.1% formic acid) from 5% to 30% for 215 min and from 30% to 60% for 5 min, followed by washing with 95% of solvent B for 5 min and re-equilibration with 5% of solvent B. Full MS spectra were acquired in a mass range of 300 to 1750 m/z with a resolution of 60,000 at 200 m/z. The ion injection target was set to 3×10^6 and the maximum injection time limited to 20 ms. Ions were fragmented by high-energy collision dissociation (HCD) using a normalized collision energy of 27 and an ion injection target of 5×10^5 with a maximum injection time of 20 ms. The resulting tandem mass spectra (MS/MS) were acquired with a resolution of 15,000 at 200 m/z using data dependent mode with a loop count of 15 (top 15). MS raw data were processed by MaxQuant (1.5.3.12) [54] using the Uniprot database for *R. capsulatus* containing 4290 entries (release date July 2016). The following parameters were used for data processing: maximum of two miss cleavages, mass tolerance of 4.5 ppm for main search, trypsin as digesting enzyme, carbamidomethylation of cysteines as fixed modification, oxidation of methionine, and acetylation of the protein N-terminus as variable modifications. For protein quantification, the LFQ function of MaxQuant was used. Peptides with a minimum of seven amino acids and at least one unique peptide were required for protein identification. Only proteins with at least two peptides and at least one unique peptide were considered to have been identified and were used for further data analysis. LFQ intensities for all identified proteins can be found in Table S4.

2.8. Search for Orthologous Rhodobacter Genes and Synteny Analysis

To find orthologous genes in *R. capsulatus* and *R. sphaeroides*, the Genome Gene Best Homologs tool from the IMG web resources was used [55]. The pBLAST-based search of orthologous genes used a 30% amino acid identity as a cutoff value for homology. PHYRE^2 was applied for a structural homology search [56]. Synteny analysis was performed using EDGAR 2.3, a software platform for comparative gene content analyses [57].

2.9. Gene Ontology Enrichment Analysis

Significantly enriched functional groups were determined with the program Cytoscape version 3.6.0 [58] according to Gene Ontology (GO) terms using the BiNGO tool [59]. Overrepresented GO categories in the data sets were determined with a hypergeometric test with Benjamini-Hochberg false discovery rate correction and a significance level of 0.05. The whole *R. capsulatus* genome served as a reference. The selected ontology file was *gb.obo*, format-version 1.2, released 06/10/2017 [60,61]. The resulting networks were searched for overrepresented GO categories.

3. Results

3.1. Adaptation of R. sphaeroides and R. capsulatus to Different Growth Conditions

Based on the different pigment content of the two *Rhodobacter* species, we hypothesized differences in their adaptation to phototrophic growth. When cultures of the two species were kept under microaerobic conditions in the dark, the growth behavior was nearly identical (doubling time t_d of 4 h ± 3 min for *R. capsulatus* and 4 h 10 min ± 15 min for *R. sphaeroides*, Figure 1A). a shift from high oxygen to phototrophic conditions with 60 W·m^{-2} white light revealed a remarkably faster adaption process for *R. capsulatus*, i.e., entering the exponential growth after the shift took ~4 h for *R. capsulatus* but ~21 h for *R. sphaeroides* (Figure 1B). As seen by the doubling time, *R. capsulatus* also grew faster in exponential phase under phototrophic conditions than *R. sphaeroides* (t_d of 3 h 35 min ± 4 min for *R. capsulatus* and 6 h 50 min ± 30 min for *R. sphaeroides*). This supported our hypothesis that the higher pigment content of *R. capsulatus* would allow a faster switch to occur to phototrophic growth. Interestingly, carotenoids (especially SE and SO) strongly contributed to the growth benefit of *R. capsulatus*, as shown by experiments with transposon mutants lacking these carotenoids (Figure S1).

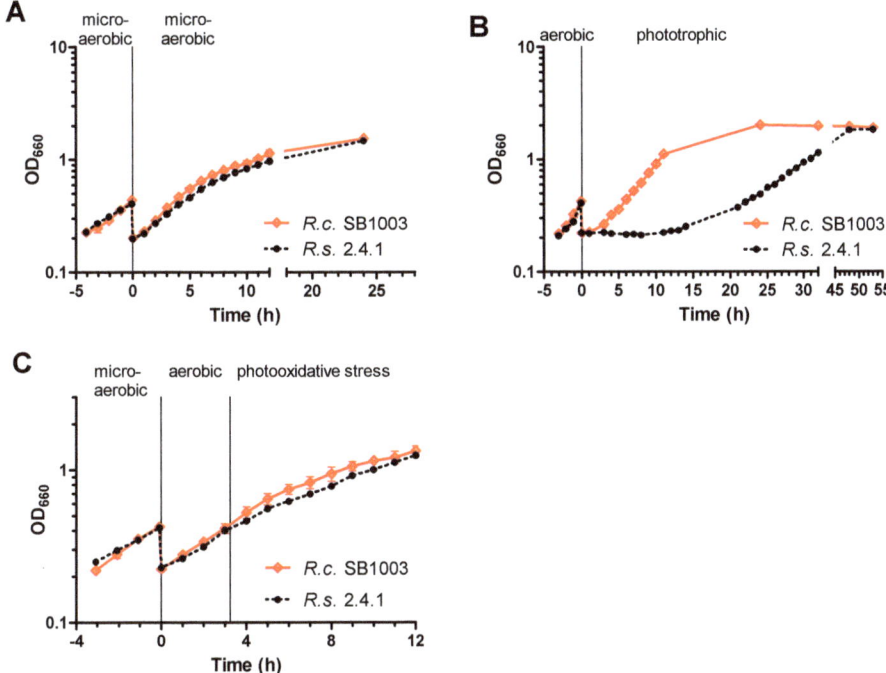

Figure 1. Growth of *R. capsulatus* and *R. sphaeroides* under different oxygen and light conditions. Exponential phase cultures of *R. capsulatus* (*R.c.* SB1003) and *R. sphaeroides* (*R.s.* 2.4.1) were diluted to an OD_{660} of 0.2 at time point 0 h. The OD_{660} was plotted semi-logarithmically against the time. Data points represent the mean of biological triplicates and error bars depict the standard deviation (standard deviations might not be visible if they are too small). (**A**) Microaerobically growing cultures. (**B**) Aerobically growing cultures shifted to phototrophic growth. (**C**) Microaerobically growing cultures were exposed to photooxidative stress when an OD_{660} of ~0.4 was reached. Figure S2 shows growth under dark and light conditions with or without methylene blue.

We also asked the question of how a strong increase of photooxidative stress caused by $^1O_2$ would affect growth of the two *Rhodobacter* species. To test this, pigmented cultures (after microaerobic cultivation) were cultivated under aerobic conditions in the dark and shifted to high light conditions (800 W·m^{-2}) in the presence of methylene blue (0.2 µM) when an OD_{660} of 0.4 was reached. These conditions were previously shown to produce $^1O_2$ and to induce a specific response in *R. sphaeroides* [27,28]. Importantly, neither methylene blue in the dark nor high light without methylene blue resulted in a strong growth retardation (Figure S2). After initiating photooxidative stress, *R. capsulatus* showed faster growth than *R. sphaeroides*, but slowed down earlier. As a consequence, both strains reached the same OD_{660} after 12 h (Figure 1C).

Although a stronger pigmentation of *R. capsulatus* cultures under high oxygen tension compared to *R. sphaeroides* was obvious, we wanted to quantify the differences and also analyze the ratio of carotenoids and bacteriochlorophylls throughout growth at different conditions. While bacteriochlorophyll functions as a photosensitizer that promotes the production of $^1O_2$, carotenoids can quench $^1O_2$, and are thus part of the defense system against $^1O_2$. In general, *R. capsulatus* showed a higher amount of bacteriochlorophyll *a* (Bchl *a*) and carotenoids, e.g., under microaerobic conditions (Figure S3) at an OD_{660} of ~0.4: *R. capsulatus* had ~3.0 µM per OD_{660} carotenoids and ~2.8 µM per OD_{660} Bchl *a*, whereas *R. sphaeroides* had ~0.7 µM per OD_{660} carotenoids and ~2.0 µM per OD_{660} Bchl *a*. After a shift to phototrophic growth, carotenoid and Bchl *a* levels steadily increased in *R. capsulatus*, reaching levels that were much higher than under aerobic conditions (Figure 2A). Since *R. sphaeroides* stopped growing after this transition for nearly 13 h, the carotenoid and Bchl *a* levels stayed low during this time period and increased only slowly. Thirty-two hours after the shift, carotenoid and Bchl *a* levels were about 3.6-fold and 4.1-fold higher, respectively, in *R. capsulatus* (Figure 2A). After a shift from low to high oxygen tension and addition of methylene blue in the dark, the pigment level steadily dropped in both strains (Figure 2B). When illumination was started to generate $^1O_2$, the Bchl *a* level dropped further. The carotenoid level in *R. capsulatus*, however, remained fairly constant (~1.4 µM per OD_{660}), with a small peak (~1.6 µM per OD_{660}) after four hours of stress. By contrast, *R. sphaeroides* showed a steady decline of carotenoids.

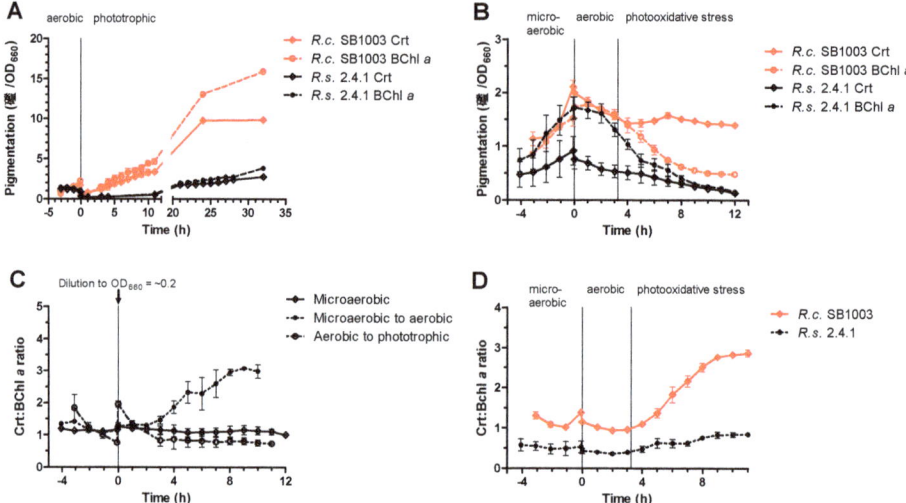

Figure 2. *R. capsulatus* has a stronger pigmentation and a higher carotenoid to bacteriochlorophyll *a* ratio than *R. sphaeroides*. Cultures in the exponential phase were diluted to an OD_{660} of 0.2 at time point 0 h. The content of carotenoids (Crt) and bacteriochlorophyll *a* (Bchl *a*) was normalized to the respective OD_{660}-values and plotted against the time for (**A**) aerobically growing cultures shifted to phototrophic growth, and (**B**) microaerobically growing cultures shifted to aerobic dark conditions followed by exposure to photooxidative stress. The Crt:Bchl *a* ratio was plotted against the time for (**C**) *R. capsulatus* either continuously grown under microaerobic conditions or shifted to different oxygen and light conditions, and for (**D**) *R. capsulatus* and *R. sphaeroides* after a shift from microaerobic to aerobic dark conditions followed by photooxidative stress. Data points represent the mean of biological triplicates and error bars depict the standard deviation (standard deviations might not be visible if they are too small).

Not only the total amount of pigments may be important for adaptation of *Rhodobacter* species to changing conditions, but also the ratio of the $^1O_2$-quenching carotenoids (Crt) to the $^1O_2$-producing BChl *a*. In *R. capsulatus*, the Crt:BChl *a* ratio did not change much during continuous cultivation under microaerobic conditions (ratio of ~1.1) or after a shift from aerobic to phototrophic conditions (ratio of ~0.8; Figure 2C). However, a shift from microaerobic to aerobic conditions resulted in a strong increase of the Crt:Bchl *a* ratio from ~1.2 to ~3.0 in *R. capsulatus* (Figure 2C). Figure 2D compares the change in the Crt:Bchl *a* ratio between *R. sphaeroides* and *R. capsulatus* upon exposure to $^1O_2$. The ratios remained fairly constant in both species during microaerobic growth and also after the shift to aerobic dark conditions in the presence of methylene blue. The ratios were ~1.1 in *R. capsulatus* and ~0.5 in *R. sphaeroides*. After the start of illumination and the production of $^1O_2$, the Crt:Bchl *a* ratio increased in

R. capsulatus from ~1.1 to nearly 3.0. In *R. sphaeroides* this ratio was below one under all conditions, and only increased from ~0.5 to ~0.8 after the initiation of photooxidative stress.

3.2. Generation of ROS in R. sphaeroides and R. capsulatus upon Photooxidative Stress

To see whether the different Crt:BChl *a* ratios observed in the two *Rhodobacter* species under photooxidative stress conditions (Figure 2D) would affect $^1O_2$ levels, the fluorescent probe Singlet Oxygen Sensor Green (SOSG) was used for in vivo $^1O_2$ measurements. Since white light itself can affect SOSG fluorescence, red light was used for illumination [62]. In cell-free reactions, the combination of oxygen, red light, and methylene blue caused enhanced SOSG fluorescence due to photosensitized formation of $^1O_2$ (Figure S4). When pigmented cultures were incubated under aerobic conditions in the presence of methylene blue, the ratio of SOSG fluorescence between illuminated samples and dark controls was significantly higher for *R. sphaeroides* than for *R. capsulatus* (Figure 3A), indicating enhanced $^1O_2$ levels in *R. sphaeroides*. In addition, general ROS formation was measured by applying 2,7-dihydrodichlorofluorescein diacetate (H_2DCFDA) before and after starting the white light illumination of aerobic cultures in the presence of methylene blue. The fluorogenic probe H_2DCFDA is mainly specific for hydrogen peroxide, peroxynitrite anions, and peroxyl radicals [63], ROS that are partly generated downstream of $^1O_2$ [15,64]. An increase in DCF fluorescence indicates elevated ROS levels. We found significant differences in fluorescence levels between the two species before and 10 min after initiating photooxidative stress (Figure 3B). The relative fluorescence intensity indicated higher ROS levels in *R. sphaeroides* compared to *R. capsulatus* by a factor of 1.6 and 1.5 for 0 and 10 min of photooxidative stress, respectively (Figure 3B).

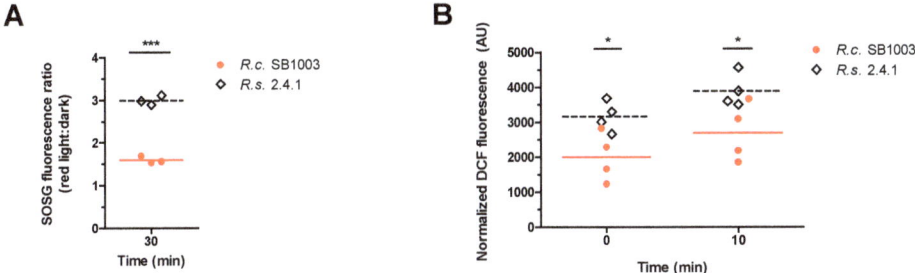

Figure 3. Determination of intracellular $^1O_2$ and ROS levels in *R. capsulatus* and *R. sphaeroides* under photooxidative stress conditions. (**A**) The fluorogenic probe SOSG was used for $^1O_2$ detection. Fluorescence intensities were normalized to BChl *a* levels. Ratios between illuminated samples (800 W·m^{-2} red light) and dark controls were calculated. (**B**) The fluorogenic probe H_2DCFDA was used for ROS detection. Fluorescence intensities were normalized to the OD_{660} and displayed in arbitrary units (AU). Data points indicate individual measurements and bars represent the mean. Two-way ANOVA followed by Bonferroni posttest was used to compare results from *R. capsulatus* (R.c. SB1003) and *R. sphaeroides* (R.s. 2.4.1) (* p-value < 0.05, *** p-value <0.001).

3.3. The rpoE-chrR Locus of R. capsulatus Shows a Unique Genetic Context in Comparison to Other Bacteria within the Rhodobacteraceae

Detailed work in *R. sphaeroides* identified the sigma factor RpoE as the master regulator of the response to $^1O_2$ [1,65]. RpoE is primarily controlled by its cognate antisigma factor ChrR, but full activation of RpoE requires the RpoE regulon members RSP_1090/91 (putative cyclopropane/cyclopropene fatty acid synthesis proteins) and the cyclopropane-fatty-acyl-phospholipid synthase CfaS [22,23]. In addition, DegS and RseP homologous proteases are involved in the degradation of ChrR [22]. In *R. sphaeroides*, the RSP_1090/91 genes are located immediately upstream of the *rpoE-chrR* locus, a genetic arrangement that is conserved among many bacteria within the *Rhodobacteraceae* (e.g., *Roseobacter denitrificans, Dinoroseobacter shibae, Jannaschia rubra, Ruegeria litorea,* and *Oceanicola litoreus*).

Interestingly, RpoE and ChrR homologs were not found in *R. capsulatus* using simple sequence alignment tools, but were revealed here by a structural homology search using Phyre[2] [56]. The RpoE proteins of *R. sphaeroides* and *R. capsulatus* only share 24% identity, but the confidence in the Phyre[2] structural homology analysis is high (99.9%). Importantly, the gene adjacent to the *rpoE* gene of *R. capsulatus* encodes a putative antisigma factor with 13% identity to ChrR (confidence of 99.4%). Although the identity values are relatively low, the high confidence in the Phyre[2] analysis strongly suggests that *R. capsulatus* has true RpoE (RCAP_rcc00699) and ChrR (RCAP_rcc00698) homologs. Moreover, and similar to *R. sphaeroides*, the *rpoE-chrR* locus of *R. capsulatus* is induced by $^1O_2$, as revealed by RNA-seq (Figure 4 and Table 1). However, microsynteny analysis using Edgar 2.3 [57] showed that the genetic context of the *rpoE-chrR* locus in *R. capsulatus* is different from *R. sphaeroides* (Figure 4). The RpoE-dependent operon RSP_1087-1091, which is located upstream of *rpoE-chrR* in *R. sphaeroides*, cannot be found next to *rpoE-chrR* in *R. capsulatus*. Moreover, respective homologs seem to be completely absent from the *R. capsulatus* genome (Table 1). Instead, the *rpoE-chrR* locus is located next to an operon, which encodes the glutathione peroxidase BsaA1, the cryptochrome/photolyase CryB, and two hypothetical proteins. All four genes are clearly induced by $^1O_2$ (Figure 4). Interestingly, *cryB* belongs to the RpoH$_{II}$ regulon in *R. sphaeroides* [66]. Microsynteny analysis further revealed that *rpoE-chrR* is in close proximity to the photosynthetic gene cluster of *R. capsulatus* (*bch* and *puf* genes in Figure 4). This genetic arrangement cannot be found in closely related members within the *Rhodobacteraceae* (155 genomes analyzed in total), and it remains speculative whether this unique gene colocalization is purely coincidental or represents a strong functional relationship between photosynthesis and the response to $^1O_2$.

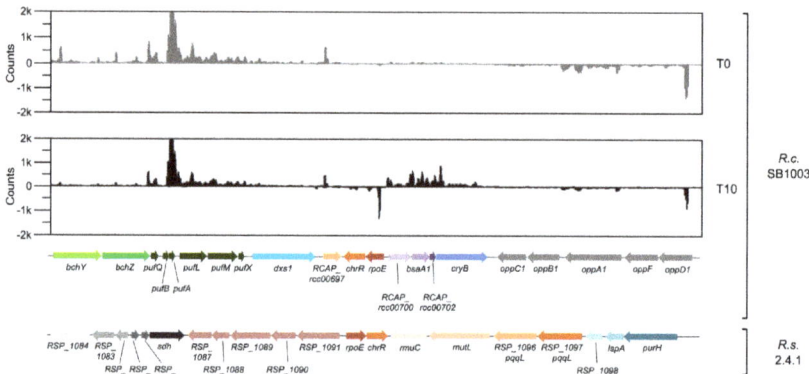

Figure 4. Microsynteny analysis of the *rpoE-chrR* locus in *R. capsulatus* and *R. sphaeroides*. The two upper panels show normalized read count distributions of a representative RNA-seq experiment with *R. capsulatus* (R.c. SB1003) before (T0) and 10 min after the onset of photooxidative stress (T10). The lower panels show the microsynteny analysis of the *rpoE-chrR* locus in *R. capsulatus* and its homologous genes in *R. sphaeroides* (R.s. 2.4.1). Homologs are indicated by identical colors.

Table 1. Transcriptome changes upon photooxidative stress of the RpoE regulon in *R. sphaeroides* compared to homologs in *R. capsulatus*.

R. sphaeroides Gene	Description	Log$_2$ FC [1] (7 min)	Log$_2$ FC [1] (45 min)	R. capsulatus Gene [2]	Log$_2$ FC [1] (10 min)
RSP_1092 *rpoE*	RNA polymerase sigma-70 factor	2.2 **	2.0 ***	RCAP_rcc00699 *rpoE*	3.7 ***
RSP_1093 *chrR*	Antisigma factor ChrR	2.0 **	2.2 ***	RCAP_rcc00698 *chrR*	3.4 ***
RSP_2144 *cfaS*	Cyclopropane-fatty-acyl -phospholipid synthase CfaS	1.4 *	1.0 ***	RCAP_rcc00273 *rsmB1*	0.3
RSP_2143 *phrA*	DNA photolyase	1.6 **	1.5 ***	RCAP_rcc02958 *phrB*	2.7 ***
RSP_1091	Putative cyclopropane or cyclopropene fatty acid synthesis protein	2.2 ***	2.0 ***	No homolog	
RSP_1090	Putative cyclopropane/cyclopropene fatty acid synthesis protein	2.4	1.9 ***	No homolog	
RSP_1089	Sugar/cation symporter, GPH family	1.9 *	1.8 ***	No homolog	
RSP_1088	Hypothetical protein	1.1 *	0.8 **	No homolog	
RSP_1087	Short-chain dehydrogenase/reductase family member	0.9	0.7 **	No homolog	
RSP_0601 *rpoH$_{II}$*	RNA polymerase sigma factor	2.0 *	2.1 *	RCAP_rcc00458 *rpoH$_{II}$* [3]	2.1 ***
RSP_1409	Beta-Ig-H3/fasciclin	2.8 **	4.5 *	No homolog	
RSP_1852 *folE2*	Hypothetical protein	2.2 ***	2.3 ***	RCAP_rcc01493 *folE2*	3.3 ***
RSP_0296 *cycA*	Cytochrome c2	−0.3	0.5 **	RCAP_rcc01240 *cycA1*	0.2
RSP_3336	ABC spermidine/putrescine transporter, inner membrane subunit	0.0	0.2 *	RCAP_rcc01895 *potH1*	−0.6 *
RSP_6222	Hypothetical protein	0.1	0.3	No homolog	

[1] It is indicated whether log$_2$ fold changes (log$_2$ FC; stressed versus unstressed) were statistically significant (* *p*-value < 0.05, ** *p*-value < 0.01, and *** *p*-value < 0.001). Data for *R. sphaeroides* were retrieved from Berghoff and colleagues [28]. [2] Homologs found either by pBLAST (30% protein identity) or Phyre². [3] Annotated as *rpoH$_I$* in public databases.

3.4. Transcriptome Analysis of the Response to Singlet Oxygen in R. capsulatus

To learn more about the response to $^1O_2$ in *R. capsulatus*, RNA-seq was performed for samples collected before (T0) and 10 min after the onset of $^1O_2$ stress (T10). The reproducibility of biological replicates is shown by correlation analysis with highly significant Pearson's *r*-values of ≥ 0.985 for all possible interreplicate comparisons (Figure S5). Principal component analysis (PCA) further revealed a clear separation between T0 and T10 samples along the first dimension (Figure S6A). In total, 3441 transcripts were quantified in the RNA-seq analysis. Four hundred and seventy-one transcripts were up- and 261 transcripts were down-regulated upon $^1O_2$ stress (log$_2$ fold change ≥ 1 or ≤ −1 and *p*-value < 0.05). To validate the RNA-seq approach for *R. capsulatus* and to further confirm the microarray results for *R. sphaeroides* [28], quantitative RT-PCR (qRT-PCR) was performed for selected genes. The qRT-PCR data were in good agreement with both RNA-seq and microarray results (Figure 5). However, for *cbbM*, *gltD*, *cysP*, *crtI*, and *cbiX*, transcript levels were only increased in *R. capsulatus*. In *R. sphaeroides*, three independent primer pairs were unable to detect *cbiX* transcripts.

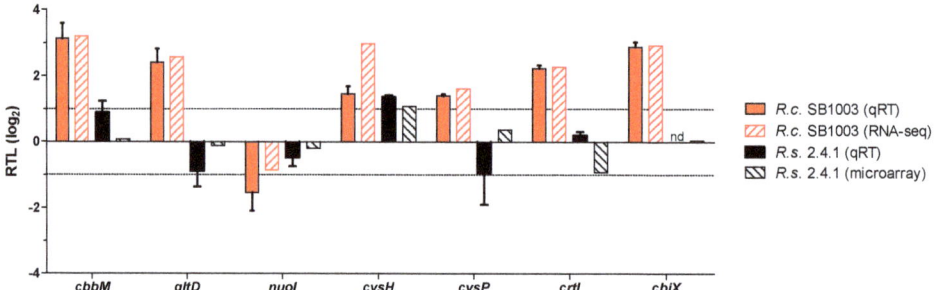

Figure 5. Gene expression changes for selected genes upon photooxidative stress in *Rhodobacter*. Relative transcript levels (RTL) were calculated after the onset of photooxidative stress in comparison to a non-stressed control in *R. capsulatus* (*R.c.* SB1003) and *R. sphaeroides* (*R.s.* 2.4.1). RNA-seq (this study) and microarray data [28] are shown for comparison. For qRT-PCR (qRT), bars represent the mean of biological triplicates, and error bars depict the standard deviation. The *cbiX* transcript was not detected (nd) in *R. sphaeroides*.

From *R. sphaeroides*, it is known that 13 out of 15 genes of the RpoE regulon are induced by $^1O_2$ (Table 1; [28]). In *R. capsulatus*, protein homologs were only found for seven of the 15 RpoE regulon members, which applies to RpoE, ChrR, photolyase PhrA (PhrB), sigma factor RpoH$_{II}$, cytochrome c2 CycA (CycA1), a polyamine transporter subunit (PotH1), and cyclohydrolase FolE2 (RCAP_rcc01493) (Table 1). It is worth noting that the aforementioned *R. capsulatus* sigma factors RpoH$_I$ and RpoH$_{II}$ are wrongly annotated in public databases: our analyses (see below) clearly show that RCAP_rcc00458 (annotated as *rpoH$_I$*) represents *rpoH$_{II}$*, and vice versa, RCAP_rcc02811 (annotated as *rpoH$_{II}$*) corresponds to *rpoH$_I$* from *R. sphaeroides*. Thus, we refer to RCAP_rcc00458 as *rpoH$_{II}$* and to RCAP_rcc02811 as *rpoH$_I$*. Among the seven homologs of the RpoE regulon, only *rpoE*, *chrR*, *phrB*, *rpoH$_{II}$*, and *folE2* were induced upon $^1O_2$ stress in *R. capsulatus*, indicating that similarities are probably limited to the most important features, which has also been observed for *Roseobacter denitrificans* [67]. One of the conserved $^1O_2$-related features includes sigma factor RpoH$_{II}$, which shares a partially overlapping regulon with heat-shock sigma factor RpoH$_I$ in *R. sphaeroides* [10,19,20]. In contrast to *rpoE-chrR*, the genetic context of both *rpoH$_I$* and *rpoH$_{II}$* is partly conserved between *R. sphaeroides* and *R. capsulatus* (Figure 6A,B). From *R. sphaeroides* it is known that RpoH$_{II}$ is more important for $^1O_2$ stress resistance than RpoH$_I$ [20]; the same was observed here for *R. capsulatus* (Figure 6C).

To identify the most prominently enriched functional groups in *R. capsulatus*, we conducted a Gene Ontology (GO) term enrichment analysis using BiNGO in Cytoscape [58,59] for the 100 transcripts with the strongest increase. Transcripts contributing to oxidation–reduction processes were significantly increased and formed the largest group (24 transcripts; Figure 7). Furthermore, we could identify several stress-related functional groups. The first group comprises nine transcripts encoding proteins with a role in protein turnover and repair, which applies to the peptide methionine sulfoxide reductases MsrA1, MsrA2, MsrB1, and MsrB2, the chaperones GroS and GroL, and the peptidases/proteases Dcp, TldD, and RCAP_rcc03333 (Figure 7). The second group comprises five transcripts encoding proteins with a known role in the (photo-) oxidative stress response, including RpoE and ChrR, RpoE-dependent GTP cyclohydrolase FolE2, peroxidase BsaA1, and glutathione-disulfide reductase Gor (Figure 7). The third and last group is formed by five transcripts encoding proteins with a function in DNA damage repair, which applies to photolyases PhrB and CryB, two components of the UvrABC complex, and the A/G-specific adenine glycosylase MutY (Figure 7).

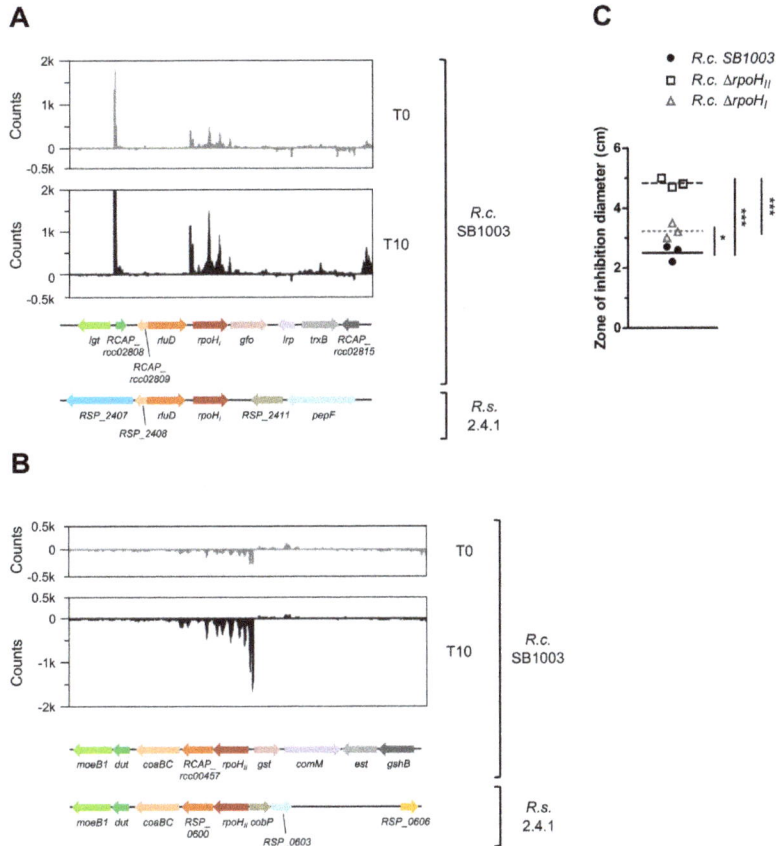

Figure 6. Analysis of $rpoH_I$ and $rpoH_{II}$ in *R. capsulatus*. RNA-seq and microsynteny analysis for (**A**) $rpoH_I$ and (**B**) $rpoH_{II}$ in *R. capsulatus*. The two upper panels show read count distributions of a representative RNA-seq experiment with *R. capsulatus* (*R.c.* SB1003) before (T0) and 10 min after the onset of photooxidative stress (T10). The lower panels show the microsynteny analysis of the *rpoH* locus in *R. capsulatus* and its homologous genes in *R. sphaeroides* (*R.s.* 2.4.1). Homologs are indicated by identical colors. (**C**) Zone of inhibition assay showing the sensitivity of *R. capsulatus* strains to photooxidative stress. Data points indicate individual measurements and bars represent the mean. One-way ANOVA followed by Bonferroni posttest was used to compare results from different *R. capsulatus* strains (* p-value < 0.05, *** p-value < 0.001).

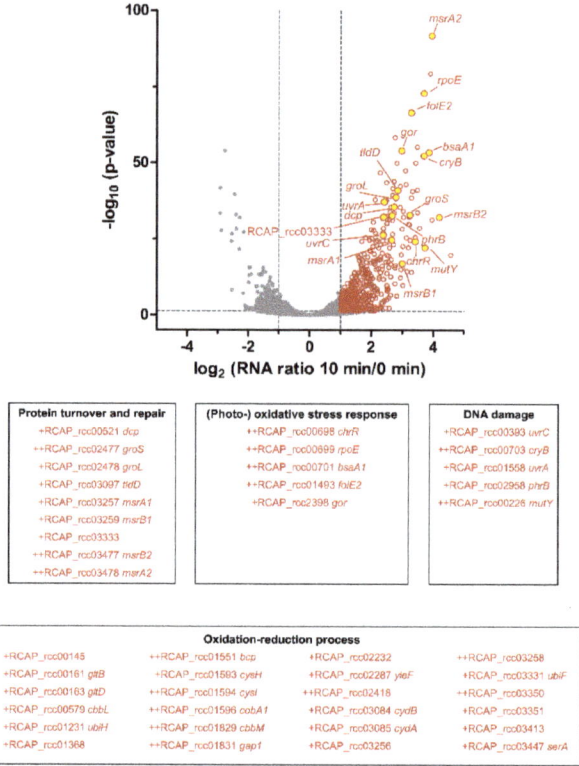

Figure 7. Transcriptome analysis of R. capsulatus reveals functional groups with importance to photooxidative stress. Changes in transcript abundance after 10 min of photooxidative stress were determined by RNA-seq of biological triplicates. The volcano plot depicts $\log_2$ fold changes (10 min versus 0 min) of all quantified transcripts and the corresponding p-values (as negative $\log_{10}$). The horizontal dashed line indicates the cutoff for statistical significance ($p < 0.05$), and the vertical dashed lines indicate $\log_2$ fold changes ≤ -1 and ≥ 1. Significantly increased transcripts with a $\log_2$ fold change ≥ 1 are indicated as red open circles. Transcripts encoding proteins with known stress-related functions are highlighted in yellow. Boxes below the volcano plot depict functional groups as determined by GO term enrichment analysis of the 100 transcripts with the strongest increase. $\log_2$ fold changes of ≥ 2 or ≥ 3 are indicated with + and ++, respectively.

3.5. Similar Proteins Fulfill Important Functions in Response to Singlet Oxygen in R. sphaeroides and R. capsulatus

It is known that changes on the RNA level are not necessarily reflected by changes in protein abundance. We therefore complemented our dataset by proteomic analysis of samples collected before (T0) and 90 min after the onset of $^1O_2$ stress (T90). Correlation between the transcriptome (T10) and proteome (T90) was fairly low (Pearson's r-value = 0.49). Proteins were analyzed by LC-MS/MS and applied to a label-free quantification (LFQ) approach [68]. Pearson's r-values of ≥ 0.975 for interreplicate comparisons demonstrated high reproducibility of the LFQ approach (Figure S7). Principal component analysis (PCA) further revealed a clear separation between T0 and T90 samples along the first dimension (Figure S6B). LFQ intensities, reflecting protein abundance, were subsequently used to calculate fold changes between conditions. In total, 1507 proteins were quantified, revealing 46 increased and 35 decreased proteins upon $^1O_2$ stress ($\log_2$ fold change ≥ 1 or ≤ -1 and p-value < 0.05). Increased proteins were subjected to GO term enrichment analysis using BiNGO in Cytoscape [58,59], and compared to proteome data from *R. sphaeroides* [28]. Seven homologous proteins were identified as increased in both organisms (Figure 8), including the three methionine sulfoxide reductases MsrA, MsrB1, and MsrB2, GTP cyclohydrolase FolE2, ATP-dependent protease ClpA, a putative protease (RCAP_rcc03333/RSP_1490), and an uncharacterized protein (RCAP_rcc00543/RSP_1760). All other proteins were only found to be increased in one of the two organisms (39 proteins in *R. capsulatus* and 43 proteins in *R. sphaeroides*), and only one functional group was clearly enriched in both organisms by means of GO terms, i.e., proteins with a function in oxidation–reduction processes. However, we identified several proteins with stress-related functions and grouped them accordingly (Figure 8). a prominent group relates to protein turnover and repair, including the aforementioned methionine sulfoxide reductases and several proteases, that are either increased in both organisms (ClpA and RCAP_rcc03333/RSP_1490) or only in one of the organisms (Lon, HslV, and ClpB in *R. capsulatus*; PqqL, MoxR, and ClpS in *R. sphaeroides*). Other proteins are directly related to the (photo-) oxidative stress response, like glutathione-disulfide reductase Gor and glutathione peroxidase BsaA1 in *R. capsulatus*, and a thioredoxin (RSP_0725), a peroxiredoxin (RSP_2973), and several RpoE regulon members in *R. sphaeroides* (Figure 8). a last group includes proteins involved in DNA damage repair, like components of the UvrABC complex and photolyase PhrB in *R. capsulatus*. Regarding other noteworthy increases, the protein CbiX caught our interest. As the terminal enzyme of the siroheme biosynthesis [69], CbiX does not belong to any of the described groups; however, its increase was one of the strongest. As *R. sphaeroides* lacked a comparable increase of the CbiX-homolog, this could hint at a core difference in the photooxidative stress responses.

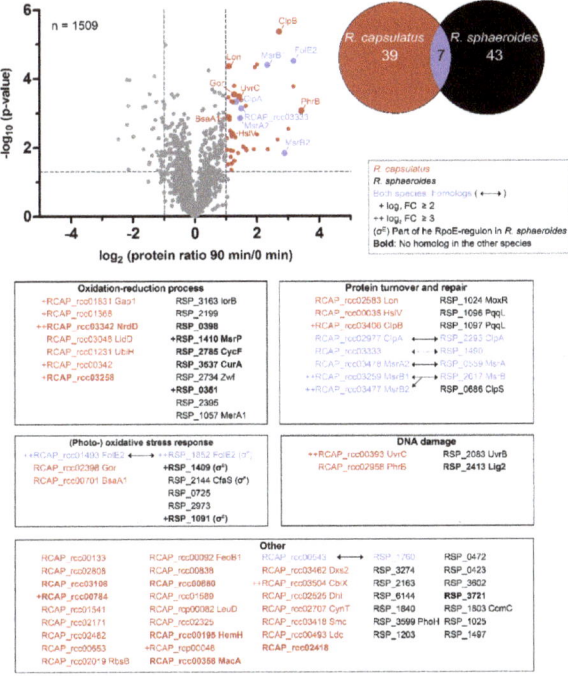

Figure 8. Functional characterization of proteins with increased abundance upon photooxidative stress. Changes in protein abundance after 90 min of photooxidative stress in *R. capsulatus* were determined by a label-free approach using LC-MS/MS of biological triplicates.

The volcano plot depicts $\log_2$ fold changes (90 min versus 0 min) of all quantified proteins and the corresponding p-values (as negative $\log_{10}$). The horizontal dashed line indicates the cutoff for statistical significance ($p < 0.05$), and the vertical dashed lines indicate $\log_2$ fold changes ≤ -1 and ≥ 1. Significantly increased proteins with a $\log_2$ fold change ≥ 1 are highlighted. Proteome data were compared to *R. sphaeroides* [28] and increased proteins illustrated in an Euler diagram. Boxes below the volcano plot depict functional groups as determined by GO term enrichment analysis. $\log_2$ fold changes of ≥ 2 or ≥ 3 are indicated with + and ++, respectively. Colors indicate whether proteins were found only in *R. capsulatus* (red), only in *R. sphaeroides* (black), or in both species (blue). See legend for details.

4. Discussion

A photosynthetic lifestyle allows organisms to use light as an energy source for growth and proliferation. However, this benefit comes at a price, that is, the risk of $^1O_2$ generation by energy transfer from (bacterio-) chlorophyll to molecular oxygen ($^3O_2$) within photosynthetic complexes. It is suspected that purple bacteria from the genus *Rhodobacter* address this problem by avoiding strong pigmentation under high light and/or high oxygen conditions, a response that is mainly regulated by light- and oxygen-sensing proteins [2,3]. Despite regulation of pigmentation in response to oxygen tension and light, *Rhodobacter* species cannot completely avoid $^1O_2$ stress in their natural aquatic environments. They have therefore evolved strategies to counteract $^1O_2$ and to deal with the resulting damages [1,65]. *R. sphaeroides* is a well-studied model organism with regard to the photooxidative stress response [70], and was compared here to *R. capsulatus* using transcriptomics and proteomics. Intriguingly, the two *Rhodobacter* species elicit very similar responses in the light of functional categories. Many proteins with an increased abundance upon $^1O_2$ exposure are involved in oxidation–reduction processes and DNA damage repair (Figure 8). Furthermore, several methionine sulfoxide reductases (MsrA and MsrB orthologs) and a variety of proteases were increased, which was also confirmed on the transcript level (Figure 7). Since proteins are the main targets of $^1O_2$ [15], we conclude that the repair of proteins and the removal of damaged proteins is essential to survive this particular stress, and that efficient protein maintenance likely represents a key feature of the response to $^1O_2$ in many organisms. In *R. sphaeroides*, the $^1O_2$ stress response is mainly controlled by the alternative sigma factors RpoE and RpoH$_{II}$ [10,18–21]; we have reason to believe that this also holds true for *R. capsulatus*. Firstly, both *rpoE* and *rpoH$_{II}$* have elevated transcript levels upon $^1O_2$ exposure (Table 1), and secondly, an *rpoH$_{II}$* deletion strain is more sensitive to $^1O_2$ than the *R. capsulatus* wild type (Figure 6). Hence, the basic regulatory principles of the photooxidative stress response might be very similar in both *Rhodobacter* species.

Despite the aforementioned similarities, there are remarkable differences between *R. sphaeroides* and *R. capsulatus*. Even though both species fall into the same clade within a phylogenetic tree based on a core genome of 580 orthologous proteins [71], synteny analysis revealed that the genetic organization of orthologous genes on the chromosome is considerably different, with the exception of, e.g., the photosynthetic gene cluster (Figure S8). This pronounced genomic rearrangement is not observed between different *R. sphaeroides* strains (e.g., strains 2.4.1, KD131, and WS8N), but between different genera within the *Rhodobacteraceae* (e.g., between *Rhodobacter* and *Roseobacter*). Hence, our analyses suggest that *R. capsulatus* and *R. sphaeroides* are, from a genomic point of view, more distantly related than expected from their common lifestyle and the arrangement of the photosynthetic gene cluster [71]. Another remarkable finding concerns the genetic context of the *rpoE-chrR* locus in *R. capsulatus*, which is in close proximity to the photosynthetic gene cluster (Figure 4). This genetic arrangement was not found in other *Rhodobacteraceae* species, and can therefore be considered as unique. It is worth noting that RpoE from *R. capsulatus* shares higher identity with the extra-cytoplasmic function sigma factor SigK from *Mycobacterium tuberculosis* (34%), compared to only 24% identity with *R. sphaeroides* RpoE. The same is true for *R. capsulatus* ChrR, which shares 25% identity with *M. tuberculosis* SigK antisigma factor RskA, but only 13% with *R. sphaeroides* ChrR. Phylogenetic trees further support the special position of the *R. capsulatus* RpoE and ChrR proteins within the *Rhodobacteraceae* family (data not shown). These findings raise the question of whether the *rpoE-chrR* locus in *R. capsulatus* and *R. sphaeroides* originated from a common ancestor and was then intensely remodeled in *R. capsulatus*, or whether *R. capsulatus* has received a *sigK-rskA*-like locus from another bacterial lineage (e.g., Gram-positives like *M. tuberculosis*) via horizontal gene transfer.

    R. capsulatus not only displays a unique colocalization of the photosynthetic gene cluster near the *rpoE-chrR* locus; it also differs to *R. sphaeroides* on the physiological level with regard to its pigmentation. a higher pigmentation of *R. capsulatus* under high oxygen conditions obviously allows much faster adaptation to occur to phototrophic conditions compared to *R. sphaeroides* (Figure 1B). Despite this higher pigmentation, *R. capsulatus* has no disadvantage when exposed to sudden photooxidative stress (Figure 1C). Low expression of BChl *a* and carotenoid biosynthesis genes under aerobic conditions is reflected by low BChl *a* and carotenoid levels in both *R. capsulatus* and *R. sphaeroides* (Figure 2A). Upon photooxidative stress, Bchl *a* and carotenoid levels decreased in *R. sphaeroides*. By contrast, in *R. capsulatus*, only BChl *a* levels declined, while carotenoid levels remained fairly constant (Figure 2B), resulting in an increasing Crt:BChl *a* ratio in the course of $^1O_2$ exposure (Figure 2D). These differences in Crt:BChl *a* ratio may be the reason for lower $^1O_2$ levels in *R. capsulatus* under photooxodative stress considering the $^1O_2$ quenching ability of carotenoids (Figure 3A). Increased carotenoid biosynthesis under conditions of photooxidative stress is a strategy that is used by some microorganisms, including the deltaproteobacterium *Myxococcus xanthus* and the yeast *Phaffia rhodozyma*, in order to avoid extensive cellular damages by direct quenching [72,73]. Accumulation of carotenoids upon $^1O_2$ exposure was, however, not observed in *R. sphaeroides* [27], and omics data even revealed declining transcript and protein levels of genes from the photosynthetic gene cluster, including genes for carotenoid biosynthesis [28]. Interestingly, RNA-seq revealed that three genes within the photosynthetic gene cluster had increased transcript levels in *R. capsulatus*, which applies to *crtI* ($\log_2$ fold change of ~2.3), *crtB* ($\log_2$ fold change of ~2.0), and *tspO* ($\log_2$ fold change of ~1.3). Induction of *crtI* was validated by qRT-PCR (Figure 5). The three genes form an operon between the carotenoid biosynthesis genes *crtA* and *crtC*. TspO is an outer membrane protein, which controls efflux of porphyrin intermediates, and thereby negatively modulates expression of photosynthesis genes, likely through AppA [74]. The increased expression of *tspO* might enhance porphyrin efflux under photooxidative stress conditions, limiting both expression of photosynthesis genes and accumulation of potential porphyrin-derived photosensitizers, which may support adaptation to this particular stress. CrtB is a phytoene synthase and CrtI is a phytoene desaturase, which catalyze the reaction of precursors to phytoene (CrtB) and from phytoene to zeta-carotene and neurosporene (CrtI). The carotenoids spheroidene (SE) and spheroidenone (SO) are then synthesized in subsequent steps from neurosporene. Even though a switch from SE to SO (catalyzed by CrtA) might be an important adaption to photooxidative stress, we could not detect a strong increase for *crtA* mRNA in the RNA-seq data. It is conceivable that enhanced CrtB and CrtI levels are needed to provide sufficient amounts of neurosporene as a precursor for SE and SO biosynthesis to maintain carotenoid levels in *R. capsulatus* (Figure 2B). By contrast, in *R. sphaeroides crtI* and *crtB* transcript levels do not increase upon photooxidative stress (Figure 5) [28], which even coincides with a decrease in carotenoid levels (Figure 2B). Obviously, both *Rhodobacter* species use different strategies to adapt to high light regimes. Since *R. capsulatus* initially grows better than *R. sphaeroides* both after a shift to phototrophic conditions (Figure 1B) and upon photooxidative stress (Figure 1C), elevated carotenoid levels are expected to be advantageous to photosynthetic bacteria in rapidly changing environments.

    Besides the protective function of carotenoids, *cbiX* might also play a more crucial role in defending against photooxidative stress in *R. capsulatus* than in *R. sphaeroides*. Our omics data show that *cbiX* is strongly increased on both transcript and protein level in *R. capsulatus* ($\log_2$ fold change of ~3). The RSP_1566 protein of *R. sphaeroides* shows 47% identity to CbiX, but neither protein nor transcript levels change significantly in response to $^1O_2$ [28], and *cbiX* transcripts were not even detected by qRT-PCR (Figure 5). The *cbiX* gene is annotated as a cobaltochelatase which incorporates cobalt into sirohydrochlorin to form cosirohydrochlorin, an early precursor of vitamin B_{12} [75]. B_{12} has special functions in the formation of the photosynthetic apparatus in *Rhodobacter*, i.e., it is required for the conversion of protoporphyrin IX to Mg-protoporphyrin monomethyl ester [76]. Furthermore, it is needed by the antirepressor AerR to efficiently bind to repressor CrtJ, thereby inducing *bch* gene expression in *R. capsulatus* [77,78]. Theoretically, increased CbiX levels favor B_{12} biosynthesis and, consequently, derepression of *bch* genes via AerR. However, increased *bch* expression would then increase photooxidative stress, but this was not observed. a structural homology search for *R. capsulatus* CbiX using PHYRE² suggests that it might rather act as a ferrochelatase involved in the biosynthesis of siroheme. This assumption is supported by a study by Bali and colleagues [69], showing that CbiX functionally replaces characteristic siroheme biosynthesis enzymes, which are missing in alphaproteobacteria, including *R. sphaeroides*. As siroheme is a cofactor of nitrite and sulfite reductases, an increased siroheme production might also explain the high accumulation of the mRNA for sulfite reductase CysI ($\log_2$ fold change of ~3.2). Increased sulfur assimilation might help to counteract the $^1O_2$-caused depletion of glutathione and abundant damages on sulfur-containing amino acids. More importantly, however, the insertion of Fe^{2+} into sirohydrochlorin reduces the concentration of free iron, which would otherwise be available for both the Fenton reaction and the formation of the photosensitizer protoporphyrin IX. Hence, additional stress is prevented by changing the flux through tetrapyrrole pathways [79–81]. As a conclusion, the induction of both the *crtIB-tspO* operon and *cbiX* might represent a successful strategy to respond to photooxidative stress, which was specifically invented in *R. capsulatus* to support adaptation. The question of whether other phototrophs have evolved similar strategies will be an exciting subject for future studies.

Supplementary Materials: The following are available online at http://www.mdpi.com/2076-2607/8/2/283/s1, Figure S1: Analysis of *Rhodobacter* transposon mutants (*R.c.* Tn5 and *R.s.* Tn5) lacking important carotenoids, Figure S2: Growth of *R. capsulatus* and *R. sphaeroides* with and without methylene blue in the dark or under high light conditions, Figure S3: Pigmentation of *R. capsulatus* and *R. sphaeroides* under microaerobic conditions, Figure S4: SOSG fluorescence in cell-free reactions, Figure S5: Correlation analysis of RNA-seq replicates, Figure S6: PCA of RNA-seq and LC-MS/MS replicate samples, Figure S7: Correlation analysis of LC-MS/MS replicates, Figure S8: Synteny plot comparing *Rhodobacter* genomes, Table S1: Strains used in this study, Table S2: Transcriptome data for *Rhodobacter* upon photooxidative stress, Table S3: Primers used for qRT-PCR in this study, Table S4: LC-MS/MS data for *R. capsulatus* upon photooxidative stress.

Author Contributions: Conceptualization, G.K.; methodology, M.K.L., A.M.N., A.K., M.B., B.A.B. and G.K.; software, M.B.; validation, M.K.L.; formal analysis, M.K.L., A.K., M.B and B.A.B.; investigation, M.K.L., A.M.N., M.V., A.K.; data curation, M.K.L., A.K. and M.B.; writing—original draft preparation, M.K.L., B.A.B. and G.K.; writing—review and editing, M.K.L., B.A.B. and G.K.; visualization, M.K.L. and B.A.B.; supervision, B.A.B. and G.K.; project administration, G.K. All authors have read and agreed to the published version of the manuscript.

Funding: This research received no external funding.

Acknowledgments: We thank Petra Dersch (University of Münster, Germany) for generous support with RNA-seq and corrections on the manuscript, and Andrew Lang (Memorial University of Newfoundland, Canada) for providing strains.

Conflicts of Interest: The authors declare no conflict of interest.

References

1. Glaeser, J.; Nuss, A.M.; Berghoff, B.A.; Klug, G. Singlet oxygen stress in microorganisms. In *Advances in Microbial Physiology*; Poole, R.K., Ed.; Elsevier Ltd. Academic Press: Amsterdam, The Netherlands, 2011; Volume 58, pp. 141–173.
2. Braatsch, S.; Gomelsky, M.; Kuphal, S.; Klug, G. a single flavoprotein, AppA, integrates both redox and light signals in *Rhodobacter sphaeroides*. *Mol. Microbiol.* **2002**, *45*, 827–836. [CrossRef] [PubMed]
3. Han, Y.; Meyer, M.H.F.; Keusgen, M.; Klug, G. a haem cofactor is required for redox and light signalling by the AppA protein of *Rhodobacter sphaeroides*. *Mol. Microbiol.* **2007**, *64*, 1090–1104. [CrossRef] [PubMed]
4. Mank, N.N.; Berghoff, B.A.; Hermanns, Y.N.; Klug, G. Regulation of bacterial photosynthesis genes by the small noncoding RNA PcrZ. *Proc. Natl. Acad. Sci. USA* **2012**, *109*, 16306–16311. [CrossRef] [PubMed]
5. Eisenhardt, K.M.H.; Reuscher, C.M.; Klug, G. PcrX, an sRNA derived from the 3′-UTR of the *Rhodobacter sphaeroides puf* operon modulates expression of *puf* genes encoding proteins of the bacterial photosynthetic apparatus. *Mol. Microbiol.* **2018**, *110*, 325–334. [CrossRef]

6. Ryter, S.W.; Tyrrell, R.M. Singlet molecular oxygen ($^1O_2$): a possible effector of eukaryotic gene expression. *Free Radic. Biol. Med.* **1998**, *24*, 1520–1534. [CrossRef]
7. Davies, M.J. Singlet oxygen-mediated damage to proteins and its consequences. *Biochem. Biophys. Res. Commun.* **2003**, *305*, 761–770. [CrossRef]
8. Frimer, A.A. The reaction of singlet oxygen with olefins: The question of mechanism. *Chem. Rev.* **1979**, *79*, 359–387. [CrossRef]
9. Cavalcante, A.K.D.; Martinez, G.R.; Di Mascio, P.; Menck, C.F.M.; Agnez-Lima, L.F. Cytotoxicity and mutagenesis induced by singlet oxygen in wild type and DNA repair deficient *Escherichia coli* strains. *DNA Repair* **2002**, *1*, 1051–1056. [CrossRef]
10. Anthony, J.R.; Warczak, K.L.; Donohue, T.J. a transcriptional response to singlet oxygen, a toxic byproduct of photosynthesis. *Proc. Natl. Acad. Sci. USA* **2005**, *102*, 6502–6507. [CrossRef]
11. Devasagayam, T.P.A.; Steenken, S.; Obendorf, M.S.W.; Schulz, W.A.; Sies, H. Formation of 8-hydroxy (deoxy) guanosine and generation of strand breaks at guanine residues in DNA by singlet oxygen. *Biochemistry* **1991**, *30*, 6283–6289. [CrossRef]
12. Floyd, R.A.; Schneider, J.E., Jr.; Dittmer, D.P. Methylene blue photoinactivation of RNA viruses. *Antiviral Res.* **2004**, *61*, 141–151. [CrossRef]
13. Joshi, S.G.; Cooper, M.; Yost, A.; Paff, M.; Ercan, U.K.; Fridman, G.; Friedman, G.; Fridman, A.; Brooks, A.D. Non-thermal dielectric-barrier discharge (DBD) plasma-induced inactivation involves oxidative-DNA damage and membrane lipid peroxidation in *Escherichia coli*. *Antimicrob. Agents Chemother.* **2011**, *55*, 1053–1062. [CrossRef] [PubMed]
14. Riske, K.A.; Sudbrack, T.P.; Archilha, N.L.; Uchoa, A.F.; Schroder, A.P.; Marques, C.M.; Baptista, M.S.; Itri, R. Giant vesicles under oxidative stress induced by a membrane-anchored photosensitizer. *Biophys. J.* **2009**, *97*, 1362–1370. [CrossRef] [PubMed]
15. Davies, M.J. Reactive species formed on proteins exposed to singlet oxygen. *Photochem. Photobiol. Sci.* **2004**, *3*, 17–25. [CrossRef]
16. Wilkinson, F.; Helman, W.P.; Ross, A.B. Rate constants for the decay and reactions of the lowest electronically excited singlet state of molecular oxygen in solution. An expanded and revised compilation. *J. Phys. Chem. Ref. Data* **1995**, *24*, 663–677. [CrossRef]
17. Pattison, D.I.; Rahmanto, A.S.; Davies, M.J. Photo-oxidation of proteins. *Photochem. Photobiol. Sci.* **2012**, *11*, 38–53. [CrossRef]
18. Glaeser, J.; Zobawa, M.; Lottspeich, F.; Klug, G. Protein synthesis patterns reveal a complex regulatory response to singlet oxygen in *Rhodobacter*. *J. Proteome Res.* **2007**, *6*, 2460–2471. [CrossRef]
19. Nuss, A.M.; Glaeser, J.; Klug, G. RpoH$_{II}$ activates oxidative-stress defense systems and is controlled by RpoE in the singlet oxygen-dependent response in Rhodobacter sphaeroides. *J. Bacteriol.* **2009**, *191*, 220–230. [CrossRef]
20. Nuss, A.M.; Glaeser, J.; Berghoff, B.A.; Klug, G. Overlapping alternative sigma factor regulons in the response to singlet oxygen in *Rhodobacter sphaeroides*. *J. Bacteriol.* **2010**, *192*, 2613–2623. [CrossRef]
21. Dufour, Y.S.; Imam, S.; Koo, B.-M.; Green, H.A.; Donohue, T.J. Convergence of the transcriptional responses to heat shock and singlet oxygen stresses. *PLoS Genet.* **2012**, *8*, e1002929. [CrossRef]
22. Nuss, A.M.; Adnan, F.; Weber, L.; Berghoff, B.A.; Glaeser, J.; Klug, G. DegS and RseP homologous proteases are involved in singlet oxygen dependent activation of RpoE in *Rhodobacter sphaeroides*. *PLoS ONE* **2013**, *8*, e79520. [CrossRef] [PubMed]
23. Nam, T.-W.; Ziegelhoffer, E.C.; Lemke, R.A.S.; Donohue, T.J. Proteins needed to activate a transcriptional response to the reactive oxygen species singlet oxygen. *MBio* **2013**, *4*, e00541-12. [CrossRef] [PubMed]
24. Hendrischk, A.-K.; Braatsch, S.; Glaeser, J.; Klug, G. The *phrA* gene of *Rhodobacter sphaeroides* encodes a photolyase and is regulated by singlet oxygen and peroxide in a σE-dependent manner. *Microbiology* **2007**, *153*, 1842–1851. [CrossRef] [PubMed]
25. Dufour, Y.S.; Landick, R.; Donohue, T.J. Organization and evolution of the biological response to singlet oxygen stress. *J. Mol. Biol.* **2008**, *383*, 713–730. [CrossRef]
26. Lemke, R.A.S.; Peterson, A.C.; Zieglhoffer, E.C.; Westphall, M.S.; Tjellström, H.; Coon, J.J.; Donohue, T.J. Synthesis and scavenging role of furan fatty acids. *Proc. Natl. Acad. Sci. USA* **2014**, *111*, E3451. [CrossRef]
27. Glaeser, J.; Klug, G. Photo-oxidative stress in *Rhodobacter sphaeroides*: Protective role of carotenoids and expression of selected genes. *Microbiology* **2005**, *151*, 1927–1938. [CrossRef]

28. Berghoff, B.A.; Konzer, A.; Mank, N.N.; Looso, M.; Rische, T.; Förstner, K.U.; Krüger, M.; Klug, G. Integrative "Omics"-approach discovers dynamic and regulatory features of bacterial stress responses. *PLoS Genet.* **2013**, *9*, e1003576. [CrossRef]
29. Schmidt, K. Biosynthesis of carotenoids. In *The Photosynthetic Bacteria*; Clayton, R.K., Sistrom, W.R., Eds.; Prenum Press: New York, NY, USA, 1978; pp. 729–750.
30. Scolnik, P.A.; Walker, M.A.; Marrs, B.L. Biosynthesis of carotenoids derived from neurosporene in *Rhodopseudomonas capsulata*. *J. Biol. Chem.* **1980**, *255*, 2427–2432.
31. Zhu, Y.S.; Cook, D.N.; Leach, F.; Armstrong, G.A.; Alberti, M.; Hearst, J.E. Oxygen-regulated mRNAs for light-harvesting and reaction center complexes and for bacteriochlorophyll and carotenoid biosynthesis in *Rhodobacter capsulatus* during the shift from anaerobic to aerobic growth. *J. Bacteriol.* **1986**, *168*, 1180–1188. [CrossRef]
32. Yeliseev, A.A.; Eraso, J.M.; Kaplan, S. Differential carotenoid composition of the B875 and B800-850 photosynthetic antenna complexes in Rhodobacter sphaeroides 2.4.1: Involvement of spheroidene and spheroidenone in adaptation to changes in light intensity and oxygen availability. *J. Bacteriol.* **1996**, *178*, 5877–5883. [CrossRef]
33. Li, Z.; Kong, L.; Hui, B.; Shang, X.; Gao, L.; Luan, N.; Zhuang, X.; Wang, D.; Bai, Z. Identification and antioxidant activity of carotenoids from superfine powder of *Rhodobacter sphaeroides*. *Emirates J. Food Agric.* **2017**, *29*, 833–845. [CrossRef]
34. Šlouf, V.; Chábera, P.; Olsen, J.D.; Martin, E.C.; Qian, P.; Hunter, C.N.; Polívka, T. Photoprotection in a purple phototrophic bacterium mediated by oxygen-dependent alteration of carotenoid excited-state properties. *Proc. Natl. Acad. Sci. USA* **2012**, *109*, 8570–8575. [CrossRef]
35. Berghoff, B.A.; Glaeser, J.; Sharma, C.M.; Vogel, J.; Klug, G. Photooxidative stress-induced and abundant small RNAs in *Rhodobacter sphaeroides*. *Mol. Microbiol.* **2009**, *74*, 1497–1512. [CrossRef]
36. Berghoff, B.A.; Glaeser, J.; Sharma Cynthia, M.; Zobawa, M.; Lottspeich, F.; Vogel, J.; Klug, G. Contribution of Hfq to photooxidative stress resistance and global regulation in *Rhodobacter sphaeroides*. *Mol. Microbiol.* **2011**, *80*, 1479–1495. [CrossRef]
37. Billenkamp, F.; Peng, T.; Berghoff, B.A.; Klug, G. a cluster of four homologous small RNAs modulates C-1 metabolism and the pyruvate dehydrogenase complex in *Rhodobacter sphaeroides* under various stress conditions. *J. Bacteriol.* **2015**, *197*, 1839–1852. [CrossRef]
38. Müller, K.M.H.; Berghoff, B.A.; Eisenhardt, B.D.; Remes, B.; Klug, G. Characteristics of Pos19—A small coding RNA in the oxidative stress response of *Rhodobacter sphaeroides*. *PLoS ONE* **2016**, *11*, e0163425. [CrossRef]
39. Adnan, F.; Weber, L.; Klug, G. The sRNA SorY confers resistance during photooxidative stress by affecting a metabolite transporter in *Rhodobacter sphaeroides*. *RNA Biol.* **2015**, *12*, 569–577. [CrossRef]
40. Peng, T.; Berghoff, B.A.; Oh, J.-I.; Weber, L.; Schirmer, J.; Schwarz, J.; Glaeser, J.; Klug, G. Regulation of a polyamine transporter by the conserved 3′ UTR-derived sRNA SorX confers resistance to singlet oxygen and organic hydroperoxides in *Rhodobacter' sphaeroides*. *RNA Biol.* **2016**, *13*, 988–999. [CrossRef]
41. Klug, G.; Adams, C.W.; Belasco, J.; Doerge, B.; Cohen, S.N. Biological consequences of segmental alterations in mRNA stability: Effects of deletion of the intercistronic hairpin loop region of the *Rhodobacter capsulatus* puf operon. *EMBO J.* **1987**, *6*, 3515–3520. [CrossRef]
42. Gregor, J.; Klug, G. Regulation of bacterial photosynthesis genes by oxygen and light. *FEMS Microbiol. Lett.* **1999**, *179*, 1–9. [CrossRef]
43. Remes, B.; Berghoff, B.A.; Foerstner, K.U.; Klug, G. Role of oxygen and the OxyR protein in the response to iron limitation in *Rhodobacter sphaeroides*. *BMC Genom.* **2014**, *15*, 794. [CrossRef]
44. Shiozawa, J.A.; Welte, W.; Hodapp, N.; Drews, G. Studies on the size and composition of the isolated light-harvesting B800-850 pigment-protein complex of *Rhodopseudomonas capsulata*. *Arch. Biochem. Biophys.* **1982**, *213*, 473–485. [CrossRef]
45. Clayton, R.K. The bacterial photosynthetic reaction center. In Proceedings of the Brookhaven Symposia in Biology, New York, NY, USA, 6–9 June 1966; Volume 19, pp. 62–70.
46. Janzon, L.; Löfdahl, S.; Arvidson, S. Evidence for a coordinate transcriptional control of alpha-toxin and protein a synthesis in *Staphylococcus aureus*. *FEMS Microbiol. Lett.* **1986**, *33*, 193–198. [CrossRef]
47. Nuss, A.M.; Heroven, A.K.; Waldmann, B.; Reinkensmeier, J.; Jarek, M.; Beckstette, M.; Dersch, P. Transcriptomic profiling of *Yersinia pseudotuberculosis* reveals reprogramming of the Crp regulon by temperature and uncovers Crp as a master regulator of small RNAs. *PLoS Genet.* **2015**, *11*, e1005087. [CrossRef]

48. Aronesty, E. *ea-Utils: Command-Line Tools for Processing Biological Sequencing Data*. 2011. Available online: https://github.com/ExpressionAnalysis/ea-utils (accessed on 19 February 2020).
49. Langmead, B.; Salzberg, S.L. Fast gapped-read alignment with Bowtie 2. *Nat. Methods* **2012**, *9*, 357–359. [CrossRef]
50. Li, H.; Handsaker, B.; Wysoker, A.; Fennell, T.; Ruan, J.; Homer, N.; Marth, G.; Abecasis, G.; Durbin, R. The sequence alignment/map format and SAMtools. *Bioinformatics* **2009**, *25*, 2078–2079. [CrossRef]
51. Love, M.I.; Huber, W.; Anders, S. Moderated estimation of fold change and dispersion for RNA-seq data with DESeq2. *Genome Biol.* **2014**, *15*, 550. [CrossRef]
52. Pfaffl, M.W. a new mathematical model for relative quantification in real-time RT–PCR. *Nucleic. Acids Res.* **2001**, *29*, e45. [CrossRef]
53. Bathke, J.; Konzer, A.; Remes, B.; McIntosh, M.; Klug, G. Comparative analyses of the variation of the transcriptome and proteome of *Rhodobacter sphaeroides* throughout growth. *BMC Genom.* **2019**, *20*, 358. [CrossRef]
54. Cox, J.; Mann, M. MaxQuant enables high peptide identification rates, individualized ppb-range mass accuracies and proteome-wide protein quantification. *Nat. Biotechnol.* **2008**, *26*, 1367–1372. [CrossRef]
55. Chen, I.-M.A.; Chu, K.; Palaniappan, K.; Pillay, M.; Ratner, A.; Huang, J.; Huntemann, M.; Varghese, N.; White, J.R.; Seshadri, R.; et al. IMG/M v. 5.0: An integrated data management and comparative analysis system for microbial genomes and microbiomes. *Nucleic Acids Res.* **2019**, *47*, D666–D677. [CrossRef]
56. Kelley, L.A.; Mezulis, S.; Yates, C.M.; Wass, M.N.; Sternberg, M.J.E. The Phyre2 web portal for protein modeling, prediction and analysis. *Nat. Protoc.* **2015**, *10*, 845–858. [CrossRef]
57. Blom, J.; Kreis, J.; Spänig, S.; Juhre, T.; Bertelli, C.; Ernst, C.; Goesmann, A. EDGAR 2.0: An enhanced software platform for comparative gene content analyses. *Nucleic Acids Res.* **2016**, *44*, W22–W28. [CrossRef]
58. Shannon, P.; Markiel, A.; Ozier, O.; Baliga, N.S.; Wang, J.T.; Ramage, D.; Amin, N.; Schwikowski, B.; Ideker, T. Cytoscape: a software environment for integrated models of biomolecular interaction networks. *Genome Res.* **2003**, *13*, 2498–2504. [CrossRef] [PubMed]
59. Maere, S.; Heymans, K.; Kuiper, M. BiNGO: a Cytoscape plugin to assess overrepresentation of gene ontology categories in biological networks. *Bioinformatics* **2005**, *21*, 3448–3449. [CrossRef] [PubMed]
60. Botstein, D.; Cherry, J.M.; Ashburner, M.; Ball, C.A.; Blake, J.A.; Butler, H.; Davis, A.P.; Dolinski, K.; Dwight, S.S.; Eppig, J.T.; et al. Gene Ontology: Tool for the unification of biology. *Nat. Genet.* **2000**, *25*, 25–29.
61. Carbon, S.; Douglass, E.; Dunn, N.; Good, B.; Harris, N.L.; Lewis, S.E.; Mungall, C.J.; Basu, S.; Chisholm, R.L.; Dodson, R.J.; et al. The Gene Ontology Resource: 20 years and still going strong. *Nucleic Acids Res.* **2019**, *47*, D330–D338.
62. Prasad, A.; Sedlářová, M.; Pospíši, P. Singlet oxygen imaging using fluorescent probe Singlet Oxygen Sensor Green in photosynthetic organisms. *Sci. Rep.* **2018**, *8*, 13685. [CrossRef]
63. Dwyer, D.J.; Belenky, P.A.; Yang, J.H.; MacDonald, I.C.; Martell, J.D.; Takahashi, N.; Chan, C.T.Y.; Lobritz, M.A.; Braff, D.; Schwarz, E.G.; et al. Antibiotics induce redox-related physiological alterations as part of their lethality. *Proc. Natl. Acad. Sci. USA* **2014**, *111*, E2100–E2109. [CrossRef]
64. Davies, M.J. Protein oxidation and peroxidation. *Biochem. J.* **2016**, *473*, 805–825. [CrossRef]
65. Ziegelhoffer, E.C.; Donohue, T.J. Bacterial responses to photo-oxidative stress. *Nat. Rev. Microbiol.* **2009**, *7*, 856–863. [CrossRef] [PubMed]
66. Hendrischk, A.-K.; Frühwirth, S.W.; Moldt, J.; Pokorny, R.; Metz, S.; Kaiser, G.; Jäger, A.; Batschauer, A.; Klug, G. a cryptochrome-like protein is involved in the regulation of photosynthesis genes in *Rhodobacter sphaeroides*. *Mol. Microbiol.* **2009**, *74*, 990–1003. [CrossRef] [PubMed]
67. Berghoff, B.A.; Glaeser, J.; Nuss, A.M.; Zobawa, M.; Lottspeich, F.; Klug, G. Anoxygenic photosynthesis and photooxidative stress: a particular challenge for *Roseobacter*. *Environ. Microbiol.* **2011**, *13*, 775–791. [CrossRef] [PubMed]
68. Cox, J.; Hein, M.Y.; Luber, C.A.; Paron, I.; Nagaraj, N.; Mann, M. Accurate proteome-wide label-free quantification by delayed normalization and maximal peptide ratio extraction, termed MaxLFQ. *Mol. Cell Proteom.* **2014**, *13*, 2513–2526. [CrossRef]
69. Bali, S.; Rollauer, S.; Roversi, P.; Raux-Deery, E.; Lea, S.; Warren, M.J.; Ferguson, S.J. Identification and characterization of the "missing" terminal enzyme for siroheme biosynthesis in α-proteobacteria. *Mol. Microbiol.* **2014**, *92*, 153–163. [CrossRef]

70. Berghoff, B.A.; Klug, G. An omics view on the response to singlet oxygen. In *Stress and Environmental Regulation of Gene Expression and Adaptation in Bacteria*; de Bruijn, F.J., Ed.; John Wiley & Sons, Inc.: Hoboken, NJ, USA, 2016; Volume 1, pp. 619–631.
71. Brinkmann, H.; Göker, M.; Koblížek, M.; Wagner-Döbler, I.; Petersen, J. Horizontal operon transfer, plasmids, and the evolution of photosynthesis in *Rhodobacteraceae*. *ISME J.* **2018**, *12*, 1984–2010. [CrossRef]
72. Schroeder, W.A.; Johnson, E.A. Carotenoids protect *Phaffia rhodozyma* against singlet oxygen damage. *J. Ind. Microbiol.* **1995**, *14*, 502–507. [CrossRef]
73. Galbis-Martínez, M.; Padmanabhan, S.; Murillo, F.J.; Elías-Arnanz, M. CarF mediates signaling by singlet oxygen, generated via photoexcited protoporphyrin IX, in *Myxococcus xanthus* light-induced carotenogenesis. *J. Bacteriol.* **2012**, *194*, 1427–1436. [CrossRef]
74. Zeng, X.; Kaplan, S. TspO as a modulator of the repressor/antirepressor (PpsR/AppA) regulatory system in *Rhodobacter sphaeroides* 2.4. 1. *J. Bacteriol.* **2001**, *183*, 6355–6364. [CrossRef]
75. Leech, H.K.; Raux-Deery, E.; Heathcote, P.; Warren, M.J. Production of cobalamin and sirohaem in *Bacillus megaterium*: An investigation into the role of the branchpoint chelatases sirohydrochlorin ferrochelatase (SirB) and sirohydrochlorin cobalt chelatase (CbiX). *Biochem. Soc. Trans.* **2002**, *30*, 610–613. [CrossRef]
76. Gough, S.P.; Petersen, B.O.; Duus, J.Ø. Anaerobic chlorophyll isocyclic ring formation in *Rhodobacter capsulatus* requires a cobalamin cofactor. *Proc. Natl. Acad. Sci. USA* **2000**, *97*, 6908–6913. [CrossRef] [PubMed]
77. Cheng, Z.; Li, K.; Hammad, L.A.; Karty, J.A.; Bauer, C.E. Vitamin B 12 regulates photosystem gene expression via the CrtJ antirepressor AerR in *Rhodobacter capsulatus*. *Mol. Microbiol.* **2014**, *91*, 649–664. [CrossRef]
78. Klug, G. Beyond catalysis: Vitamin B12 as a cofactor in gene regulation. *Mol. Microbiol.* **2014**, *91*, 635–640. [CrossRef]
79. Kim, J.-G.; Back, K.; Lee, H.Y.; Lee, H.-J.; Phung, T.-H.; Grimm, B.; Jung, S. Increased expression of Fe-chelatase leads to increased metabolic flux into heme and confers protection against photodynamically induced oxidative stress. *Plant Mol. Biol.* **2014**, *86*, 271–287. [CrossRef] [PubMed]
80. Dai, J.; Wei, H.; Tian, C.; Damron, F.H.; Zhou, J.; Qiu, D. An extracytoplasmic function sigma factor-dependent periplasmic glutathione peroxidase is involved in oxidative stress response of *Shewanella oneidensis*. *BMC Microbiol.* **2015**, *15*, 34. [CrossRef] [PubMed]
81. Qiu, D.; Xie, M.; Dai, J.; An, W.; Wei, H.; Tian, C.; Kempher, M.L.; Zhou, A.; He, Z.; Gu, B.; et al. Differential regulation of the two ferrochelatase paralogues in *Shewanella loihica* PV-4 in response to environmental stresses. *Appl. Environ. Microbiol.* **2016**, *82*, 5077–5088. [CrossRef]

© 2020 by the authors. Licensee MDPI, Basel, Switzerland. This article is an open access article distributed under the terms and conditions of the Creative Commons Attribution (CC BY) license (http://creativecommons.org/licenses/by/4.0/).

Article

Multiple Sense and Antisense Promoters Contribute to the Regulated Expression of the *isc-suf* Operon for Iron-Sulfur Cluster Assembly in *Rhodobacter*

Xin Nie, Bernhard Remes and Gabriele Klug *

Institute of Microbiology and Molecular Biology, University of Giessen, Heinrich-Buff-Ring 26-32, D-35392 Giessen, Germany; Xin.Nie@bio.uni-giessen.de (X.N.); Bernhard.Remes@gmx.de (B.R.)
* Correspondence: Gabriele.Klug@mikro.bio.uni-giessen.de

Received: 5 November 2019; Accepted: 4 December 2019; Published: 10 December 2019

Abstract: A multitude of biological functions relies on iron-sulfur clusters. The formation of photosynthetic complexes goes along with an additional demand for iron-sulfur clusters for bacteriochlorophyll synthesis and photosynthetic electron transport. However, photooxidative stress leads to the destruction of iron-sulfur clusters, and the released iron promotes the formation of further reactive oxygen species. A balanced regulation of iron-sulfur cluster synthesis is required to guarantee the supply of this cofactor, on the one hand, but also to limit stress, on the other hand. The phototrophic alpha-proteobacterium *Rhodobacter sphaeroides* harbors a large operon for iron-sulfur cluster assembly comprising the *iscRS* and *suf* genes. IscR (iron-sulfur cluster regulator) is an iron-dependent regulator of *isc-suf* genes and other genes with a role in iron metabolism. We applied reporter gene fusions to identify promoters of the *isc-suf* operon and studied their activity alone or in combination under different conditions. Gel-retardation assays showed the binding of regulatory proteins to individual promoters. Our results demonstrated that several promoters in a sense and antisense direction influenced *isc-suf* expression and the binding of the IscR, Irr, and OxyR regulatory proteins to individual promoters. These findings demonstrated a complex regulatory network of several promoters and regulatory proteins that helped to adjust iron-sulfur cluster assembly to changing conditions in *Rhodobacter sphaeroides*.

Keywords: iron-sulfur cluster; *isc* genes; *suf* genes; antisense promoters; OxyR; IscR; Irr

1. Introduction

Proteins containing iron-sulfur (Fe-S) clusters are present in almost all living organisms. They have diverse and often essential functions, for example, electron carriers in redox reactions, in redox sensing, oxidative stress defense, biosynthesis of metal-containing cofactors, DNA replication and repair, regulation of gene expression, and tRNA modification. Fe-S clusters are believed to be among the first catalysts to have evolved [1,2]. With the appearance of oxygenic photosynthesis, increasing oxygen levels drastically decrease iron availability [3]. Besides molecular oxygen, different reactive oxygen species (ROS) appear, which are very harmful to living cells since they can damage proteins, lipids, and nucleic acids. Molecular oxygen and ROS also destabilize Fe-S clusters, leading to the release of Fe^{2+} ions that, in turn, potentiate oxygen toxicity by the production of hydroxyl radicals in the Fenton reaction [4]. As a consequence, organisms have to develop multicomponent systems that promote the biogenesis of Fe-S proteins while protecting the cellular surrounding from the deleterious effects of free iron.

Different systems for the assembly of Fe-S clusters into biological macromolecules have evolved: the Isc (iron-sulfur cluster) system was identified as the system for generalized Fe-S protein maturation in *Azotobacter vinelandii* [5] and other bacteria. In *Escherichia coli*, the *isc* operon encodes the regulator

IscR (iron-sulfur cluster regulator), a cysteine desulfurase (IscS), a scaffold protein (IscU), an A-type carrier protein (IscA), a DnaJ-like co-chaperone (HscB), a DnaK-like chaperone (HscA), and a ferredoxin (fdx) [6]. The Isc machinery is widely conserved from prokaryotes to higher eukaryotes. Later, another operon for Fe-S cluster biogenesis, *suf*, was discovered in *E. coli* [7]: the *suf* operon encodes an A-type protein (SufA), a heterodimeric cysteine desulfurase (SufS and SufE), and a pseudo-ABC transporter (SufB, SufC, and SufD) that could act as a scaffold. Components of the Suf system are also found in other bacteria, including cyanobacteria and chloroplasts. The number and type of *isc* and *suf* operons, as well as their composition, vary among bacterial species. In *E. coli*, most of the Fe-S cluster biogenesis under non-stress conditions is catalyzed by the housekeeping Isc pathway, while the Suf system functions primarily under oxidative stress and/or iron starvation [8–12].

Fe-S cluster biogenesis systems have to respond to environmental stimuli to maintain and repair the pool of Fe-S proteins under changing environmental conditions. The current understanding of the regulation of genes for Fe-S cluster assembly is mostly limited to the model organism *E. coli* and a few other members of the gamma-proteobacteria. The main regulator of Fe-S assembly in *E. coli* is IscR, but other regulatory proteins, such as Fur (an iron-dependent regulator) and OxyR (an oxidative stress-dependent regulator) also contribute to regulated *isc* and *suf* operon expression [9,10,12]. The first gene of the *iscRSUA-hscBA-fdx* operon encodes the DNA-binding IscR protein that can coordinate a [2Fe-2S] cluster, which is assembled by the Isc system [13,14]. In *E. coli*, IscR regulates at least 40 genes comprising the *isc* and *suf* operons, genes for other Fe-S containing proteins, and genes encoding surface structures (*fim* and *flu*) or of unknown functions [9]. Holo-IscR (IscR containing a Fe-S cluster) represses its gene and the rest of the *isc* operon. This repression is released under conditions unfavorable for Fe-S maturation of IscR [6]. Thus, IscR also functions as a sensor for Fe-S homeostasis. Apo-IscR (protein lacking the Fe-S cluster) activates *suf* operon expression [15] in *E. coli*. Induction of the *isc* and *suf* operons by apo-IscR occurs under iron-limiting conditions and oxidative stress [8,10,11]. A switch from the Isc to the Suf system when the iron is limiting is also promoted by the non-coding sRNA RhyB that is under negative control of Fur (ferric uptake regulator) [16]. RhyB base pairs with the Shine-Dalgarno sequence of *iscS*, leading to the degradation of the 3' part of the *iscRSUA-hscBA-fdx* operon mRNA. The 5' part of the polycistronic mRNA, which contains *iscR*, is stabilized and translated [17]. This favors the formation of apo-IscR, which consequently induces *suf* expression.

Regulation of the *isc* and *suf* operons by stress signals also involves the regulators Fur, OxyR, and RhyB. The iron-sensing regulator protein Fur binds to the *suf* promoter as Fur-Fe^{2+} represses the *suf* genes [10]. OxyR, a known sensor protein for oxidative stress in *E. coli* and many other bacteria, also acts as an activator of the *suf* operon [12]. It was also proposed that IscR might directly sense oxidative stress through the destabilization of its Fe-S cluster [6]. Hydrogen peroxide (H_2O_2) was shown to inactivate the *E. coli* Isc system and to activate the *suf* operon through OxyR [18].

Phototrophic bacteria have a special need for the regulation of the Fe-S cluster assembly. Fe-S clusters are required for some components involved in photosynthesis like enzymes for bacteriochlorophyll synthesis (magnesium chelatase and the dark-operative protochlorophyllide oxidoreductase) or the cytochrome bc_1 complex for photosynthetic electron transport. On the other hand, photooxidative stress caused by the formation of ROS by the light excitation of bacteriochlorophyll destroys Fe-S clusters, which in turn leads to a further increase of ROS levels by the Fenton reaction. Latifi and co-workers suggested that elevated levels of ROS upon iron starvation [19], as also determined for *R. sphaeroides* [20], might be a characteristic for photosynthetic bacteria. Therefore, it is important to study the regulation of Fe-S cluster assembly also in phototrophic bacteria that significantly differ in their lifestyle from *E. coli*.

R. sphaeroides is a facultative photosynthetic bacterium, which can use a variety of metabolic pathways for ATP production. At high oxygen tension, aerobic respiration generates ATP. When the oxygen tension in the environment drops, the synthesis of photosynthetic complexes is induced, while aerobic respiration still takes place. Under anaerobic conditions in the light, anoxygenic photosynthesis generates ATP, while under dark conditions in the presence of a suitable electron acceptor, anaerobic respiration is performed. Oxygen is a major regulatory factor for the formation of photosynthetic complexes, and several proteins involved in oxygen-mediated gene regulation have been identified [21,22]. Under iron limitation, *R. sphaeroides* loses its purple color due to the loss of photosynthetic complexes and is no longer able to grow phototrophically [23]. Since bacteriochlorophyll synthesis requires iron, and also the reaction center contains iron, no photosynthetic complexes can be formed under iron limitation.

R. sphaeroides serves as a model organism to elucidate the response of photosynthetic bacteria to singlet oxygen [24–29] and has also been analyzed in regard to its response to iron limitation in oxic and anoxic environments [20,23,30,31]. The Fur ortholog (Fur/Mur) of *R. sphaeroides* is involved in both iron and manganese homeostasis [23]. Like in other alpha-proteobacteria, the Irr (iron response regulator) protein was shown to affect genes of the iron metabolism [30]. Lack of Fur/Mur, as well as lack of Irr, results in stronger induction of *isc/suf* genes under iron limitation [23,30]. Two-thirds of the iron-dependent genes in *R. sphaeroides* showed different responses under oxic or anoxic conditions. For some of these genes, including *isc-suf* genes, induction under iron limitation under oxic conditions was mediated by the OxyR protein [20]. *R. sphaeroides* harbors a large operon comprising *isc* and *suf* genes for iron-sulfur cluster assembly. Like in *E. coli*, the product of the first gene, IscR, functions as an iron-dependent repressor of the *isc* genes [31]. To better understand the transcriptional regulation that adjusts *isc-suf* operon expression to changing environmental conditions, we identified promoters responsible for *isc-suf* expression and quantified their activities under different oxygen concentrations in response to iron limitation and oxidative stress. This study resulted in a complex model for *isc-suf* gene regulation in a phototrophic alpha-proteobacterium.

2. Materials and Methods

2.1. Bacterial Strains and Growth Conditions

Bacterial strains are listed in Table S1. All *E. coli* strains were cultivated in Standard I medium at 37 °C, either in liquid culture by shaking at 180 rpm or on a solid growth medium, which contained 1.6% (*w/v*) agar. Depending on the cultivated strain, the antibiotics kanamycin (25 µg mL^{-1}), ampicillin (200 µg mL^{-1}), or tetracycline (20 µg mL^{-1}) were added to the liquid and solid growth media.

R. sphaeroides strains were cultivated in 50 mL Erlenmeyer flasks containing 40 mL malate minimal medium [20] with continuous shaking at 32 °C (microaerobic growth with a dissolved oxygen concentration of 25–30 µM). Aerobic conditions with 160 to 180 µM dissolved oxygen were achieved by incubating 25 mL of culture in 100 mL Erlenmeyer baffled flasks. The iron-depleted medium used was malate minimal medium without the addition of Fe(III) citrate [20,23] and containing the iron chelator 2,2′-dipyridyl (30 mM, Merck, Darmstadt, Germany). The cells were grown overnight and then transferred to new iron-depleted malate minimal medium three times more before harvesting [20,23]. Cells were harvested at an OD$_{660}$ of 0.4–0.6. Antibiotics were added to the liquid and solid growth media depending on the cultivated strain at the following concentrations: spectinomycin (10 µg mL^{-1}), kanamycin (25 µg mL^{-1}), gentamicin (25 µg mL^{-1}), tetracycline (2 µg mL^{-1}). For generating oxidative stress, the cultures of *R. sphaeroides* wild type or mutants were grown in iron-repleted malate minimal medium and treated with 1 mM (final concentration) hydrogen peroxide or 100 µM (final concentration) tertiary butyl-alcohol (tBOOH) for 7 min or 30 min at an OD$_{660}$ of 0.5. After 0, 7, and 30 min, cells were harvested and used for ß-galactosidase measurements.

2.2. Construction of R. sphaeroides RirA Deletion Mutants

R. sphaeroides strain 2.4.1ΔRSP_3341 was generated by transferring the suicide plasmid pPHU2.4.1ΔRSP_3341::Sp (RirA homolog 1) into *R. sphaeroides* 2.4.1, and screening for the insertion of the spectinomycin resistance cassette into the chromosome by homologous recombination. Parts of the *rirA* gene (RSP_3341) of *R. sphaeroides*, together with upstream and downstream sequences, were amplified by using oligonucleotides 3341up_f/3341up_r and 3341dn_f/3341dn_r. The amplified PCR fragments were cloned into the *KpnI-BamH*I and *BamH*I-*Hind*III sites of the suicide plasmid pPHU281 [32], generating the plasmid pPHU2.4.1ΔRSP_3341. A 2.2 kb *BamH*I fragment containing the spectinomycin cassette from pHP45Ω [33] was inserted into the *BamH*I site of pPHU2.4.1ΔRSP_3341 to generate pPHU2.4.1ΔRSP_3341::Sp. This plasmid was transferred into *E. coli* strain S17-1 and diparentally conjugated into *R. sphaeroides* 2.4.1 wild-type strain. Conjugants were selected on malate minimal salt agar plates containing spectinomycin. By insertion of the spectinomycin cassette, 438 bp of *R. sphaeroides rirA* gene RSP_3341 was deleted. *R. sphaeroides* strain 2.4.1ΔRSP_2888 was generated by transferring the suicide plasmid pPHU2.4.1ΔRSP_2888::Km (RirA homolog 2) into *R. sphaeroides* 2.4.1, and screening for insertion of the kanamycin resistance cassette into the chromosome by homologous recombination. Parts of the *rirA* gene (RSP_2888) of *R. sphaeroides*, together with upstream and downstream sequences, were amplified by using oligonucleotides 2888up_f/2888up_r and 2888dn_f/2888dn_r. The amplified PCR fragments were cloned into the *KpnI-BamH*I and *BamH*I-*Hind*III sites of the suicide plasmid pPHU281 [32], generating the plasmid pPHU2.4.1ΔRSP_2888. A 2.2 kb *BamH*I fragment containing the kanamycin cassette from pHP45Ω [33] was inserted into the *BamH*I site of pPHU2.4.1ΔRSP_2888 to generate pPHU2.4.1ΔRSP_2888::Sp. This plasmid was transferred into *E. coli* strain S17-1 and diparentally conjugated into *R. sphaeroides* 2.4.1 wild-type strain. The conjugants were selected on malate minimal salt agar plates containing kanamycin.

2.3. Constructions of Promoter Fusion Plasmids

According to the dRNA-seq data, fragments with lengths ranging from 98 bp to 1777 bp containing one of the putative five different single promoters or combined promoters of the *isc-suf* operon, respectively, were amplified by PCR with primers listed in Table S2 The PCR product was ligated into the pJET1.2/blunt cloning vector (Thermo Fisher, Dreieich, Germany) and then transferred into *E. coli* JM109. After confirming the correct sequence, the promoter fragment was cut from the sequenced cloning vector by *Xba*I/*Pst*I or *Spe*I/*Crf*9 and subsequently ligated into the transcriptional *lacZ* fusion vector pBBR1-MCS3-LacZ [34]. For testing the effect of antisense RNA on the activity of P2, the identical promoter fragment was also ligated into the *Xba*I/*Pst*I sites of the transcriptional *lacZ* fusion vector pBBR1-MCS5-LacZ, which harbors a gentamicin cassette and can be transferred into *R. sphaeroides* together with plasmid pRK4352 [35] and its derivatives.

2.4. Construction of Altered Promoter Sequences by Site-Directed Mutagenesis

Mutations in promoters of the *isc-suf* operon were constructed by inverse PCR. All primers used are listed in Table S2. For constructions of site-directed mutant P4 and P254, the TTG in the −35 region of the P4 promoter was replaced by AAA by using the plasmids pJET1.2-P4 or pJET1.2-P254 as templates, respectively. Similarly, for the constructions of mutant P5 and P25, the TTG in the −35 region of the P5 promoter was replaced by AAA by using the plasmids pJET1.2-P5 or pJET1.2-P25 as templates, respectively. The PCR products were digested by *Dpn*I and then transferred into JM109. Subsequently, the mutated promoter fragments were cut from the cloning vectors with *Xba*I/*Pst*I and ligated into pBBR-MCS3-LacZ [34].

Overexpression of RNA antisense to promoter P2 and sense promoter 2 fusion plasmids.

To test whether the antisense mRNA affects the activity of sense promoter, a DNA fragment antisense to P2 was amplified by PCR with primers asP2_fwd/rev listed in Table S2. The PCR product was ligated into the pJET1.2/blunt cloning vector and then transferred into *E. coli* JM109. After sequencing, the cloned fragment was cut out by *Xba*I-*BamH*I and subsequently ligated into vector pRK4352, which contains the strong 16S promoter [35].

2.5. ß-Galactosidase-Measurements

The ß-galactosidase activity of transcriptional fusions was measured by the hydrolysis of O-nitrophenyl- ß-D-galactopyranoside (ONPG) (Serva, Heidelberg, Germany) and expressed as Miller Units, as described in [36].

2.6. Differential RNA-seq

RNA isolation, TEX (Terminator EXonuclease) treatment, library construction and sequencing, read mapping, and transcription start site prediction have been described in detail in [37]. The RNA-seq data sets are available at the NCBI Gene Expression Omnibus database under accession number GSE71844.

2.7. RNA Isolation and Quantitative RT-PCR

Twenty milliliters of R. sphaeroides cells were harvested by centrifugation when an OD_{660} of 0.5 was reached. Total RNA for quantitative RT-PCR was isolated by using the peqGOLDTriFast kit (Peqlab, Erlangen, Germany), as described by the manufacturer. Remaining traces of DNA were removed by the TURBO DNaseI (Invitrogen, Schwerte, Germany). To further confirm the absence of DNA, PCR was performed targeting *globB* (RSP_0799) with the primers listed in Table S2. Quantitative RT-PCR was performed in a Bio-Rad CFX96 Real-Time system, as described in our previous study [20]. The reference gene *rpoZ* encoding the ω-subunit of RNA polymerase of *R. sphaeroides* was used to normalize the mRNA expression levels [38] according to the formula given by Pfaffl [39]. Primers are listed in Table S2.

2.8. Analysis of isc-suf Operon Synteny

The synteny analysis was performed by applying the EDGAR software [40].

Purification of the IscR, Irr, and OxyR proteins and electrophoretic mobility shift assays with dsDNA and protein.

E. coli M15 (pREP4/pQE2.4.1oxyR) was used for the overexpression of His-tagged OxyR protein, and the purification of OxyR was performed, as described earlier [41]. Isolation of His-tagged Irr from *E. coli* M15 (pREP4/pQE2.4.1*irr*) is described in [30], and the purification of His-tagged IscR from *E. coli* M15 (pREP4/pQE2.4.1*iscR*) in [31]. Binding of the proteins to the promoter regions of the *isc-suf* operon was determined by an electrophoretic mobility shift assay, as described previously [30,38]. DNA fragments containing the promoter regions of the *isc-suf* operon were produced by PCR. Primers used in the PCR are listed in Table S2. The final volume of the binding reactions was 15–20 µL. The reaction mixtures contained varying amounts of protein, ~3–7 fmol γ-^{32}PATP labeled DNA fragment (2000 cpm, Hartmann Analytik, Braunschweig, Germany), salmon sperm DNA (1 µg), and binding buffer, as described elsewhere [30,42]. After the binding reactions were incubated for 30 min at room temperature or 4 °C, the samples were loaded onto a 4% (*w/v*) polyacrylamide gel in 0.5×TBE buffer (22 mM Tris-HCl, 22 mM boric acid, 0.5 mM EDTA, pH 8.3), and the electrophoresis was performed at 130–180 V for 2–5 h at room temperature or at 4 °C. Irr and IscR were isolated and tested under non-reducing conditions. OxyR was purified and tested under reducing and non-reducing conditions with the same outcome.

3. Results

3.1. Prediction of Promoters for the isc-suf Operon of R. sphaeroides Based on dRNA-Seq Analysis

R. sphaeroides harbored a cluster of genes, with *isc* and *suf* homologs (Figure 1). RNA-seq data suggested the co-transcription of these genes, and this was also confirmed by further RT-PCR experiments [31]. The first gene of the *R. sphaeroides isc-suf* operon encoded the IscR regulator, which coordinates a Fe-S cluster with a unique Fe-S ligation scheme [31]. The other genes of the operon encoded two cysteine desulfurases (IscS and SufS), the membrane component of an iron-regulated

ABC transporter (SufB), a hypothetical protein (RSP_0439), the ATPase subunit of an ATP transporter (SufC), an Fe-S cluster assembly protein (SufD), and two proteins of the Yip1 family (RSP_0433 and RSP_0432) (Figure 1). This arrangement of genes was highly conserved in the family of *Rhodobacteraceae* (Figure S1). The synteny of genes that were annotated as *isc* or *suf* genes was highly conserved among the anaerobic anoxygenic phototrophs (AnAPs) and aerobic anoxygenic phototrophs (AAPs). There was some variation in regard to the non-annotated open reading frames. The non-phototrophs *Ruegeria* and *Paracoccus* also showed a similar synteny of *isc* and *suf* genes, and the gene for the IscR homolog of *Paracoccus denitrificans* was, however, positioned at a different chromosomal locus. A similar synteny of *iscS* and *suf* genes was also found in *Rhizobiales* (Figure S1); however, these bacteria lack IscR homologs [43]. The arrangement of genes did not provide information on the organization within an operon. In *Rhodobacter capsulatus*, SB1003 RNAseq data indicated co-transcription of *isc-suf* genes (GEO GSE134200) as in *R. sphaeroides*.

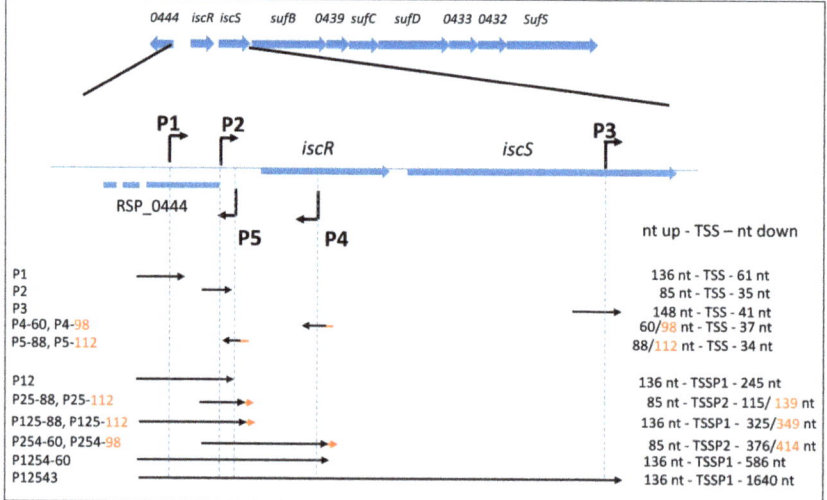

Figure 1. Schematic overview of the *isc-suf* operon of *R. sphaeroides*. The scheme at the top shows the arrangement of *isc-suf* genes and the arrows indicate the five transcriptional start sites (TSS) identified by dRNAseq (Figure S2). Fragments used in the reporter assays to examine the promoter activity for individual promoters and promoter combinations are indicated by the arrows below. The designations for the constructs are given on the left, and the number of nucleotides upstream or downstream of the TSS that are present in the reporter plasmids, are indicated on the right. Upstream or downstream regions of different sizes are marked in orange, and the number of the nucleotides (nt) that extend the shorter DNA fragment is added to the name of the shorter construct. According to the genome annotation of *R. sphaeroides*, the *isc-suf* genes are transcribed from the minus strand. We flipped this orientation in all figures to allow direct recognition of promoter sequences or other motifs.

Previous studies addressed the role of proteins with established roles in iron-dependent regulation in *isc-suf* expression in *R. sphaeroides* [23,30,31,36,37]. Different effects of Fur/Mur, Irr, and IscR on individual *iscR* and *suf* genes suggested that not all *isc-suf* genes were under the control of a single promoter and the identical regulatory elements.

Most *R. sphaeroides* promoters share little sequence identity [37], which makes sequence-based promoter identification almost impossible. Therefore, total RNA was isolated from cultures in the exponential growth phase and used for a differential RNA-seq analysis to determine transcriptional start sites (TSS) [37]. This analysis compared RNA, which was treated with terminator exonuclease (TEX) with untreated RNA samples [44]. While TEX degraded RNAs with 5'monophosphate, which were generated by processing, primary transcripts with 5′triphosphate were protected from degradation. Accumulation of the sequencing reads 5′of a gene in the TEX-treated sample, therefore, strongly supported the presence of a TSS at this position. As shown in Figure S2 (overview in Figure 1), the dRNA-seq read coverage indicated the presence of two promoters (P1 and P2) upstream of *iscR*. Another promoter (P3) was predicted within *iscS* and initiated transcripts spanning the *suf* genes. Furthermore, two promoters on the opposite DNA strand might lead to transcripts that were partially antisense to *isc-suf* transcripts. P4 was represented by a very low number of reads and led to a transcript mostly antisense to the *iscR* mRNA. P5 represented the promoter for transcription of RSP_0444. In previous studies, RSP_0444 showed only small differences in expression in the various strains or response to iron-limitation or H_2O_2 [23,30,31,36,37]. However, the 5′end of this transcript would partially be antisense to transcripts for the *isc-suf* genes initiating at P1 or P2. The dRNA-seq data did not hint to further promoters for *isc-suf* transcription [31]. Figure 1 shows a schematic overview of the promoter arrangement.

Many promoters in *R. sphaeroides* have a TTG around position −35 and an A at position −10/−11 relative to the TSS [37]. For the putative P1 promoter, an A was in position −11 in regard to the TSS, but no TTG around position −35 was present. The putative P2 promoter had an A residue at position −11 and a TTG around position −35 (Figure S3). Rodionov and co-workers predicted the presence of an Irr box and an IscR box upstream of the *iscR* gene [43]. The predicted IscR box spanned the positions −36 to −19 and the predicted Irr box the positions −13 to +1 in relation to the TSS at P2 (Figure S2, TSS indicated in red). Binding of IscR to the upstream region of the *iscR* gene was experimentally verified previously [31]. For the putative promoter within *iscS* (P3), an A residue at position −10 and a TTG around position −34 were present. Putative promoter P4, which seemed to be very weak based on the read number detected in RNA-seq, had a TTG around position −35, and the same was true for the putative promoter P5.

3.2. Activities of the isc-suf Sense and Antisense Promoters Alone or in Combination, as Determined by Transcriptional Reporter Gene Fusions

In order to confirm the activity of the predicted promoters, we constructed transcriptional fusions to the *lacZ* gene. Fragments containing 60–148 nt upstream and 34–63 nt downstream of the TSS, as determined by dRNA-seq, were cloned in front of the *lacZ* gene of plasmid pBBR1-MCS3-*lacZ* [34] (Table S3). An overview of the cloned fragments is shown in Figure 1, while the exact positions of primers together with predicted TSS are depicted in Figure S3. The plasmids were transferred to *R. sphaeroides* 2.4.1 wild type by conjugation, and the ß-galactosidase activity was determined for exponentially growing cultures. Since the oxygen levels influence the formation of photosynthetic complexes and consequently the demand for Fe-S clusters in *R. sphaeroides*, the cultures were either incubated under aerobic or microaerobic conditions. Under the latter conditions, the formation of photosynthetic complexes was strongly increased. The activity determined for the different promoter fusions showed marked differences (Figure 2). While P1 and P2 exhibited the weak activity of about 20–50 Miller Units (MU), the P3-*lacZ* and P5-88-*lacZ* fusions resulted in about 150–200 MU, the P4-98-*lacZ* fusion in around 400 MU under microaerobic conditions. The high activity of P4 was surprising, considering the fact that only a low number of reads from the P4 promoter was detected by RNA-seq, indicating that the native transcript initiating at P4 might be very unstable in contrast to the P4-*lacZ* fusion. This high activity, however, strongly decreased when the upstream region was reduced from 98 nt to 60 nt (P4–60). Likewise, the activity of P5 was dependent on the length of the upstream region: a shorter upstream region (88 nt) resulted in higher promoter activity than a longer upstream region (112 nt). We describe below that the OxyR protein binds to the P5 upstream region and represses P5 activity. Only for promoter P3, the activity was significantly (increase >1.5-fold and $p < 0.01$) but only slightly higher (slightly more than 1.5-fold) in aerobic conditions compared to microaerobic growth.

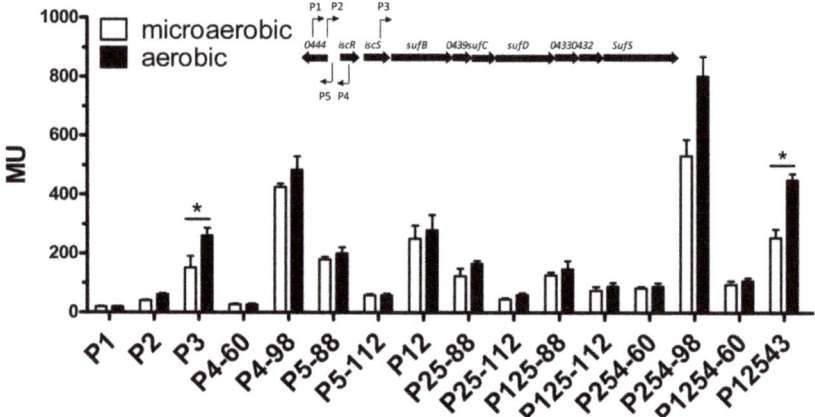

Figure 2. The activity of individual promoters and promoter combinations, as determined by *lacZ* reporter assays and quantified by measuring the ß-galactosidase activity in Miller Units (MU). Cells were cultured under aerobic or microaerobic conditions, as described in Materials and Methods. The scheme at the top shows the arrangement of *isc-suf* genes and the arrows indicate the five transcriptional start sites (TSS) identified by dRNAseq (Figure S2).The designations for all constructs are shown in Figure 1. The bars represent the average of technical duplicates from biological triplicates, and the standard deviation is indicated. *: the difference between the values for iron-replete and iron deplete conditions is >1.5-fold with a *p*-value of <0.01.

Considering the arrangement of the different promoters on the chromosome, it is conceivable that both P1, P2, as well as P3, contribute to *isc-suf* operon transcription and that the antisense transcripts initiating at P4 and P5 may also influence *isc-suf* operon transcription. To test this hypothesis, we applied the same primers as used for the single promoter fusions to also construct reporter plasmids that harbor combinations of the different promoters (Figure 1). Our results confirmed that the presence of additional promoters could influence the expression of the *lacZ* gene, which was fused to one of the promoters (Figure 2). When P1 was present together with P2, the activity was clearly higher than for the single promoter fusions. Remarkably, also the presence of the P5 promoter (RSP_0444 promoter) with 88 nt long upstream sequence, which generated transcripts that were antisense to the 28 nt at the 5'end of the *iscR* transcript, increased P2 activity, but not P12 activity. This activation did not occur, when 112 nt of the P5 upstream region were present (P5–112). The additional presence of the strong P4 promoter led to a strong increase of the activity compared to the P25-*lacZ* fusion, when 98 nt long upstream region was present (P254–98) but not with only 60 nt of the upstream region (P254–60).

When we used the 1777 nt fragment harboring all five promoter regions upstream of *lacZ* (P12543), the ß-galactosidase level was lower than for P254 and similar to P12 under microaerobic conditions (Figure 2). While the P12 activity was independent of the oxygen levels, the P12543 activity was slightly (1.8-fold) but significantly increased under aerobic conditions.

3.3. An AntiSense Promoter Stimulates Transcription of the isc-suf Operon

To study a possible influence of the antisense promoters P4 and P5 (promoter for RSP_0444) on P2 (main promoter for *iscR*) activity, we used different sized downstream fragments of P2 fused to *lacZ*. As shown in Figure 2, our data indicated that both P4 and P5 might influence P2 activity. However, we also had to consider the possibility that the prolonged DNA fragments fused to *lacZ* might contain other elements, which could affect the *lacZ* transcript level. For determining the influence of P5, the P2 downstream DNA fragment fused to *lacZ* was extended by only 80 nt (P25–88) or by 104 nt (P25–112). For testing a putative additional effect by P4, the P25–112 fragment was extended by 237 nt (P254–60) or by 275 nt (P254–98) (Figure 1). To verify that different activities of the P2-reporter versus the P25-

and P254- reporters are really due to the promoter activity of P4 and P5, we introduced point mutations into the −35 regions of P4 and P5 promoter regions (see material and methods, TTG changed to AAA) to abolish their activity. Figure 3A shows that the point mutation in the −35 region (mutP5–88) indeed almost abolished the activity of the P5 promoter. While the presence of the wild type P5–88 sequence induced P2 activity, no effect of the mutated P5–88 promoter on the P2 activity was observed, strongly suggesting that the activity of this promoter influenced transcription of the opposite DNA strand (Figure 3A). There were two possibilities, how P5 could influence P2 promoter activity: i) through the production of an antisense transcript, or ii) through changing the local DNA topology by its promoter activity. To discriminate between these possibilities, we introduced a second plasmid (pRK4352-asP2) into the strain harboring the P2-*lacZ* fusion. This plasmid allowed the production of an antisense RNA as produced by P5 from the strong 16S promoter. Real-time RT-PCR proved that in the presence of pRK4352-asP2, the amount of the antisense RNA was about 30-fold higher than in a strain lacking this plasmid (Figure 3B). As shown in Figure 3C, the production of this antisense RNA did not affect the activity of P2. The P2 promoter used in these assays (P2 Gm) was cloned into a different plasmid with gentamicin resistance to allow overexpression of the antisense RNA as P2 from a plasmid with tetracycline resistance.

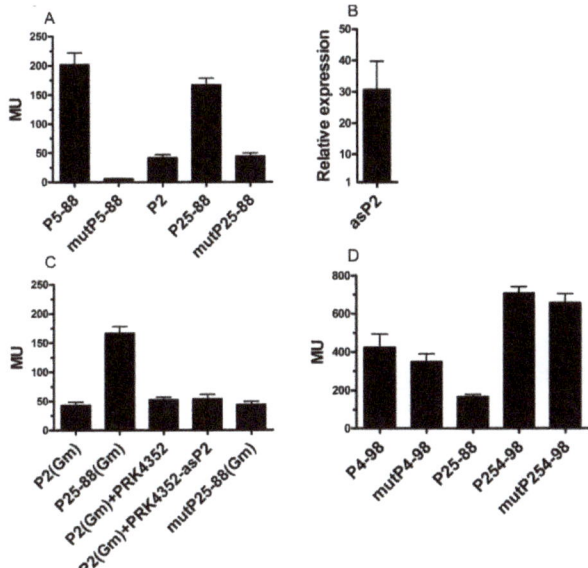

Figure 3. Effects of the antisense promoters P5 and P4 on the activity of P2, as determined by the *lacZ*-reporter assay (ß-galactosidase activity in Miller units). (**A**) The activity of individual promoters P5 or P2 or the combined P25 promoters. The numbers indicate the length of the DNA sequence upstream of the TSS, as shown in Figure 1. "mut" indicates that the TTG at position −35 of promoter P5 was changed to AAA. (**B**) Change of the level of RNA antisense to P2 in a wild type strain harboring the P2 reporter plasmid and plasmid pRK4352-asP2 (overexpression of the antisense RNA) compared to a strain just harboring the P2 promoter. The RNA level was determined by real-time RT-PCR, and the bar represents the average level of technical triplicates from biological triplicates with standard deviation. (**C**) Effect of elevated levels of RNA antisense to P2 on the activity of P5, as determined by the *lacZ*-reporter assay. (Gm) indicates that these reporter constructs carry a gentamycin resistance (all other reporters carry tetracycline resistance) to allow selection of the overexpressed plasmid pRK4352-asP2 (tetracycline resistance). (**D**) The activity of individual promoter P4 or the combined P25 and P254 promoters. The numbers indicate the length of the DNA sequence upstream of the TSS. "mut" indicates that the TTG at position −35 of promoter P4 was changed to AAA.

We applied the same strategy to test the effect of P4 on P2 activity. However, changing the TTG at the putative −35 of P4 only slightly decreased ß-galactosidase activity in the P254–98 construct. Consequently, the resulting construct mutP254–98 would still increase the activity of the P25 fusion (Figure 3D). As already shown in Figure 2, the P4–60 construct had a strongly reduced activity compared to P4–98. The low activity of P4-60 was almost abolished, when the −35 region was mutated. This indicated that the sequences in the −98 to −60 upstream region of the P4 promoter were responsible for the strong activity of P4–98 and the stimulating effect of P4–98 on P25. We were unable to recognize a particular motif that could cause this effect.

3.4. Effect of Oxidative Stress and Iron Availability on the Activity of the Promoters of the isc-suf Operon

Previous studies on global gene expression by microarrays or RNA-seq revealed that expression of the *isc-suf* genes is affected by oxidative stress [41,45] and by iron availability [20,23]. In this study, we applied the different reporter fusions to test which promoters are affected by these external factors. We tested hydrogen peroxide and tertiary butyl alcohol (tBOOH) as oxidative stress agents. tBOOH represents organic peroxides, which are generated in the cell, e.g., upon singlet oxygen exposure, and thus represent the photo-oxidative stress that *R. sphaeroides* faces in the presence of pigments, oxygen, and light. None of the promoters P1, P2, P3, P4, P5, or any combination of promoters showed marked changes in activity upon addition of hydrogen peroxide (same values for t0 as shown in Figure 4A and no significant changes at later time points), although hydrogen peroxide was previously shown to strongly induce the *isc-suf* mRNA levels [36]. We had, however, noted before that the response of *lacZ* reporters to H_2O_2 might be weak, possibly because of a negative effect on the enzymatic activity [46], while tBOOH was shown to induce the activity of certain *lacZ* fusions [47]. The addition of tBOOH resulted in significantly increased activities of the P2 promoter, and, also, the combination P25–88 showed significant tBOOH-dependent expression (all other promoter fusions did not show altered activity in response to tBOOH). The changes in activity were, however, less than 2-fold (Figure 4A). P12543 was also activated by tBOOH, but the increase was only 1.5-fold.

A previous study demonstrated that the *iscR* transcript level increases upon iron depletion [31]. A strong increase (about 5-fold) of P2 promoter activity upon iron depletion was confirmed in this study and was also observed for all other fusions containing P2, including the long fusion extending to P3 (Figure 4B). No significant effect of iron on activities of P1, P3, P4, or P5 was observed in the wild type (Figure 4B).

3.5. Protein Regulators of the isc-suf Operon

Previous bioinformatic analysis predicted IscR and Irr binding sites upstream of the *iscR* gene [43] (Figure S2), and the binding of IscR to this region was experimentally verified [31]. The role of Irr in the expression of the *suf* genes was supported by microarray analysis [30], but binding of the Irr protein to the *iscR* promoter region of *R. sphaeroides* was not reported. Microarray analyses also revealed a stronger effect of iron depletion on *isc-suf* expression in a mutant lacking the Fur/Mur regulator [23] and an effect of OxyR on *isc-suf* expression [18,20]. We tested the activity of the individual *isc-suf* promoters in different mutant strains to get more insights into the influence of these regulatory proteins on the *isc-suf* promoters.

Only the mutant strains, lacking the IscR or Irr protein, showed the altered activity of the P1 promoter (Figure 5A). The activity of P1 was almost 3-fold lower in the *irr* mutant than in the wild type under iron repletion and 1.6-fold lower under iron depletion. A sequence (TAGAAGGCATAGTGC) with similarity to the consensus Irr-box (Figure S2) was present directly upstream of the TSS of P1. However, we could not confirm binding of Irr to the P1 promoter region in vitro, while in a parallel control assay binding to the *mbfA* promoter, as described in [30], could be observed and confirmed that the isolated Irr protein was able to bind to one of its known targets (Figure S4A). Activity in the *iscR* mutant was similar to that of the wild type under iron repletion but about 2-fold higher under iron depletion. A sequence with some similarity to the *iscR* box (Figure S2) was present at the transcriptional start site (TAGACGACCTTGTTGTT, Figure S3). Indeed, gel retardation revealed that IscR showed specific interaction to the P1 promoter region (Figure 6A). Increasing amounts of IscR protein shifted the P1 containing DNA fragment, while the addition of unlabeled, specific competitor released the shift.

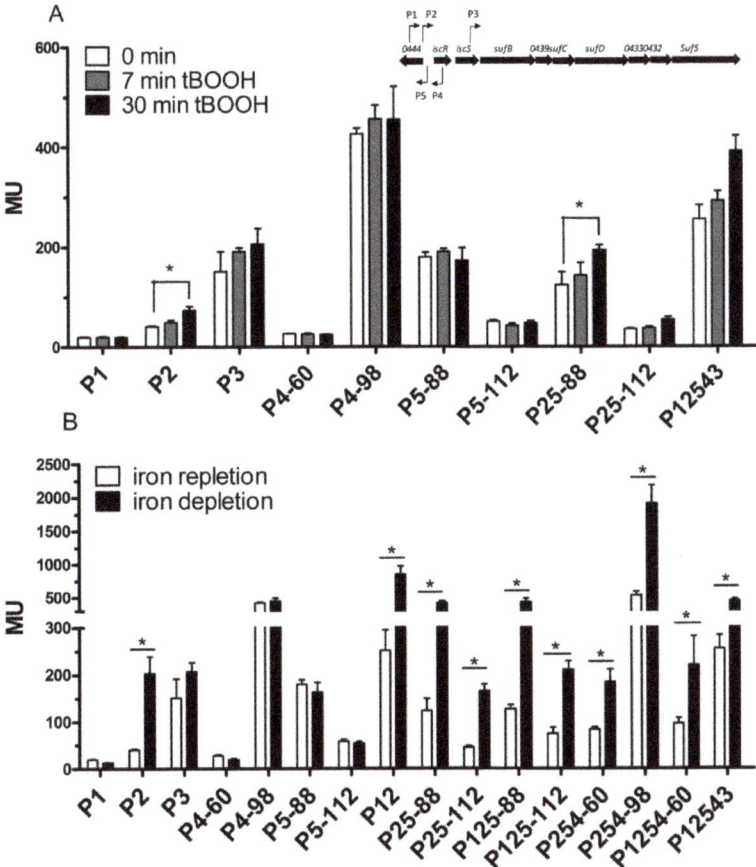

Figure 4. The activity of individual promoters and promoter combinations, as determined by *lacZ* reporter assays and quantified by measuring the ß-galactosidase activity in Miller Units (MU). (**A**) The ß-galactosidase activity was measured before, 7 min, and 30 min after the addition of tBOOH (100 µM final concentration). (**B**) The ß-galactosidase activity was determined under iron repletion and iron depletion. The designations for all constructs are shown in Figure 1. The bars represent the average of technical duplicates from biological triplicates, and the standard deviation is indicated. *: the difference is >1.5-fold with a *p*-value of <0.01.

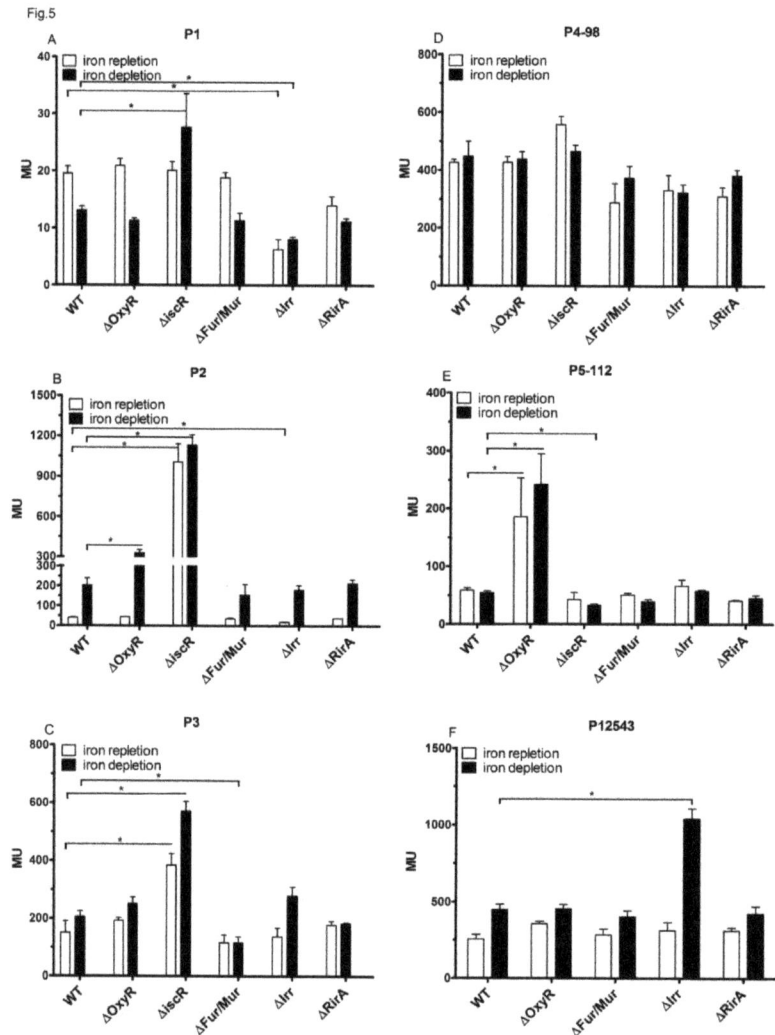

Figure 5. The activity of the individual promoters and P12543, as determined by *lacZ* reporter assays and quantified by measuring the ß-galactosidase activity in Miller Units (MU) in the wild type and mutant strains under iron repletion or iron depletion. (**A**) The activity of promoter P1, (**B**) The activity of promoter P2, (**C**) The activity of promoter P3, (**D**) The activity of promoter P4, (**E**) The activity of promoter P5, (**F**) The activity of P12543. The bars represent the average of technical duplicates from biological triplicates, and the standard deviation is indicated. *: the difference is >1.5-fold with a *p*-value of <0.01.

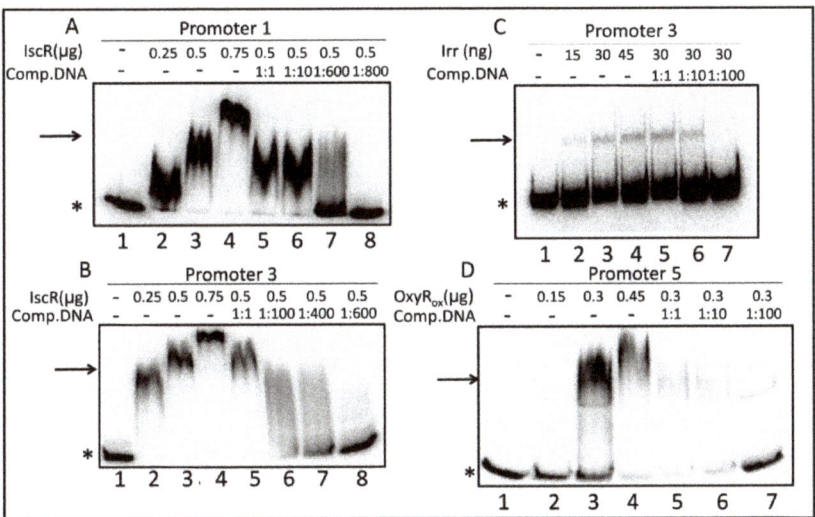

Figure 6. Electrophoretic mobility shift assays showing the interaction of (**A**) IscR to the P1 promoter region (199 bp fragment) and (**B**) IscR to the P3 promoter region (190 bp fragment) and (**C**) Irr to the P3 promoter region (190 bp fragment) and (**D**) oxidized OxyR to the P5 promoter region (147 bp fragment). The star labels the radio-labeled input DNA fragment, and the arrow points to the shifted bands of the DNA protein complexes. The amount of the protein input is given for each lane, as well as the molar ratio of specific, unlabeled competitor DNA (same DNA fragment but without a label).

The activity of P2 was strongly influenced by IscR under iron repletion and iron depletion, as demonstrated previously [31] (Figure 5B). The activity of P2 was also elevated in the presence of tBOOH in iron-replete medium (Figure 4A). This response of P2 to tBOOH was lost in the *iscR* mutant (Figure S5). Furthermore, we observed small but significant effects of OxyR and Irr on P2 activity: lack of OxyR increased P2 activity about 1.5-fold under iron depletion, the lack of Irr decreased P2 activity about 2-fold under iron repletion. Binding of IscR to P2, as well as regulation of P2 by IscR, was demonstrated before [31]. An Irr binding site was predicted for the P2 promoter region [43]. As for P1, we could not confirm binding of Irr to the P2 promoter region, while in a parallel control assay, binding to the *mbfA* promoter, as described in [30], could be observed (Figure S4B).

P3 activity was affected by IscR, both under iron repletion and iron depletion (Figure 5C). Lack of IscR resulted in higher activity (2.8-fold) in iron depletion, as well as in iron repletion (2.5-fold), indicating a repressing effect of IscR under both conditions. In the wild type, P3 showed no significant response to iron, while we observed slightly higher activity (1.5-fold) under iron depletion in the strain lacking IscR. Inspection of the sequence around the predicted TSS of P3 revealed a motif (GACTATTTCTGTCGG) with similarity to the consensus IscR box (Figure S2). Indeed, IscR showed specific binding to a DNA fragment containing the predicted binding site in a gel retardation assay (Figure 6B). Furthermore, the activity of the P3 promoter was significantly reduced in the Fur/Mur mutant under iron depletion (Figure 5C). Only in the strain lacking Irr, we had a significantly increased P3 activity (2-fold increase) in iron depletion compared to iron repletion. This agreed with our previous microarray data: the iron dependency of *suf* transcript levels is more pronounced in a strain lacking Irr than in the wild type. A putative Irr binding site (TTAGAAATATTCTAGA) was present about 50 nt upstream of the TSS of P3, which is a similar distance to the TSS as observed for the confirmed Irr target *ccpA* [30]. A gel retardation assay confirmed the specific binding of Irr to this DNA region (Figure 6C).

Differences in activity between wild type and the tested mutants for antisense promoter P4 were small (≤1.5-fold) and/or statistically not significant for the reporter construct with 98 nt upstream of the TSS (Figure 5D) and also for the construct with only 60 nt of the upstream region.

The activity of antisense promoter P5 was strongly affected by the OxyR protein when 112 nt upstream of the TSS was present. The activity in the mutant was increased by a factor of 3-4 under both iron repletion and iron depletion (Figure 5E). OxyR did not affect P5 activity when only 88 nt of the upstream sequence was present. None of the other tested mutants showed significantly altered P5 activity. Despite the strong differences in the activity of P5 in the presence or absence of OxyR, P5 did not show a response to high oxygen levels (Figure 2) or tBOOH (Figure 4A). We tested whether OxyR was able to bind to the P5 promoter by the gel retardation assay. As shown in Figure 6D, the addition of increasing amounts of oxidized OxyR resulted in retardation of a 147 nt DNA fragment harboring the P5 promoter region fragment. The same result was achieved with reduced OxyR protein. The addition of an excess of the same, but un-labeled DNA fragment led to a decrease of retardation due to specific competition with the labeled fragment for binding. This result strongly supported that OxyR was indeed binding to the P5 promoter region.

We also tested the activity of the P12543 fusion in different mutant strains (Figure 5F) to see how the action of the regulators on promoters P1, P2, P4, or P5 would influence P3 expression. Fur/Mur and the RirA proteins did not influence P12543 activity, as this was also the case for P3 alone. Since the P12543 fusion comprises an intact copy of the *iscR* gene and may, therefore, increase IscR levels in the cell, it could not be tested in conditions lacking IscR. The repressing effect of Irr under iron depletion was more pronounced when all promoters were present than in the P3-fusion alone (compare Figure 5C–F).

3.6. The RirA Proteins of R. sphaeroides Have No Effect on isc-suf Expression in R. sphaeroides

While Fur is the dominant iron regulator in gamma-proteobacteria, other proteins have important roles in iron-dependent regulation in alpha-proteobacteria [43]. Besides IscR, another transcriptional regulator of the Rrf2 protein family, RirA, was identified to have an important role in iron regulation in *Rhizobia* [48,49]. The genes RSP_2888 and RSP_3341 of *R. sphaeroides* 2.4.1 have 59%–63% identity to RirA from *Rhizobium leguminosarum* or *Agrobacterium tumefaciens* (*Rhizobium radiobacter*). We constructed knock out strains of *R. sphaeroides* lacking either RSP_2888 or RSP_3341 or both genes together. As seen in Figure 5A–F and Figure S6, the promoters of the *isc-suf* operon showed very similar activities in the wild type and the double mutant, and also the effect of iron depletion on promoter activity was very similar in both strains. We concluded that the RirA homologs of *R. sphaeroides* had no major role in the iron-dependent regulation of the *isc-suf* operon. The double mutant showed identical growth curves as the wild type in iron repletion and iron depletion (Figure S7), indicating that the RirA proteins have also no major impact on iron regulation in *R. sphaeroides* in general.

4. Discussion

The assembly of iron-sulfur clusters is an important cellular process in almost all living organisms, and stress conditions that destroy iron-sulfur clusters increase the need for iron-sulfur cluster assembly. The facultative phototrophic bacterium *R. sphaeroides* harbors a large operon that comprises *isc* and *suf* genes, which are distributed to different operons in *E. coli*. Unlike *E. coli*, *R. sphaeroides* can form photosynthetic complexes that cause an additional demand for the Fe-S cluster and can also cause photooxidative stress that destroys the Fe-S cluster. The synteny of *isc* and *suf* genes is conserved among *Rhodobacteraceae* independently of the ability to perform photosynthesis and is also similar in *Rhizobiales*. This suggests that the combined *isc-suf* operons arose early in evolution, in a common ancestor of these orders.

Previous work revealed iron- and hydrogen peroxide-dependent expression of the *isc-suf* genes of *R. sphaeroides* and the role of OxyR, Fur/Mur, Irr, and IscR regulators in *isc-suf* expression [20,23,30,31]. To verify promoter activities as predicted by dRNA-seq (Figure S2) and to understand the contribution of the individual promoters, we analyzed activities of individual promoters and of promoter combinations and the effect of iron, oxygen, and oxidative stress on the activities by using reporter gene fusions. These data were combined in a model (Figure 7) that visualizes the complex regulatory network controlling *isc-suf* expression in *R. sphaeroides*.

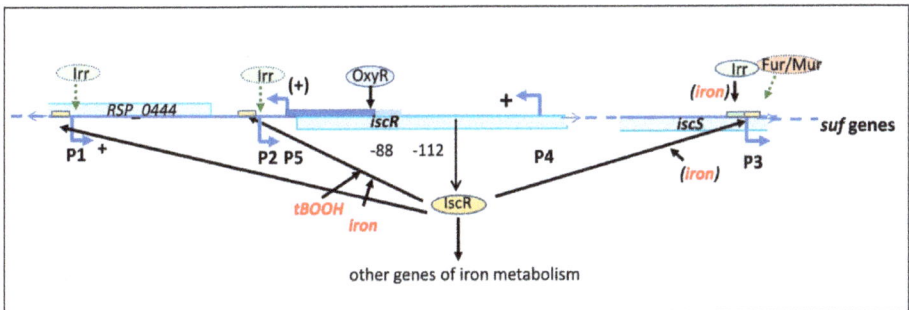

Figure 7. A schematic model, combining the influence of other promoters on P2 and the action of different proteins on the activity of the *isc-suf* promoters. + indicates a stimulating effect of P1, P4, and P5 promoter regions (dependent on the length of the upstream region) on P2 activity. The IscR protein binds to IscR boxes (yellow bars) in the promoter regions of P1, P2, and P3, and OxyR binds upstream of the P5 promoter (indicated by solid arrows), and Irr binds to an Irr-box (green bar) in the P3 promoter region. At P2, IscR mediates the response to iron and tBOOH. The iron-dependent activity of P3 is only observed in the absence of IscR or the absence of Irr. Green arrows indicate activation by the protein regulators, and black arrows indicate repression. The small effects of Irr on P1 and P2 and of Fur/Mur on P3 are most likely indirect and do not involve direct binding (indicated by dashed arrows).

Our data supported the view that both P1 and P2 contributed to *isc-suf* expression, and P3 further contributed to transcription of the *suf* genes. P3 was the strongest of the sense promoters, and extending the upstream region of P3 and including P1 and P2 elevated the activity further and conferred iron-dependent expression. An additional promoter for the *suf* genes (P3) might guarantee the high expression of the proteins required for the assembly of iron-sulfur clusters. IscR has a regulatory function and may not be required in a high amount. IscS is a cysteine desulfurase that is required for mobilization of sulfur from L-cysteine. A higher expression of *iscS* might not be required, since, with SufS, another cysteine desulfurase was encoded by the *isc-suf* operon downstream of P3. This arrangement with two cysteine desulfurases was conserved among *Rhodobacteraceae* and also found in *Rhizobiaceae* (Figure S1).

Two antisense promoters were present downstream of P2. P4 was located within the *iscR* gene, while P5 was located upstream of *iscR* and responsible for transcription of RSP_0444. The gene product of RSP_0444 is annotated as a putative hydrolase, but no experimental data on its function is available. The position of this gene on the chromosome was also found in most *Rhodobacteraceae* and *Rhizobiaceae*, although in *Rhizobiaceae, iscR* was not located upstream of IscS (Figure S1). Surprisingly, both antisense promoter regions, P5 and P4, positively affected transcription in sense direction. Transcription of *iscR* from P2 was higher, when, at the same time, P5 was present, and even higher, when, also, the region containing P4 was present. RNA-seq analyses revealed the presence of remarkably high numbers of antisense promoters in many bacterial species, and different effects of antisense transcripts have been reported [50–52]. In many cases, antisense transcription produces non-coding sRNAs that either interfere with translation or influence stability of the sense RNA. We could not confirm the existence of a distinct, small RNA originating at P4 by Northern blot analysis. Alternatively, the formation of an open complex during initiation of transcription might also allow more efficient transcription in the opposite direction. An effect of promoters on superhelicity-dependent processes is well documented [53]. The exact mechanisms by which P5 and the P4 region affect *isc-suf* expression remain elusive, but our data emphasized that it is important to consider the effect of such antisense promoters for sense promoter activity.

P2 was the main promoter for transcription of *iscRS*, and binding of IscR to P2 conferred iron-dependent expression to P2 and to the downstream promoter P3 that initiated further *suf* transcripts. Our present study revealed that IscR was also required for the tBOOH-dependent activity of P2. Thus, IscR could function as a sensor for iron availability and organic peroxide stress in *R. sphaeroides*. Furthermore, we could demonstrate the binding of IscR to the other sense promoters—P1 and P3. In contrast to its effect on P2 activity, IscR binding to P1 or P3 did not mediate tBOOH-dependent expression. Two types of IscR binding sites—Type 1 and Type 2—were identified in *E. coli*: holo-IscR (containing the Fe-S cluster) shows a higher affinity to Type 1 sites than apo-IscR, while both IscR forms bind with similar affinity to Type 2 sites [9]. Interestingly, this regulatory mechanism is even conserved in the Gram-positive *Thermincola potens* [54]. Our in vivo data revealed stronger repression of the P2 promoter by holo-IscR, while the repressing effect on P3 was similar in iron repletion and iron depletion, and only apo-IscR had a repressing effect on P1. This indicated the presence of different types of IscR binding sites also in *R. sphaeroides*.

Despite the presence of an Irr box around the P2 promoter, we detected only a small effect of Irr on P2 (1.8-fold) and did not detect binding of Irr to the P2 region in vitro. A small effect of Irr on P1 activity (1.5-fold) was observed, but no binding of Irr to the promoter region in vitro was observed. Our results implied that Irr did not make a major contribution to P1 and P2 regulation under the tested conditions and suggested that the influence of Irr might be indirect. There was also an influence of Irr on P3 that was even more pronounced when all upstream promoters were present, and binding of Irr to P3 was demonstrated. Since the P12543 reporter also carries a complete copy of *iscR*, we could not exclude that elevated levels of IscR influence activity of the P3 promoter in this strain. Higher IscR levels could increase the effect of iron in the *irr* deletion strain (compare 5C to 5F). The P3 promoter was the only promoter that showed a significant but small effect in the mutant lacking Fur/Mur, indicating an activating function of Fur/Mur under iron depletion. Johnston and co-workers suggested that iron-dependent regulation in alpha-proteobacteria mainly occurs by regulators different from Fur [55]. Some Fur homologs in *Rhizobia* and the Fur homolog of *R. sphaeroides* were shown to affect the expression of the *sit* operon in an Mn^{2+}-dependent manner and were consequently named Mur [23,56–58]. The deletion of the *fur/mur* gene resulted in stronger expression of many iron-dependent genes in *R. sphaeroides* [23], suggesting that Fur/Mur has a role in regulating iron metabolism in this bacterium and has a repressing function under iron depletion. Rodionov and co-workers suggested a Fur-box and a slightly differing Mur-box as consensus sequences for DNA binding sites in alpha-proteobacteria [43]. Such sequences are not present in the vicinity of P3 of the *isc-suf* operon, while an almost perfect Mur box is located between the −10 and −35 region of the *sitA* promoter. Fur/Mur is required for a strong induction (about 50-fold) of *sitA* expression in response to Mn^{2+} limitation [23]. It is likely that the small effects of Fur/Mur on many genes of iron metabolism, including P3 of the *isc-suf* operon, does not include direct binding but is rather mediated indirectly.

While none of the sense promoters was influenced by the OxyR protein, antisense promoter P5 was strongly repressed by this regulator, under iron-replete and iron deplete conditions, and binding of OxyR to the P5 upstream region was confirmed. The repressing effect of OxyR on the P5 promoter required 112 nt of the upstream region and was not present with only 88 nt of the upstream region (Figure 5E). An effect of OxyR on the iron-dependent levels of *isc* and *suf* genes was reported previously [20,41,45]. Since the activity of P12543 was not influenced by OxyR (Figure 5F), it was likely that other mechanisms than binding to P5 were responsible for the OxyR effect on the *suf* genes, which might also be indirect. Since P5 is the promoter for RSP_0444, OxyR could also affect the expression of RSP_0444, but previous microarray studies revealed hydrogen peroxide- and OxyR-independent expression of RSP_0444 [41,45]. OxyR is mostly known as an activator of gene expression in response to oxidative stress. A repressor function of OxyR under non-stress conditions, as observed for P5, was, however, also described in *R. sphaeroides* [41].

5. Conclusions

Figure 7 summarizes the effects of external factors and regulatory proteins on *isc-suf* expression and also indicates the influence of the P1, P5, and the P4 promoter region on the activity of P2, which is the main promoter for *iscRS* expression but also regulates other genes of iron metabolism. Our study revealed IscR as the main regulator for iron- and tBOOH-dependent expression of the *isc-suf* operon. IscR (under iron repletion and iron depletion) and Irr (under iron depletion) also repressed P3 activity by direct binding. The presence of the upstream region, including P1, allowed higher transcription rates of *isc-suf* than P2 alone, and both antisense promoters stimulated P2 activity. The influence of Fur/Mur was observed for P3, but the effects were small and most likely indirect. Thus, multiple promoters and multiple regulators were involved in adjusting *isc-suf* expression to environmental conditions, and by regulating *iscR* expression indirectly, other genes of the iron metabolism that were controlled by IscR were also affected.

Supplementary Materials: The following are available online at http://www.mdpi.com/2076-2607/7/12/671/s1.

Author Contributions: X.N. contributed to acquisition, analysis, interpretation of the data, preparation of figures, and writing of the part of the manuscript. B.R. contributed to the design of the study. G.K. managed the project, contributed to the conception and design of the study, to the interpretation of the data, and writing of the manuscript.

Funding: The work was supported by Deutsche Forschungsgemeinschaft [grant Kl563/34-1], and Xin Nie was supported by a fellowship of the Chinese Scholarship Council (CSC).

Acknowledgments: We thank Katrin Werler (former employee of the Institute of Microbiology and Molecular Biology, University of Giessen) for constructing some of the reporter fusions and the RirA single mutants; Kerstin Haberzettl (Institute of Microbiology and Molecular Biology, University of Giessen) for excellent technical assistance; and Jochen Blom (Institute of Bioinformatics and Systems Biology, University of Giessen) for his kind help with the synteny analysis.

Conflicts of Interest: The authors declare no conflict of interest.

References

1. Py, B.; Barras, F. Building Fe-S proteins: Bacterial strategies. *Nat. Rev. Microbiol.* **2010**, *8*, 436–446. [CrossRef] [PubMed]
2. Johnson, D.C.; Dean, D.R.; Smith, A.D.; Johnson, M.K. Structure, function, and formation of biological iron-sulfur clusters. *Annu. Rev. Biochem.* **2005**, *74*, 247–281. [CrossRef] [PubMed]
3. Andrews, S.C.; Robinson, A.K.; Rodriguez-Quinones, F. Bacterial iron homeostasis. *FEMS Microbiol. Rev.* **2003**, *27*, 215–237. [CrossRef]
4. Touati, D. Iron and oxidative stress in bacteria. *Arch. Biochem. Biophys.* **2000**, *373*, 1–6. [CrossRef] [PubMed]
5. Zheng, L.M.; Cash, V.L.; Flint, D.H.; Dean, D.R. Assembly of iron-sulfur clusters—Identification of an *iscSUA-hscBA-fdx* gene cluster from Azotobacter vinelandii. *J. Biol. Chem.* **1998**, *273*, 13264–13272. [CrossRef] [PubMed]
6. Roche, B.; Aussel, L.; Ezraty, B.; Mandin, P.; Py, B.; Barras, F. Reprint of: Iron/sulfur proteins biogenesis in prokaryotes: Formation, regulation and diversity. *Biochim. Biophys. Acta Bioenerg.* **2013**, *1827*, 923–937. [CrossRef]
7. Takahashi, Y.; Tokumoto, U. A third bacterial system for the assembly of iron-sulfur clusters with homologs in archaea and plastids. *J. Biol. Chem.* **2002**, *277*, 28380–28383. [CrossRef]
8. Outten, F.W.; Djaman, O.; Storz, G. A *suf* operon requirement for Fe-S cluster assembly during iron starvation in Escherichia coli. *Mol. Microbiol.* **2004**, *52*, 861–872. [CrossRef]
9. Giel, J.L.; Rodionov, D.; Liu, M.Z.; Blattner, F.R.; Kiley, P.J. IscR-dependent gene expression links iron-sulphur cluster assembly to the control of O-2-regulated genes in Escherichia coli. *Mol. Microbiol.* **2006**, *60*, 1058–1075. [CrossRef]
10. Lee, K.C.; Yeo, W.S.; Roe, J.H. Oxidant-Responsive Induction of the *suf* Operon, Encoding a Fe-S Assembly System, through Fur and IscR in Escherichia coli. *J. Bacteriol.* **2008**, *190*, 8244–8247. [CrossRef]
11. Yeo, W.S.; Lee, J.H.; Lee, K.C.; Roe, J.H. IscR acts as an activator in response to oxidative stress for the *suf* operon encoding Fe-S assembly proteins. *Mol. Microbiol.* **2006**, *61*, 206–218. [CrossRef] [PubMed]

12. Lee, J.H.; Yeo, W.S.; Roe, J.H. Induction of the sufA operon encoding Fe-S assembly proteins by superoxide generators and hydrogen peroxide: Involvement of OxyR, IHF and an unidentified oxidant-responsive factor. *Mol. Microbiol.* **2004**, *51*, 1745–1755. [CrossRef] [PubMed]
13. Giel, J.L.; Nesbit, A.D.; Mettert, E.L.; Fleischhacker, A.S.; Wanta, B.T.; Kiley, P.J. Regulation of iron-sulphur cluster homeostasis through transcriptional control of the Isc pathway by [2Fe-2S]-IscR in *Escherichia coli*. *Mol. Microbiol.* **2013**, *87*, 478–492. [CrossRef] [PubMed]
14. Schwartz, C.J.; Giel, J.L.; Patschkowski, T.; Luther, C.; Ruzicka, F.J.; Beinert, H.; Kiley, P.J. IscR, an Fe-S cluster-containing transcription factor, represses expression of *Escherichia coli* genes encoding Fe-S cluster assembly proteins. *Proc. Natl. Acad. Sci. USA* **2001**, *98*, 14895–14900. [CrossRef]
15. Nesbit, A.D.; Giel, J.L.; Rose, J.C.; Kiley, P.J. Sequence-Specific Binding to a Subset of IscR-Regulated Promoters Does Not Require IscR Fe-S Cluster Ligation. *J. Mol. Biol.* **2009**, *387*, 28–41. [CrossRef]
16. Masse, E.; Vanderpool, C.K.; Gottesman, S. Effect of RyhB small RNA on global iron use in *Escherichia coli*. *J. Bacteriol.* **2005**, *187*, 6962–6971. [CrossRef]
17. Desnoyers, G.; Morissette, A.; Prevost, K.; Masse, E. Small RNA-induced differential degradation of the polycistronic mRNA iscRSUA. *Embo J.* **2009**, *28*, 1551–1561. [CrossRef]
18. Jang, S.J.; Imlay, J.A. Hydrogen peroxide inactivates the *Escherichia coli* Isc iron-sulphur assembly system, and OxyR induces the Suf system to compensate. *Mol. Microbiol.* **2010**, *78*, 1448–1467. [CrossRef]
19. Latifi, A.; Jeanjean, R.; Lemeille, S.; Havaux, M.; Zhang, C.C. Iron starvation leads to oxidative stress in Anabaena sp strain PCC 7120. *J. Bacteriol.* **2005**, *187*, 6596–6598. [CrossRef]
20. Remes, B.; Berghoff, B.A.; Forstner, K.U.; Klug, G. Role of oxygen and the OxyR protein in the response to iron limitation in *Rhodobacter sphaeroides*. *BMC Genom.* **2014**, *15*, 794. [CrossRef]
21. Zeilstra-Ryalls, J.H.; Kaplan, S. Oxygen intervention in the regulation of gene expression: The photosynthetic bacterial paradigm. *Cell. Mol. Life Sci.* **2004**, *61*, 417–436. [CrossRef] [PubMed]
22. Elsen, S.; Swem, L.R.; Swem, D.L.; Bauer, C.E. RegB/RegA, a highly conserved redox-responding global two-component regulatory system. *Microbiol. Mol. Biol. Rev.* **2004**, *68*, 263–279. [CrossRef] [PubMed]
23. Peuser, V.; Metz, S.; Klug, G. Response of the photosynthetic bacterium *Rhodobacter sphaeroides* to iron limitation and the role of a Fur orthologue in this response. *Environ. Microbiol. Rep.* **2011**, *3*, 397–404. [CrossRef] [PubMed]
24. Glaeser, J.; Klug, G. Photo-oxidative stress in Rhodobacter sphaeroides: Protective role of carotenoids and expression of selected genes. *Microbiology* **2005**, *151*, 1927–1938. [CrossRef] [PubMed]
25. Anthony, J.R.; Warczak, K.L.; Donohue, T.J. A transcriptional response to singlet oxygen, a toxic byproduct of photosynthesis. *Proc. Natl. Acad. Sci. USA* **2005**, *102*, 6502–6507. [CrossRef] [PubMed]
26. Glaeser, J.; Zobawa, M.; Lottspeich, F.; Klug, G. Protein synthesis patterns reveal a complex regulatory response to singlet oxygen in *Rhodobacter*. *J. Proteome Res.* **2007**, *6*, 2460–2471. [CrossRef]
27. Glaeser, J.; Nuss, A.M.; Berghoff, B.A.; Klug, G. Singlet Oxygen Stress in Microorganisms. *Adv. Microb. Physiol.* **2011**, *58*, 141–173. [CrossRef]
28. Berghoff, B.A.; Glaeser, J.; Sharma, C.M.; Vogel, J.; Klug, G. Photooxidative stress-induced and abundant small RNAs in *Rhodobacter sphaeroides*. *Mol. Microbiol.* **2009**, *74*, 1497–1512. [CrossRef]
29. Berghoff, B.A.; Konzer, A.; Mank, N.N.; Looso, M.; Rische, T.; Forstner, K.U.; Kruger, M.; Klug, G. Integrative "Omics"-Approach Discovers Dynamic and Regulatory Features of Bacterial Stress Responses. *PLoS Genet.* **2013**, *9*, e1003576. [CrossRef]
30. Peuser, V.; Remes, B.; Klug, G. Role of the Irr Protein in the Regulation of Iron Metabolism in *Rhodobacter sphaeroides*. *PLoS ONE* **2012**, *7*, e42231. [CrossRef]
31. Remes, B.; Eisenhardt, B.D.; Srinivasan, V.; Klug, G. IscR of *Rhodobacter sphaeroides* functions as repressor of genes for iron-sulfur metabolism and represents a new type of iron-sulfur-binding protein. *Microbiologyopen* **2015**, *4*, 790–802. [CrossRef] [PubMed]
32. Hubner, P.; Willison, J.C.; Vignais, P.M.; Bickle, T.A. Expression of Regulatory Nif Genes in *Rhodobacter-Capsulatus*. *J. Bacteriol.* **1991**, *173*, 2993–2999. [CrossRef] [PubMed]
33. Prentki, P.; Binda, A.; Epstein, A. Plasmid vectors for selecting IS1-promoted deletions in cloned DNA: Sequence analysis of the omega interposon. *Gene* **1991**, *103*, 17–23. [CrossRef]
34. Kovach, M.E.; Elzer, P.H.; Hill, D.S.; Robertson, G.T.; Farris, M.A.; Roop, R.M., 2nd; Peterson, K.M. Four new derivatives of the broad-host-range cloning vector pBBR1MCS, carrying different antibiotic-resistance cassettes. *Gene* **1995**, *166*, 175–176. [CrossRef]

35. Mank, N.N.; Berghoff, B.A.; Hermanns, Y.N.; Klug, G. Regulation of bacterial photosynthesis genes by the small noncoding RNA PcrZ. *Proc. Natl. Acad. Sci. USA* **2012**, *109*, 16306–16311. [CrossRef]
36. Klug, G.; Jager, A.; Heck, C.; Rauhut, R. Identification, sequence analysis, and expression of the *lepB* gene for a leader peptidase in *Rhodobacter capsulatus*. *Mol. Gen. Genet. MGG* **1997**, *253*, 666–673. [CrossRef]
37. Remes, B.; Rische-Grahl, T.; Muller, K.M.H.; Forstner, K.U.; Yu, S.H.; Weber, L.; Jager, A.; Peuser, V.; Klug, G. An RpoHI-Dependent Response Promotes Outgrowth after Extended Stationary Phase in the Alphaproteobacterium *Rhodobacter sphaeroides*. *J. Bacteriol.* **2017**, *199*. [CrossRef]
38. Zeller, T.; Klug, G. Detoxification of hydrogen peroxide and expression of catalase genes in *Rhodobacter*. *Microbiology* **2004**, *150*, 3451–3462. [CrossRef]
39. Pfaffl, M.W. A new mathematical model for relative quantification in real-time RT-PCR. *Nucleic Acids Res.* **2001**, *29*, e45. [CrossRef]
40. Blom, J.; Kreis, J.; Spanig, S.; Juhre, T.; Bertelli, C.; Ernst, C.; Goesmann, A. EDGAR 2.0: An enhanced software platform for comparative gene content analyses. *Nucleic Acids Res.* **2016**, *44*, W22–W28. [CrossRef]
41. Zeller, T.; Mraheil, M.A.; Moskvin, O.V.; Li, K.Y.; Gomelsky, M.; Klug, G. Regulation of hydrogen peroxide-dependent gene expression in *Rhodobacter sphaeroides*: Regulatory functions of OxyR. *J. Bacteriol.* **2007**, *189*, 3784–3792. [CrossRef] [PubMed]
42. Wu, Y.; Outten, F.W. IscR controls iron-dependent biofilm formation *in Escherichia coli* by regulating type I fimbria expression. *J. Bacteriol.* **2009**, *191*, 1248–1257. [CrossRef] [PubMed]
43. Rodionov, D.A.; Gelfand, M.S.; Todd, J.D.; Curson, A.R.J.; Johnston, A.W.B. Computational reconstruction of iron- and manganese-responsive transcriptional networks in alpha-proteobacteria. *PLoS Comput. Biol.* **2006**, *2*, 1568–1585. [CrossRef] [PubMed]
44. Sharma, C.M.; Hoffmann, S.; Darfeuille, F.; Reignier, J.; Findeiss, S.; Sittka, A.; Chabas, S.; Reiche, K.; Hackermuller, J.; Reinhardt, R.; et al. The primary transcriptome of the major human pathogen *Helicobacter pylori*. *Nature* **2010**, *464*, 250–255. [CrossRef]
45. Zeller, T.; Moskvin, O.V.; Li, K.Y.; Klug, G.; Gomelsky, M. Transcriptome and physiological responses to hydrogen peroxide of the facultatively phototrophic bacterium *Rhodobacter sphaeroides*. *J. Bacteriol.* **2005**, *187*, 7232–7242. [CrossRef]
46. Hendrischk, A.K.; Braatsch, S.; Glaeser, J.; Klug, G. The *phrA* gene of *Rhodobacter sphaeroides* encodes a photolyase and is regulated by singlet oxygen and peroxide in a sigma(E)-dependent manner. *Microbiology* **2007**, *153*, 1842–1851. [CrossRef]
47. Li, H.S.; Zhou, Y.N.; Li, L.; Li, S.F.; Long, D.; Chen, X.L.; Zhang, J.B.; Feng, L.; Li, Y.P. HIF-1alpha protects against oxidative stress by directly targeting mitochondria. *Redox Biol.* **2019**, 101109. [CrossRef]
48. Imam, S.; Noguera, D.R.; Donohue, T.J. Global Analysis of Photosynthesis Transcriptional Regulatory Networks. *PLoS Genet.* **2014**, *10*, e1004837. [CrossRef]
49. Imam, S.; Noguera, D.R.; Donohue, T.J. An Integrated Approach to Reconstructing Genome-Scale Transcriptional Regulatory Networks. *PLoS Comput. Biol.* **2015**, *11*, e1004103. [CrossRef]
50. Thomason, M.K.; Storz, G. Bacterial antisense RNAs: How many are there, and what are they doing? *Annu. Rev. Genet.* **2010**, *44*, 167–188. [CrossRef]
51. Pernitzsch, S.R.; Sharma, C.M. Transcriptome complexity and riboregulation in the human pathogen *Helicobacter pylori*. *Front. Cell. Infect. Microbiol.* **2012**, *2*, 14. [CrossRef] [PubMed]
52. Sesto, N.; Wurtzel, O.; Archambaud, C.; Sorek, R.; Cossart, P. The excludon: A new concept in bacterial antisense RNA-mediated gene regulation. *Nat. Rev. Microbiol.* **2013**, *11*, 75–82. [CrossRef] [PubMed]
53. Beaucage, S.L.; Miller, C.A.; Cohen, S.N. Gyrase-dependent stabilization of pSC101 plasmid inheritance by transcriptionally active promoters. *EMBO J.* **1991**, *10*, 2583–2588. [CrossRef] [PubMed]
54. Santos, J.A.; Alonso-Garcia, N.; Macedo-Ribeiro, S.; Pereira, P.J.B. The unique regulation of iron-sulfur cluster biogenesis in a Gram-positive bacterium. *Proc. Natl. Acad. Sci. USA* **2014**, *111*, E2251–E2260. [CrossRef] [PubMed]
55. Johnston, A.W.; Todd, J.D.; Curson, A.R.; Lei, S.; Nikolaidou-Katsaridou, N.; Gelfand, M.S.; Rodionov, D.A. Living without Fur: The subtlety and complexity of iron-responsive gene regulation in the symbiotic bacterium *Rhizobium* and other alpha-proteobacteria. *Biometals* **2007**, *20*, 501–511. [CrossRef] [PubMed]
56. Chao, T.C.; Becker, A.; Buhrmester, J.; Puhler, A.; Weidner, S. The *Sinorhizobium meliloti fur* gene regulates, with dependence on Mn(II), transcription of the sitABCD operon, encoding a metal-type transporter. *J. Bacteriol.* **2004**, *186*, 3609–3620. [CrossRef]

57. Diaz-Mireles, E.; Wexler, M.; Sawers, G.; Bellini, D.; Todd, J.D.; Johnston, A.W. The Fur-like protein Mur of Rhizobium leguminosarum is a Mn(2+)-responsive transcriptional regulator. *Microbiology* **2004**, *150*, 1447–1456. [CrossRef]
58. Platero, R.; Peixoto, L.; O'Brian, M.R.; Fabiano, E. Fur is involved in manganese-dependent regulation of *mntA* (*sitA*) expression in *Sinorhizobium meliloti*. *Appl. Environ. Microbiol.* **2004**, *70*, 4349–4355. [CrossRef]

© 2019 by the authors. Licensee MDPI, Basel, Switzerland. This article is an open access article distributed under the terms and conditions of the Creative Commons Attribution (CC BY) license (http://creativecommons.org/licenses/by/4.0/).

Article

Interactions among Redox Regulators and the CtrA Phosphorelay in *Dinoroseobacter shibae* and *Rhodobacter capsulatus*

Sonja Koppenhöfer and Andrew S. Lang *

Department of Biology, Memorial University of Newfoundland, St John's, NL A1B 3X9, Canada; skoppenhofer@mun.ca
* Correspondence: aslang@mun.ca

Received: 11 March 2020; Accepted: 10 April 2020; Published: 14 April 2020

Abstract: Bacteria employ regulatory networks to detect environmental signals and respond appropriately, often by adjusting gene expression. Some regulatory networks influence many genes, and many genes are affected by multiple regulatory networks. Here, we investigate the extent to which regulatory systems controlling aerobic–anaerobic energetics overlap with the CtrA phosphorelay, an important system that controls a variety of behavioral processes, in two metabolically versatile alphaproteobacteria, *Dinoroseobacter shibae* and *Rhodobacter capsulatus*. We analyzed ten available transcriptomic datasets from relevant regulator deletion strains and environmental changes. We found that in *D. shibae*, the CtrA phosphorelay represses three of the four aerobic–anaerobic Crp/Fnr superfamily regulator-encoding genes (*fnrL*, *dnrD*, and especially *dnrF*). At the same time, all four Crp/Fnr regulators repress all three phosphorelay genes. Loss of *dnrD* or *dnrF* resulted in activation of the entire examined CtrA regulon, regardless of oxygen tension. In *R. capsulatus* FnrL, in silico and ChIP-seq data also suggested regulation of the CtrA regulon, but it was only with loss of the redox regulator RegA where an actual transcriptional effect on the CtrA regulon was observed. For the first time, we show that there are complex interactions between redox regulators and the CtrA phosphorelays in these bacteria and we present several models for how these interactions might occur.

Keywords: Alphaproteobacteria; Rhodobacteraceae; nitric oxide; quorum sensing; gene transfer agent; motility; Crp/Fnr; Dnr; RegA; ChpT

---

1. Introduction

Bacteria sense and process environmental signals in order to adapt to changes in their surroundings. These signals are relayed through regulatory networks that adjust the cells' behavior, often through changes in gene expression. The alphaproteobacterium *Dinoroseobacter shibae* is a member of the marine roseobacter group and an aerobic anoxygenic photoheterotrophic bacterium, capable of both aerobic and anaerobic respiration [1]. It can be free-living or an algal symbiont [1] and is a metabolically versatile bacterium able to adapt to changes in its highly dynamic environment. For example, at the end of an algal bloom when the oxygen concentration drops, an alternative terminal electron acceptor such as nitrate can be used for respiration [1,2].

The response to the change from aerobic to anaerobic conditions is controlled by four Crp/Fnr transcriptional regulators in *D. shibae* [3]. Crp/Fnr regulators are widely distributed among bacteria and form a superfamily consisting of 14 phylogenetic subgroups [4]. The versatility of this family is reflected by both the wide range of signals that are sensed, such as temperature [5], oxygen [6], and nitric oxide (NO) [7], and the range of metabolic processes regulated upon activation, which include respiration-related processes and especially the transition between aerobic and anaerobic lifestyles [3,8].

Two well-studied members of this family are the Dnr and Fnr proteins. Dnr proteins bind a heme cofactor that allows for sensing of NO [4,9], while Fnr proteins react to low oxygen tension [4,6]. In *D. shibae*, FnrL and DnrD regulate DnrE and DnrF in a cascade-type network that controls the transition from aerobic to anaerobic growth, heme and carotenoid synthesis, multiple other metabolic processes, and flagellar synthesis [3]. The importance of these regulators in *D. shibae* is well illustrated by the observation that loss of FnrL affects the transcript levels of over 400 genes [3].

Another important regulatory system in *D. shibae* is the CtrA phosphorelay [10]. Like the Crp/Fnr regulators, this phosphorelay integrates an environmental signal, in this case, the autoinducer concentration as an indicator of cell density, and adjusts gene expression in response [11]. This phosphorelay is conserved within the majority of alphaproteobacterial lineages and consists of the histidine kinase CckA, the phosphotransferase ChpT and the transcriptional regulator CtrA [10]. In *D. shibae*, the CtrA phosphorelay is activated by the quorum sensing (QS) signal of the main acyl-homoserine lactone (AHL) synthase (LuxI$_1$) with subsequent regulation of genes for flagellar motility, recombination and competence proteins, a tight adherence (tad) pilus involved in attachment to carbohydrates on the host cells [12], cell cycle control, gene transfer agent (GTA) production, bis-(3'-5')-cyclic dimeric guanosine monophosphate (c-di-GMP) signaling, the NO-sensing heme-nitric oxide/oxygen binding domain (HNOX) protein, and the AHL synthases LuxI$_2$ and LuxI$_3$ [11,13,14]. Deletion of *cckA* has been found to abolish the mutualistic interaction between *D. shibae* and its algal host, demonstrating that the CtrA phosphorelay is essential for establishment of this symbiosis, at least partly due to the requirement for flagella [15]. The Crp/Fnr and CtrA phosphorelay networks are connected by their shared regulation of flagellar gene expression and due to their involvement in symbiosis with the host dinoflagellate.

There are three ways bacteria can be exposed to NO. Some bacteria generate NO during denitrification, and this is considered the activator for DnrD in *D. shibae* [3,16]. NO can be produced intracellularly through the oxidization of L-arginine to NO and L-citrulline [17] or via a nitric oxide synthase (NOS) [17,18]. NO released by some eukaryotic organisms can be a form of communication with their symbiotic bacteria and is then typically sensed by HNOX proteins [19]. The HNOX genes are often located adjacent to genes encoding c-di-GMP signaling proteins or histidine kinases. In the context of symbioses, only a few NO-detecting systems have been found that do not involve c-di-GMP signaling but instead directly integrate into QS systems [20–22]. In *D. shibae*, an HNOX protein detects NO and thereupon inhibits the c-di-GMP synthesizing enzyme Dgc1 [23].

The potential for overlap between Crp/Fnr-based regulation and the CtrA phosphorelay also exists in the purple non-sulfur alphaproteobacterium *Rhodobacter capsulatus*. Its CtrA phosphorelay was originally discovered due to its regulation of GTA production [24], but it also affects many other genes such as those associated with flagellar motility, gas vesicles, and c-di-GMP signaling [24,25]. Like *D. shibae*, *R. capsulatus* can switch between aerobic and anaerobic lifestyles, which involves Crp/Fnr regulation, the RegA/B two-component system, and CrtJ [26–28]. Loss of FnrL affects the transcript levels of 20% of *R. capsulatus* genes [29], including 42 that are directly regulated and encode c-di-GMP signaling, gas vesicle, and flagellar proteins, among others [29].

These initial surveys of the activities of redox regulators and the CtrA phosphorelays in *D. shibae* and *R. capsulatus* indicated a potential connection of the regulons. Therefore, we were interested in exploring in more detail the extent to which these regulatory systems interact. We re-analyzed ten available transcriptomic datasets for the two species. Deletion mutants, including those of redox regulators and the CtrA phosphorelay/QS networks, were analyzed to examine the regulon overlap of these systems and to evaluate their potential integration. We also included further analyses of available transcriptomic datasets of wild type strains undergoing physiological changes related to the environmental signals integrated by these regulatory systems.

2. Materials and Methods

2.1. Datasets Analyzed in this Study

Ten published and accessible microarray and RNA-seq transcriptomic datasets for chosen gene knockout strains and experiments monitoring responses to changes in environmental conditions were obtained from the NCBI GenBank database (Table 1).

Table 1. Description of the transcriptomic datasets analyzed in this study.

Species	Strains and Culture Conditions	Type of Data	Accession Number	Reference
D. shibae	Time-resolved response to addition of AHL to $\Delta luxI_1$	RNA-seq	GSE122111	[13]
	Time resolved co-cultivation with *Prorocentrum minimum*	RNA-seq	GSE55371	[15]
	Knockouts of *ctrA*, *chpT*, and *cckA* in exponential and stationary phases of growth	Agilent dual-color microarray	GSE47451	[11]
	Knockouts of *fnrL*, *dnrD*, *dnrE*, and *dnrF* under aerobic conditions and 60 min after shift to anaerobic, denitrifying conditions	Agilent dual-color microarray	GSE93652	[3]
	Time-resolved growth of wild type and $\Delta luxI_1$ strains from OD_{600} 0.1 to stationary phase	Agilent dual-color microarray	GSE42013	[14]
	$\Delta luxI_2$ growth to OD_{600} of 0.4	RNA-seq	PRJEB20656	[30]
	Time-resolved shift of the wild type from aerobic to anaerobic growth conditions	Agilent single-color microarray	GSE47445	[31]
R. capsulatus	Knockouts of *regA*, *crtJ*, and *fnrL* in mid-exponential growth phase	RNA-seq	PRJNA357604	[32]
	Knockouts of *ctrA* and *cckA* in mid-exponential growth phase	Affymetrix microarray	GSE53636	[33]
	Knockout of *ctrA* during exponential and stationary growth phases	Affymetrix microarray	GSE18149	[34]

2.2. Processing and Analysis of Datasets

This study includes four different types of transcriptomic data (Table 1) that could not be processed and analyzed as one dataset. We therefore used the changes in transcript levels ($\log_2$ fold change) compared to the controls used in the respective studies (e.g., wild type or time point before changes in the environmental conditions) for each dataset. RNA-seq data from *D. shibae* (reads per gene) and *R. capsulatus* ($\log_2$ fold change) were obtained from the respective publications (Table 1).

Agilent microarray datasets were processed using the LIMMA package in R [35]. Background correction was performed with the "normexp" method and an offset of 10. Two-color microarrays were normalized with the "loess" method before quantile normalization. Signals/intensities from spots were averaged.

Affymetrix microarray datasets were processed using the R packages LIMMA, makecdfenv, and affy [35–37]. The CDF environment for GSE18149 was generated using the corresponding CDF file downloaded from GEO (accession GPL9198). Data were normalized with the rma function. A linear fit model was generated for comparison.

In order to analyze the CckA and ChpT regulons, thresholds were set that allowed definition of regulated and non-regulated genes. These thresholds were applied to the $\log_2$ fold change in transcript level values in the *cckA* and *chpT* deletion mutants. A gene was not considered regulated when its $\log_2$ fold change was between 1 and −1 while a $\log_2$ fold change value above 1 or below −1 indicated an affected gene. The analyzed genes were grouped based on published information about their functional categories as described (Supplementary Table S1).

3. Results

3.1. Overlap of the Crp/Fnr and CtrA Regulons in Dinorosebacter shibae

The possible interaction between the Crp/Fnr regulator and CtrA phosphorelay networks was first assessed using transcriptomic datasets for regulator deletion mutants. The changes of transcript levels of known Crp/Fnr- and CtrA-controlled traits revealed a strong overlap of both regulons, with the regulator-encoding genes themselves affected by losses of the other regulators (Figure 1). Under both aerobic and anaerobic conditions, the loss of *dnrD* or *dnrF* resulted in increased transcript levels of the CtrA phosphorelay, QS, flagellar motility, tad pilus, competence and recombination, gene transfer agent (GTA), *divL* and c-di-GMP signaling genes (Figure 1A). In all datasets, the GTA genes showed comparatively small changes in transcript levels (Figure 1), probably as a result of a small subpopulation actually expressing these genes [13]. Only the loss of *dnrF* led to a change in gene expression between aerobic and anaerobic conditions, since a greater increase in the transcript levels could be observed under anaerobic conditions for most of its regulon (Figure 1C). The loss of *fnrL* or *dnrE* resulted in increased transcript levels of *ctrA*, *cckA*, *chpT*, *luxI$_1$*, *luxR$_1$*, and *luxR$_2$* but had little to no effect on the downstream CtrA regulon (Figure 1B).

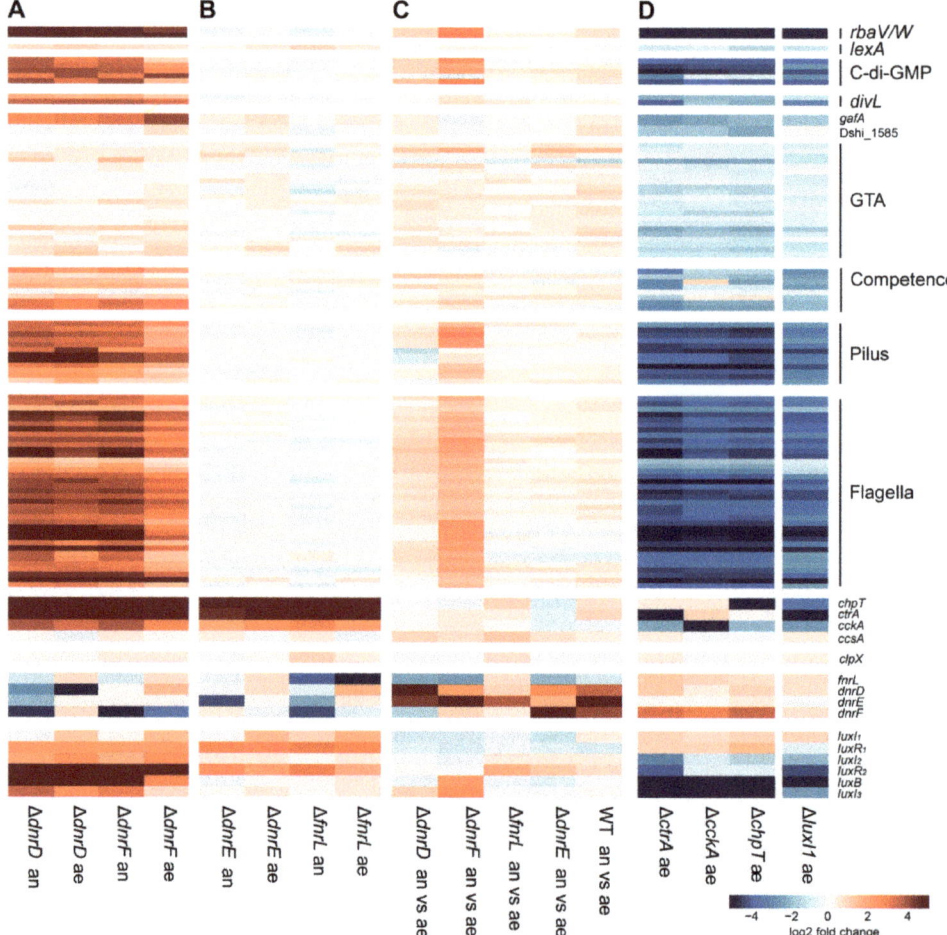

Figure 1. Transcriptomic data for genes in selected functional groups in different knockout strains. The four Crp/Fnr regulator knockouts were grown under aerobic (ae) or anaerobic (an) conditions. The log$_2$ fold changes compared to the respective wild type (WT) (**A**,**B**) or against themselves grown at different conditions, are shown (**C**). The CtrA phosphorelay and quorum sensing system knockouts were grown aerobically to the stationary phase and compared to the WT (**D**). The functional group assignments on the right are based on published information as described in Supplementary Table S1. Note: the Δ*luxI$_1$* strain retains a portion of the gene that can therefore result in mapped reads.

Almost all examined genes showed an opposite pattern in the CtrA phosphorelay and *luxI₁* mutants (Figure 1D) compared to *dnrD* and *dnrF* (Figure 1A). Most of the genes showed decreased transcript levels in strains lacking any of the CtrA phosphorelay genes, with the exceptions of the Crp/Fnr regulators where the largest increase was found for *dnrF* (Figure 1D). Loss of *luxI₁* resulted in increased transcript levels for *fnrL*, *dnrD*, and *dnrF*, but no changes were observed for *dnrE* (Figure 1D).

3.2. The Role of ChpT in Signal Integration

In *D. shibae*, deletion of neither *ctrA* nor *cckA* had an influence on expression of *chpT*, whereas the loss of either *ctrA* or *chpT* resulted in decreased expression of *cckA* (Figure 1D) [11]. However, all three CtrA phosphorelay component genes showed reduced transcript levels in the absence of the AHL synthase *luxI₁* (Figure 1D) [13], whereas loss of the Crp/Fnr regulators resulted in increased transcript levels of these genes (Figure 1A,B). Therefore, in contrast to *ctrA* and *ccka*, *chpT* is not regulated by the CtrA phosphorelay itself, but by other factors that can thereby control the phosphorylation state of CtrA. These findings also suggest that *chpT* transcription is regulated oppositely by QS and the Crp/Fnr regulators.

This is supported by binding site predictions for FnrL [3] that suggest it binds at the promoter of *chpT* and *clpX*, which encodes a protease known to cleave CtrA [3,38,39]. Deletion of *fnrL* strongly increased the expression of *chpT* but only resulted in minimal changes for *clpX* (Figure 1B). Binding site prediction for the Dnr regulators did not find any binding sites near *clpX* or the CtrA phosphorelay genes [3].

It was previously found that more genes were affected by the loss of *chpT* than *cckA* [11], suggesting ChpT regulates some genes independent of CckA and that a different kinase might regulate its activity and thereby affect downstream gene expression. Among the genes affected by the loss of *chpT* but not *cckA*, *dnrF* was the most upregulated gene during exponential growth while *lexA* and *recA* were among those most downregulated genes in both exponential and stationary phases (Figure 2). Although there was a small increase in transcript levels of *dnrF* in the *cckA* deletion strain during exponential growth, it did not pass the threshold we defined (see Materials and Methods). These findings suggest a link between *dnrF* and *chpT*.

Additional discrepancies between CckA and ChpT are apparent from their opposing effects on the *nap* gene cluster during exponential growth (Figure 2A), although this is not maintained in stationary phase (Figure 2B). In exponential phase, loss of *cckA* led to decreased transcript levels of the *nap* gene cluster, while the loss of *ctrA* and *chpT* led to increased levels (Figure 2A). This cluster is the only denitrification cluster activated by FnrL but repressed by the three Dnr regulators [3]. Interestingly, transcript levels of all four denitrification gene clusters were increased in the AHL synthase knockout ΔluxI₂ but were unaffected in ΔluxI₁ (Figure 2C).

3.3. Time-Resolved Evaluation of Environmental Changes and the Regulation of c-di-GMP Signaling Genes

Interactions between the networks in *D. shibae* were further analyzed using time-resolved transcriptomic datasets. These were collected following the switch from aerobic to anaerobic conditions in wild type cells (Figures 3A and 4A) [31], following the external addition of AHL autoinducer to the AHL synthase mutant ΔluxI₁ (Figures 3B and 4B) [13], and through the culture growth phases for ΔluxI₁ in the absence of AHL (Figure 4C) [14].

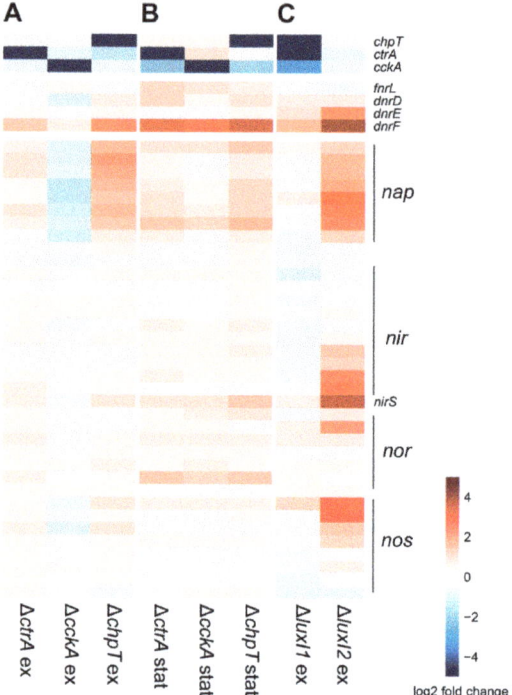

Figure 2. Comparison of CtrA phosphorelay, Crp/Fnr regulator, and denitrification gene expression control by CtrA phosphorelay and LuxI$_{1/2}$ synthases during exponential and stationary growth phases. Samples for the *ctrA*, *cckA*, and *chpT* knockout mutants were analyzed at mid-exponential (OD 0.4) (**A**) and stationary (six hours after onset of stationary phase) (**B**) phases of growth. The Δ*luxI$_1$* data were obtained during stationary phase, six hours after the onset of stationary phase, and the Δ*luxI$_2$* data were obtained during the mid-exponential growth phase (**C**).

Upon the shift to anaerobic conditions, all three *dnr* genes showed an immediate increase in transcript levels for 30 min and then stayed constant, whereas those of *fnrL* decreased (Figure 4A). These changes corresponded with increased transcript levels of the denitrification gene clusters, with the *nap* cluster showing a slightly different pattern than the *nir* and *nos* clusters (Figure 3A). Slight increases were observed for the c-di-GMP signaling, flagellar, tad pilus, and QS genes (Figure 3A). Four of the five c-di-GMP signaling genes showed increased transcript levels following the transfer to an anaerobic environment, whereas *dgc2* showed a slight decrease (Figure 4A).

The addition of AHL to the Δ*luxI$_1$* strain led to increased transcript levels for all CtrA- and QS-controlled genes (Figure 3B). This included the CtrA phosphorelay and c-di-GMP signaling genes, with *dgc2* showing the largest increase (Figure 4B). No effect was visible for the Crp/Fnr regulator-encoding genes (Figure 4B) and only a minor increase of the *nap* gene cluster was observed among the denitrification genes (Figure 3B).

Due to the increased transcript levels observed for CtrA regulon genes in the *dnrD* and *dnrF* deletion strains, it was expected that the same genes would also be decreased under anaerobic conditions. Instead, it turned out that the change from aerobic to anaerobic conditions (Figure 3) resulted in increased transcript levels for these genes. However, this increase was small, and effects were not observed for some genes that appeared to be controlled by the individual regulators based on the knockout transcriptomic data (Figure 1). This included the regulation of the CtrA phosphorelay genes by the Crp/Fnr regulators. Vice versa, loss of the CtrA phosphorelay genes indicated their repression of Crp/Fnr regulator gene expression (Figure 1D), but the contrary was observed in the respective physiological datasets where the Crp/Fnr regulators seem to be upregulated (Figure 3B). Notably however, in both physiological datasets, *dgc2* stands out as distinctly affected compared to other c-di-GMP signaling genes (Figure 4A,B). Also, in the non-induced Δ*luxI₁* culture, no influence of the QS null mutant on the Crp/Fnr regulators was observed, but the CtrA phosphorelay and c-di-GMP signaling genes were down-regulated (Figure 4C).

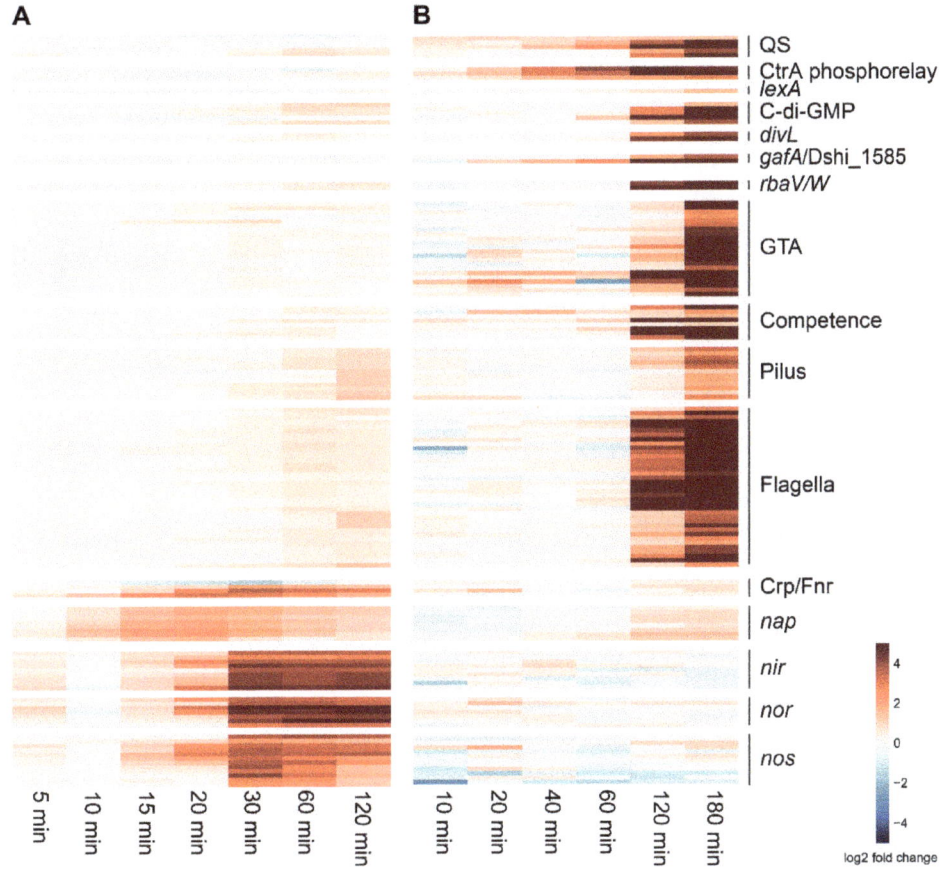

Figure 3. Time-resolved transcriptomic analysis for genes in selected groups in response to environmental changes. (**A**) Gene expression changes after the shift to anaerobic growth compared to aerobic conditions. (**B**) Gene expression after external addition of autoinducer 3-oxo C14 HSL to the QS synthase null mutant (Δ*luxI₁*).

Interestingly, in contrast to *fnrL*, *dgc2*, and *chpT*, the other Crp/Fnr regulators, c-di-GMP signaling, and CtrA phosphorelay genes all decreased at the onset of the stationary phase (Figure 4C). Moreover, analysis of the Crp/Fnr knockout data showed that the loss of *dnrF* or *dnrD* resulted in increased transcript levels of four of the c-di-GMP signaling genes under anaerobic growth conditions, with only *dgc2* being unaffected (Figure S1A). Loss of *luxI₁* and the CtrA phosphorelay genes resulted in decreased transcripts for all five genes (Figure S1B,C), although the effects on *dgc2* were smaller than for the other genes in the stationary phase (Figure S1C).

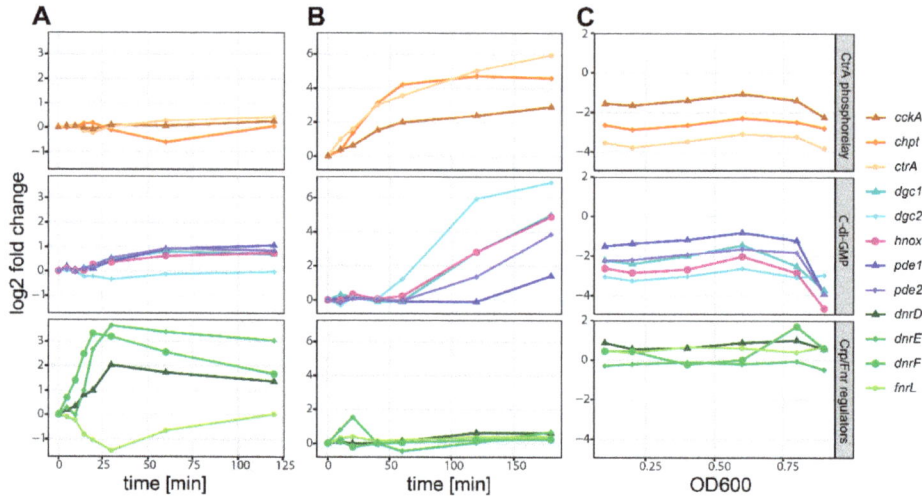

Figure 4. Time- and density-resolved transcript levels in three different conditions for three groups of regulators. The expression profiles of the CtrA phosphorelay genes (top), c-di-GMP signaling genes (middle), and four Crp/Fnr regulator-encoding genes (bottom) are plotted. The changes in transcript levels were monitored after the switch from aerobic to anaerobic growth over a time period of 120 min (**A**), after the external addition of autoinducer (3-oxo C14 HSL) to the synthase null mutant ($\Delta luxI_1$) over a period of 180 min (**B**), and during logarithmic (samples 1–5) and stationary (sample 6) phases of growth as determined by optical density (**C**).

3.4. Effects on the CtrA Regulon during Coculture of Dinoroseobacter shibae and Its Algal Host

In the two-phase interaction of *D. shibae* with its dinoflagellate host *Prorocentrum minimum* [14,40], a mutualistic growth phase (0 to 21 days of cocultivation) is followed by a pathogenic growth phase (21 to 30 days of cocultivation) that leads to death of the algae [15]. Analysis of the transcriptomic data of *D. shibae* during cocultivation showed an overall increase in the transcription for the CtrA regulon genes during the transition between the two phases (day 24 compared to day 18), followed by a decrease during the late-pathogenic phase, after 30 days (Supplementary Figure S2).

Of the CtrA phosphorelay genes, only *chpT* remained upregulated during the pathogenic interaction. Evaluation of the denitrification gene clusters showed strong variation among these genes (Supplementary Figure S2), likely arising from overall low expression levels, and this made it difficult to draw any conclusions.

3.5. RegA Activates the CtrA Regulon in Rhodobacter capsulatus

Next, we asked if the observed overlap between redox regulators and the CtrA phosphorelay system is conserved in another member of the family *Rhodobacteraceae*. For *R. capsulatus*, transcriptomic data were available for knockout mutants of *ctrA*, *cckA*, and the known redox regulator-encoding genes *fnrL*, *regA*, and *crtJ*. We identified three additional Crp/Fnr regulator-encoding genes in this bacterium based on blast searches (RCAP_rcp00107, RCAP_rcc01561, RCAP_rcc03255), but these genes showed no evidence of differential regulation in any of the analyzed datasets and we did not consider them further. A blast search also identified a homologue (RCAP_rcc02630) of the HNOX-encoding gene of *D. shibae* (Dshi_2815). This gene encodes a protein with a predicted heme nitric oxide/oxygen binding (HNOB) domain and is located adjacent to a c-di-GMP signaling gene (RCAP_rcc02629) that was recently demonstrated to affect GTA production and motility in *R. capsulatus* [41]. When bound to NO, the HNOX homologue in *D. shibae* inhibits the activity of the diguanylate cyclase Dgc1, which is encoded by the neighboring gene [23].

FnrL is the only Crp/Fnr regulator that has been studied in *R. capsulatus* [29]. Its loss did not result in any large changes in transcript levels for the examined traits under anaerobic phototrophic conditions (Figure 5), and the same was observed for the loss of *crtJ*, which encodes a transcription factor that controls numerous photosynthesis and cytochrome genes [32] (Figure 5). RegA is the response regulator of the RegB/A two-component system that controls photosynthesis, nitrogen and carbon fixation, denitrification, and respiration genes in response to oxygen availability [26]. In contrast to *fnrL* and *crtJ*, we found that the loss of *regA* resulted in a strong decrease in transcript levels of the CtrA regulon genes (Figure 5), indicating that RegA acts as a direct or indirect activator of these genes. Like the genes involved in regulation of photosynthesis and the change between aerobic/anaerobic lifestyle in *D. shibae*, loss of *regA* affected *chpT* the most among the CtrA phosphorelay genes in *R. capsulatus*. Loss of the CtrA phosphorelay genes had no influence on transcription of *fnrL*, *regA*, *regB*, or the other putative Crp/Fnr regulator-encoding genes (Supplementary Figure S3). A comparison of photosynthetic anaerobic growth and aerobic cultivation in *R. capsulatus* showed the CtrA-regulated traits have reduced transcript levels under anaerobic conditions (Figure 5).

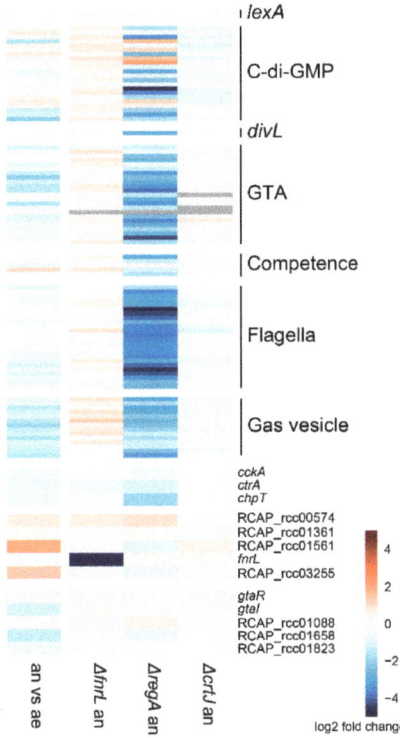

Figure 5. Effects of growth conditions and three regulator knockouts on the transcript levels of eight categorized groups of genes in *Rhodobacter capsulatus*. The microarray-based transcriptomic data for aerobic versus anaerobic growth in the wild type and for three mutants, *fnrL*, *regA*, and *crtJ*, compared to the wild type are shown.

4. Discussion

4.1. The Crp/Fnr and CtrA/QS Regulons Overlap in Dinoroseobacter shibae

Our analysis revealed an inverse regulatory crosstalk between the Crp/Fnr and CtrA systems in *D. shibae*. We found the denitrification gene clusters and Crp/Fnr regulator genes, especially *dnrF*, to be part of the CtrA phosphorelay and LuxI$_2$ regulons. *DnrE* was affected exclusively by loss of LuxI$_2$, whereas loss of LuxI$_1$ only had minor effects on *fnrL*, *dnrD*, and *dnrF* and no effect on *dnrE*. In addition to their regulation by LuxI$_1$, which signals cell density, the Crp/Fnr regulators integrate oxygen and NO levels and affect all three CtrA phosphorelay genes.

Until now, overlapping regulation by the Crp/Fnr and CtrA systems has only been noted in *D. shibae* for flagellar genes and *recA* [3,12,13], and to our knowledge this level of regulatory interaction has not been reported for alphaproteobacteria. However, a comparable connection between QS and Crp/Fnr regulators has been documented for the gammaproteobacterium *Pseudomonas aeruginosa* where the regulons of the FnrL homolog Anr and QS synthase LasR overlap. Here, denitrification genes are induced by Anr and inhibited by LasR. Additionally, in the absence of *lasR*, Anr regulates production of the QS molecule 4-hydroxy-2-alkylquinoline [42]. At the protein level, nitrite reductase (NirS), a flagellar protein (FliC), and the chaperone DnaK form a complex that influences flagellar formation and motility and thus creates a link between denitrification and motility [43]. In cystic fibrosis infections, *P. aeruginosa* is exposed to ambient conditions with low oxygen tension. The intracellular levels of c-di-GMP increase, which leads to biofilm formation. These conditions also lead to an increase in mutations in the QS transcriptional regulator-encoding gene *lasR*. As *lasR* deletion strains grow to higher cell densities and have higher denitrification rates, it has been suspected that these mutations increase the fitness of the population during infection [44–46].

Combined, these observations indicate that there may be a more widely conserved interaction of Crp/Fnr regulators and QS in proteobacteria. The CtrA phosphorelay is unique to alphaproteobacteria, indicating a potential independent evolution of this regulatory crosstalk in this lineage.

4.2. Inverse Control of the CtrA Regulon by RegA and Anaerobic Photosynthetic Growth Conditions in Rhodobacter capsulatus

In *R. capsulatus*, the regulons of the redox-responsive two-component system RegA/B [47] and the CtrA phosphorelay overlap. Interestingly, *chpT* stands out because it is the only CtrA phosphorelay gene that is regulated by RegA. Similar to Dnr and Fnr in *D. shibae*, RegA controls the expression of photosynthesis and respiration genes [26]. ChIP-seq with RegA identified binding sites adjacent to several genes also targeted by CtrA: RCAP_rcc02857 (a c-di-GMP signaling gene involved in GTA production) and its divergently transcribed neighbor (RCAP_rcc02856), RCAP_rcc02683 (a chemotaxis gene), and *dksA* (a *dnaK* suppressor gene) [34].

As in *D. shibae*, transcriptomic data from a *fnrL* deletion strain showed no effects on the CtrA-controlled traits outside of the CtrA phosphorelay genes themselves. However, ChIP-seq and in silico binding site predictions [29] suggest FnrL binding adjacent to *divL*, *dnaK*, *recA*, flagellar gene clusters, the RcGTA capsid protein-encoding gene, and c-di-GMP signaling genes (including those affecting RcGTA production [41]). Similarly, ChIP-seq with CrtJ [48], a regulator controlling expression of multiple genes involved in photosynthesis, also revealed a connection to the CtrA phosphorelay. Even though the observed transcript level changes in the *crtJ* mutant were small, binding was found adjacent to *ctrA*, *clpX*, a *luxR* family gene, *dnaA*, *spoT*, *ftsZ*, and the first gene in the GTA structural gene cluster (RCAP_rcc01682) under aerobic and anaerobic cultivation. Binding sites adjacent to *dnaK* and two flagellar genes (*flgB* and *flaA*) were identified under aerobic and anaerobic conditions, respectively.

In *D. shibae*, deletions of the Crp/Fnr regulator-encoding genes indicated an inhibition of the CtrA regulon, but the physiological changes detected by these regulators (switch from aerobic to anaerobic conditions) showed a tendency towards activation of the CtrA regulon. The same was observed for the deletion mutants of the CtrA phosphorelay components and their regulation of the Crp/Fnr regulator genes. In *R. capsulatus*, we could observe a similar pattern but in reverse for regulation of the CtrA regulon by RegA. While the *regA* knockout indicated activation of the CtrA regulon, the switch to anaerobic photosynthetic growth conditions showed an inhibition. This is probably indicative of a more complex interaction among these regulatory systems. However, the *regA* deletion transcriptomic data are supported by in vivo motility tests that showed reduced swimming ability of the Δ*regA* strain [26].

4.3. Integration of Crp/Fnr Regulation into the CtrA Phosphorelay and Regulon

In *D. shibae*, CtrA binding site predictions and expression data for *ctrA* and *cckA* suggest that CtrA directly regulates its own expression and that of *cckA*, but not *chpT* [13]. Therefore, *chpT* transcription must be regulated from outside of the CtrA phosphorelay and upstream of CtrA. Both, regulatory control of *chpT* and signal integration upstream of CtrA is known for LuxI$_1$ [11]. A similar situation might be possible for Crp/Fnr signal integration due to their regulation of *chpT* (Figure 6A). Since *chpT* is the only RegA-regulated CtrA phosphorelay gene in *R. capsulatus* (and it has a RegA binding site), it seems to play a central role here, too. However, there are also RegA binding sites associated with *clpX* and other genes of the CtrA regulon [26]. Interestingly, the Dnr/Fnr binding site in the *nosR2* promoter in *D. shibae* has the sequence 5′-TTAAC-N4-GTCAA-3′ [3], which shares a half-site binding motif with CtrA 5′-TTAAC-N5-GTTAAC-3′ [11]. Previously, comparison between transcriptional regulation and the presence of full and half-site motifs revealed the potential importance of half-site motifs for transcriptional control by CtrA in *R. capsulatus* [34]. Thus, CtrA and Fnr regulators might interact with some of the same/overlapping sequences (Figure 6B).

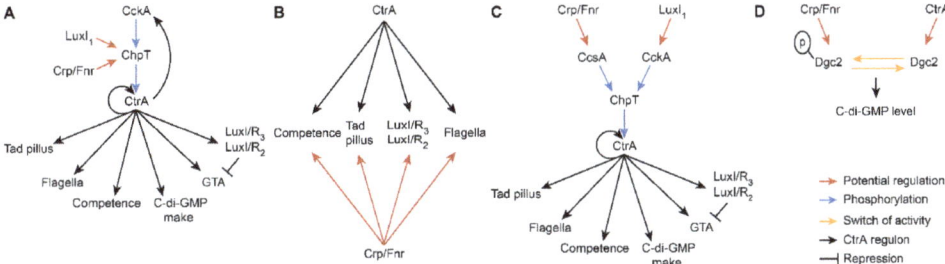

Figure 6. Possible mechanisms of integration of the Crp/Fnr and CtrA systems. (**A**) The LuxI$_1$ and Crp/Fnr signals could be integrated into the CtrA phosphorelay via *chpT* regulation, which does not happen via CckA or CtrA. (**B**) Shared binding site motifs for Crp/Fnr regulators and CtrA might allow direct integration of the NO/oxygen signal into the CtrA regulon. (**C**) An additional histidine kinase (CcsA) has been reported to phosphorylate ChpT in another bacterium, and this could integrate the Crp/Fnr signals and disconnect CckA from the integration. (**D**) Phosphorylation of the Dgc2 receiver domain likely regulates the enzyme's diguanylate cyclase activity and thereby alters the intracellular levels of c-di-GMP, which are known to affect the CtrA regulon.

A distinct role for ChpT is supported by the observation that loss of *chpT* or *ctrA* but not *cckA* results in decreased transcript levels of *dnrF*. It is possible that ChpT integrates signals from more than one kinase into its regulation of CtrA. To our knowledge, the only other instance of a histidine kinase affecting phosphorylation of CtrA via ChpT is CcsA from *Sphingomonas melonis* [49]. Potential homologues of CcsA are encoded in *D. shibae* (Dshi_1893) and *R. capsulatus* (RCAP_rcc02545), but effects on transcript levels of these genes were not observed in any of the analyzed datasets. This does not exclude their involvement but also does not allow us to draw further conclusions (Figure 6C).

4.4. Crp/Fnr Regulation of the CtrA Regulon is Largely Independent of Oxygen Tension

Among the Crp/Fnr regulators, only loss of the NO-sensing DnrF resulted in higher inhibition activity of the CtrA system under anaerobic conditions. In *P. aeruginosa*, swimming motility is controlled anaerobically and aerobically and it was suggested that NirS promotes motility in multiple ways, at the protein level or via signaling pathways, depending on oxygen availability [50]. Regulation of QS traits by both NO and oxygen was also found in the interaction of *Vibrio fischeri* with the light organ of its squid host. Here, NO released by the host's immune system regulates the symbionts' settlement via biofilm production while the host's control of oxygen availability regulates bacterial bioluminescence in a circadian manner [51–53]. However, since the Crp/Fnr knockout and physiological change transcriptomic data have opposite effects on the CtrA phosphorelay (inhibition indicated by the knockouts and activation by the shift to anaerobic growth), it is difficult to determine the role of oxygen on the CtrA phosphorelay. In *R. capsulatus*, the knockout transcriptomic data were supported by in vivo experiments, so if the knockout transcriptomic data also reflect the actual CtrA regulon in *D. shibae*, the Crp/Fnr regulators have an inhibitory effect on the CtrA phosphorelay and its regulon. It is known that *Dinoroseobacter* establishes a mutualistic symbiosis with its dinoflagellate host via the CtrA phosphorelay and by means of flagella. It is possible this interaction is repressed towards the end of an algal bloom when oxygen concentrations change, resulting in downregulation of flagella (and other CtrA-regulated traits) via Crp/Fnr regulation.

4.5. The Role of c-di-GMP

Multiple eukaryotic hosts are known to use NO for communication with microbial symbionts. In some of the characterized systems, NO is sensed by HNOX proteins, which then control c-di-GMP signaling proteins or histidine kinases encoded by genes adjacent to the HNOX-encoding gene. For example, in *Vibrio harveyi*, the HNOX-neighboring histidine kinase phosphorylates the QS transcription regulator LuxU [20], and in *D. shibae*, HNOX inhibits the c-di-GMP signaling enzyme Dgc1 [23]. However, *D. shibae* also has a second c-di-GMP synthesizing enzyme, Dgc2. During adaptation to anaerobic cultivation and at the onset of stationary phase, *dgc2* transcriptional patterns were similar to *chpT* and *fnrL*. The transcript levels of these three genes plateaued, whereas those of the other c-di-GMP signaling, CtrA phosphorelay and Crp/FnrL genes decreased. A unique regulation of *dgc2* was also observed in the *dnrF*, *dnrD*, *cckA*, and *chpT* knockout strains. Thus, both networks (Crp/Fnr and CtrA phosphorelay) regulate *dgc2* and affect its expression in a similar manner as a response to the onset of stationary phase.

The role of *dgc2* in the CtrA phosphorelay and FnrL networks and how it might connect both remain to be clarified. For example, it is possible that phosphorylation of the receiver domain of Dgc2 regulates its c-di-GMP synthase activity. As a result, regulation by the Crp/Fnr or CtrA phosphorelay systems could have different effects on the shared traits (Figure 6D).

5. Conclusions

In this study we show that regulation of the CtrA regulon, including traits related to phenotypic heterogeneity, is additionally controlled by the aerobic–anerobic regulators Crp/Fnr in *D. shibae* and by FnrL/RegA in *R. capsulatus*. This finding is especially important for the understanding of the metabolically flexible lifestyles of these bacteria. The analysis of the available transcriptomic datasets revealed multiple possible integration sites of the Crp/Fnr signal into the CtrA phosphorelay, but a final explanation is still elusive based on these data. Nevertheless, this investigation provides the first insights into the integration of a second environmental signal into the CtrA phosphorelay and demonstrates a strong transcriptional connection between QS, CtrA-regulated traits, and Crp/Fnr regulators in alphaproteobacteria, which has an interesting parallel with QS and Crp/Fnr regulators in a second class of proteobacteria. To our knowledge, *D. shibae* and *R. capsulatus* are the first two organisms where both Dnr and HNOX NO-sensor proteins have been studied. Further investigation is necessary to clarify the interaction between the CtrA phosphorelay and the Crp/Fnr regulators. For example, it would be helpful to confirm if an additional kinase is indeed regulating ChpT in these bacteria.

Supplementary Materials: The following are available online at http://www.mdpi.com/2076-2607/8/4/562/s1: Figure S1. Transcript level changes of *D. shibae* c-di-GMP signaling genes; Figure S2. Comparison of changes in transcript levels during different stages of the "Jekyll and Hyde" interaction between *Dinoroseobacter shibae* and the dinoflagellate *Prorocentrum minimum*; Figure S3. Transcript level changes of various FnrL- and RegA-related genes in CtrA phosphorelay mutants during exponential and stationary phases of growth in *Rhodobacter capsulatus*; Table S1. Assignment of genes into functional categories.

Author Contributions: Conceptualization, S.K.; methodology, formal analysis, and data curation S.K.; writing—original draft preparation, S.K.; writing—review and editing, A.S.L.; visualization, S.K.; supervision, A.S.L.; funding acquisition, A.S.L. All authors have read and agree to the published version of the manuscript.

Funding: This research was funded by the Natural Sciences and Engineering Research Council of Canada (NSERC) (grant number 341561 to A.S.L.). S.K. was supported in part by funding from the Memorial University of Newfoundland School of Graduate Studies.

Conflicts of Interest: The authors declare no conflict of interest. The funders had no role in the design of the study; in the collection, analyses, or interpretation of data; in the writing of the manuscript, or in the decision to publish the results.

References

1. Wagner-Döbler, I.; Ballhausen, B.; Berger, M.; Brinkhoff, T.; Buchholz, I.; Bunk, B.; Cypionka, H.; Daniel, R.; Drepper, T.; Gerdts, G.; et al. The complete genome sequence of the algal symbiont *Dinoroseobacter shibae*: A hitchhiker's guide to life in the sea. *ISME J.* **2010**, *4*, 61–77. [CrossRef] [PubMed]
2. Pitcher, G.C.; Probyn, T.A. Suffocating phytoplankton, suffocating waters-red tides and anoxia. *Front. Mar. Sci.* **2016**, *3*, 186. [CrossRef]
3. Ebert, M.; Laaß, S.; Thürmer, A.; Roselius, L.; Eckweiler, D.; Daniel, R.; Härtig, E.; Jahn, D. FnrL and three Dnr regulators are used for the metabolic adaptation to low oxygen tension in *Dinoroseobacter shibae*. *Front. Microbiol.* **2017**, *8*, 642. [CrossRef] [PubMed]
4. Körner, H.; Sofia, H.J.; Zumft, W.G. Phylogeny of the bacterial superfamily of Crp-Fnr transcription regulators: Exploiting the metabolic spectrum by controlling alternative gene programs. *FEMS Microbiol. Rev.* **2003**, *27*, 559–592. [CrossRef]
5. Leimeister-Wachter, M.; Domann, E.; Chakraborty, T. The expression of virulence genes in *Listeria monocytogenes* is thermoregulated. *J. Bacteriol.* **1992**, *174*, 947–952. [CrossRef]
6. Volbeda, A.; Darnault, C.; Renoux, O.; Nicolet, Y.; Fontecilla-Camps, J.C. The crystal structure of the global anaerobic transcriptional regulator FNR explains its extremely fine-tuned monomer-dimer equilibrium. *Sci. Adv.* **2015**, *1*, e1501086. [CrossRef]
7. Poole, R.K.; Anjum, M.F.; Membrillo-Hernandez, J.; Kim, S.O.; Hughes, M.N.; Stewart, V. Nitric oxide, nitrite, and Fnr regulation of *hmp* (flavohemoglobin) gene expression in *Escherichia coli* K-12, *J. Bacteriol.* **1996**, *178*, 5487–5492. [CrossRef]
8. Beliaev, A.S.; Beliaev, A.S.; Thompson, D.K.; Thompson, D.K.; Fields, M.W.; Fields, M.W.; Wu, L.; Wu, L.; Lies, D.P.; Lies, D.P.; et al. Microarray transcription profiling of a *Shewanella oneidensis etrA* mutant. *J. Bacteriol.* **2002**, *184*, 4612–4616. [CrossRef]
9. Ebert, M.; Schweyen, P.; Bröring, M.; Laass, S.; Härtig, E.; Jahn, D. Heme and nitric oxide binding by the transcriptional regulator DnrF from the marine bacterium *Dinoroseobacter shibae* increases *napD* promoter affinity. *J. Biol. Chem.* **2017**, *292*, 15468–15480. [CrossRef]
10. Poncin, K.; Gillet, S.; De Bolle, X. Learning from the master: Targets and functions of the CtrA response regulator in *Brucella abortus* and other alpha-proteobacteria. *FEMS Microbiol. Rev.* **2018**, *019*, 500–513.
11. Wang, H.; Ziesche, L.; Frank, O.; Michael, V.; Martin, M.; Petersen, J.; Schulz, S.; Wagner-Döbler, I.; Tomasch, J. The CtrA phosphorelay integrates differentiation and communication in the marine alphaproteobacterium *Dinoroseobacter shibae*. *BMC Genomics* **2014**, *15*, 130. [CrossRef] [PubMed]
12. Motherway, C.; Zomer, A.; Leahy, S.C.; Reunanen, J.; Bottacini, F.; Claesson, M.J.; Flynn, K.; Casey, P.G.; Antonio Moreno Munoz, J.; Kearney, B.; et al. Functional genome analysis of *Bifidobacterium breve* UCC2003 reveals type IVb tight adherence (Tad) pili as an essential and conserved host-colonization factor. *Proc. Natl. Acad. Sci. USA* **2011**, *108*, 11217–11222. [CrossRef]
13. Koppenhöfer, S.; Wang, H.; Scharfe, M.; Kaever, V.; Wagner-Döbler, I.; Tomasch, J. Integrated transcriptional regulatory network of quorum sensing, replication control, and SOS response in *Dinoroseobacter shibae*. *Front. Microbiol.* **2019**, *10*, 803. [CrossRef] [PubMed]
14. Patzelt, D.; Wang, H.; Buchholz, I.; Rohde, M.; Gröbe, L.; Pradella, S.; Neumann, A.; Schulz, S.; Heyber, S.; Münch, K.; et al. You are what you talk: Quorum sensing induces individual morphologies and cell division modes in *Dinoroseobacter shibae*. *ISME J.* **2013**, *7*, 2274–2286. [CrossRef] [PubMed]

15. Wang, H.; Tomasch, J.; Michael, V.; Bhuju, S.; Jarek, M.; Petersen, J.; Wagner-Döbler, I. Identification of genetic modules mediating the Jekyll and Hyde interaction of *Dinoroseobacter shibae* with the dinoflagellate *Prorocentrum minimum*. Front. Microbiol. **2015**, *6*, 1262. [CrossRef] [PubMed]
16. Zumft, W.G. Nitric oxide signaling and NO dependent transcriptional control in bacterial denitrification by members of the FNR-CRP regulator family, J. Mol. Microbiol. Biotechnol. **2002**, *4*, 277–286.
17. Crane, B.R.; Sudhamsu, J.; Patel, B.A. Bacterial nitric oxide synthases. Annu. Rev. Biochem **2010**, *79*, 445–470. [CrossRef]
18. Rao, M.; Smith, B.C.; Marletta, M.A. Nitric oxide mediates biofilm formation and symbiosis in *Silicibacter* sp. strain TrichCH4B. *mBio* **2015**, *6*, e00206-15. [CrossRef]
19. Wang, Y.; Ruby, E.G. The roles of NO in microbial symbioses. Cell. Microbiol. **2013**, *13*, 518–526. [CrossRef]
20. Henares, B.M.; Higgins, K.E.; Boon, E.M. Discovery of a nitric oxide responsive quorum sensing circuit in *Vibrio harveyi*. ACS Chem. Biol **2012**, *7*, 28. [CrossRef]
21. Hossain, S.; Heckler, I.; Boon, E.M. Discovery of a nitric oxide responsive quorum sensing circuit in *Vibrio cholerae*. ACS Chem. Biol **2018**, *13*, 56. [CrossRef] [PubMed]
22. Nisbett, L.-M.; Boon, E.M. Nitric oxide regulation of H-NOX signaling pathways in bacteria. *Biochemistry* **2016**, *55*, 32. [CrossRef] [PubMed]
23. Bedrunka, P.; Olbrisch, F.; Rüger, M.; Zehner, S.; Frankenberg-Dinkel, N. Nitric oxide controls c-di-GMP turnover in *Dinoroseobacter shibae*. Microbiology **2018**, *164*, 1405–1415. [CrossRef] [PubMed]
24. Lang, A.S.; Beatty, J.T. Genetic analysis of a bacterial genetic exchange element: The gene transfer agent of *Rhodobacter capsulatus*. Proc. Natl. Acad. Sci. **2000**, *97*, 859–864. [CrossRef] [PubMed]
25. Lang, A.S.; Beatty, J.T. A bacterial signal transduction system controls genetic exchange and motility a bacterial signal transduction system controls genetic exchange and motility. J. Bacteriol. **2002**, *184*, 913–918. [CrossRef]
26. Schindel, H.S.; Bauer Biochemistry, C.E.; Bauer, C.E. The RegA regulon exhibits variability in response to altered growth conditions and differs markedly between *Rhodobacter* species. Microb. Genomics **2016**, *2*, e000081. [CrossRef]
27. Smart, J.L.; Willett, J.W.; Bauer, C.E. Regulation of *hem* gene expression in *Rhodobacter capsulatus* by redox and photosystem regulators RegA, CrtJ, FnrL, and AerR. J. Mol. Biol. **2004**, *342*, 1171–1186. [CrossRef]
28. Ponnampalam, S.N.; Bauer, C.E. DNA binding characteristics of RegA. J. Biol. Chem. **1998**, *273*, 18509–18513.
29. Kumka, J.E.; Bauer, C.E. Analysis of the FnrL regulon in *Rhodobacter capsulatus* reveals limited regulon overlap with orthologues from *Rhodobacter sphaeroides* and *Escherichia coli*. BMC Genomics **2015**, *16*. [CrossRef]
30. Tomasch, J.; Wang, H.; Hall, A.T.K.; Patzelt, D.; Preuße, M.; Brinkmann, H.; Bhuju, S.; Jarek, M.; Geffers, R.; Lang, A.S. Packaging of *Dinoroseobacter shibae* DNA into Gene Transfer Agent particles is not random. Genome Biol. Evol. **2018**, *10*, 359–369. [CrossRef]
31. Laass, S.; Kleist, S.; Bill, N.; Drüppel, K.; Kossmehl, S.; Wöhlbrand, L.; Rabus, R.; Klein, J.; Rohde, M.; Bartsch, A.; et al. Gene regulatory and metabolic adaptation processes of *Dinoroseobacter shibae* DFL12 T during oxygen depletion. J. Biol. Chem. **2014**, *289*, 13219–13231. [CrossRef] [PubMed]
32. Kumka, J.E.; Schindel, H.; Fang, M.; Zappa, S.; Bauer, C.E. Transcriptomic analysis of aerobic respiratory and anaerobic photosynthetic states in *Rhodobacter capsulatus* and their modulation by global redox regulators RegA, FnrL and CrtJ. Microb. Genomics **2017**, *3*, e000125. [CrossRef] [PubMed]
33. Peña-Castillo, L.; Mercer, R.G.; Gurinovich, A.; Callister, S.J.; Wright, A.T.; Westbye, A.B.; Beatty, J.T.; Lang, A.S. Gene co-expression network analysis in *Rhodobacter capsulatus* and application to comparative expression analysis of *Rhodobacter sphaeroides*. BMC Genomics **2014**, *15*, 730. [CrossRef] [PubMed]
34. Mercer, R.G.; Callister, S.J.; Lipton, M.S.; Pasa-tolic, L.; Strnad, H.; Paces, V.; Beatty, J.T.; Lang, A.S. Loss of the response regulator CtrA causes pleiotropic effects on gene expression but does not affect growth phase regulation in *Rhodobacter capsulatus*. J. Bacteriol. **2010**, *192*, 2701–2710. [CrossRef]
35. Smyth, G.K. Limma: Linear models for microarray data. In *Bioinformatics and Computational Biology Solutions Using R and Bioconductor*; Gentleman, R., Carey, V.J., Huber, W., Irizarry, R.A., Dudoit, S., Eds.; Springer: New York, NY, USA, 2005; pp. 397–420.
36. Irizarry, R.A.; Gautier, L.; Huber, W.; Bolstad, B.M. makecdfenv: CDF Environment Maker. R Packag. Version 1.62.0. 2019. Available online: https://rdrr.io/bioc/makecdfenv/ (accessed on 1 March 2019).
37. Gautier, L.; Cope, L.; Bolstad, B.M.; Irizarry, R.A. Affy-analysis of affymetrix GeneChip data at the probe level. *Bioinformatics* **2004**, *20*, 307–315. [CrossRef]

38. Westbye, A.; Kater, L.; Wiesmann, C.; Ding, H.; Yip, C.; Beatty, J. The protease ClpXP and the PAS-domain protein DivL regulate CtrA and gene transfer agent production in *Rhodobacter capsulatus*. *Appl. Environ. Microbiol.* **2018**, *84*, e00275-18. [CrossRef]
39. Brilli, M.; Fondi, M.; Fani, R.; Mengoni, A.; Ferri, L.; Bazzicalupo, M.; Biondi, E.G. The diversity and evolution of cell cycle regulation in alpha-proteobacteria: A comparative genomic analysis. *BMC systems biology* **2010**, *4*, 52. [CrossRef]
40. Wang, H.; Tomasch, J.; Jarek, M.; Wagner-Döbler, I. A dual-species co-cultivation system to study the interactions between *Roseobacters* and dinoflagellates. *Front. Microbiol.* **2014**, *5*, 311. [CrossRef]
41. Pallegar, P.; Peña-Castillo, L.; Evan, L.; Mark, G.; Lang, A.S. Cyclic-di-GMP-mediated regulation of gene transfer and motility in *Rhodobacter capsulatus*. *J. Bacteriol.* **2020**, *202*, e00554-19. [CrossRef]
42. Hammond, J.H.; Dolben, E.F.; Smith, T.J.; Bhuju, S.; Hogan, D.A. Links between Anr and quorum sensing in *Pseudomonas aeruginosa* biofilms. *J. Bacteriol.* **2015**, *197*, 2810–2820. [CrossRef] [PubMed]
43. Heylen, K.; Gevers, D.; Vanparys, B.; Wittebolle, L.; Geets, J.; Boon, N.; De Vos, P. The incidence of *nirS* and *nirK* and their genetic heterogeneity in cultivated denitrifiers. *Environ. Microbiol.* **2006**, *8*, 2012–2021. [CrossRef] [PubMed]
44. Wang, Y.; Gao, L.; Rao, X.; Wang, J.; Yu, H.; Jiang, J.; Zhou, W.; Wang, J.; Xiao, Y.; Li, M.; et al. Characterization of *lasR*-deficient clinical isolates of *Pseudomonas aeruginosa*. *Sci. Rep.* **2018**, *8*, 13344. [CrossRef] [PubMed]
45. Barraud, N.; Hassett, D.J.; Hwang, S.-H.; Rice, S.A.; Kjelleberg, S.; Webb, J.S. Involvement of nitric oxide in biofilm dispersal of *Pseudomonas aeruginosa*. *J. Bacteriol.* **2006**, *188*, 7344–7353. [CrossRef]
46. Toyofuku, M.; Nomura, N.; Fujii, T.; Takaya, N.; Maseda, H.; Sawada, I.; Nakajima, T.; Uchiyama, H. Quorum sensing regulates denitrification in *Pseudomonas aeruginosa* PAO1. *J. Bacteriol.* **2007**, *189*, 4969–4972. [CrossRef]
47. Elsen, S.; Swem, L.R.; Swem, D.L.; Bauer, C.E. RegB/RegA, a highly conserved redox-responding global two-component regulatory system. *Microbiol. Mol. Biol. Rev.* **2004**, *68*, 263–279. [CrossRef]
48. Cheng, Z.; Li, K.; Hammad, L.A.; Karty, J.A.; Bauer, C.E. Vitamin B12 regulates photosystem gene expression via the CrtJ antirepressor AerR in *Rhodobacter capsulatus*. *Mol. Microbiol.* **2014**, *91*, 649–664. [CrossRef]
49. Francez-Charlot, A.; Kaczmarczyk, A.; Vorholt, J.A. The branched CcsA/CckA-ChpT-CtrA phosphorelay of *Sphingomonas melonis* controls motility and biofilm formation. *Mol. Microbiol.* **2015**, *97*, 47–63. [CrossRef]
50. Cutruzzolà, F.; Frankenberg-Dinkel, N. Origin and impact of nitric oxide in *Pseudomonas aeruginosa* biofilms. *J. Bacteriol.* **2015**, *198*, 55–65. [CrossRef]
51. Boettcher, K.J.; Ruby, E.G.; Mcfall-Ngai, M.J. Bioluminescence in the symbiotic squid *Euprymna scolopes* is controlled by a daily biological rhythm. *J. Comp. Physiol. A* **1996**, *179*, 65–73. [CrossRef]
52. Wang, Y.; Dufour, Y.S.; Carlson, H.K.; Donohue, T.J.; Marletta, M.A.; Ruby, E.G. H-NOX-mediated nitric oxide sensing modulates symbiotic colonization by *Vibrio fischeri*. *Proc. Natl. Acad. Sci. USA* **2010**, *107*, 8375–8380. [CrossRef] [PubMed]
53. Bouchard, J.N.; Yamasaki, H. Heat stress stimulates nitric oxide production in *Symbiodinium microadriaticum*: A possible linkage between nitric oxide and the coral bleaching phenomenon. *Plant. Cell Physiol.* **2008**, *49*, 641–652. [CrossRef] [PubMed]

© 2020 by the authors. Licensee MDPI, Basel, Switzerland. This article is an open access article distributed under the terms and conditions of the Creative Commons Attribution (CC BY) license (http://creativecommons.org/licenses/by/4.0/).

Article

Characterization of the Aerobic Anoxygenic Phototrophic Bacterium *Sphingomonas* sp. AAP5

Karel Kopejtka [1], Yonghui Zeng [1,2], David Kaftan [1,3], Vadim Selyanin [1], Zdenko Gardian [3,4], Jürgen Tomasch [5,†], Ruben Sommaruga [6] and Michal Koblížek [1,*]

[1] Centre Algatech, Institute of Microbiology, Czech Academy of Sciences, 379 81 Třeboň, Czech Republic; kopejk00@alga.cz (K.K.); marinezeng@gmail.com (Y.Z.); kaftan@alga.cz (D.K.); selyaninvv@gmail.com (V.S.)
[2] Department of Plant and Environmental Sciences, University of Copenhagen, Thorvaldsensvej 40, 1871 Frederiksberg C, Denmark
[3] Faculty of Science, University of South Bohemia, 370 05 České Budějovice, Czech Republic; zdenogardian@gmail.com
[4] Institute of Parasitology, Biology Centre, Czech Academy of Sciences, 370 05 České Budějovice, Czech Republic
[5] Research Group Microbial Communication, Technical University of Braunschweig, 38106 Braunschweig, Germany; Juergen.Tomasch@helmholtz-hzi.de
[6] Laboratory of Aquatic Photobiology and Plankton Ecology, Department of Ecology, University of Innsbruck, 6020 Innsbruck, Austria; Ruben.Sommaruga@uibk.ac.at
* Correspondence: koblizek@alga.cz
† Present Address: Department of Molecular Bacteriology, Helmholtz-Centre for Infection Research, 38106 Braunschweig, Germany.

Citation: Kopejtka, K.; Zeng, Y.; Kaftan, D.; Selyanin, V.; Gardian, Z.; Tomasch, J.; Sommaruga, R.; Koblížek, M. Characterization of the Aerobic Anoxygenic Phototrophic Bacterium *Sphingomonas* sp. AAP5. *Microorg* **2021**, *9*, 768. https://doi.org/10.3390/microorganisms9040768

Academic Editor: Johannes F. Imhoff

Received: 9 March 2021
Accepted: 5 April 2021
Published: 6 April 2021

Publisher's Note: MDPI stays neutral with regard to jurisdictional claims in published maps and institutional affiliations.

Copyright: © 2021 by the authors. Licensee MDPI, Basel, Switzerland. This article is an open access article distributed under the terms and conditions of the Creative Commons Attribution (CC BY) license (https://creativecommons.org/licenses/by/4.0/).

Abstract: An aerobic, yellow-pigmented, bacteriochlorophyll *a*-producing strain, designated AAP5 (=DSM 111157=CCUG 74776), was isolated from the alpine lake Gossenköllesee located in the Tyrolean Alps, Austria. Here, we report its description and polyphasic characterization. Phylogenetic analysis of the 16S rRNA gene showed that strain AAP5 belongs to the bacterial genus *Sphingomonas* and has the highest pairwise 16S rRNA gene sequence similarity with *Sphingomonas glacialis* (98.3%), *Sphingomonas psychrolutea* (96.8%), and *Sphingomonas melonis* (96.5%). Its genomic DNA G + C content is 65.9%. Further, in silico DNA-DNA hybridization and calculation of the average nucleotide identity speaks for the close phylogenetic relationship of AAP5 and *Sphingomonas glacialis*. The high percentage (76.2%) of shared orthologous gene clusters between strain AAP5 and *Sphingomonas paucimobilis* NCTC 11030^T, the type species of the genus, supports the classification of the two strains into the same genus. Strain AAP5 was found to contain $C_{18:1}\omega7c$ (64.6%) as a predominant fatty acid (>10%) and the polar lipid profile contained phosphatidylglycerol, diphosphatidylglycerol, phosphatidylethanolamine, sphingoglycolipid, six unidentified glycolipids, one unidentified phospholipid, and two unidentified lipids. The main respiratory quinone was ubiquinone-10. Strain AAP5 is a facultative photoheterotroph containing type-2 photosynthetic reaction centers and, in addition, contains a xathorhodopsin gene. No CO_2-fixation pathways were found.

Keywords: aerobic anoxygenic phototrophic bacteria; bacteriochlorophyll a; photosynthesis genes; rhodopsin; Sphingomonadaceae

1. Introduction

The genus *Sphingomonas* (Alphaproteobacteria) was originally proposed by Yabuuchi and coworkers [1] as a genus accommodating Gram-negative, strictly aerobic, non-sporulating, non-motile or motile, non-fermenting, chemoheterotrophic bacteria [2,3]. Later, this genus was divided into four genera and genus *Sphingomonas* was redefined in *sensu stricto* [4]. Over time, several photoheterotrophic representatives of *Sphingomonas* were cultivated [5–8]. *Sphingomonas* are found in a wide range of environmental niches, such as soils [6,9,10], fresh and marine waters [11–13], plants [14,15], airborne dust [16,17], and clinical samples [1,18,19]. Some representatives of *Sphingomonas* have a potential for biotechnological applications [20–22].

We previously isolated a novel *Sphingomonas* sp. strain designated AAP5 from the alpine lake Gossenköllesee located in the Tyrolean Alps, Austria. This aerobic yellow-pigmented strain contains bacteriochlorophyll-containing reaction centers [23].

Anoxygenic photosynthesis is relatively common among members of the order Sphingomonadales. Indeed, the first cultured aerobic anoxygenic phototrophic (AAP) bacterium was *Erythrobacter longus* isolated from the Bay of Tokyo [24]. Many AAP species have been cultured from freshwater, namely from the genera *Porphyrobacter*, *Erythromicrobium*, *Erythromonas*, *Sandarakinorhabdus*, and *Blastomonas* [25]. However, the unique feature of the strain AAP5 is that, together with genes for bacterial reaction center it contains also gene for another light harvesting protein xanthorhodopsin [23].

Here, we report the detailed phenotypic, phylogenetic, genomic, physiological, and biochemical characterization of the AAP5 strain. Furthermore, we compared it with its closest relative, *Sphingomonas glacialis*, and with the type species of the genus, *S. paucimobilis*.

2. Materials and Methods

Sampling site and strain isolation. The sampling was conducted in the clear alpine lake Gossenköllesee, Tyrolean Alps, Austria in September 2012. The lake is situated in a siliceous catchment area at 2417 m above sea level (47.2298° N, 11.0140° E). Details on sampling site and sampling procedure were described previously [12].

1 µl of the lake water sample was diluted into 100 µl of sterile half-strength R2A medium, and the dilution spread onto half-strength standard R2A agar plates (DSMZ medium 830). The plates were incubated aerobically at 25 °C under 12-h-light/dark cycles until colonies were visible, which were then screened for the presence of bacteriochlorophyll *a* (BChl *a*) using an infrared (IR) imaging system [26]. IR positive colonies were repeatedly streaked onto new agar plates until pure cultures were obtained.

Cultivation conditions. For all analyses conducted, the strain was grown either on R2A solid medium (DSMZ medium 830) or in R2A liquid medium. Cultures were incubated aerobically in 100 mL of the liquid medium in 250-mL flasks with cotton plugs on an orbital shaker (150 RPM). Illumination was provided by a bank of Dulux L 55W/865 luminescent tubes (Osram, Germany, spectral temperature of 6500 K) delivering ca. 100 µmol photons $m^{-2} s^{-1}$. Unless stated otherwise, the cultures were grown under 12-h dark/12-h light regime and at 22°C. The growth of cultures was followed by turbidity measurements at 650 nm.

Microscopy. Samples for epifluorescence microscopy were diluted 1000-fold in a sterile medium, fixed with sterile-filtered formalin to a final concentration of 1%, filtered onto white polycarbonate filters (Nuclepore, pore size 0.2 µm, diameter 25 mm, Whatman) and stained with 4′,6-diamidino-2-phenylindole at final concentration of 1 mg l^{-1} [27]. The cells were visualized under UV/blue excitation emission, and the autofluorescence of BChl *a* was visualized under white light/IR emission, using a Zeiss Axio Imager.D2 microscope equipped with a Plan-Apochromat 63x/1.46 Oil Corr objective, a Hamamatsu EMCCD camera C9100-02Min and Collibri2 LED illumination, as described previously [28].

Samples for transmission electron microscopy (TEM) and scanning electron microscopy (SEM) were fixed with 2.5% glutaraldehyde in 0.1 M phosphate buffer (pH = 7.2) for 2 days at 4°C. TEM samples were post-fixed in osmium tetroxide for 2 h, at 4°C, washed, dehydrated through an acetone series and embedded in Spurr's resin. A series of ultrathin sections were cut using a Leica UCT ultramicrotome (Leica Microsystems), counterstained with uranyl acetate and lead citrate, then examined in a JEOL TEM 1010 operated at 80 kV. SEM samples were dehydrated through an acetone series and dried by means of a critical point dryer CPD 2 (Pelco TM). Dry samples were attached to an aluminum target by means of carbon tape, coated with gold using a sputter coater E5100 (Polaron Equipment Ltd.) and examined with JEOL SEM JSM 7401F. Images were digitally recorded for the determination of morphological parameters.

Samples for atomic force microscopy (AFM) were resuspended in a buffer containing 20 mM Tris-HCl pH 8.0, 50 mM NaCl and adsorbed onto a clean glass coverslip functionalized by Corning™ Cell-Tak Cell and Tissue Adhesive (Corning Inc, USA). Cells were imaged by NanoWizard4 BioAFM (Bruker, USA) atop of an inverted optical microscope (IX73P2F, Olympus, Japan) placed on active vibration isolation system (Halcyonics) inside a custom-made acoustic enclosure. Cells were imaged with lever 3 of qp-BioAC AFM probe (Nanosensors, Switzerland). Cantilevers were calibrated using thermal noise, non-contact Sader method [29] providing resonant frequency f_o = 8.94 kHz and spring constant k = 0.076 N m^{-1}. Quantitative Imaging maps (5 × 5 µm^2, 10 × 10 µm^2) were recorded with resolution of 256 × 256 pixels^2. The force-distance curves recorded over the sample's surface were baseline corrected and the vertical tip position was estimated by fitting the position of the contact point. The Young's Modulus was calculated by fitting the processed curves using the Herz/Sneddon model [30,31] according to Rico and coworkers [32].

Phylogenetic analyses. The 16S rRNA gene sequence (GenBank accession number MW410774) of strain AAP5 was retrieved from its genome sequence (GenBank accession number GCA_004354345.1). Reference sequences were obtained either from the SILVA database [33] or NCBI GenBank (May 2020), and aligned using ClustalW [34]. Ambiguously aligned regions and gaps were manually excluded from further analysis. The 16S rRNA phylogenetic tree was computed using both neighbor-joining (NJ) [35] and maximum likelihood (ML) [36] algorithms included in the MEGA 6.06 software [37]. The Tamura-Nei model [38] was used for inferring the NJ tree. The ML tree was constructed using GTR nucleotide substitution model [39]. A uniform rate of nucleotide substitution was used. For the PufLM concatenated tree, amino acid sequences were retrieved from GenBank™ (December 2020) and aligned using ClustalX version 2.1. Sites containing gaps and ambiguously aligned regions were manually excluded. Amino acid sequence alignments of PufL and PufM were concatenated with Geneious version 8.1.2 (Biomatters Ltd.). Phylogenomic trees were inferred by MEGA 6.0 software using the ML algorithm with LG model [40] and 1000 bootstrap replicates.

In silico DNA-DNA hybridization (iDDH), average nucleotide identity (ANI), DNA base composition analysis, and orthologous gene cluster analysis. Genome sequences of AAP5 and reference strains (*S. glacialis* C16y^T, *S. psychrolutea* MDB1-A^T, and *S. paucimobilis* NCTC 11030^T) were retrieved from GenBank with accession numbers GCA_004354345.1, GCA_014653575.1, GCF_014636175.1, and GCA_900457515.1, respectively. iDDH was performed between AAP5 and reference strains using the Genome-to-Genome Distance Calculator (GGDC 2.1) web server [41]. To support the iDDH results, ANI was calculated using the EzBiocloud web server [42]. The genomic G + C content of the AAP strain was taken from the published genome record (GCA_004354345.1). Orthologous gene cluster analysis was performed using the OrthoVenn2 web server [43]. Pairwise sequence similarities between all input protein sequences were calculated with an E-value cut-off of 1e^{-2}. An inflation value (-I) of 1.5 was used to define orthologous cluster structure.

Identification of DNA methylation sites. PacBio-sequencing was performed as previously described [44]. Methylome analysis was performed using the "RS_Modification_and_Motif_Analysis.1" protocol included in SMRT Portal version 2.3.0. Only modifications with an identification phred score > 30 were considered.

Physiological and biochemical characterization. For physiological experiments with the AAP5 standard R2A liquid or solid medium (DSMZ medium 830) was used. Growth was monitored by measuring colony size on plates and optical density (λ = 650 nm) in the culture. Growth at 0, 0.5, 1, 5, 10, 50 and 70 g NaCl l^{-1} was examined in R2A liquid medium under dark conditions. The pH range for growth was investigated at pH 6–10 in increments of 1 pH unit, with an additional test at pH 7.3. The following pH buffer solutions were used: acetate buffer solutions (acetic acid/sodium acetate) for pH 6, NaH$_2$PO$_4$/Na$_2$HPO$_4$ for pH 6–8, Tris-HCl buffer for pH 9, and Tris buffer for pH 10. Cell motility was tested using motility assay as described previously [45]. Growth at 8, 23, 25, 27, and 37 °C was examined on R2A agar plates under dark conditions with an incubation time of

1 week. Catalase activity was determined by observing bubble production in a 3% H_2O_2 solution. Oxidase activity was determined by monitoring the oxidation of tetramethyl *p*-phenylenediamine dichloride on filter paper. Antibiotic susceptibility tests were performed using the disc diffusion method with commercially available discs (BioRad, CA, USA). Nutrient source utilization was assayed using Phenotype MicroArrays (BIOLOG, Inc., Hayward, CA, USA). The system was modified for use with an organic medium containing [L^{-1}] 0.05 g glucose, 0.05 g peptone, 0.05 g yeast extract, 0.03 g sodium pyruvate, 0.3 g K_2HPO_4, and 1 g NaCl. Phenotype MicroArrays were incubated at 22 °C under aerobic and dark conditions. OD_{750} was measured using an Infinite F200 spectrophotometer (Tecan Trading AG, Mannendorf, Switzerland) after 1–4 days.

Other analyses. For spectroscopic analyses, the cells harvested directly from agar plates were resuspended in 70% glycerol to reduce scattering. The in vivo absorption spectra were recorded using a Shimadzu UV 2600 spectrophotometer equipped with an integrating sphere. The same cells were also used to record fluorescence emission spectra. Fluorescence was excited by a single Cyan (505 nm) Luxeon Rebel light-emitting diode (Quadica Developments Inc., Canada). The emission spectra were recorded by a QEPro high-sensitivity fiber optics spectrometer (OceanOptics, FL, USA). Pigments were analyzed using high-performance liquid chromatography as described previously [23]. BChl *a* peaks were detected at 770 nm and its content was normalized on a per total protein basis. Lowry assay was used for protein extraction (Total Protein Kit, Micro Lowry, Peterson's Modification, Sigma-Aldrich). Protein absorption was determined with UV-500 Thermo Scientific spectrophotometer at 650 nm. Respiratory quinones were extracted and analyzed as described previously [46].

Analyses of fatty acids and polar lipids were carried out by the Identification Service of the Deutsche Sammlung für Mikroorganismen und Zellkulturen (DSMZ), Germany. For this purpose, the strain was grown aerobically in full-strength R2A medium at 22 °C under 12-h dark/12-h light regime. Cells were harvested by centrifugation at $6000 \times g$ after reaching the late exponential phase (approx. OD_{650} = 0.8), freeze-dried, and sent to DSMZ.

3. Results and Discussion

Cultivation and physiology. The strain AAP5 forms yellow colonies on R2A agar. In liquid culture, growth of the strain occurred at a wide temperature range 8–37 °C, with an optimum temperature of 25–27 °C. The pH range for growth was 7.0–8.0, with an optimum at pH 7.0. Under optimal salinity (1 g NaCl l^{-1}) the value for pH optimum shifted to 7.54. AAP5 did not require NaCl for growth, but it tolerated it up to 5 g NaCl l^{-1}. The highest protein content was measured at 1 g NaCl l^{-1}. Under optimal (1 g NaCl l^{-1}) and higher than optimal salinity the cells of strain AAP5 were more abundant but smaller, which can be explained by a shorter division time.

Morphology. Epifluorescence microscopy showed that AAP5 cells are rods with a clear IR autofluorescence signal of BChl *a* (Figure 1A left). The shape of the cells corresponds to the AAP morphotype C reported from the Gossenköllesee [12]. SEM images (Figure 1A center) confirmed the rod-like shape of AAP5 cells with length of 1.8 ± 0.3 μm in solitary cells and up to 3 μm in cells prior their dividing. TEM images (Figure 1A right) revealed a corrugated surface of a thin dense lipopolysaccharide layer above the outer membrane of the G-cell wall. The cell is additionally surrounded by fine mucilaginous sheaths of 140 ± 10 nm thickness resulting in an overall cell width of 0.8 ± 0.1 μm. AFM images documented that mean cell length was 2.9 ± 0.5 μm varying between 2.2–3.6 μm, occasionally forming short chains containing up to four cells. Two conjoined cells were usually longer with mean length of 4.6 ± 0.1 μm (min-max 4.4–4.7 μm). Mean cell width reported by AFM was 1.5 ± 0.2 μm, ranging from 1.2 to 1.8 μm. The wider cell dimensions apparent in AFM scans is due to the cell surface is coated by a soft diffusive mucilage (Figure 1B center) that was also seen in the TEM images. Map of Young's modulus shows soft surface of YM = 710 ± 260 kPa. Several narrow stiffer areas (YM = 3.3 ± 1.5 MPa) orthogonal to the longer cell axis marking evidence of a site of a future cell division.

The cell's surface exhibits generally low adhesion of 95 ± 27 pN (Figure 1B right). RMS topography roughness was low over most of the cell surface (20–30 nm) but consistently higher above the division regions and at the cell terminus (50–70 nm).

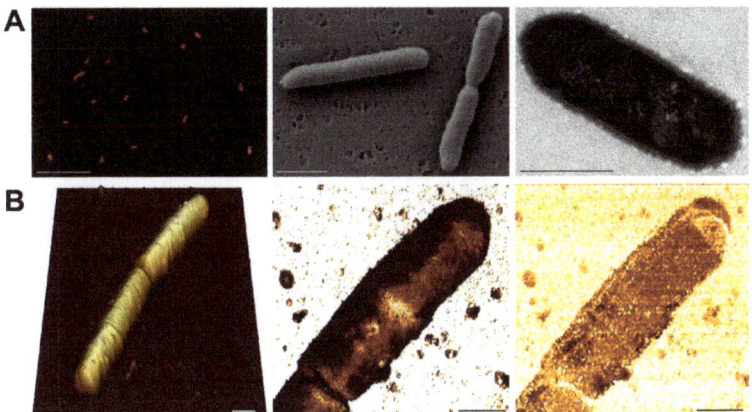

Figure 1. Microscopy images of the studied AAP5 strain. (**A**, Left) Infrared epifluorescence microscopy image (false color) showing autofluorescence from BChl *a*. *(Center)* Scanning electron microscopy image. (Right) Transmission electron microscopy image. Scale bars represent 10 µm, 1 µm, and 0.5 µm, respectively. (**B**) Atomic force microscopy images of AAP5 cells. (Left) 3-D topography image of two conjoined cells. Full image vertical range is 1.5 um. (Center) Map of Young's modulus of the upper cell shown on right reveals elastic surface (YM = 710 ± 260 kPa) with a narrow stiffer area (YM = 3.3 ± 1.5 MPa) orthogonal to the longer cell axis. Full vertical range of the image is 4 GPa. (Right) Cell surface is decorated by a soft diffusive mucilage responsible for the low stiffness and generally low adhesion of 95 ± 27 pN. Full image vertical range is 300 pN. Cell terminus exhibits both higher roughness (50–70 nm) and adhesion (156 ± 20 pN) than the rest of the cell. Scale bars represent 1 µm.

Physiological and biochemical characteristics. Strain AAP5 and *S. glacialis* C16y[T] could be differentiated by cell properties, NaCl concentration tolerance, temperature range and optimum for growth, utilization of carbon sources, as well as susceptibility to antibiotics (Table 1). Strain AAP5 was able to utilize the following compounds as a carbon source: pyruvic acid, succinic acid, L-malic acid, citramalic acid, α-keto-butyric acid, β-hydroxy butyric acid, D-tartaric acid, D-glucuronic acid, D-galacturonic acid, L-arabinose, D-glucose, D-galactose, maltose, D-mannose, D-melibiose, L-rhamnose, glycerol, sucrose, lactulose, uridine, L-glutamine, α/β/γ-cyclodextrin, dextrin, pectin, and amygdalin. It was not able to utilize citric acid, acetic acid, D-gluconic acid, adipic acid, capric acid, phenylacetic acid, D-sorbitol, D-mannitol, and inositol. The ability to utilize ammonia, N-amylamine, N-butylamine, ethylamine, ethanolamine, putrescine, β-phenylethylamine, acetamine, glucuronamide, N-acetyl-D-glucosamine, N-acetyl-D-mannosamine, cytidine, guanosine, uridine, xanthine, ε-amino-N-caproic acid, and δ-amino-N-valeric acid as a nitrogen source was also observed. From 22 proteinogenic amino acids, only L-aspartic acid, L-glutamic acid, L-serine, L-tyrosine, L-valine, and D/L-glutamic acid were utilized. It did not utilize nitrite, nitrate, and urea.

Table 1. The comparison of strain AAP with other three *Sphingomonas* strains: 1, *Sphingomonas* sp. AAP5 (data were taken from this study); 2, *Sphingomonas glacialis* C16y[T] [47,48]; 3, *Sphingomonas psychrolutea* MDB1-A[T] [48]; 4, *Sphingomonas melonis* LMG 19484[T] [14]; 5, *Sphingomonas paucimobilis* NCTC 11030[T] [1,49].

Parameter	1	2	3	4	5
Colony color	Yellow	Yellow	Orange-yellow	Deep yellow	Yellow
Cell width [µm]	0.7–0.8	0.5	0.5–0.6	0.68–0.85	0.7
Cell length [µm]	1.1–2.3	0.8	1.8–2.2	1.2–1.9	1.4
Motility	No	No	No	No	Yes
Genome characteristics [‡]					
G + C content [%]	65.9	65.7	64.2	67.1	65.7
PGC	Yes	Yes	Yes	No	No
Xanthorhodopsin	Yes	Yes	Yes	No	No
Utilization of					
D-mannose	Yes	No	Yes	Yes	No
D-melibiose	Yes	No	No	No	Yes
L-rhamnose	Yes	No	No	No	No
Phenylacetate	No	No	No	Yes	No
Antibiotics resistance					
Penicillin G 100 µg ml^{-1}	No	Yes	n.d.	Yes	Yes
Tetracycline 30 µg ml^{-1}	Yes	No	n.d.	No	No

[‡] Based on published genome sequences (accession numbers GCA_004354345.1, GCA_014653575.1, GCA_014636175.1, GCA_001761345.1, and GCA_900457515.1, respectively). n.d., no data.

AAP5 exhibited natural resistance to (µg ml^{-1}) ciprofloxacin (5), erythromycin (15), neomycin (6), ofloxacin (5), and tetracycline (30), but it was sensitive to cefoxitin (30), gentamicin (10), and penicillin G (100). The major respiratory quinone was ubiquinone-10. The predominant (>10%) fatty acid was $C_{18:1}\omega 7c$ (64.6%). The polar lipid profile of AAP5 contained phosphatidylglycerol, diphosphatidylglycerol, phosphatidylethanolamine, sphingoglycolipid, six unidentified glycolipids, one unidentified phospholipid, and two unidentified lipids. The fatty acid and polar lipid profiles agreed with those of *S. glacialis* C16y[T] [47]. The cells were positive for oxidase and catalase.

The main carotenoid was nostoxanthin. The in vivo absorption spectrum of AAP5 cells displayed a clear single near IR BChl *a* absorption band at 872 nm. In the blue part of the spectrum there was an intense absorption of carotenoids, with main absorption peaks at 433, 458, and 489 nm (Figure 2). As reported previously BChl *a* was detected in cells grown on R2A agar plates, but not in cells grown in full-strength R2A broth [23].

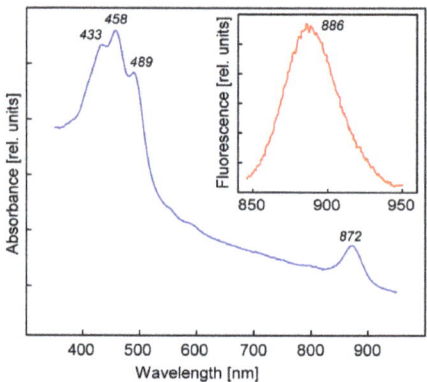

Figure 2. In vivo absorption spectra of *Sphingomonas* sp. AAP5 cells. (*Inset*) The fluorescence emission spectrum.

Although containing genes (E2E30_RS14635–E2E30_14725) necessary for flagellar biosynthesis and assembly, under the experimental conditions we used AAP5 was non-motile. Moreover, no structures similar to the flagella were found by neither of the microscopic techniques use to visualize the cells.

Phylogenomy and genomic traits. AAP5 genome contains three copies of the 16S rRNA gene. Two copies are identical (at position 3117387..3118878 and 3704698..3706189), the third one (at position 905790..907281) differs in one nucleotide position. For the alignment we used sequence from the two identical copies. The 16S rRNA tree (Figure 3) showed that AAP5 strain grouped with the genus *Sphingomonas* and formed a distinct cluster with *S. glacialis* (98.3% pairwise 16S rRNA sequence similarity), *S. psychrolutea* (96.8%), and *S. melonis* (96.5%). Further evidence provided *in silico* comparison of genomic sequences of strain AAP5 and reference strains of three species, *S. glacialis* C16y^T, *S. psychrolutea* MDB1-A^T, and the type species of the genus *S. paucimobilis* NCTC 11030^T. In silico DNA-DNA hybridization values between strain AAP5 and *S. glacialis* C16y^T, between strain AAP5 and *S. psychrolutea* MDB1-A^T, and between strain AAP5 and *S. paucimobilis* NCTC 11030^T were 74.50 ± 2.95%, 23.60 ± 2.4%, and 20.0 ± 2.3%, respectively. The cut-off value for species delineation is 70%. Average nucleotide identity values, representing mean identity values between a given pair of genomes, were 97.09% between strain AAP5 and *S. glacialis* C16y^T, 80.35% between strain AAP5 and *S. psychrolutea* MDB1-A^T, and 74.26% between strain AAP5 and *S. paucimobilis* NCTC 11030^T. The proposed boundary for defining a novel species is 95–96% [50]. In addition, we calculated the numbers of shared orthologous gene clusters among strain AAP5 and the two reference strains. Strain AAP5 shares 2,174 orthologous gene clusters with the type strain of *S. pauci-mobilis* NCTC 11030T. The number of unique orthologous gene clusters (620) shared by AAP5 and *S. glacialis* C16y^T was much greater than the number shared by either AAP5 and *S. psychrolutea* MDB1-A^T (44) or AAP5 and *S. paucimobilis* NCTC 11030^T (85), indicating a close relationship between AAP5 and *S. glacialis* C16y^T also at the genomic level. In summary, phylogenetic and genomic analyses imply that AAP5 belongs to the same genus as the three reference *Sphingomonas* strains with *S. glacialis* C16y^T representing its closest relative. The genome of AAP5 consists of four replicons (Figure 4) including one circular chromosome (3,987,367 bp; GC content 66.2%), and three plasmids p32 (31,551 bp; GC content 63.3%), p150 (150,273 bp; GC content 63.3%), and p213 (213,080 bp; GC content 62.7%) with a total length of 4,382,271 bp encoding 4,128 genes. Overall, GC content is 65.9%. Similar GC content of all four replicons suggests their common origin. The chromosome contains three copies of rRNA operons (5S, 16S, 23S), 51 tRNA genes, and 4,065 protein-coding sequences. It is notable that the AAP5 chromosome contains one continuous 38.6-kb-long photosynthesis gene cluster (E2E30_16220–E2E30_16405) and a gene coding for xanthorhodopsin (E2E30_05030), a transmembrane protein with (presumably) proton-pumping activity [22]. Furthermore, AAP5 contains a complete and contiguous gene cluster (E2E30_17610–E2E30_17700) with genes coding for gene transfer agents (GTA), virus-like particles transferring pieces of genomic DNA between prokaryotic cells. The gene organization is similar to the GTA cluster in *Rhodobacter capsulatus* [51]. The *pufL* and *pufM* genes, coding proteins of the type 2 reaction centers, represent standard genetic markers of anoxygenic phototrophs and were used to infer the phylogenetic relationship of strain AAP5 with other phototrophic *Proteobacteria*. The ML phylogenetic tree shows a split between basic alphaproteobacterial orders (Rhodobacterales, Sphingomonadales, Rhizobiales) (Figure 5). The AAP5 strain clearly clusters with other photosynthetic *Sphingomonas*. Similarly, as in the 16S rRNA tree, *Sphingomonas glacialis* C16y^T represents its closest relative. Interestingly, inside of the Sphingomonadales branch lays also one species from a distinct order, *Polynucleobacter duraquae* DSM 21495^T (Burkholderiales) (Figure 5). The presented tree agrees well with recently published PufLM phylogeny of phototrophic Proteobacteria [52,53].

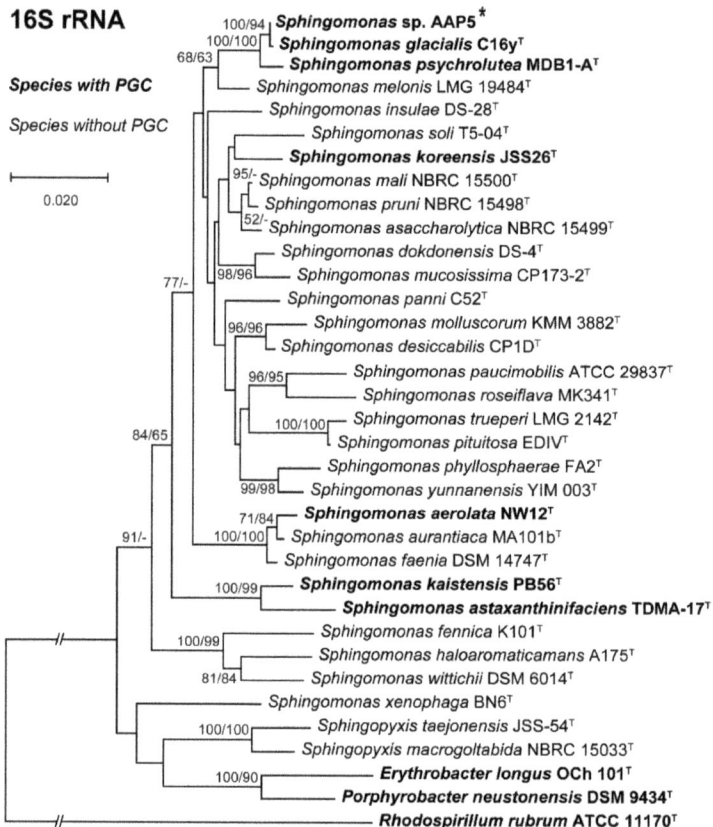

Figure 3. 16S rRNA phylogenetic tree showing position of the AAP5 strain (marked by the asterisk) within the genus *Sphingomonas* and some representatives of other related taxa. Strains with photosynthesis gene cluster (PGC) are in bold. Phylogenetic tree was based on 16S rRNA gene sequences downloaded from the SILVA database and NCBI GenBank (June, 2020). Nucleotide sequences were aligned using ClustalW resulting in alignment with 1,313 conserved nucleotide positions (after ambiguously aligned regions and gaps being manually excluded). The phylogenetic tree was calculated using both neighbor-joining (NJ) and maximum likelihood (ML) algorithms and 1,500× bootstrap replicates. *Rhodospirillum rubrum* ATCC 11170^T was used as an outgroup organism. Scale bar represents changes per position. NJ/ML bootstrap values >50% are shown.

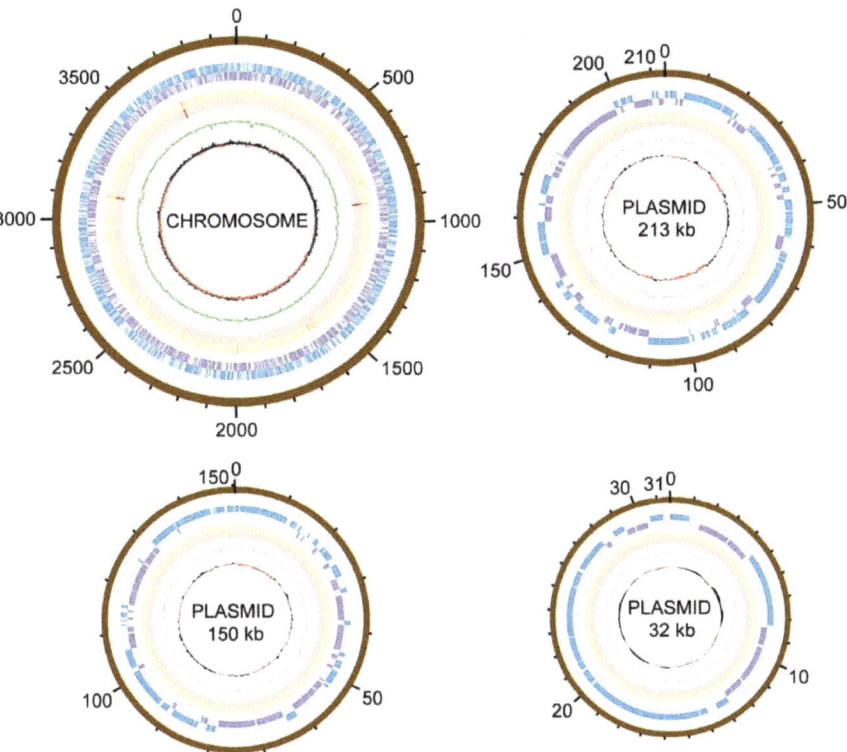

Figure 4. Circular representations of the AAP5 chromosome and three plasmids. The outer to inner rings represent: scale of replicon size in kb; position of open reading-frames encoded on the plus/minus strand (in blue/purple); tRNA (orange); rRNA (red); GC content (in green); GC skew (in red/black).

DNA methylation. Single-molecule real-time sequencing allows for detection of any DNA modifications [54]. Three different methylation sites are commonly found in bacteria [55]. We identified 19502 N^6-Adenosine (m6A) and 420 N^4-Cytosine (m4C) bases but no N^5-Cytosine (m5C) motifs on the chromosome and plasmids of AAP5 (Figure 6). The only present m4C-motif was restricted to the chromosome and plasmid p213, whereas the five m6A motifs were found on all four replicons. By far, the most abundant was the GANTC motif that is m6A-methylated by the methyltransferase CcrM [56]. This enzyme is highly conserved in Alphaproteobacteria. The GANTC motif is overrepresented in intergenic regions and underrepresented in genes [57]. In *Caulobacter vibrioides* CcrM-activity is restricted to the pre-divisional cell, thus newly replicated DNA stays hemimethylated until replication has finished. The activity of several *Caulobacter vibrioides* promoters is dependent on the methylation state and, thereby, coupled to replication [58,59]. The density of GANTC-motifs is similar for the chromosome and plasmid p32 but lower for the plasmids p213 and p150. The other five m6A-motifs as well as the m4C motif might represent targets for restriction-modification systems [60]. The motif CATGAG is underrepresented on the chromosome compared to the plasmids and might, therefore, have a regulatory function for extrachromosomal elements.

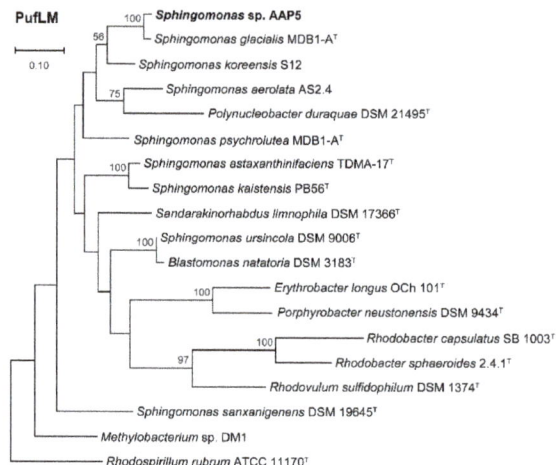

Figure 5. Maximum likelihood (ML) phylogenetic tree based on concatenated alignments of amino acid sequences of the photosynthetic reaction center subunit L and M (PufLM; 573 common amino acid positions). The ML tree was calculated using LG model and bootstrap 1,000x. *Rhodospirillum rubrum* ATCC 11170^T was used as an outgroup organism. Scale bars represent changes per position. Bootstrap values >50% are shown. Studied strain is in bold.

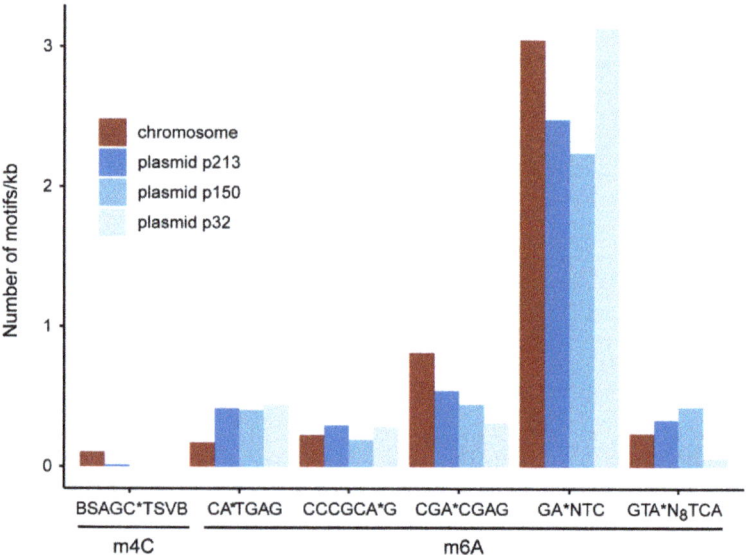

Figure 6. DNA methylation motifs inferred from SMART sequencing. The number of motifs has been normalized to one kb sequence length. The methylation site is indicated by an asterisk.

4. Conclusions

On the basis of phylogenetic and genomic evidence, strain AAP5 belongs to the species *S. glacialis*. This organism contains genes for anoxygenic photosynthesis as well as xanthorhodopsin. Although phototrophy is common among Alphaproteobacteria, the common presence of two phototrophic systems in *Sphingomonas* represents a unique phenomenon, which deserves further attention.

Author Contributions: Conceptualization, M.K.; Data curation, J.T.; Formal analysis, K.K. and J.T.; Funding acquisition, M. K.; Investigation, K.K., Y.Z., D.K., V.S., Z.G., R.S., and M.K.; Supervision, Y.Z., and M.K.; Writing – original draft, K.K., D.K., and J.T. All authors have read and agreed to the published version of the manuscript.

Funding: This research received no external funding.

Institutional Review Board Statement: Not applicable.

Informed Consent Statement: Not applicable.

Data Availability Statement: The complete genome sequence is deposited at NCBI GenBank under the accession numbers: CP037913 (chromosome) and CP037914-CP037916 (plasmids). The GenBank accesion number for the 16S rRNA sequence is MW410774.

Acknowledgments: KK was partially funded by the Czech Science Foundation project no. 18-14095Y, DK was supported by the European Regional Development Fund-Project No. CZ.02.1.01/0.0/0.0/15_003/00004 The authors thank Sonja Koppenhöfer for help with the motility assay, Kasia Piwosz for conducting the IR epifluorescence microscopy, and Alastair T. Gardiner for English proof-reading.

Conflicts of Interest: Authors declare no conflict of interests.

Abbreviations

AAP: aerobic anoxygenic phototrophic; AFM, atomic force microscopy; ANI, average nucleotide identity; BChl a, bacteriochlorophyll a; DSMZ, Deutsche Sammlung für Mikroorganismen und Zellkulturen; GTA, gene transfer agents; iDDH, in silico DNA-DNA hybridization; IR, infrared; ML, maximum likelihood; NJ, neighbor-joining; OD, optical density; PGC, photosynthesis gene cluster; SEM, scanning electron microscopy; TEM, transmission electron microscopy.

References

1. Yabuuchi, E.; Yano, I.; Oyaizu, H.; Hashimoto, Y.; Ezaki, T.; Yamamoto, H. Proposals of *Sphingomonas paucimobilis* gen. nov. and comb. nov., *Sphingomonas parapaucimobilis* sp. nov., *Sphingomonas yanoikuyae* sp. nov., *Sphingomonas adhaesiva* sp. nov., *Sphingomonas capsulata* comb, nov., and Two Genospecies of the Genus. *Sphingomonas. Microbiol. Immunol.* **1990**, *34*, 99–119. [CrossRef] [PubMed]
2. Takeuchi, M.; Sawada, H.; Oyaizu, H.; Yokota, A. Phylogenetic evidence for *Sphingomonas* and *Rhizomonas* as nonphotosynthetic members of the alpha-4 subclass of the Proteobacteria. *Int. J. Syst. Evol. Microbiol.* **1994**, *44*, 308–314. [CrossRef]
3. White, D.C.; Sutton, S.D.; Ringelberg, D.B. The genus *Sphingomonas*: Physiology and ecology. *Curr. Opin. Biotechnol.* **1996**, *7*, 301–306. [CrossRef]
4. Takeuchi, M.; Hamana, K.; Hiraishi, A. Proposal of the genus Sphingomonas sensus stricto and three new genera, Sphingobium, Novosphingobium Sphingopyxis, on the basis of phylogenetic and chemotaxonomic analyses. *Int. J. Syst. Evol. Microbiol.* **2001**, *51*, 1405–1417. [CrossRef]
5. Hiraishi, A.; Kuraishi, H.; Kawahara, K. Emendation of the description of *Blastomonas natatoria* (Sly 1985) Sly and Cahill 1997 as an aerobic photosynthetic bacterium and reclassification of *Erythromonas ursincola* Yurkov et al. 1997 as *Blastomonas ursincola* comb. nov. *Int. J. Syst. Evol. Microbiol.* **2000**, *50*, 1113–1118. [CrossRef]
6. Kim, M.K.; Schubert, K.; Im, W.T.; Kim, K.H.; Lee, S.T.; Overmann, J.; Affiliations, V. *Sphingomonas kaistensis* sp. nov., a novel alphaproteobacterium containing pufLM genes. *Int. J. Syst. Evol. Microbiol.* **2007**, *57*, 1527–1534. [CrossRef]
7. Salka, I.; Srivastava, A.; Allgaier, M.; Grossart, H.P. The draft genome sequence of *Sphingomonas* sp. strain FukuSWIS1, obtained from acidic Lake Grosse Fuchskuhle, indicates photoheterotrophy and a potential for humic matter degradation. *Genome Announc.* **2014**, *2*, e01183-14. [CrossRef]
8. Tahon, G.; Willems, A. Isolation and characterization of aerobic anoxygenic phototrophs from exposed soils from the Sør Rondane Mountains, East Antarctica. *Syst. Appl. Microbiol.* **2017**, *40*, 357–369. [CrossRef]
9. Huang, H.D.; Wang, W.; Ma, T.; Li, G.Q.; Liang, F.L.; Liu, R.-L. *Sphingomonas sanxanigenens* sp. nov., isolated from soil. *Int. J. Syst. Evol. Microbiol.* **2009**, *59*, 719–723. [CrossRef]
10. Manandhar, P.; Zhang, G.; Lama, A.; Liu, F.; Hu, Y. *Sphingomonas montana* sp. nov., isolated from a soil sample from the Tanggula Mountain in the Qinghai Tibetan Plateau. *Antonie Leeuwenhoek* **2017**, *110*, 1659–1668. [PubMed]
11. Asker, D.; Beppu, T.; Ueda, K. *Sphingomonas jaspsi* sp. nov., a novel carotenoid-producing bacterium isolated from Misasa, Tottori, Japan. *Int. J. Syst. Evol. Microbiol.* **2007**, *57*, 1435–1441. [CrossRef]
12. Čuperová, Z.; Holzer, E.; Salka, I.; Sommaruga, R.; Koblížek, M. Temporal changes and altitudinal distribution of aerobic anoxygenic phototrophs in mountain lakes. *Appl. Environ. Microbiol.* **2013**, *79*, 6439–6446.

13. Marizcurrena, J.J.; Morales, D.; Smircich, P.; Castro-Sowinski, S. Draft Genome Sequence of the UV-Resistant Antarctic Bacterium *Sphingomonas* sp. Strain UV9. *Microbiol. Resour. Announc.* **2019**, *8*, e01651-18. [CrossRef] [PubMed]
14. Buonaurio, R.; Stravato, V.M.; Kosako, Y.; Fujiwara, N.; Naka, T.; Kobayashi, K.; Cappelli, C.; Yabuuchi, E. Sphingomonas melonis sp. nov., a novel pathogen that causes brown spots on yellow Spanish melon fruits. *Int. J. Syst. Evol. Microbiol.* **2002**, *52*, 2081–2087.
15. Shin, S.C.; Ahn, D.H.; Lee, J.K.; Kim, S.J.; Hong, S.G.; Kim, E.H.; Park, H. Genome sequence of *Sphingomonas* sp. strain PAMC 26605, isolated from arctic lichen (*Ochrolechia* sp.). *J. Bacteriol.* **2012**, *194*, 1607. [CrossRef]
16. Busse, H.J.; Denner, E.B.; Buczolits, S.; Salkinoja-Salonen, M.; Bennasar, A.; Kämpfer, P. *Sphingomonas aurantiaca* sp. nov., *Sphingomonas aerolata* sp. nov. and *Sphingomonas faeni* sp. nov., air-and dustborne and Antarctic, orange-pigmented, psychrotolerant bacteria, and emended description of the genus *Sphingomonas*. *Int. J. Syst. Evol. Microbiol.* **2003**, *53*, 1253–1260. [CrossRef]
17. Amato, P.; Parazols, M.; Sancelme, M.; Laj, P.; Mailhot, G.; Delort, A.-M. Microorganisms isolated from the water phase of tropospheric clouds at the Puy de Dôme: Major groups and growth abilities at low temperatures. *FEMS Microbiol. Ecol.* **2007**, *59*, 242–254. [CrossRef] [PubMed]
18. Angelakis, E.; Roux, V.; Raoult, D. *Sphingomonas mucosissima* bacteremia in patient with sickle cell disease. *Emerg. Infect. Dis.* **2009**, *15*, 133. [PubMed]
19. Ryan, M.P.; Adley, C.C. *Sphingomonas paucimobilis*: A persistent Gram-negative nosocomial infectious organism. *J. Hosp. Infect.* **2010**, *75*, 153–157. [CrossRef] [PubMed]
20. Denner, E.B.; Paukner, S.; Kämpfer, P.; Moore, E.R.; Abraham, W.R.; Busse, H.J.; Wanner, G.; Lubitz, W. *Sphingomonas pituitosa* sp. nov., an exopolysaccharide-producing bacterium that secretes an unusual type of sphingan. *Int. J. Syst. Evol. Microbiol.* **2001**, *51*, 827–841. [CrossRef] [PubMed]
21. Gai, Z.; Wang, X.; Zhang, X.; Su, F.; Tang, H.; Tai, C.; Tao, F.; Ma, C.; Xu, P. Genome sequence of *Sphingomonas elodea* ATCC 31461, a highly productive industrial strain of gellan gum. *J. Bacteriol.* **2011**, *193*, 7015–7016. [CrossRef]
22. Kera, Y.; Abe, K.; Kasai, D.; Fukuda, H.; Takahashi, S. Draft genome sequences of *Sphingobium* sp. strain TCM1 and *Sphingomonas* sp. strain TDK1, haloalkyl phosphate flame retardant-and plasticizer-degrading bacteria. *Genome Announc.* **2016**, *4*, e00668-16. [CrossRef]
23. Kopejtka, K.; Tomasch, J.; Zeng, Y.; Selyanin, V.; Dachev, M.; Piwosz, K.; Tichý, M.; Bína, D.; Gardian, Z.; Bunk, B.; et al. Simultaneous presence of bacteriochlorophyll and xanthorhodopsin genes in a freshwater bacterium. *mSystems* **2020**, *5*, e01044-20. [CrossRef]
24. Shiba, T.; Simidu, U. *Erythrobacter longus* gen. nov., sp. nov., an aerobic bacterium which contains bacteriochlorophyll *a*. *Int. J. Syst. Bacteriol.* **1982**, *32*, 211–217. [CrossRef]
25. Yurkov, V.V.; Csotonyi, J.T. New light on aerobic anoxygenic phototrophs. In *The Purple Phototrophic Bacteria. Advances in Photosynthesis and Respiration*; Hunter, C.N., Daldal, F., Thurnauer, M.C., Beatty, J.T., Eds.; Springer Verlag: Dordrecht, The Netherlands, 2009; Volume 28, pp. 31–35.
26. Zeng, Y.; Feng, F.; Medová, H.; Dean, J.; Koblížek, M. Functional type 2 photosynthetic reaction centers found in the rare bacterial phylum *Gemmatimonadetes*. *Proc. Natl. Acad. Sci. USA* **2014**, *111*, 7795–7800. [CrossRef]
27. Coleman, A.W. Enhanced detection of bacteria in natural environments by fluorochrome staining of DNA. *Limnol. Oceanogr.* **1980**, *25*, 948–951. [CrossRef]
28. Fecskeová, L.K.; Piwosz, K.; Hanusová, M.; Nedoma, J.; Znachor, P.; Koblížek, M. Diel changes and diversity of *pufM* expression in freshwater communities of anoxygenic phototrophic bacteria. *Sci. Rep.* **2019**, *9*, 1–12. [CrossRef]
29. Sader, J.E.; Chon, J.W.; Mulvaney, P. Calibration of rectangular atomic force microscope cantilevers. *Rev. Sci. Instrum.* **1999**, *70*, 3967–3969. [CrossRef]
30. Hertz, H. Über die Berührung fester elastischer Körper. *J. Reine Angew. Math.* **1882**, *92*, 156–171.
31. Sneddon, I.N. The relation between load and penetration in the axisymmetric Boussinesq problem for a punch of arbitrary profile. *Int. J. Eng. Sci.* **1965**, *3*, 47–57. [CrossRef]
32. Rico, F.; Roca-Cusachs, P.; Gavara, N.; Farré, R.; Rotger, M.; Navajas, D. Probing mechanical properties of living cells by atomic force microscopy with blunted pyramidal cantilever tips. *Phys. Rev. E* **2005**, *72*, 021914. [CrossRef] [PubMed]
33. Quast, C.; Pruesse, E.; Yilmaz, P.; Gerken, J.; Schweer, T.; Yarza, P.; Peplies, J.; Glöckner, F.O. The SILVA ribosomal RNA gene database project: Improved data processing and web-based tools. *Nucleic Acids Res.* **2012**, *41*, D590–D596. [CrossRef] [PubMed]
34. Thompson, J.D.; Higgins, D.G.; Gibson, T.J. CLUSTAL W: Improving the sensitivity of progressive multiple sequence alignment through sequence weighting, position-specific gap penalties and weight matrix choice. *Nucleic Acids Res.* **1994**, *22*, 4673–4680. [CrossRef] [PubMed]
35. Saitou, N.; Nei, M. The neighbor-joining method: A new method for reconstructing phylogenetic trees. *Mol. Biol. Evol.* **1987**, *4*, 406–425. [PubMed]
36. Felsenstein, J. Evolutionary trees from DNA sequences: A maximum likelihood approach. *J. Mol. Evol.* **1981**, *17*, 368–376. [CrossRef]
37. Tamura, K.; Stecher, G.; Peterson, D.; Filipski, A.; Kumar, S. MEGA6: Molecular evolutionary genetics analysis version 6.0. *Mol. Biol. Evol.* **2013**, *30*, 2725–2729. [CrossRef] [PubMed]
38. Tamura, K.; Nei, M. Estimation of the number of nucleotide substitutions in the control region of mitochondrial DNA in humans and chimpanzees. *Mol. Biol. Evol.* **1993**, *10*, 512–526.

39. Tavaré, S. Some probabilistic and statistical problems in the analysis of DNA sequences. Some Mathematical Questions in Biology: DNA Sequence Analysis. In *Lectures on Mathematics in the Life Sciences*, 2nd ed.; Miura, R.M., Ed.; The American Mathematical Society: Providence, RI, USA,, 1986; Volume 17, pp. 57–86.
40. Le, S.Q.; Gascuel, O. An improved general amino acid replacement matrix. *Mol. Biol. Evol.* **2008**, *25*, 1307–1320. [CrossRef] [PubMed]
41. Meier-Kolthoff, J.P.; Göker, M.; Spröer, C.; Klenk, H.P. When should a DDH experiment be mandatory in microbial taxonomy? *Arch. Microbiol.* **2013**, *195*, 413–418. [CrossRef] [PubMed]
42. Yoon, S.H.; Ha, S.M.; Lim, J.; Kwon, S.; Chun, J. A large-scale evaluation of algorithms to calculate average nucleotide identity. *Antonie Leeuwenhoek* **2017**, *110*, 1281–1286.
43. Xu, L.; Dong, Z.; Fang, L.; Luo, Y.; Wei, Z.; Guo, H.; Zhang, G.; Gu, Y.Q.; Coleman-Derr, D.; Xia, Q.; et al. OrthoVenn2: A web server for whole-genome comparison and annotation of orthologous clusters across multiple species. *Nucleic Acids Res.* **2019**, *47*, W52–W58. [CrossRef]
44. Bartling, P.; Brinkmann, H.; Bunk, B.; Overmann, J.; Göker, M.; Petersen, J. The Composite 259-kb Plasmid of *Martelella mediterranea* DSM 17316^T–A Natural Replicon with Functional RepABC Modules from *Rhodobacteraceae* and *Rhizobiaceae*. *Front. Microbiol.* **2017**, *8*, 1787. [CrossRef]
45. Bartling, P.; Vollmers, J.; Petersen, J. The first world swimming championships of roseobacters—Phylogenomic insights into an exceptional motility phenotype. *Syst. Appl. Microbiol.* **2018**, *41*, 544–554. [CrossRef]
46. Zeng, Y.; Nupur Wu, N.; Madsen, A.M.; Chen, X.; Gardiner, A.T.; Koblížek, M. Gemmatimonas groenlandica sp. nov. is an aerobic anoxygenic phototroph in the phylum Gemmatimonadetes. *Front. Microbiol.* **2021**, *11*, 606612. [CrossRef]
47. Zhang, D.C.; Busse, H.J.; Liu, H.C.; Zhou, Y.G.; Schinner, F.; Margesin, R. Sphingomonas glacialis sp. nov., a psychrophilic bacterium isolated from alpine glacier cryoconite. *Int. J. Syst. Evol. Microbiol.* **2011**, *61*, 587–591. [CrossRef] [PubMed]
48. Liu, Q.; Liu, H.C.; Zhang, J.L.; Zhou, Y.G.; Xin, Y.H. Sphingomonas psychrolutea sp. nov., a psychrotolerant bacterium isolated from glacier ice. *Int. J. Syst. Evol. Microbiol.* **2015**, *65*, 2955–2959. [CrossRef] [PubMed]
49. Holmes, B.; Owen, R.J.; Evans, A.; Malnick, H.; Willcox, W.R. Pseudomonas paucimobilis, a new species isolated from human clinical specimens, the hospital environment, and other sources. *Int. J. Syst. Evol. Microbiol.* **1977**, *27*, 133–146. [CrossRef]
50. Goris, J.; Konstantinidis, K.T.; Klappenbach, J.A.; Coenye, T.; Vandamme, P.; Tiedje, J.M. DNA–DNA hybridization values and their relationship to whole-genome sequence similarities. *Int. J. Sys. Evol. Microbiol.* **2007**, *57*, 81–91. [CrossRef] [PubMed]
51. Lang, A.S.; Beatty, J.T. Genetic analysis of a bacterial genetic exchange element: The gene transfer agent of Rhodobacter capsulatus. *Proc. Nat. Acad. Sci. USA* **2000**, *97*, 859–864. [CrossRef] [PubMed]
52. Imhoff, J.F.; Rahn, T.; Künzel, S.; Neulinger, S.C. Photosynthesis is widely distributed among Proteobacteria as demonstrated by the phylogeny of PufLM reaction center proteins. *Front. Microbiol.* **2018**, *8*, 2679. [CrossRef]
53. Imhoff, J.F.; Rahn, T.; Künzel, S.; Neulinger, S.C. Phylogeny of anoxygenic photosynthesis based on sequences of photosynthetic reaction center proteins and a key enzyme in bacteriochlorophyll biosynthesis, the chlorophyllide reductase. *Microorganisms* **2019**, *7*, 576. [CrossRef] [PubMed]
54. Flusberg, B.; Webster, D.; Lee, J.; Travers, K.J.; Olivares, E.C.; Clark, T.A.; Korlach, J.; Turner, S.W. Direct detection of DNA methylation during single-molecule, real-time sequencing. *Nat. Methods* **2010**, *7*, 461–465. [CrossRef] [PubMed]
55. Blow, M.J.; Clark, T.A.; Daum, C.G.; Deutschbauer, A.M.; Fomenkov, A.; Fries, R.; Froula, J.; Kang, D.D.; Malmstrom, R.R.; Morgan, R.D.; et al. The epigenomic landscape of prokaryotes. *PLoS Genet.* **2016**, *12*, e1005854. [CrossRef] [PubMed]
56. Zweiger, G.; Marczynski, G.; Shapiro, L. A Caulobacter DNA methyltransferase that functions only in the predivisional cell. *J. Mol. Biol.* **1994**, *235*, 472–485. [CrossRef]
57. Gonzalez, D.; Kozdon, J.B.; McAdams, H.H.; Shapiro, L.; Collier, J. The functions of DNA methylation by CcrM in *Caulobacter crescentus*: A global approach. *Nucleic Acids Res.* **2014**, *42*, 3720–3735. [CrossRef]
58. Domian, I.J.; Reisenauer, A.; Shapiro, L. Feedback control of a master bacterial cell-cycle regulator. *Proc. Nat. Acad. Sci.* **1999**, *96*, 6648–6653. [CrossRef]
59. Gonzalez, D.; Collier, J. DNA methylation by CcrM activates the transcription of two genes required for the division of *Caulobacter crescentus*. *Mol. Microbiol.* **2013**, *88*, 203–218. [CrossRef]
60. Loenen, W.A.; Dryden, D.T.; Raleigh, E.A.; Wilson, G.G.; Murray, N.E. Highlights of the DNA cutters: A short history of the restriction enzymes. *Nucleic Acids Res.* **2014**, *42*, 3–19. [CrossRef]

Article

In-Situ Metatranscriptomic Analyses Reveal the Metabolic Flexibility of the Thermophilic Anoxygenic Photosynthetic Bacterium *Chloroflexus aggregans* in a Hot Spring Cyanobacteria-Dominated Microbial Mat

Shigeru Kawai [1,2,*], Joval N. Martinez [1,3], Mads Lichtenberg [4], Erik Trampe [4], Michael Kühl [4], Marcus Tank [1,5], Shin Haruta [1], Arisa Nishihara [1,6], Satoshi Hanada [1] and Vera Thiel [1,5,*]

Citation: Kawai, S.; Martinez, J.N.; Lichtenberg, M.; Trampe, E.; Kühl, M.; Tank, M.; Haruta, S.; Nishihara, A.; Hanada, S.; Thiel, V. In-Situ Metatranscriptomic Analyses Reveal the Metabolic Flexibility of the Thermophilic Anoxygenic Photosynthetic Bacterium *Chloroflexus aggregans* in a Hot Spring Cyanobacteria-Dominated Microbial Mat. *Microorg* 2021, *9*, 652. https://doi.org/10.3390/microorganisms9030652

Academic Editor: Johannes F. Imhoff

Received: 28 January 2021
Accepted: 17 March 2021
Published: 21 March 2021

Publisher's Note: MDPI stays neutral with regard to jurisdictional claims in published maps and institutional affiliations.

Copyright: © 2021 by the authors. Licensee MDPI, Basel, Switzerland. This article is an open access article distributed under the terms and conditions of the Creative Commons Attribution (CC BY) license (https://creativecommons.org/licenses/by/4.0/).

[1] Department of Biological Sciences, Tokyo Metropolitan University, Hachioji, Tokyo 192-0397, Japan; j.martinez@usls.edu.ph (J.N.M.); mat19@dsmz.de (M.T.); sharuta@tmu.ac.jp (S.H.); arisa.nishihara@aist.go.jp (A.N.); satohana@tmu.ac.jp (S.H.)
[2] Institute for Extra-cutting-edge Science and Technology Avant-garde Research (X-star), Japan Agency for Marine-Earth Science and Technology (JAMSTEC), Yokosuka, Kanagawa 237-0061, Japan
[3] Department of Natural Sciences, College of Arts and Sciences, University of St. La Salle, Bacolod City, Negros Occidental 6100, Philippines
[4] Department of Biology, Marine Biological Section, University of Copenhagen, Strandpromenaden 5, 3000 Helsingør, Denmark; mlichtenberg@sund.ku.dk (M.L.); etrampe@bio.ku.dk (E.T.); mkuhl@bio.ku.dk (M.K.)
[5] DSMZ—German Culture Collection of Microorganisms and Cell Culture, GmbH Inhoffenstraße 7B, 38124 Braunschweig, Germany
[6] Bioproduction Research Institute, National Institute of Advanced Industrial Science and Technology (AIST), Ibaraki 305-8566, Japan
* Correspondence: kawais@jamstec.go.jp (S.K.); Vera.Thiel@DSMZ.de (V.T.)

Abstract: *Chloroflexus aggregans* is a metabolically versatile, thermophilic, anoxygenic phototrophic member of the phylum *Chloroflexota* (formerly *Chloroflexi*), which can grow photoheterotrophically, photoautotrophically, chemoheterotrophically, and chemoautotrophically. In hot spring-associated microbial mats, *C. aggregans* co-exists with oxygenic cyanobacteria under dynamic micro-environmental conditions. To elucidate the predominant growth modes of *C. aggregans*, relative transcription levels of energy metabolism- and CO_2 fixation-related genes were studied in Nakabusa Hot Springs microbial mats over a diel cycle and correlated with microscale in situ measurements of O_2 and light. Metatranscriptomic analyses indicated two periods with different modes of energy metabolism of *C. aggregans*: (1) phototrophy around midday and (2) chemotrophy in the early morning hours. During midday, *C. aggregans* mainly employed photoheterotrophy when the microbial mats were hyperoxic (400–800 µmol L^{-1} O_2). In the early morning hours, relative transcription peaks of genes encoding uptake hydrogenase, key enzymes for carbon fixation, respiratory complexes as well as enzymes for TCA cycle and acetate uptake suggest an aerobic chemomixotrophic lifestyle. This is the first in situ study of the versatile energy metabolism of *C. aggregans* based on gene transcription patterns. The results provide novel insights into the metabolic flexibility of these filamentous anoxygenic phototrophs that thrive under dynamic environmental conditions.

Keywords: filamentous anoxygenic phototroph; microbial mats; hot springs; metatranscriptomics; energy metabolism; carbon fixation

1. Introduction

Members of the genus *Chloroflexus* are thermophilic, filamentous anoxygenic phototrophs (FAPs) in the phylum *Chloroflexota* (formerly *Chloroflexi*). They are well known to have the ability to grow photoheterotrophically under anaerobic conditions and chemoheterotrophically under aerobic conditions in the laboratory [1–3]. While photoautotrophic growth in the laboratory has been observed only in a small number of isolated strains (e.g., *Chloroflexus aurantiacus* strain OK-70-fl [4–6], *Chloroflexus* sp. strain MS-G [7], and *Chloroflexus aggregans* strains NA9-6 [8,9] and ACA-12 [10]), the genes necessary for the 3-hydroxypropionate (3-OHP) bi-cycle, which is a carbon fixation pathway found only in members of the order *Chloroflexales* among bacteria, are present in all of the available *Chloroflexus* spp. genomes [11]. *C. aggregans* strains NA9-6 and ACA-12, which were isolated from Nakabusa Hot Springs (Nagano Prefecture, Japan), can grow photoautotrophically with hydrogen gas (H_2) [8] and sulfide [10] as the electron donors in pure culture, respectively. In addition to the long known phototrophic and chemoheterotrophic metabolism in *Chloroflexus* spp., chemoautotrophic growth has recently been shown in lab studies of *C. aggregans* strain NA9-6 [8]. In addition, fermentative growth has been shown in two isolates of *C. aurantiacus*, strains B3 and UZ [12].

Microbial mats in the slightly alkaline, sulfidic Nakabusa Hot Springs have been intensively studied with regard to their microbial diversity and functions [13–21]. At water temperatures of 63–70 °C, olive-green microbial mats ("*Chloroflexus* mats") are dominated by *C. aggregans* [14,15], and oxygenic cyanobacteria are not found. At lower temperatures of 45–62°C, *Chloroflexus* spp. co-exist with cyanobacteria in dark blue-green microbial mats ("cyanobacterial mats"). These blue-green mats are stratified with a green upper layer dominated by the thermophilic cyanobacteria on top of an orange-colored layer that is frequently inhabited by *C. aggregans* [13].

In the anoxygenic, cyanobacteria-free phototrophic mats, *C. aggregans* is considered to be the main primary producer, using sulfide as the major electron source [9,10,14–16]. The metabolic repertoire of *C. aggregans* in the blue-green cyanobacterial mats has remained unstudied. In situ isotopic studies of similar cyanobacterial mats colonizing the effluent channels of Mushroom Spring and Octopus Spring in Yellowstone National Park (YNP; WY, U.S.) suggested that filamentous phototrophic *Chloroflexota* vary their carbon metabolisms over a diel cycle [22]. Based on transcriptomic data, Klatt et al. (2013) suggested photomixotrophic growth of a member of FAPs—i.e., *Roseiflexus* spp.—in Mushroom Spring during daytime and fermentative growth during the night [23]. Compared to the microbial mats in YNP, Nakabusa Hot Spring cyanobacterial mats are rich in *C. aggregans*, at a relative abundance of approximately 21–22% compared to only 1% in Mushroom Spring cyanobacterial mats [13,24]. This suggests an important ecological role and potential function of *C. aggregans* as a primary producer in the Nakabusa mats.

In this study, the in situ metabolic lifestyle of *C. aggregans* in the blue-green microbial mats of Nakabusa Hot Springs was analyzed by using a metatranscriptomic approach. Light is the main energy source during daytime, supporting photoautotrophic, photomixotrophic and photoheterotrophic growth of *C. aggregans*, while chemotrophic growth is prevalent during the afternoon and night. During the afternoon, under microaerobic low-light conditions chemoheterotrophic growth is based on O_2 respiration, while at night fermentation is conducted under anaerobic conditions. Unexpectedly, chemoautotrophic growth using O_2 as the terminal electron acceptor appeared to take place during early morning hours before sunrise, which suggests a vertical migration of *C. aggregans* cells to the microaerobic surface layers of the mats.

2. Materials and Methods
2.1. Field Site and Sample Collection

Blue-green cyanobacterial mat samples were collected from a small pool at 56 °C with slightly alkaline (pH 8.5–8.9) and sulfidic (46–138 µM) hot spring water [18,25–27] at Nakabusa Hot Springs, Nagano Prefecture, Japan (36°23′33″ N, 137°44′52″ E) [20]. Microbial mat samples of approximately 3 mm thickness with two distinct vertical layers, a green top layer and an orange-colored bottom layer (Figure 1), were randomly collected in triplicate using a size 4 cork borer (8 mm diameter) as previously described [13,14]. Samples were placed in 2 mL screw-cap tubes and snap-frozen in a dry-ice cooled, 70% (v/v) ethanol bath on site. Samples were taken at 12 different time points over a diel cycle on 3 to 4 November, 2016 (19:00 and 23:00 on 3 November; 02:10, 05:00, 06:00, 07:00, 11:00, 15:00, 16:00, 17:00, 18:00, and 19:00 on 4 November) and were brought back to the laboratory on dry ice and stored in a −80 °C freezer until further processing for metatranscriptomic analyses.

 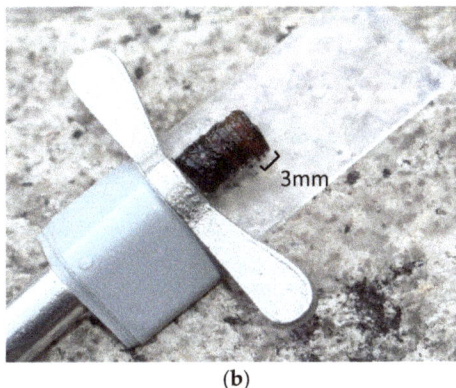

(a) (b)

Figure 1. Photographs of the sampling site and cyanobacteria-dominated microbial mat. (**a**) The dark blue-green microbial mats developed at the "Stream Site" of Nakabusa Hot Springs, Japan [19]. (**b**) The microbial mat core samples collected at each time point were 8 mm in diameter and approx. 15 mm thick. The upper 3 mm of the core samples was used in this study.

2.2. RNA Extraction

RNA extraction from microbial mat samples was performed as previously described [18]. Briefly, 0.10–0.21 g wet weight samples were used for RNA extraction with an RNeasy PowerBiofilm Kit (Qiagen, Valencia, CA, USA) following the manufacturer's protocol. The RNA was treated with DNase I and eluted with RNase-free water. Purity and concentration of the RNA were determined using an RNA High Sensitivity (HS) assay with a Qubit 3.0 fluorometer (Life Technologies, Grand Island, NY, USA).

2.3. RNA Sequencing

Library preparation and sequencing of the RNA samples were conducted at DNALink Inc. (Seoul, Korea) as described previously [18]. RNA purity was determined by assaying 1 µL of total RNA extract on a NanoDrop8000 spectrophotometer (Thermo Fisher Scientific, Waltham, MA, USA). Total RNA integrity was assessed by the RNA integrity number (RIN) using a 2100 Bioanalyzer (Agilent Technologies, Palo Alto, CA, USA). Total RNA sequencing libraries were prepared using a Truseq Stranded Total RNA Library prep kit and Ribo-Zero bacteria kit (both from Illumina, San Diego, CA, USA) according to the manufacturer's instructions.

First, 0.5 µg of total RNA was subjected to ribosomal RNA depletion with Ribo-Zero bacteria reagent using biotinylated probes that selectively bind rRNA species. Following purification, the rRNA-depleted total RNA was fragmented into small pieces using divalent cations under elevated temperature. The cleaved RNA fragments were copied into first-strand cDNA using random primers and reverse transcriptase, followed by second-strand cDNA synthesis using DNA polymerase I and RNase H. A single 'A' base was then added to these cDNA fragments, and the adapter was ligated. The products were purified and enriched by polymerase chain reaction (PCR) to create the final cDNA library.

The quality of the amplified libraries was verified by capillary electrophoresis using the 2100 Bioanalyzer (Agilent Technologies, Palo Alto, CA, USA). After a quantitative (q)PCR using SYBR Green PCR Master Mix (Applied Biosystems, Carlsbad, CA, USA), index-tagged libraries were combined in equimolar amounts. RNA sequencing was performed using an Illumina NextSeq 500 system following the provided protocols for 2 × 150 sequencing.

2.4. Sequence Data Analyses

Raw RNA reads were pre-processed using FastQC [28]. Adapter sequence and low-quality reads were trimmed by Cutadapt ver. 1.12 [29]. Quality-checked reads were mapped against the complete genome of *C. aggregans* DSM 9485^T (RefSeq acc. No. NC_011831.1) [2] with bowtie2 ver. 2.3.0 [30] with default settings allowing no mismatches. The reads were then aligned using the EDGE-pro algorithm [31] with the rRNA depletion option.

Transcriptomic analyses were conducted as described previously [18]. In short, read counts were normalized for each time point by the total number of reads retrieved for the target organism. The relative transcription of each gene during the cycle was then calculated and normalized against the mean of all of the reads at each time point for that particular gene over the diel cycle. This method allows comparison of the relative transcription abundance levels (rather than the absolute values) for each gene across the diel cycle.

2.5. Statistical Analyses

As described above, diel transcriptomic data in this study lacked replication. In the following Results and Discussion sections, the authors carefully interpreted and described and intentionally averaged the normalized transcriptional patterns of several genes related in a single pathway to recognize those gene transcription patterns as the pathway-level metabolic dynamics. However, some important genes function in an important enzyme reaction solely, the statistical analyses of each gene were performed to discuss the transcriptional changes over a diel cycle. For each gene in dual datasets, and for every possible pair-wise comparison of the 11 sets of adjacent samples (November 3 19:00–23:00; 3 November 23:00–4 November 2:10; 4 November 02:10–05:00, 05:00–06:00, 06:00–07:00, 07:00–11:00, 11:00–15:00, 15:00–16:00, 16:00–17:00, 17:00–18:00, 18:00–19:00), "exactTest" program in edgeR with dispersion set at 0.1 was used to determine the probability that the gene was differentially transcribed in a statistically significant manner [32,33].

2.6. Microsensor Analyses

The profiles of the O_2 concentration as a function of depth in the microbial mat were measured in situ by using a Clark-type O_2 microsensor (OX25; Unisense, Aarhus, Denmark) with a tip diameter of <25 µm, low stirring sensitivity (<1–2%) and fast response time (t_{90} < 0.5 s). The O_2 microsensor was mounted on a motorized micromanipulator (Unisense, Aarhus, Denmark) and connected to a PC-interfaced pA-meter (Unisense, Aarhus, Denmark), both of which were controlled by dedicated data acquisition, profiling, and positioning software (SensorTrace Pro, Unisense, Aarhus, Denmark). The micromanipulator was mounted on a metal stand placed next to the hot spring, allowing for vertical insertion of the microsensor tip into the microbial mat under natural flow, temperature and light conditions. The microsensor tip was carefully positioned at the mat surface (defined as 0 µm) by manual operation of the micromanipulator. Subsequently, O_2 microprofiles were recorded automatically every 15 min for 24 h starting at 18:00 on 3 November 2016. In each profile, O_2 measurements were made in 100 µm increments from the water-phase and into the mat. One measurement was taken per depth and, for each measurement a 10 s wait period was applied, to ensure steady O_2 signal, and the O_2 signal was then recorded averaged over a 1 s period.

2.7. Irradiance Measurements

Downwelling solar photon irradiance (400–700 nm) at the water surface next to the mat was logged every 5 min throughout the 24 h diel sampling cycle with a calibrated light meter connected to a cosine-corrected photon irradiance sensor (ULM-500, MQS-B; Walz, Effeltrich, Germany).

3. Results
3.1. Irradiance and In Situ Oxygen Dynamics in the Microbial Mat

The O_2 concentration and penetration from the surface green layer to the deeper orange layer in the microbial mat varied dramatically with irradiance. The whole mat was anoxic during the night, and the O_2 concentration started to increase at the mat surface at around 06:00, correlating with the time of sunrise and thus the onset of cyanobacterial oxygenic photosynthesis under diffuse light (Figures 1 and 2). However, O_2 did not accumulate in deeper mat layers until later in the morning at around 09:00, when the microbial mats were exposed to direct sunlight as the sun rose over the surrounding mountains. Supersaturating O_2 levels were observed in the uppermost mat layers at ~12:00 (noon) during the highest solar irradiance (1531 µmol photons m^{-2} s^{-1}; 400–700 nm). The maximum O_2 concentration at the microbial mat surface reached >900 µmol O_2 L^{-1} and with a maximal O_2 penetration of >2 mm depth under the highest irradiance between 10:00 and 14:00. Thus, there was sufficient O_2 available for aerobic microbes in the upper 2 mm of the mat during this period of the day. In the afternoon, after 14:00, O_2 concentration started to decrease gradually (from approx. 500 µmol O_2 L^{-1}) and no O_2 was detected at a depth of 1–2 mm shortly after 15:00. The upper layer remained oxic (100–200 µmol O_2 L^{-1}) until 16:00. At this time, the microbial mats experienced a substantial decrease in solar irradiance (see Figures 3–8 and S1–S5) as the sun set behind the mountains. However, low levels of diffuse sunlight (<100 µmol photons m^{-2} s^{-1}; 400–700 nm) hit the microbial mats until complete darkness was observed at 17:00. Anoxic conditions started to become established in the lower parts of the microbial mats ~2 h before sunset, potentially enabling anaerobic and microaerophilic metabolism under low-light conditions during this time interval.

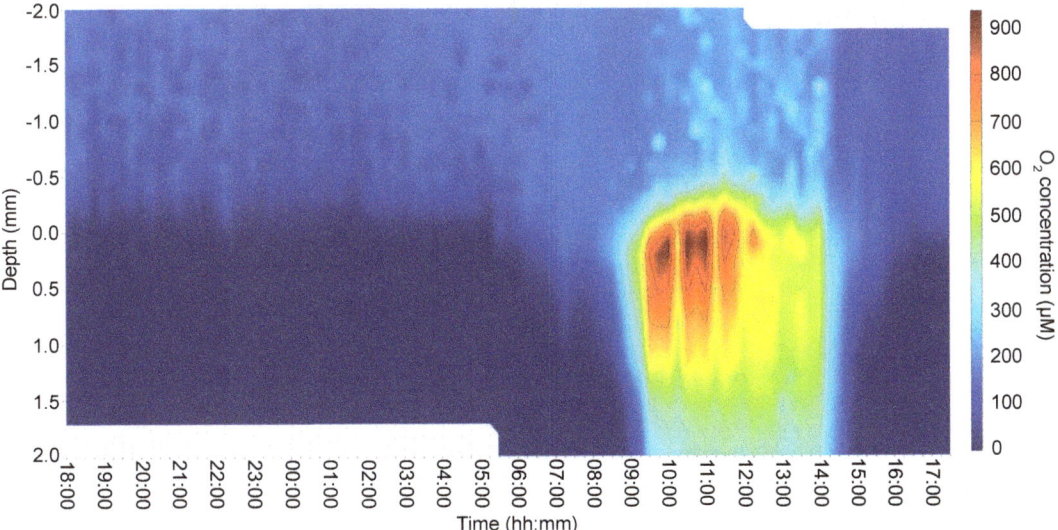

Figure 2. A heat map of the vertical O_2 concentration profiles in the cyanobacteria-dominated microbial mat of Nakabusa Hot Springs as measured over a diel cycle. The O_2 concentration (µmol L^{-1}) was measured as a function of depth in the microbial mat at 15-min intervals for 24 h from 18:00 on 3 November to 18:00 on 4 November. The mat surface is indicated by 0 mm. Positive depth values indicate the depth below the mat surface, and negative values indicate the depth above the mat surface, i.e., the distance into the overlying water column of the hot spring.

3.2. Transcriptome Profiles and Differentially Transcribed Genes

Approx. 0.26–5.23 million reads of transcripts were assigned to the *C. aggregans* genome throughout the day (Table S1). Among 3848 CDSs contained in the genome of *C. aggregans* DSM9485^T, the number of genes in which more than 10 transcripts were detected in all timepoints was 2542–3506 genes. Statistical significance of transcription level changes of each gene was determined using the p-value ($p < 0.05$) based on the "exactTest" function in edgeR [32]. Thousands of genes were differentially transcribed during the period from 19:00 on 3 November to 15:00 on 4 November indicating a versatile and changing transcriptional activity between the different time points. During 15:00–16:00 as well as 17:00–19:00, the numbers of significantly differentially transcribed genes were considerably lower (less than 10% of all CDSs), indicating transcriptional activity of *C. aggregans* during this period was relatively stable compared with former 20 h of the day. Those times were taken together as 'afternoon' and 'evening', respectively, in which both transcriptional activity as well as environmental conditions are expected to be rather similar. Overall the statistics support the changes in relative transcriptional levels and activity of *C. aggregans* to be significant. Detailed information on the differential transcription of the genes discussed is given in Tables S2 and S3.

3.3. Transcription of Photosynthesis-Related Genes

Chloroflexus aggregans contains a type 2 photosynthetic reaction center complex (RC) and light-harvesting chlorosomes; the main photosynthetic pigments are bacteriochlorophylls (BChls) *c* and *a* [2]. Transcripts of *pufLMC* genes (Cagg_1639–1640 and Cagg_2631) encoding RC proteins in *C. aggregans* showed significant nocturnal patterns and were most abundant in the evening and at night (Figure 3). Similarly, nocturnal transcriptional patterns of chlorosome proteins encoded by *csmAMNOPY* genes (Cagg_1222, Cagg_1209, Cagg_1208, Cagg_2486, Cagg_1206, and Cagg_1296) were detected (Figure 3).

BChl synthesis-related *bch* genes in *C. aggregans* in the mats showed inconsistent transcription patterns, with the highest relative transcription levels either under oxic conditions during the daytime (11:00) and/or under microoxic conditions in the afternoon (15:00; Table S4). In total, four different groups of transcription patterns could be distinguished for *bch* genes. Group I (*bchH*-III, *bchI*-I and II, *bchY*, *bchZ*, *acsF*, *bchL*, *bchB*, *bchN*, *bchM*) showed highest relative transcription during the daytime (midday, 11:00). Group II (*bchH*-I and II, *bchJ*, *bchF*, *bchX*, *bchC*) showed maximal values in the afternoon around 15:00, with more or less pronounced lows at 11:00. The two other groups showed highest relative transcription levels under anaerobic conditions. Four genes (*bchG*, *bchK*, *bchU*, and *bchP*) showed peaks at 18:00. *bchD* and *bchI*-III had maximal relative transcription in the early morning at 05:00.

Paralogs with different patterns were present for the genes encoding Mg-chelatase, i.e., *bchH* (three paralogs; Cagg_0239, 0575, 1286) and *bchI* (three paralogs; Cagg_1192, 2319, 3123). In all three cases, the paralogs showed differences in absolute read abundance as well as different temporal peaks in relative transcription over the day (Table S4).

Two different genes encoding for Mg-protoporphyrin monomethylester cyclase are present in *C. aggregans*. AcsF and BchE both catalyze the synthesis of divinylprotochlorophyllide from Mg-protoporphyrin IX 13-monomethyl ester (one of the intermediates in the BChl synthesis pathway) under aerobic and anaerobic conditions, respectively [34–36]. *acsF* (Cagg_1285) was transcribed throughout the diel cycle with relative transcription levels about four times higher during day, when the mat was (super)oxic than under anoxic conditions at night (Figure 3). In contrast to *acsF*, the *bchE* (Cagg_0316) showed significantly lower relative transcription levels (only 1/32 of the diel average) under high-light/high-O_2 conditions in the mat and higher transcription levels during the night as well as during low-light transition times in the morning and afternoon (Figure 3).

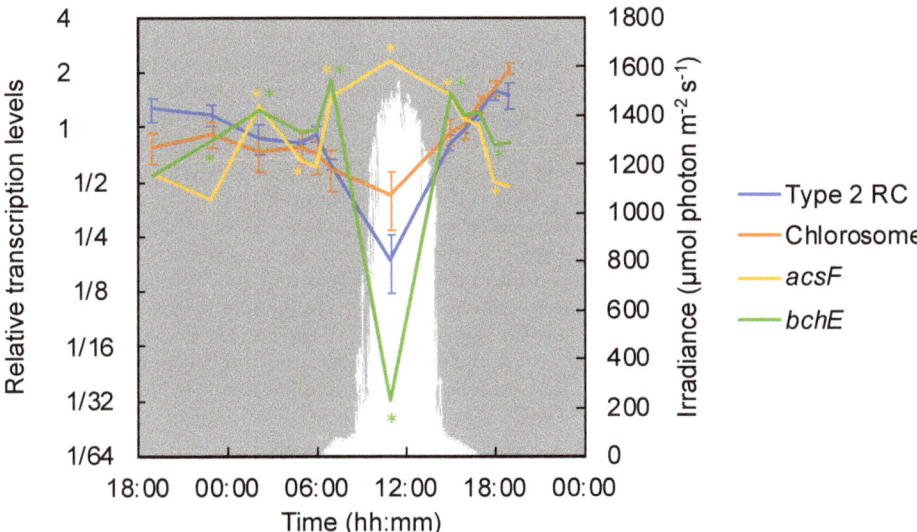

Figure 3. Relative transcription levels of genes encoding photosynthetic reaction center and chlorosome proteins, as well as genes involved in bacteriochlorophyll biosynthesis. Mean values of genes encoding the type 2 reaction center (RC) (*pufLMC*: Cagg_1639–1640 and Cagg_2631) and chlorosome proteins (*csmAMNOPY*: Cagg_1222, Cagg_1209, Cagg_1208, Cagg_2486, Cagg_1206, and Cagg_1296) are represented by a blue line and an orange line, respectively, with standard deviations. The values of Mg-protoporphyrin IX monomethyl ester aerobic cyclase (*acsF*, Cagg_1285, yellow line) and anaerobic cyclase (*bchE*, Cagg_0316, green line) are shown. The downwelling photon irradiance (photosynthetically active radiation [PAR]; 400–700 nm) is indicated in white. The asterisk indicates the transcription of a particular gene corresponding to the color in a timepoint differed significantly ($p < 0.05$) from that in the previous timepoint.

3.4. Phototrophic and Respiratory Electron Transport

Electron transport chains are involved in both phototrophic and chemotrophic (respiratory) metabolism. *Chloroflexus* spp. contain paralogs of some of the major enzyme complexes involved in the electron transport chain, namely, NADH:menaquinone oxidoreductase (Complex I) [11,37,38] alternative complex III (ACIII) [39–41] and the soluble electron carrier auracyanin [42–45] in their genomes. Single-copy genes are present encoding succinate dehydrogenase (Complex II) [46,47] and F-type ATP synthase (Complex V) [48].

The respiratory complex I carries out the transfer of electrons between soluble cytoplasmic electron carriers and membrane-bound electron carriers coupled to proton translocation generating a transmembrane proton motive force. *C. aggregans* DSM9485^T contains two sets of genes encoding Complex I (NADH:menaquinone oxidoreductase, *nuo*): one (Cagg_1620–1631) represents a cluster comprising 12 genes with an additional *nuoM* (2-M complex) inserted between the original *nuoM1* and *nuoN* genes (*nuoABCDHJKLMMN*, 3); the second is a complete cluster comprising 14 genes (Cagg_1036–1049). The additional proton-pumping subunit NuoM has been speculated to lead to a higher stoichiometry of protons translocated per 2e$^-$ reaction cycle [49]. Both sets of *nuo* genes were transcribed and showed the highest relative transcription during the daytime (Figure S1). The 14-gene set showed a transcriptional peak at 11:00 and the 12-gene peak came slightly later (in the afternoon at 15:00–16:00). Both *nuo* gene sets showed a small transcription peak in the early morning at 05:00.

Alternative complex III (ACIII) transfers electrons from menaquinol to water-soluble proteins such as auracyanin, the blue copper electron carrier protein found in *Chloroflexus* spp. [44]. Two types of ACIII have been reported in *Chloroflexus* spp.: Cp, which is thought to be involved in cyclic phototrophic electron transport, and Cr, which is predicted to be related to the reduction of oxygen in respiratory electron transport [11,23,39]. In the present study, clear diel patterns as well as differences between the two sets of genes were observed. The genes encoding Cp showed high relative transcription levels all day with the highest level at 11:00 and a significant decrease in the afternoon as the sunlight vanished. The Cr genes were not highly transcribed under the daytime high O_2 conditions in the mat, but they significantly increased and exhibited maximal relative transcription levels in the microoxic low-light afternoon hours at 15:00 and 16:00 (Figure 4). Genes for both Cp and Cr were highly transcribed in the early morning at 5:00.

Two homologs encoding the soluble electron carrier protein auracyanin, *aurA* and *aurB*, are present in the genome of *C. aggregans*. The transcriptional profile of *aurA* in *C. aggregans* in the cyanobacterial mats in the present investigation showed patterns similar to those of Cr and other genes involved in the respiratory electron transport chain, with one significant peak in the early morning at 05:00 and another under microoxic and anaerobic conditions in the afternoon and evening (15:00 until 18:00; Figure 4). In contrast, *aurB* was significantly higher transcribed during a high-light period (11:00) as well as at 05:00.

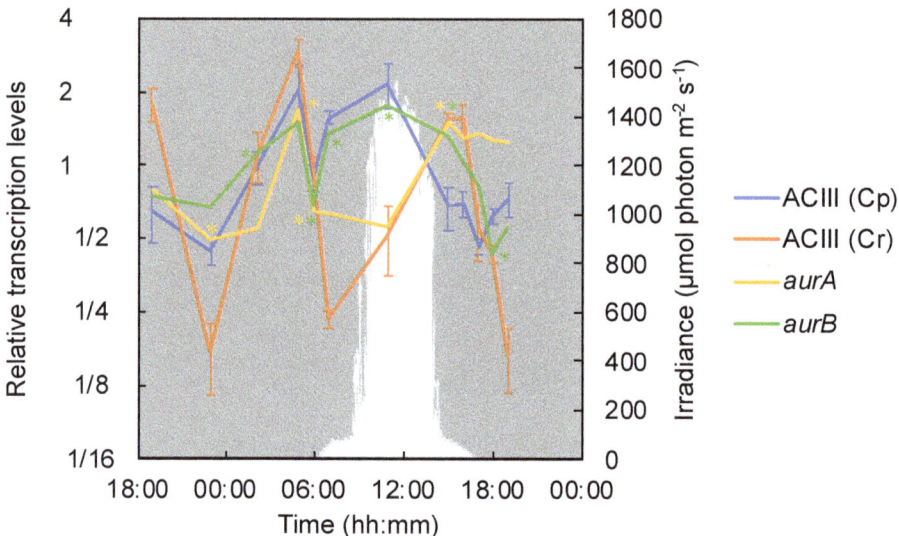

Figure 4. Relative transcriptional levels for alternative complex III (ACIII) and auracyanin. The mean values of ACIII (Cp, Cagg_3382–3383, and 3385–3387) for phototrophic electron transfer and ACIII (Cr, Cagg_1523–1527) for chemotrophic electron transfer are represented by a blue line and an orange line, respectively, with standard deviations. The values of *aurA* and *aurB* (Cagg_0327 and 1833) encoding auracyanin are respectively displayed as yellow (*aurA*) and green (*aurB*) lines. The downwelling photon irradiance (PAR; 400–700 nm) is indicated in white. The asterisk indicates the transcription of a particular gene corresponding to the color in a timepoint differed significantly from that in the previous timepoint.

Respiratory Complex IV—i.e., the cytochrome *c* oxidase complex—plays a key role in the reduction of O_2 to H_2O in the respiratory electron transport chain. The oxygen profiles over the diel cycle indicated high O_2 concentrations in the mats during the daytime (Figure 2). In the laboratory, *Chloroflexus* spp. can grow chemoheterotrophically using respiration under aerobic dark conditions. In *Chloroflexus* spp., the cytochrome c oxidase (COX, or Complex IV; EC 1.9.3.1) genes are clustered with the Cr operon (Cagg_1519–1522). Similar to Cr, the average relative transcription levels of these COX genes reached their highest values in the early morning at 05:00 and in the afternoon at 15:00 and 16:00 (Figure 5).

Chloroflexus aggregans possesses a type-B succinate dehydrogenase (Complex II) which comprises one polypeptide and two hemes for a transmembrane cytochrome *b* (*sdhC*) in addition to a flavoprotein subunit (*sdhA*) and iron-sulfur subunit (*sdhB*) [50,51]. Complex II encoded by *sdhCAB* (Cagg_1576–1578) is involved in electron transport as well as in the TCA cycle and the 3-OHP bi-cycle. They showed significant high relative transcription levels in the morning at 05:00 and during the daytime, and were significantly low throughout the night (Figure 5).

F-type ATP synthase is involved in the production of ATP based on the proton motive force obtained by phototrophic as well as respiratory electron chain activity. The ATP-synthase consists of two parts: F^1, which is a catalytic part, and F^0, which is a transmembrane proton channel part [48]; *C. aggregans* DSM9485^T contains a complete gene set for both parts in the genome [11]. Similar to Complex I-1, II, and ACIII Cp, the relative transcription of ATP-synthase showed a diurnal pattern with two peaks, one at 05:00 and the other at 11:00 (Figure 5).

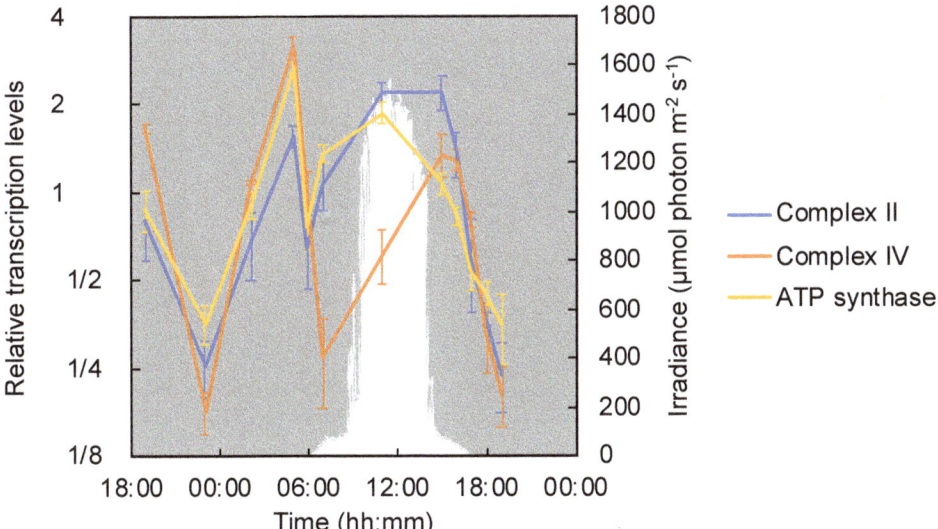

Figure 5. Relative transcription levels of respiratory complex II, IV and ATP synthase. The mean values of the relative transcripts of respiratory complex II (Cagg_1576–1578, blue line), complex IV (Cagg_1519–1522, orange line) and ATP synthase (Cagg_0984—991, yellow line) are shown with standard deviations. The downwelling photon irradiance (PAR; 400–700 nm) is indicated in white.

3.5. 3-Hydroxypropionate Bi-Cycle and Anaplerotic Carbon Fixation

Chloroflexus spp. contain all genes for the 3-hydroxypropionate (3-OHP) bi-cycle, a carbon fixation pathway found only in members of filamentous phototrophic Chloroflexota [11,52–57]. The number of transcripts per million (TPM) for genes encoding key enzymes of the 3-OHP bi-cycle, i.e., malonyl-CoA reductase (Cagg_1256) and 3-hydroxypropionyl-CoA synthase (Cagg_3394) [23], were considerably higher than the average of all genes and appeared relatively stable over the diel cycle (Table S5). Although transcription was detected at all times, the relative transcripts of the two key enzyme genes on average peaked at 15:00. The second highest peak of 3-OHP bi-cycle key enzyme genes was detected at 05:00 before sunrise, and again at 07:00. Under the oxic, high-light conditions, at 11:00, the relative transcriptional levels of the genes encoding key enzymes of 3-OHP bi-cycle were the lowest (Figure 6).

Filamentous anoxygenic phototrophs also contain anaplerotic pathways for incorporating inorganic carbon, such as phosphoenolpyruvate carboxylase (*ppc*, Cagg_0399), which catalyzes the unidirectional production of oxaloacetate from phosphoenolpyruvate [11,58,59]. Transcripts of *ppc* were abundant during the daytime under aerobic, highlight conditions (Figure 6).

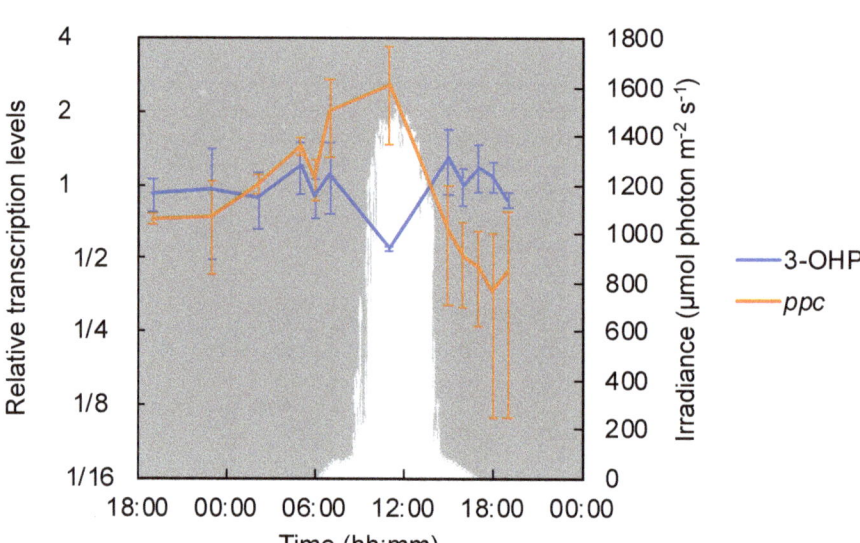

Figure 6. Relative transcription levels of genes encoding key enzymes of the 3-OHP bi-cycle and related enzymes of the anaplerotic pathway in cyanobacterial mats. The mean values of the relative transcription levels of key 3-OHP enzymes, i.e., malonyl-CoA reductase (Cagg_1256) and propionyl-CoA synthase (Cagg_3394), plus that of phosphoenolpyruvate carboxylase (*ppc*, Cagg_0058 and 0399) are represented by a blue line and an orange line, respectively, with standard deviations. The downwelling photon irradiance (PAR; 400–700 nm) is indicated in white.

3.6. Electron Donors: Hydrogenase, Sulfide: Quinone Reductase and CO-Dehydrogenase

Recent studies demonstrated that *C. aggregans* has the capability to use sulfide as well as H_2 as an electron donor for photoautotrophic growth [9,10]. Genome analyses have suggested that carbon monoxide can also serve as a potential electron donor [11]. The correlation between the relative transcriptions of genes encoding hydrogenases, sulfide:quinone reductase and carbon monoxide dehydrogenase and the genes encoding key enzymes in the 3-OHP bi-cycle was analyzed in order to predict autotrophic metabolism.

C. aggregans contains two Ni-Fe hydrogenases: a bidirectional hydrogenase (Cagg_2476–2480) and an uptake hydrogenase (*hyd*, Cagg_0470–0471)–that can provide electrons for autotrophic growth [60,61]. Relative transcription levels of *hyd* genes encoding the uptake hydrogenase significantly increased in the afternoon (at 15:00) shortly after the direct solar illumination of the mats ended around 14:00 and anaerobic conditions were established in the deeper layers of the mat, as well as in the early morning at 05:00 (Figures 2 and 7). The relative transcription levels for genes encoding the bidirectional hydrogenase (Cagg_2476–2480) peaked a little later in the evening after sunset (17:00, PAR=0), stayed high throughout the night, and then decreased during the day. Nickel transporter genes showed the same high relative transcription pattern as the *hyd* genes, with peaks in the early morning and the afternoon (Figure 7).

Chloroflexus spp. contain type-II sulfide:quinone oxidoreductase (SQR), which oxidizes sulfide to elemental sulfur, but lack the *dsr* genes, which encode genes involved in the oxidation of elemental sulfur to sulfate as observed in green and purple sulfur bacteria [11,62], and also lack the *sox* system, which oxidizes elemental sulfur and thiosulfate to sulfate and is widespread in chemoautotrophic sulfur oxidizers [63]. In the present study, a significant increase of relative transcription levels with the highest peak of *sqr* was detected in the afternoon, at 15:00 (Figure 7) under microaerobic to anaerobic low-light conditions in the mat.

As a third possibility, based on the presence of genes encoding carbon monoxide dehydrogenase (*coxGSML*, Cagg_0971–0974) in the genome, the capability of *Chloroflexus* spp. to utilize CO as an electron donor and/or carbon source during aerobic or microaerobic growth has been discussed [11]. In the present investigation, the *coxGSML* genes were significantly higher transcribed during high-light conditions around noon (Figure 7) as well as during the afternoon as the mats turned anoxic. A significant increase in the relative transcription of *cox* genes was seen under anaerobic, low-light conditions in the morning at 07:00 together with a spike in the relative transcription for genes encoding key enzymes of the 3-OHP bi-cycle. The relative transcription of hydrogenase and *sqr* genes were clearly and significantly decreased at that time.

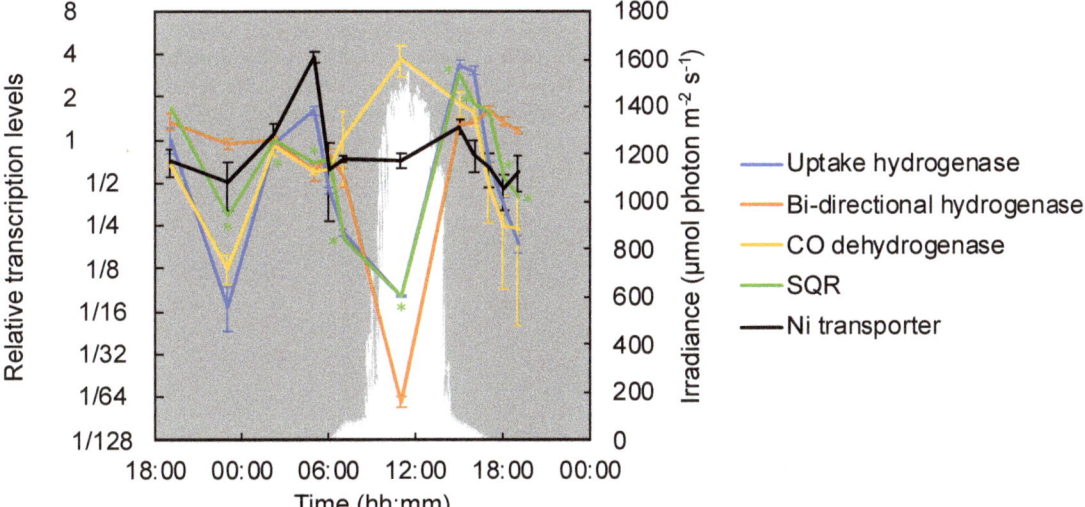

Figure 7. Relative transcription levels of genes encoding hydrogenases, sulfide:quinone oxidoreductase and nickel transporter in cyanobacterial mats. The mean values of the relative transcription levels for type-I uptake Ni-Fe hydrogenase genes *hydAB* (Cagg_0470–0471, blue line), bi-directional Ni-Fe hydrogenase genes homologous to *frhA*, *frhG*, *hoxU*, *nuoF* and *hoxE* (Cagg_2476–2480, orange line), carbon monoxide (CO) dehydrogenase genes *coxGSML* (Cagg_0971–0974, gray line), and nickel transporter genes (Cagg_1273–1276, light blue line) are shown with standard deviations. The relative transcription levels of the type-II sulfide:quinone oxidoreductase gene (*sqr*, Cagg_0045) are represented by the black line. The downwelling photon irradiance (PAR; 400–700 nm) is indicated in *white*. The asterisk indicates the transcription of a particular gene corresponding to the color in a timepoint differed significantly from that in the previous timepoint.

3.7. Carbohydrate Metabolism and the TCA Cycle

Chloroflexus aggregans contains the gene set for the pentose phosphate pathway (PP), including the key enzymes for the oxidative phase involved in anabolic pathways, i.e., glucose-6-phosphate dehydrogenase (Cagg_3190) and 6-phosphogluconate dehydrogenase (Cagg_3189) [23]. Herein, the transcription levels of the two key enzyme genes showed a significant diurnal pattern with the highest relative transcription under high-light conditions (11:00) and an additional peak at 05:00 (Figure 8). Similar to the genes of the oxidative pentose phosphate pathway, high relative transcription levels were also detected at those times for other genes that are indicative of active metabolism and growth, such as DNA gyrase, DNA polymerase, and RNA polymerase (Figure S2).

Because many of the enzymes involved in glycolysis are bi-directional and similarly used in gluconeogenesis, the transcriptional patterns of genes related to glycolysis were analyzed by focusing on the unidirectional enzyme, 6-phosphofructokinase (Cagg_3643), which irreversibly catalyzes the reaction from fructose-1,6-phosphate to fructose-6-bisphosphate to predict chemoheterotrophic growth and catabolism. In *C. aggregans*, two genes are annotated as genes encoding 6-phosphofructokinase (Cagg_3643 and Cagg_2702). These two genes differ considerably in length, with Cagg_2702 encoding a protein identified as a member of the 6PF1K_euk superfamily, the eukaryotic type of the 6-phosphofructokinase, which is almost twice as long as the 'bacterial' 6-phosphofructokinase version, represented by Cagg_3643 (747 aa vs. 356 aa) [64]. Homologs of the Cagg_2702 gene are present in many *Chloroflexota* genomes as identified by a BLAST search, but are not generally present in many other bacteria. The two genes differed in the relative transcription levels and patterns. Cagg_2702 showed the same diel transcription pattern as other genes involved in glycolysis, with its highest relative transcription during high-light conditions at 11:00 (see Table S5). In contrast, Cagg_3643 showed the highest relative transcription under anaerobic conditions in the evening and the lowest transcription levels during superoxic, high-light conditions (Figure 8), but it showed considerably higher absolute transcription levels (TPM average of 1312.12 vs. 9.28 for Cagg_2702). Because Cagg_3643 represents the 'bacterial' type of the enzyme (with higher similarity to the 6-phosphofructokinase in *E. coli*) and since it showed higher absolute transcription levels, it is hypothesized that it represents the unidirectional gene involved in the oxidative activity of glycolysis in *C. aggregans*.

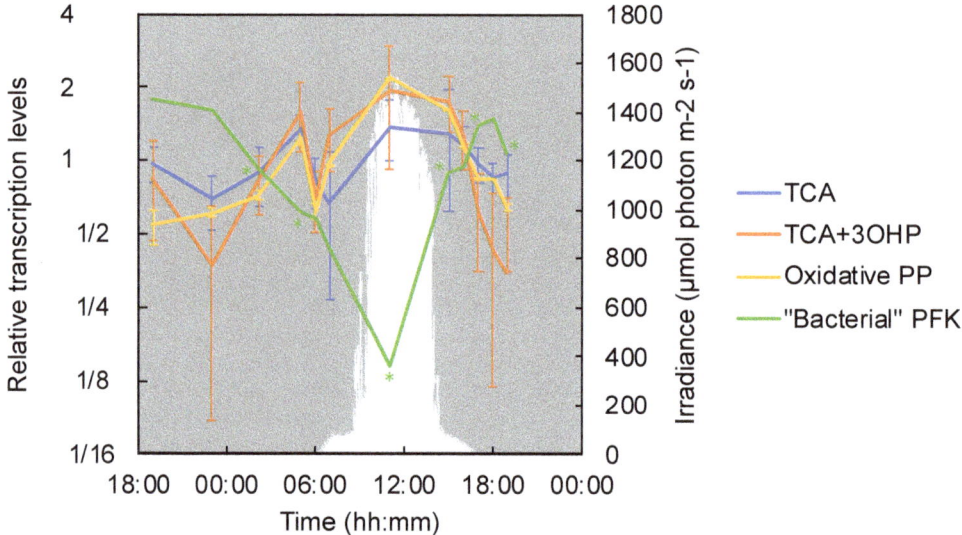

Figure 8. Relative transcription levels of genes for carbohydrate metabolism. The mean values of the relative transcriptional levels for the TCA cycle (Cagg_3738, 3721, 2500, and 2290) and common enzymes of the TCA cycle and the 3-OHP bi-cycle that is labeled as 'TCA+3-OHP' (Cagg_2086, 2819, and 1576–1578) are represented by a blue line and an orange line with standard deviations. The mean values of the relative transcripts of genes encoding key enzymes of the pentose phosphate pathway (Oxidative PP)—i.e., glucose-6-phosphate dehydrogenase (Cagg_3190) and 6-phosphogluconate dehydrogenase (Cagg_3189)—are represented by the yellow line with standard deviation. The values of the relative transcription levels are displayed for genes encoding the 'bacterial' type of the 6-phosphofructokinase that is labeled as "Bacterial" PFK (Cagg_3643, green line). The downwelling photon irradiance (PAR; 400–700 nm) is indicated in white. The asterisk indicates the transcription of a particular gene corresponding to the color in a timepoint differed significantly from that in the previous timepoint.

The oxidative TCA cycle is important for oxygen-respiring heterotrophic organisms, and all *Chloroflexus* species are known to have the ability to grow chemoheterotrophically [1–3] with oxygen as the terminal electron acceptor. Some of the reactions involved in the TCA cycle—e.g., the conversions from succinyl-CoA to malate—are also part of the 3-OHP bi-cycle [65]. Therefore, the transcriptional patterns of succinyl-CoA synthase (Cagg_2086 and

Cagg_2819), succinate dehydrogenase (Cagg_1576–1578), and fumarate lyase (Cagg_2500) are labeled as 'TCA+3-OHP' and the genes that are exclusively present in the TCA cycle are labeled as 'TCA-only' (Figure 8). Genes involved in the TCA cycle were transcribed relatively evenly over the diel cycle, with two peaks: one in the early morning at 05:00 and the other increasing during the day from 07:00 to 15:00. At 05:00, genes encoding acetate/CoA ligase (Cagg_3789), which catalyzes the production of acetyl-CoA from acetate, also showed significantly higher relative transcription levels (Table S5). The two TCA-affiliated gene groups showed only small differences in their relative transcription patterns. After a small peak at 06:00 for all TCA-related genes, the relative transcription of the 'TCA+3-OHP' genes had already increased at 07:00, whereas the relative transcription levels of the 'TCA-only' genes were low at that time and showed a small increase later in the morning (Figure 8).

3.8. Transcription of Oxygen Protection Genes

Genes for two oxidative stress-protection enzymes present in the *C. aggregans* genome were analyzed in this study: superoxide dismutase (Cagg_2494) and two copies of a glutathione peroxidase (1: Cagg_0324 and 2: Cagg_0446). Transcripts for the enzymes showed their highest relative transcription levels during the daytime, when both O_2 and light were present (Figure S3). The gene encoding superoxide dismutase, i.e., an enzyme-detoxifying reactive oxygen species, exhibited a second peak of relative transcripts in the early morning (at 05:00), when cytochrome c oxidase genes were also highly transcribed (see Section 3.4 above, "Phototrophic and Respiratory Electron Transport"). Glutathione peroxidase reduces lipid hydroperoxides to their corresponding alcohols and reduces free hydrogen peroxide to water. Significant high relative transcription levels of glutathione peroxidase-encoding genes in *C. aggregans* were observed at 11:00 (Figure S3), thus indicating the possible presence of not only O_2 but also hydrogen peroxide.

Hydrogen peroxide is produced as a by-product in the oxidation of glycolate to glyoxylate by glycolate oxidase [66]. The encoding genes (*glcDEF*, Cagg_1528, Cagg_1530–1531, and Cagg_1892–1893) showed patterns similar to that of glutathione peroxidase, with the highest relative transcription levels during midday, indicating both the presence of glycolate in the mat environment and its oxidation by *C. aggregans* (Figure S3). The oxidation of glycolate may be linked to a photoheterotrophic metabolism, which is further supported by the high relative transcript levels for genes encoding glycoside, sugar, and amino acid transporter genes during this time under high-light conditions (Figures S4 and S5).

4. Discussion

4.1. Light and O_2 Dynamics Shape the Environmental Conditions for C. aggregans

The results of microsensor analyses revealed a strong correlation between solar irradiance and the O_2 concentration and penetration depth in the cyanobacterial microbial mats from Nakabusa Hot Springs (Figure 2). Since no increase in O_2 concentration was observed in the bottom layer lower than 1mm in depth, oxygenic cyanobacteria was supposed to be absent in the undermat as supported by the Martinez et al. [13]. The diel changes between anoxia and hyperoxic conditions driven by the oxygenic activity of cyanobacteria lead to drastic changes in the conditions for their microbial metabolism. Consequently, mat-inhabiting microbes may be under optimal conditions during only part of the diel cycle, and may need to endure unfavorable conditions at other times. Under such conditions, a versatile metabolism is thought to be advantageous in terms of ensuring continuous energy production under dynamic environmental conditions. In the following sections it is discussed how *Chloroflexus aggregans* use their metabolic flexibility to thrive in the highly variable conditions in the microbial mat over a diel cycle.

4.2. Low Light and Low O_2 Dominated the Morning Hours (07:00)

After sunrise, although no direct sunlight hit the mats, the irradiance from diffuse light increased and stimulated cyanobacterial oxygenic photosynthesis, as indicated by the increasing O$_2$ concentrations at the mat surface as well as the significant increase in the relative transcription levels of genes for protection against reactive oxygen species in *C. aggregans*. However, deeper mat layers were still anoxic, which in combination with low-light conditions seems to provide suitable conditions for anoxygenic photosynthesis by FAPs such as *Chloroflexus spp.* [67]. The increasing transcription of genes encoding housekeeping enzymes suggested increasingly active metabolism (anabolism) in *C. aggregans* (Figure S2). In the morning, the transcription of the phototrophy-affiliated ACIII Cp gene increased slowly and the transcription levels of genes for ACIII Cr, *aurA* and cytochrome *c* oxidase decreased, which in combination indicate active phototrophy. Photoheterotrophy of *C. aggregans* is suggested by increases in the transcription of TCA-related genes, probably using fermentation products in the mats that accumulated during the nighttime, as reported in similar hot spring systems in Yellowstone National Park [68–72]. At the same time, photoautotrophy (and thus photomixotrophy) is indicated by the increased transcription of key 3-OHP genes, probably supported by anaplerotic carbon fixation as indicated by the increased transcription of phosphoenolpyruvate carboxylase genes (Figure 6).

Unexpectedly, neither molecular hydrogen nor sulfide seems to function as an electron donor for autotrophic growth at this time, as neither *hyd* nor *sqr* genes were highly transcribed. In contrast, a significant increase in the relative transcription levels of *cox* gene transcription was seen starting at 07:00, indicating that carbon monoxide is a potential electron source for photoautotrophic metabolism in the early morning, as hypothesized for the CO utilization of *Roseiflexus* spp. and *C. aurantiacus* based on the genomic analysis [11,37].

4.3. High-Light and Super-Oxygenated Midday Hours (11:00)

Oxygen started to accumulate in the upper mat layers from around 09:00 and with time also accumulated in the deeper mat layers. Hyperoxic O$_2$ levels (>800 µmol O$_2$ L^{-1}) were observed in the uppermost mat layers and O$_2$ penetration reached a >2 mm depth under high irradiance (approx. 1000–1500 µmol photons m^{-2} s^{-1}; 400–700 nm) between 10:00 and 14:00. The relative transcription levels of genes encoding DNA gyrase, DNA/RNA polymerase and ATP synthase peaked at the same time (Figure 5 and Figure S2), indicating active growth and energy production of *C. aggregans* during midday high-light and oxic conditions.

In the laboratory, *C. aggregans* and other FAPs have long been known to grow chemoheterotrophically via oxic respiration under aerobic dark conditions [2]. Additionally, aerobic growth in the light has been shown for *C. aurantiacus* only very recently [73]. In the present study, despite the presence of O$_2$ in the upper 2 mm of the mat during this period of the day, there was no indication of chemoheterotrophic growth or aerobic respiration. These results suggest the absence of both active glycolysis (as indicated by low relative transcription of the 'bacterial' type of the 6-phosphofructokinase gene) and aerobic respiration (as inferred from the transcription data of cytochrome *c* oxidase genes and the ACIII Cr genes). In contrast, active phototrophy in the presence of O$_2$ is suggested by the high relative transcription levels of genes involved in electron transport, including ACIII Cp, *aurB*, and the TCA cycle. Structural analyses indicated that AurA and AurB in *C. aurantiacus* are active during phototrophic growth and chemotrophic growth, respectively [74]. In contrast, the transcription of *aurB* under phototrophic conditions in this study is consistent with the data obtained in a proteomic study of *C. aurantiacus* in which AurA was more abundant during chemoheterotrophic growth, while AurB was observed under photoheterotrophic conditions [45]. While the phototrophic growth under aerobic conditions correlates with earlier laboratory studies demonstrating that O$_2$ did not inhibit energy transfer between the chlorosomes and the reaction center in *C. aurantiacus* [75].

However, bacteriochlorophyll biosynthesis in *C. aurantiacus* was long believed to be inhibited by O$_2$, and enzymes involved in the biosynthesis were only detected or significantly increased in cultures grown under anaerobic phototrophic conditions [45]. Accordingly, metatranscriptomic studies of alkaline hot spring microbial mats in Yellowstone National Park showed that the transcripts of most of the genes involved in the biosynthesis of

BChls in different FAPs, including *Chloroflexus* spp., were the most abundant at night [23]. Inhibitory effects on the biosynthesis of the photosynthetic apparatus under aerobic conditions in *C. aurantiacus* were also shown in the laboratory [76]. In contrast, transcription of phototrophy-related genes under aerobic light conditions were shown for *C. aurantiacus* only very recently [73].

In accordance with the expectation that oxygen represses the biosynthesis of phototrophic apparatus and pigments, nocturnal transcription patterns of the photosynthetic apparatus-related *puf* and *csm* genes were also observed in the present study and might be negatively correlated with the increasing O_2 concentrations in the environment. In contrast, many of the *bch* genes were observed to be transcribed during the day, with the majority showing the highest relative transcription levels at 11:00. This correlates with a recent study of *C. aurantiacus* showing biosynthesis of BChl *a* and *c* under anaerobic as well as aerobic conditions in the light, while being suppressed in the presence of O_2 in the dark [73]. This finding supports the hypothesis that *C. aggregans* in Nakabusa Hot Springs cyanobacterial mats can produce BChls for phototrophic growth under aerobic conditions during the daytime.

The high relative transcription levels for glutathione peroxidase-encoding genes during the midday not only reflect high O_2 levels in the environment; they might also indicate high H_2O_2 levels, since the main function of this enzyme is to reduce lipid hydroperoxides to their corresponding alcohols and to reduce free hydrogen peroxide to water [77,78]. Hydrogen peroxide can be formed via photochemical reactions with dissolved organic carbon in hot spring waters [79], and H_2O_2 is produced as a byproduct of the oxidation of glycolate (to glyoxylate) [66], which has been shown to be present under high-light conditions in hot spring microbial mats, presumably produced by photoinhibited cyanobacteria [69]. The presence of glycolate and its oxidation at this time of day (11:00) is further suggested by the gene transcription pattern of glycolate oxidase genes, which showed patterns similar to those of the glutathione peroxidase, with highest relative transcription levels during the midday (Figure S3). Glycolate oxidase transcription activity supports the hypothesis that *Chloroflexus* species photoassimilate the glycolate supplied by the cyanobacteria in such microbial mats [80]. This indicates a photoheterotrophic metabolism for *C. aggregans*, which is further supported by the high relative transcript levels for glycoside, sugar, and amino acid transporter genes during high-light conditions (Figures S4 and S5).

Carbon monoxide (CO) might function as an electron and/or carbon source under aerobic conditions during this time of day, as aerobic carbon monoxide dehydrogenase genes are found to peak at midday [23]. The transcription profile of carbon monoxide dehydrogenase genes was related to the transcription pattern of the phosphoenolpyruvate carboxylase gene (Figure 6), which indicates that CO might be converted to CO_2, which then is incorporated by the phosphoenolpyruvate carboxylase-catalyzed anaplerotic reaction (phosphoenolpyruvate to oxaloacetate). In *Thermomicrobium roseum*, an obligately aerobic chemoheterotroph in the phylum *Chloroflexota*, carbon monoxide dehydrogenase is utilized to produce ATP and NADPH under aerobic conditions [81]. It is speculated that *C. aggregans* may use CO for both anaplerotic carbon fixation and supplemental energy production in the presence of O_2 due to the limitation of available CO_2 caused by active cyanobacterial photosynthetic carbon fixation.

Two gene clusters encoding the "respiratory" complex I (NADH:menaquinone oxidoreductase, *nuo* genes) are present in *C. aggregans* and *C. aurantiacus* [54,82], as well as in the red FAP species *Roseiflexus castenholzii* and *Roseiflexus* sp. RS-1 (acc. nos. CP000804 and CP000686, respectively [23,37]). To our knowledge, neither of these clusters has been affiliated with phototrophic electron transport. The different relative transcription patterns obtained in the present study might indicate that the 14-gene set is used primarily in photosynthesis and the 12-gene set is used primarily in respiratory electron transport. However, two sets of *nuo* operons have also been described for other, non-phototrophic bacteria such as *Ignavibacterium album* [83] and "*Candidatus* Thermonerobacter thiotrophicus" [18], which might indicate specification to different O_2 levels rather than phototrophic versus respiratory electron transport. If true, this might indicate a potential higher O_2 tolerance for the 14-gene set and a higher oxygen affinity for the 12-gene set. Because these results indicate that chemoheterotrophic, respiratory metabolism does not take place during highlight and superoxic conditions at midday, when the 14-gene set is highly transcribed, it is hypothesized that this NADH:menaquinone oxidoreductase plays a role in phototrophic electron transport, perhaps by donating electrons from NADH similar to as it has been proposed to be the case in the cyclic electron transport chain in heliobacteria [84]. However, further biochemical analyses are required to precisely determine the different functions of the two NADH: menaquinone oxidoreductases in *Chloroflexus*.

4.4. Low Light and Low O_2 Dominated the Afternoon Hours (15:00–16:00)

In the afternoon, a substantial decrease in solar irradiance and O_2 was observed after 14:00, as the sun set behind the surrounding mountains. Between 15:00 and 16:00, O_2 was still detected only in the upper layer of the mat (Figure 2), and *C. aggregans* is hypothesized to experience microaerobic conditions enabling aerobic chemoheterotrophic metabolism, as indicated by the high relative transcription levels of TCA cycle, ACIII Cr, *aurA*, and cytochrome *c* oxidase-encoding genes (Figures 4, 5 and 8).

Simultaneously, high relative transcription levels of the gene encoding the O_2-sensitive version of Mg-protoporphyrin monomethylester cyclase, *bchE*, indicated that part of the *C. aggregans* population was exposed to anoxic conditions, at least in the deeper mat layers, as supported by the microsensor data of the vertical O_2 distribution in the mat (Figures 2 and 3). High relative transcription levels of genes involved in the 3-OHP bi-cycle suggest autotrophic growth, especially under anaerobic conditions, where sulfide and/or H_2 is available [85]. Sulfide concentrations are expected to rise under anaerobic conditions due to biological sulfate-reduction, as was similarly shown for the bacterial community of Mushroom Spring in Yellowstone National Park [86]. The sulfide oxidation capabilities of *C. aggregans* in the Nakabusa mats are supported by high relative transcription levels of *sqr* for utilization of sulfide as an electron donor [87]. At the same time, uptake hydrogenase genes are transcribed, pointing to the use of molecular hydrogen for autotrophic growth (Figure 7). However, although anoxic low-light conditions were prevalent in the deeper layers of the microbial mats for approx. 2 h before sunset, phototrophy does not seem to be the predominating metabolic growth mode during this time of day, as the decreasing relative transcription levels of ACIII Cp suggest. Thus, sulfide- and H_2-oxidizing enzymes of *C. aggregans* may be employed for aerobic chemoautotrophic metabolism instead, or additionally for photoautotrophic metabolism at dusk, as suggested by the transcriptional peaks of ACIII Cr, *aurA* and cytochrome *c* oxidase (Figures 4 and 5).

4.5. Dark and Anoxic Nighttime Hours (17:00–19:00, 23:00, 02:10)

After 17:00, the microbial mat community experienced dark and anoxic conditions. The low relative transcription levels for housekeeping genes such as DNA gyrase and DNA/RNA polymerases indicate low metabolic activity of *C. aggregans* during this period. The low transcription values for transporter genes support this conclusion. The electron transport gene transcripts for both phototrophy and respiration as well as for the ATP-synthase genes were low. The unidirectional glycolysis enzyme 6-phosphofructokinase gene showed significant changes towards high relative transcription levels at the beginning of the night while TCA cycle-related gene transcriptions were low, suggesting fermentative metabolism and possibly the degradation of internal glycogen storage. This has been shown for *C. aurantiacus* in the laboratory [56] and has been suggested for FAPs inhabiting hot spring microbial mats in Yellowstone National Park [23]. High relative transcription levels of a bi-directional hydrogenase further support the fermentative growth mode and the potential production of H_2 by *C. aggregans* during the night. This is in accordance with the recent detection of H_2 production under fermentative conditions (dark, anaerobic) in *C. aggregans* strain NA9-6 (unpublished data).

4.6. Early Morning Hours (05:00)

Under dark, anoxic conditions in the early morning hours, an unexpected significant increase in the relative transcription of genes encoding the respiratory chain components (respiratory complexes I, II, and IV and ACIII Cr) as well as ATP-synthase was observed. It is suggested that the transcription of these genes is indicative of the occurrence of chemotrophic growth involving O_2 respiration at that time of day. Although the microsensor measurements showed no presence of O_2 in the mats until later in the morning, the transcription of genes encoding enzymes involved in O_2 protection, such as superoxide dismutase and glutathione peroxidase, indicated a (micro)aerobic environment for *C. aggregans* at that time of day, and trace amounts of O_2 were detected at the very surface throughout the night. As *C. aggregans* is known to have gliding motility and chemotaxis toward reduced O_2 concentrations [2,88–91], it is speculated that *C. aggregans* migrates from anaerobic deeper layers to the micro-oxic surface layers in the early morning, in which a diffusive supply of O_2 from the overlying water leads to microaerobic conditions during the nighttime, as has been suggested previously [23,92].

Similar to the ACIII Cr operon, the genes of the ACIII Cp operon, presumed to be involved in phototrophy, were also highly transcribed at 05:00. However, active phototrophy is ruled out due to the lack of light. Because the red filamentous anoxygenic phototrophic members of *Chloroflexota*—i.e., *Roseiflexus* spp.—contain only one copy of Cp-like ACIII genes, which are predicted to work under both phototrophic and chemotrophic conditions [23], the Cp-related genes in *C. aggregans* might also function under chemotrophic growth in the mats.

A capacity for chemoautotrophic growth was very recently observed in *Chloroflexus* spp. isolates obtained from Nakabusa Hot Springs microbial mats [8]. The high relative transcription levels of *hyd* uptake hydrogenase and Ni-transporter genes further suggest the use of H_2 as an electron donor for the aerobic chemoautotrophic growth in the microoxic surface layers of the cyanobacterial mats around 05:00. Additionally, the high transcription levels of genes for the TCA cycle and acetate/CoA ligase—the latter of which catalyzes the production of acetyl-CoA from acetate—indicate that acetate, supplied mainly from the fermentation of co-existing microbes as shown in similar mats in Yellowstone National Park, might be taken up at this time of day [22,68,93,94] (Table S5). This points to the possibility of an assimilation of acetate in addition to the purely autotrophic metabolism, suggesting the chemomixotrophic lifestyle of *C. aggregans* during predawn.

5. Conclusions

This study suggests that *C. aggregans* uses its metabolic flexibility and capability for both phototrophic and chemotrophic growth to optimize its performance under the varying environmental conditions in its natural habitat, the microbial mat community at Nakabusa Hot Springs. The main ATP-generating and thus metabolically most-active times are not only the high-light hours around midday (phototrophy), but—most notably—also the early morning hours around 05:00, when the cells are hypothesized to conduct chemomixotrophic growth (Figure 9).

Genes for the biosynthesis of the photosynthetic apparatus are predominantly transcribed during the night; however, photosynthesis is active during the light hours in the morning, midday and afternoon. Under low O_2 concentrations in the dim morning light (≤ 100 µmol photons m^{-2} s^{-1}), photoauto/mixotrophic metabolism potentially using CO as an electron donor is suggested to be the major energy source for *C. aggregans* in the cyanobacteria-dominated mats. Later on, under midday high-light conditions, intense oxygenic photosynthesis by cyanobacteria renders the upper millimeters of the microbial mat highly oxic. However, O_2 respiration in *C. aggregans* does not seem to take place under these conditions. Instead, photoheterotrophic growth (and the assimilation of glycolate) is most likely the dominant lifestyle, supplemented with a certain degree of anaplerotic CO_2 fixation. In the afternoon, under anaerobic light conditions, photoautotrophic or photomixotrophic growth with sulfide and/or H_2 as the electron donor takes place in the deeper mat layers, and aerobic respiration and chemoheterotrophic growth are hypothesized for the cells in the upper layers. At nighttime, chemoheterotrophic fermentative growth and the production of H_2 may take place. In the late-night/early morning hours, at around 05:00, *Chloroflexus* migrates to the mat surface and undergoes mixotrophic growth with H_2 and O_2 prior to sunrise, after which *C. aggregans* switches back to phototrophy.

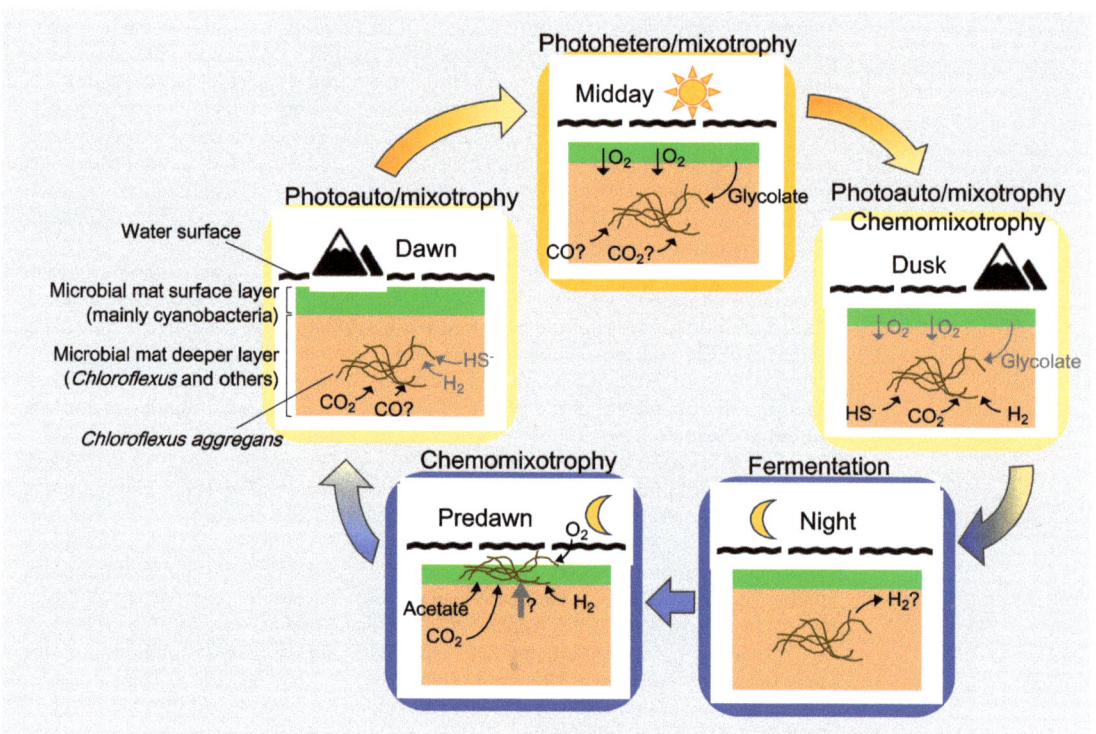

Figure 9. Proposed diel growth modes of *C. aggregans* (indicated by brown filaments) in cyanobacterial mats based on in situ metatranscriptomic and microsensor analyses in the cyanobacterial mats of Nakabusa Hot Springs. The green area represents the green upper layer containing oxygenic phototrophs (i.e., cyanobacteria), while the orange area corresponds to the orange colored undermat. The black curvy line indicated the overflowing water surface.

Supplementary Materials: The following are available online at https://www.mdpi.com/2076-2607/9/3/652/s1, Figure S1: Relative transcription levels of paralogous respiratory complex I; Figure S2: Relative transcription levels of housekeeping genes; Figure S3: Relative transcription levels of genes related to oxidative stresses; Figure S4: Relative transcription levels of amino acid and oligopeptide transporter genes; Figure S5: Relative transcription levels of glycoside and sugar transporter genes; Table S1: Transcriptomic profiles and differentially transcribed genes; Table S2: Differential transcription analyses of genes involved in energy metabolisms of *C. aggregans*; Table S3: Summary of differentially transcribed genes; Table S4: Transcription patterns of *bch* genes; Table S5: Transcription patterns of genes involved in energy metabolisms of *C. aggregans*.

Author Contributions: Conceptualization, S.K. and V.T.; Transcriptomic data analysis, S.K.; RNA extraction, J.N.M.; Microsensor measurements and analysis, M.L., E.T., and M.K.; Sample collection, J.N.M., M.T., A.N., and V.T.; Writing—original draft preparation, S.K. and V.T.; Writing—review and editing, S.K., V.T., M.T., M.K., M.L., E.T., S.H. (Shin Haruta), A.N., and J.N.M.; Supervision, V.T. and S.H. (Shin Haruta); Funding acquisition, S.H. (Satoshi Hanada), M.K., and A.N. All authors have read and agreed to the published version of the manuscript.

Funding: This study was supported by the Institute for Fermentation, Osaka (IFO) to S.H. (Satoshi Hanada), grants from the Independent Research Fund Denmark (DFF-1323-00065B; DFF-8021-00308B) to M.K., and the fund from international practice course of Tokyo Metropolitan University to A.N.

Institutional Review Board Statement: Not applicable.

Informed Consent Statement: Not applicable.

Data Availability Statement: The sequence collected in this study is available under NCBI BioProject accession number PRJNA715822.

Acknowledgments: The authors would like to thank the owner of Nakabusa Hot Spring (Takahito Momose) for the permission to collect samples from the hot spring.

Conflicts of Interest: The authors declare no conflict of interest.

References

1. Pierson, B.K.; Castenholz, R.W. A phototrophic gliding filamentous bacterium of hot springs, *Chloroflexus aurantiacus*, gen. and sp. nov. *Arch. Microbiol.* **1974**, *100*, 5–24. [CrossRef]
2. Hanada, S.; Hiraishi, A.; Shimada, K.; Matsuura, K. *Chloroflexus aggregans* sp. nov., a filamentous phototrophic bacterium which forms dense cell aggregates by active gliding movement. *Int. J. Syst. Bacteriol.* **1995**, *45*, 676–681. [CrossRef] [PubMed]
3. Gaisin, V.A.; Kalashnikov, A.M.; Grouzdev, D.S.; Sukhacheva, M.V.; Kuznetsov, B.B.; Gorlenko, V.M. *Chloroflexus islandicus* sp. nov., a thermophilic filamentous anoxygenic phototrophic bacterium from a geyser. *Int. J. Syst. Evol. Microbiol.* **2017**, *67*, 1381–1386. [CrossRef] [PubMed]
4. Madigan, M.T.; Petersen, S.R.; Brock, T.D. Nutritional studies on *Chloroflexus*, a filamentous photosynthetic, gliding bacterium. *Arch. Microbiol.* **1974**, *100*, 97–103. [CrossRef]
5. Sirevag, R.; Castenholz, R. Aspects of carbon metabolism in *Chloroflexus*. *Arch. Microbiol.* **1979**, *120*, 151–153. [CrossRef]
6. Holo, H.; Sirevag, R. Autotrophic growth and CO_2 fixation of *Chloroflexus aurantiacus*. *Arch. Microbiol.* **1986**, *145*, 173–180. [CrossRef]
7. Thiel, V.; Hamilton, T.L.; Tomsho, L.P.; Burhans, R.; Gay, S.E.; Schuster, S.C.; Ward, D.M.; Bryant, D.A. Draft genome sequence of a sulfide-oxidizing, autotrophic filamentous anoxygenic phototrophic bacterium, *Chloroflexus sp.* strain MS-G (*Chloroflexi*). *Genome Announc.* **2014**, *2*, 9–10. [CrossRef] [PubMed]
8. Kawai, S.; Nishihara, A.; Matsuura, K.; Haruta, S. Hydrogen-dependent autotrophic growth in phototrophic and chemolithotrophic cultures of thermophilic bacteria, *Chloroflexus aggregans* and *Chloroflexus aurantiacus*, isolated from Nakabusa hot springs. *FEMS Microbiol. Lett.* **2019**, *366*, 1–6. [CrossRef]
9. Kawai, S.; Kamiya, N.; Matsuura, K.; Haruta, S. Symbiotic growth of a thermophilic sulfide-oxidizing photoautotroph and an elemental sulfur-disproportionating chemolithoautotroph and cooperative dissimilatory oxidation of sulfide to sulfate. *Front. Microbiol.* **2019**, *10*, 1–8. [CrossRef] [PubMed]
10. Kanno, N.; Haruta, S.; Hanada, S. Sulfide-dependent photoautotrophy in the filamentous anoxygenic phototrophic bacterium, *Chloroflexus aggregans*. *Microbes Environ.* **2019**, *34*, 304–309. [CrossRef] [PubMed]
11. Tang, K.H.; Barry, K.; Chertkov, O.; Dalin, E.; Han, C.S.; Hauser, L.J.; Honchak, B.M.; Karbach, L.E.; Land, M.L.; Lapidus, A.; et al. Complete genome sequence of the filamentous anoxygenic phototrophic bacterium *Chloroflexus aurantiacus*. *BMC Genom.* **2011**, *12*, 1–21. [CrossRef] [PubMed]
12. Kondratieva, E.N.; Ivanovsky, R.N.; Krasilnikova, E.N. Carbon metabolism in *Chloroflexus aurantiacus*. *FEMS Microbiol. Lett.* **1992**, *100*, 269–271. [CrossRef]
13. Martinez, J.N.; Nishihara, A.; Lichtenberg, M.; Trampe, E.; Kawai, S.; Tank, M.; Kühl, M.; Hanada, S.; Thiel, V. Vertical distribution and diversity of phototrophic bacteria within a hot spring microbial mat (Nakabusa hot springs, Japan). *Microbes Environ.* **2019**, *34*, 374–387. [CrossRef]
14. Everroad, R.C.; Otaki, H.; Matsuura, K.; Haruta, S. Diversification of bacterial community composition along a temperature gradient at a thermal spring. *Microbes Environ.* **2012**, *27*, 374–381. [CrossRef] [PubMed]
15. Kubo, K.; Knittel, K.; Amann, R.; Fukui, M.; Matsuura, K. Sulfur-metabolizing bacterial populations in microbial mats of the Nakabusa hot spring, Japan. *Syst. Appl. Microbiol.* **2011**, *34*, 293–302. [CrossRef] [PubMed]
16. Otaki, H.; Everroad, R.C.; Matsuura, K.; Haruta, S. Production and consumption of hydrogen in hot spring microbial mats dominated by a filamentous anoxygenic photosynthetic bacterium. *Microbes Environ.* **2012**, *27*, 293–299. [CrossRef] [PubMed]
17. Tamazawa, S.; Yamamoto, K.; Takasaki, K.; Mitani, Y.; Hanada, S.; Kamagata, Y.; Tamaki, H. In situ gene expression responsible for sulfide oxidation and CO_2 fixation of an uncultured large sausage-shaped *Aquificae* bacterium in a sulfidic hot spring. *Microbes Environ.* **2016**, *31*, 194–198. [CrossRef] [PubMed]
18. Thiel, V.; Garcia Costas, A.M.; Fortney, N.W.; Martinez, J.N.; Tank, M.; Roden, E.E.; Boyd, E.S.; Ward, D.M.; Hanada, S.; Bryant, D.A. "Candidatus thermonerobacter thiotrophicus," a non-phototrophic member of the Bacteroidetes/Chlorobi with dissimilatory sulfur metabolism in hot spring mat communities. *Front. Microbiol.* **2019**, *10*, 3159. [CrossRef] [PubMed]

19. Nishihara, A.; Matsuura, K.; Tank, M.; McGlynn, S.E.; Thiel, V.; Haruta, S. Nitrogenase activity in thermophilic chemolithoautotrophic bacteria in the phylum *Aquificae* isolated under nitrogen-fixing conditions from Nakabusa hot springs. *Microbes Environ.* **2018**, *33*, 394–401. [CrossRef] [PubMed]
20. Nishihara, A.; Haruta, S.; McGlynn, S.E.; Thiel, V.; Matsuura, K. Nitrogen fixation in thermophilic chemosynthetic microbial communities depending on hydrogen, sulfate, and carbon dioxide. *Microbes Environ.* **2018**, *33*, 10–18. [CrossRef]
21. Nishihara, A.; Thiel, V.; Matsuura, K.; McGlynn, S.E.; Haruta, S. Phylogenetic diversity of nitrogenase reductase genes and possible nitrogen-fixing bacteria in thermophilic chemosynthetic microbial communities in Nakabusa hot springs. *Microbes Environ.* **2018**, *33*, 357–365. [CrossRef] [PubMed]
22. van der Meer, M.T.J.; Schouten, S.; Bateson, M.M.; Nübel, U.; Wieland, A.; Kühl, M.; De Leeuw, J.W.; Damsté, J.S.S.; Ward, D.M. Diel variations in carbon metabolism by green nonsulfur-like bacteria in alkaline siliceous hot spring microbial mats from Yellowstone National Park. *Appl. Environ. Microbiol.* **2005**, *71*, 3978–3986. [CrossRef] [PubMed]
23. Klatt, C.G.; Liu, Z.; Ludwig, M.; Kühl, M.; Jensen, S.I.; Bryant, D.A.; Ward, D.M. Temporal metatranscriptomic patterning in phototrophic *Chloroflexi* inhabiting a microbial mat in a geothermal spring. *ISME J.* **2013**, *7*, 1775–1789. [CrossRef] [PubMed]
24. Thiel, V.; Wood, J.M.; Olsen, M.T.; Tank, M.; Klatt, C.G.; Ward, D.M.; Bryant, D.A. The dark side of the mushroom spring microbial mat: Life in the shadow of chlorophototrophs. I. Microbial diversity based on 16S rRNA gene amplicons and metagenomic sequencing. *Front. Microbiol.* **2016**, *7*, 1–25. [CrossRef]
25. Hiroyuki, K.; Mori, K.; Nashimoto, H.; Hanada, S.; Kato, K. In situ biomass production of a hot spring sulfur-turf microbial mat. *Microbes Environ.* **2010**, *25*, 140–143.
26. Kato, K.; Kobayashi, T.; Yamamoto, H.; Nakagawa, T.; Maki, Y.; Hoaki, T. Microbial mat boundaries between chemolithotrophs and phototrophs in geothermal hot spring effluents. *Geomicrobiol. J.* **2004**, *21*, 91–98. [CrossRef]
27. Nakagawa, T.; Fukui, M. Phylogenetic characterization of microbial mats and streamers from a Japanese alkaline hot spring with a thermal gradient. *J. Gen. Appl. Microbiol.* **2002**, *48*, 211–222. [CrossRef]
28. Andrews S FastQC. A Quality Control Tool for High Throughput Sequence Data. Available online: http://www.bioinformatics.babraham.ac.uk/projects/fastqc/ (accessed on 29 March 2018).
29. Martin, M. Cutadapt removes adapter sequences from high-throughput sequencing reads. *EMBnet. J.* **2011**, *17*, 10–12. [CrossRef]
30. Langmead, B.; Salzberg, S. Fast gapped-read alignment with Bowtie 2. *Nat. Methods* **2013**, *9*, 357–359. [CrossRef] [PubMed]
31. Magoc, T.; Wood, D.; Salzberg, S.L. EDGE-pro: Eestimated Degree of Gene Expression in prokaryotic genomes. *Evol. Bioinform.* **2013**, *9*, 127–136. [CrossRef] [PubMed]
32. Robinson, M.D.; McCarthy, D.J.; Smyth, G.K. edgeR: A Bioconductor package for differential expression analysis of digital gene expression data. *Bioinformatics* **2009**, *26*, 139–140. [CrossRef]
33. Liu, Z.; Koid, A.E.; Terrado, R.; Campbell, V.; Caron, D.A.; Heidelberg, K.B. Changes in gene expression of Prymnesium parvum induced by nitrogen and phosphorus limitation. *Front. Microbiol.* **2015**, *6*, 1–13. [CrossRef] [PubMed]
34. Tang, K.H.; Wen, J.; Li, X.; Blankenship, R.E. Role of the AcsF protein in *Chloroflexus aurantiacus*. *J. Bacteriol.* **2009**, *191*, 3580–3587. [CrossRef]
35. Frigaard, N.U.; Maqueo Chew, A.G.; Maresca, J.A.; Bryant, D.A. Bacteriochlorophyll biosynthesis in green bacteria. *Chlorophylls Bacteriochlorophylls* **2007**, *25*, 201–221.
36. Chew, A.G.M.; Bryant, D.A. Chlorophyll biosynthesis in bacteria: The origins of structural and functional diversity. *Annu. Rev. Microbiol.* **2007**, *61*, 113–129. [CrossRef] [PubMed]
37. van der Meer, M.T.J.; Klatt, C.G.; Wood, J.; Bryant, D.A.; Bateson, M.M.; Lammerts, L.; Schouten, S.; Sinninghe Damste, J.S.; Madigan, M.T.; Ward, D.M. Cultivation and genomic, nutritional, and lipid biomarker characterization of *Roseiflexus* strains closely related to predominant in situ populations inhabiting Yellowstone hot spring microbial mats. *J. Bacteriol.* **2010**, *192*, 3033–3042. [CrossRef] [PubMed]
38. Garcia Costas, A.M.; Tsukatani, Y.; Rijpstra, W.I.C.; Schouten, S.; Welander, P.V.; Summons, R.E.; Bryant, D.A. Identification of the bacteriochlorophylls, carotenoids, quinones, lipids, and hopanoids of "candidatus chloracidobacterium thermophilum". *J. Bacteriol.* **2012**, *194*, 1158–1168. [CrossRef] [PubMed]
39. Yanyushin, M.F.; Del Rosario, M.C.; Brune, D.C.; Blankenship, R.E. New class of bacterial membrane oxidoreductases. *Biochemistry* **2005**, *44*, 10037–10045. [CrossRef]
40. Gao, X.; Xin, Y.; Blankenship, R.E. Enzymatic activity of the alternative complex III as a menaquinol:auracyanin oxidoreductase in the electron transfer chain of *Chloroflexus aurantiacus*. *FEBS Lett.* **2009**, *583*, 3275–3279. [CrossRef]
41. Gao, X.; Xin, Y.; Bell, P.D.; Wen, J.; Blankenship, R.E. Structural analysis of alternative complex III in the photosynthetic electron transfer chain of *Chloroflexus aurantiacus*. *Biochemistry* **2010**, *49*, 6670–6679. [CrossRef]
42. McManus, J.D.; Brune, D.C.; Han, J.; Sanders-Loehr, J.; Meyer, T.E.; Cusanovich, M.A.; Tollin, G.; Blankenship, R.E. Isolation, characterization, and amino acid sequences of auracyanins, blue copper proteins from the green photosynthetic bacterium *Chloroflexus aurantiacus*. *J. Biol. Chem.* **1992**, *267*, 6531–6540. [CrossRef]
43. Van Driessche, G.; Hu, W.; Van De Werken, G.; Selvaraj, F.; Mcmanus, J.D.; Blankenship, R.E.; Van Beeumen, J.J. Auracyanin a from the thermophilic green gliding photosynthetic bacterium *Chloroflexus aurantiacus* represents an unusual class of small blue copper proteins. *Protein Sci.* **1999**, *8*, 947–957. [CrossRef]

44. Tsukatani, Y.; Nakayama, N.; Shimada, K.; Mino, H.; Itoh, S.; Matsuura, K.; Hanada, S.; Nagashima, K.V.P. Characterization of a blue-copper protein, auracyanin, of the filamentous anoxygenic phototrophic bacterium *Roseiflexus castenholzii*. *Arch. Biochem. Biophys.* **2009**, *490*, 57–62. [CrossRef] [PubMed]
45. Cao, L.; Bryant, D.A.; Schepmoes, A.A.; Vogl, K.; Smith, R.D.; Lipton, M.S.; Callister, S.J. Comparison of *Chloroflexus aurantiacus* strain J-10-fl proteomes of cells grown chemoheterotrophically and photoheterotrophically. *Photosynth. Res.* **2012**, *110*, 153–168. [CrossRef]
46. Xin, Y.; Lu, Y.K.; Fromme, R.; Fromme, P.; Blankenship, R.E. Purification, characterization and crystallization of menaquinol:fumarate oxidoreductase from the green filamentous photosynthetic bacterium *Chloroflexus aurantiacus*. *Biochim. Biophys. Acta Bioenerg.* **2009**, *1787*, 86–96. [CrossRef] [PubMed]
47. Schmidt, K. A comparative study on the composition of chlorosomes (*Chlorobium* vesicles) and cytoplasmic membranes from *Chloroflexus aurantiacus* strain Ok-70-fl and *Chlorobium limicola* f. *thiosulfatophilum* strain 6230. *Arch. Microbiol.* **1980**, *124*, 21–31. [CrossRef]
48. Yanyushin, M.F. Subunit structure of ATP synthase from *Chloroflexus aurantiacus*. *FEBS Lett.* **1993**, *335*, 85–88. [CrossRef]
49. Chadwick, G.L.; Hemp, J.; Fischer, W.W.; Orphan, V.J. Convergent evolution of unusual complex I homologs with increased proton pumping capacity: Energetic and ecological implications. *ISME J.* **2018**, *12*, 2668–2680. [CrossRef] [PubMed]
50. Hägerhäll, C. Succinate: Quinone oxidoreductases. *Biochim. Biophys. Acta Bioenerg.* **1997**, *1320*, 107–141. [CrossRef]
51. Lemos, R.S.; Fernandes, A.S.; Pereira, M.M.; Gomes, C.M.; Teixeira, M. Quinol:fumarate oxidoreductases and succinate:quinone oxidoreductases: Phylogenetic relationships, metal centres and membrane attachment. *Biochim. Biophys. Acta Bioenerg.* **2002**, *1553*, 158–170. [CrossRef]
52. Eisenreich, W.; Strauss, G.; Werz, U.; Fuchs, G.; Bacher, A. Retrobiosynthetic analysis of carbon fixation in the phototrophic eubacterium *Chloroflexus aurantiacus*. *Eur. J. Biochem.* **1993**, *215*, 619–632. [CrossRef] [PubMed]
53. Strauss, G.; Fuchs, G. Enzymes of a novel autotrophic CO_2 fixation pathway in the phototrophic bacterium *Chloroflexus aurantiacus*, the 3- hydroxypropionate cycle. *Eur. J. Biochem.* **1993**, *215*, 633–643. [CrossRef] [PubMed]
54. Klatt, C.G.; Bryant, D.A.; Ward, D.M. Comparative genomics provides evidence for the 3-hydroxypropionate autotrophic pathway in filamentous anoxygenic phototrophic bacteria and in hot spring microbial mats. *Environ. Microbiol.* **2007**, *9*, 2067–2078. [CrossRef] [PubMed]
55. Herter, S.; Fuchs, G.; Bacher, A.; Eisenreich, W. A bicyclic autotrophic CO_2 fixation pathway in *Chloroflexus aurantiacus*. *J. Biol. Chem.* **2002**, *277*, 20277–20283. [CrossRef]
56. Zarzycki, J.; Fuchs, G. Coassimilation of organic substrates via the autotrophic 3-hydroxypropionate bi-cycle in *Chloroflexus aurantiacus*. *Appl. Environ. Microbiol.* **2011**, *77*, 6181–6188. [CrossRef] [PubMed]
57. van der Meer, M.T.J.; Schouten, S.; De Leeuw, J.W.; Ward, D.M. Autotrophy of green non-sulphur bacteria in hot spring microbial mats: Biological explanations for isotopically heavy organic carbon in the geological record. *Environ. Microbiol.* **2000**, *2*, 428–435. [CrossRef]
58. Dunn, M.F. Anaplerotic function of phosphoenolpyruvate carboxylase in *Bradyrhizobium japonicum* USDA110. *Curr. Microbiol.* **2011**, *62*, 1782–1788. [CrossRef]
59. Dunn, M.F. Tricarboxylic acid cycle and anaplerotic enzymes in rhizobia. *FEMS Microbiol. Rev.* **1998**, *22*, 105–123. [CrossRef] [PubMed]
60. Vignais, P.M.; Billoud, B.; Meyer, J. Classifcation and phylogeny of hydrogenases. *FEMS Microbiol. Rev.* **2001**, *25*, 455–501. [CrossRef]
61. Vignais, P.M.; Billoud, B. Occurrence, classification, and biological function of hydrogenases: An overview. *Chem. Rev.* **2007**, *107*, 4206–4272. [CrossRef] [PubMed]
62. Gregersen, L.H.; Bryant, D.A.; Frigaard, N.U. Mechanisms and evolution of oxidative sulfur metabolism in green sulfur bacteria. *Front. Microbiol.* **2011**, *2*, 1–14. [CrossRef]
63. Friedrich, C.G.; Bardischewsky, F.; Rother, D.; Quentmeier, A.; Fischer, J. Prokaryotic sulfur oxidation. *Curr. Opin. Microbiol.* **2005**, *8*, 253–259. [CrossRef]
64. Lu, S.; Wang, J.; Chitsaz, F.; Derbyshire, M.K.; Geer, R.C.; Gonzales, N.R.; Gwadz, M.; Hurwitz, D.I.; Marchler, G.H.; Song, J.S.; et al. CDD/SPARCLE: The conserved domain database in 2020. *Nucleic Acids Res.* **2020**, *48*, D265–D268. [CrossRef] [PubMed]
65. Hügler, M.; Sievert, S.M. Beyond the calvin cycle: Autotrophic carbon fixation in the ocean. *Ann. Rev. Mar. Sci.* **2011**, *3*, 261–289. [CrossRef] [PubMed]
66. Zimmermann, P.; Zentgraf, U. The correlation between oxidative stress and leaf senescence during plant development. *Cell. Mol. Biol. Lett.* **2005**, *10*, 515–534.
67. Oostergetel, G.T.; van Amerongen, H.; Boekema, E.J. The chlorosome: A prototype for efficient light harvesting in photosynthesis. *Photosynth. Res.* **2010**, *104*, 245–255. [CrossRef]
68. van der Meer, M.T.J.; Schouten, S.; Damsté, J.S.S.; Ward, D.M. Impact of carbon metabolism on ^{13}C signatures of cyanobacteria and green non-sulfur-like bacteria inhabiting a microbial mat from an alkaline siliceous hot spring in Yellowstone National Park (USA). *Environ. Microbiol.* **2007**, *9*, 482–491. [CrossRef] [PubMed]
69. Kim, Y.M.; Nowack, S.; Olsen, M.; Becraft, E.D.; Wood, J.M.; Thiel, V.; Klapper, I.; Kühl, M.; Fredrickson, J.K.; Bryant, D.A.; et al. Diel metabolomics analysis of a hot spring chlorophototrophic microbial mat leads to new hypotheses of community member metabolisms. *Front. Microbiol.* **2015**, *6*, 209. [CrossRef] [PubMed]

70. Bhaya, D.; Grossman, A.R.; Steunou, A.S.; Khuri, N.; Cohan, F.M.; Hamamura, N.; Melendrez, M.C.; Bateson, M.M.; Ward, D.M.; Heidelberg, J.F. Population level functional diversity in a microbial community revealed by comparative genomic and metagenomic analyses. *ISME J.* **2007**, *1*, 703–713. [CrossRef]
71. Steunou, A.S.; Jensen, S.I.; Brecht, E.; Becraft, E.D.; Bateson, M.M.; Kilian, O.; Bhaya, D.; Ward, D.M.; Peters, J.W.; Grossman, A.R.; et al. Regulation of *nif* gene expression and the energetics of N_2 fixation over the diel cycle in a hot spring microbial mat. *ISME J.* **2008**, *2*, 364–378. [CrossRef] [PubMed]
72. Steunou, A.S.; Bhaya, D.; Bateson, M.M.; Melendrez, M.C.; Ward, D.M.; Brecht, E.; Peters, J.W.; Kühl, M.; Grossman, A.R. In situ analysis of nitrogen fixation and metabolic switching in unicellular thermophilic cyanobacteria inhabiting hot spring microbial mats. *Proc. Natl. Acad. Sci. USA* **2006**, *103*, 2398–2403. [CrossRef] [PubMed]
73. Izaki, K.; Haruta, S. Aerobic production of bacteriochlorophylls in the filamentous anoxygenic photosynthetic bacterium, *Chloroflexus aurantiacus* in the light. *Microbes Environ.* **2020**, *35*, 1–5. [CrossRef]
74. Lee, M.; Del Rosario, M.C.; Harris, H.H.; Blankenship, R.E.; Guss, J.M.; Freeman, H.C. The crystal structure of auracyanin A at 1.85 Å resolution: The structures and functions of auracyanins A and B, two almost identical "blue" copper proteins, in the photosynthetic bacterium *Chloroflexus aurantiacus*. *J. Biol. Inorg. Chem.* **2009**, *14*, 329–345. [CrossRef]
75. Frigaard, N.U.; Tokita, S.; Matsuura, K. Exogenous quinones inhibit photosynthetic electron transfer in *Chloroflexus aurantiacus* by specific quenching of the excited bacteriochlorophyll *c* antenna. *Biochim. Biophys. Acta Bioenerg.* **1999**, *1413*, 108–116. [CrossRef]
76. Oelze, J. Light and oxygen regulation of the synthesis of bacteriochlorophylls *a* and *c* in *Chloroflexus aurantiacus*. *J. Bacteriol.* **1992**, *174*, 5021–5026. [CrossRef]
77. Mishara, S.; Imlay, J. Why do bacteria use so many enzymes to scavenge hydrogen peroxide? *Arch. Biochem. Biophys.* **2012**, *525*, 145–160. [CrossRef] [PubMed]
78. Cabiscol, E.; Tamarit, J.; Ros, J. Oxidative stress in bacteria and protein damage by reactive oxygen species. *Int. Microbiol.* **2000**, *3*, 3–8. [PubMed]
79. Wilson, C.L.; Hinman, N.W.; Cooper, W.J.; Brown, C.F. Hydrogen peroxide cycling in surface geothermal waters of yellowstone national park. *Environ. Sci. Technol.* **2000**, *34*, 2655–2662. [CrossRef]
80. Bateson, M.M.; Ward, D.M. Photoexcretion and fate of glycolate in a hot spring cyanobacterial mat. *Appl. Environ. Microbiol.* **1988**, *54*, 1738–1743. [CrossRef] [PubMed]
81. Wu, D.; Raymond, J.; Wu, M.; Chatterji, S.; Ren, Q.; Graham, J.E.; Bryant, D.A.; Robb, F.; Colman, A.; Tallon, L.J.; et al. Complete genome sequence of the aerobic CO-oxidizing thermophile *Thermomicrobium roseum*. *PLoS ONE* **2009**, *4*, e4207. [CrossRef] [PubMed]
82. Zannoni, D.; Fuller, R.C. Functional and spectral characterization of the respiratory chain of *Chloroflexus aurantiacus* grown in the dark under oxygen-saturated conditions. *Arch. Microbiol.* **1988**, *150*, 368–373. [CrossRef]
83. Liu, Z.; Frigaard, N.U.; Vogl, K.; Iino, T.; Ohkuma, M.; Overmann, J.; Bryant, D.A. Complete genome of *Ignavibacterium album*, a metabolically versatile, flagellated, facultative anaerobe from the phylum *Chlorobi*. *Front. Microbiol.* **2012**, *3*, 1–15. [CrossRef]
84. Kramer, D.M.; Schoepp, B.; Liebl, U.; Nitschke, W. Cyclic electron transfer in *Heliobacillus mobilis* involving a menaquinol-oxidizing cytochrome *bc* complex and an RCI-type reaction center. *Biochemistry* **1997**, *36*, 4203–4211. [CrossRef] [PubMed]
85. Revsbech, N.P.; Trampe, E.; Lichtenberg, M.; Ward, D.M.; Kühl, M. In situ hydrogen dynamics in a hot spring microbial mat during a diel cycle. *Appl. Environ. Microbiol.* **2016**, *82*, 4209–4217. [CrossRef] [PubMed]
86. Dillon, J.G.; Fishbain, S.; Miller, S.R.; Bebout, B.M.; Habicht, K.S.; Webb, S.M.; Stahl, D.A. High rates of sulfate reduction in a low-sulfate hot spring microbial mat are driven by a low level of diversity of sulfate-respiring microorganisms. *Appl. Environ. Microbiol.* **2007**, *73*, 5218–5226. [CrossRef] [PubMed]
87. Marcia, M.; Ermler, U.; Peng, G.; Michel, H. The structure of *Aquifex aeolicus* sulfide:quinone oxidoreductase, a basis to understand sulfide detoxification and respiration. *Proc. Natl. Acad. Sci. USA* **2009**, *106*, 9625–9630. [CrossRef] [PubMed]
88. Morohoshi, S.; Matsuura, K.; Haruta, S. Secreted protease mediates interspecies interaction and promotes cell aggregation of the photosynthetic bacterium *Chloroflexus aggregans*. *FEMS Microbiol. Lett.* **2015**, *362*, 1–5. [CrossRef]
89. Hanada, S.; Shimada, K.; Matsuura, K. Active and energy-dependent rapid formation of cell aggregates in the thermophilic photosynthetic bacterium *Chloroflexus aggregans*. *FEMS Microbiol. Lett.* **2002**, *208*, 275–279. [CrossRef] [PubMed]
90. Fukushima, S. Analysis of gliding motility of the filamentous bacterium *Chloroflexus aggregans*. Ph.D. Thesis, Tokyo Metropolitan University, Tokyo, Japan, 2016.
91. Fukushima, S.I.; Morohoshi, S.; Hanada, S.; Matsuura, K.; Haruta, S. Gliding motility driven by individual cell-surface movements in a multicellular filamentous bacterium *Chloroflexus aggregans*. *FEMS Microbiol. Lett.* **2016**, *363*, 1–5. [CrossRef] [PubMed]
92. Doemel, W.N.; Brock, T.D. Structure, growth, and decomposition of laminated algal-bacterial mats in alkaline hot springs. *Appl. Environ. Microbiol.* **1977**, *34*, 433–452. [CrossRef] [PubMed]
93. Anderson, K.L.; Tayne, T.A.; Ward, D.M. Formation and fate of fermentation products in hot spring cyanobacterial mats. *Appl. Environ. Microbiol.* **1987**, *53*, 2343–2352. [CrossRef] [PubMed]
94. Nold, S.C.; Ward, D.M. Photosynthate partitioning and fermentation in hot spring microbial mat communities. *Appl. Environ. Microbiol.* **1996**, *62*, 4598–4607. [CrossRef] [PubMed]

MDPI
St. Alban-Anlage 66
4052 Basel
Switzerland
Tel. +41 61 683 77 34
Fax +41 61 302 89 18
www.mdpi.com

Microorganisms Editorial Office
E-mail: microorganisms@mdpi.com
www.mdpi.com/journal/microorganisms